Applied Mathematical Sciences
Volume 168

For further volumes:
http://www.springer.com/series/34

George W. Bluman
Alexei F. Cheviakov
Stephen C. Anco

Applications of Symmetry Methods to Partial Differential Equations

Springer

George W. Bluman
Department of Mathematics
The University of British Columbia
Vancouver, British Columbia V6T 1Z2
Canada
bluman@math.ubc.ca

Alexei F. Cheviakov
Department of Mathematics and Statistics
University of Saskatchewan
Saskatoon SK S7N 5E6
Canada
cheviakov@math.usask.ca

Stephen C. Anco
Department of Mathematics
Brock University
St. Catharines, Ontario L2S 3A1
Canada
sanco@brocku.ca

ISSN 0066-5452
ISBN 978-0-387-98612-8 e-ISBN 978-0-387-68028-6
DOI 10.1007/978-0-387-68028-6
Springer New York Dordrecht Heidelberg London

Library of Congress Control Number: 2009937642

Mathematics Subject Classification (2000): 58J70, 70H33, 35A30, 58J72, 35A20, 35A10, 35A15, 35A25, 70H06

Printed on acid-free paper

Springer is part of Springer Science+Business Media (www.springer.com)

Contents

Preface

This book is a sequel to *Symmetries and Integration Methods* (2002), by George W. Bluman and Stephen C. Anco. It includes a significant update of the material in the last three chapters of *Symmetries and Differential Equations* (1989; reprinted with corrections, 1996), by George W. Bluman and Sukeyuki Kumei. The emphasis in the present book is on how to find systematically symmetries (local and nonlocal) and conservation laws (local and nonlocal) of a given PDE system and how to use systematically symmetries and conservation laws for related applications. In particular, for a given PDE system, it is shown how systematically (1) to find higher-order and nonlocal symmetries of the system; (2) to construct by direct methods its conservation laws through finding sets of conservation law multipliers and formulas to obtain the fluxes of a conservation law from a known set of multipliers; (3) to determine whether it has a linearization by an invertible mapping and construct such a linearization when one exists from knowledge of its symmetries and/or conservation law multipliers, in the case when the given PDE system is nonlinear; (4) to use conservation laws to construct equivalent nonlocally related systems; (5) to use such nonlocally related systems to obtain nonlocal symmetries, nonlocal conservation laws and non-invertible mappings to linear systems; and (6) to construct specific solutions from reductions arising from its symmetries as well as from extensions of symmetry methods to find such reductions.

This book is aimed at applied mathematicians, scientists and engineers interested in finding solutions of partial differential equations and is written in the style of the above-mentioned 1989 book by Bluman and Kumei. There are numerous examples involving various well-known physical and engineering PDE systems.

The preceding book by Bluman and Anco includes comprehensive treatments of dimensional analysis, Lie groups of transformations, the discovery and use of symmetries to construct solutions of ordinary differential equations, and also shows how to construct conservation laws (first integrals) of ordinary differential equations through multipliers (integrating factors) as well as how to construct invariant solutions of partial differential equations from their point symmetries.

Chapter 1 reviews essential material from the Bluman and Anco book on one-parameter Lie groups of point transformations and how to find point symmetries of PDE systems and extends this material to the consideration of one-parameter higher-order local transformations and the finding of higher-order symmetries of PDE systems. This is followed by a comprehensive treatment on how to construct directly the local conservation laws essentially for any given PDE system. This treatment is based on first finding conservation law multipliers. It is shown how this treatment is related to and subsumes the classical Noether's theorem (which only holds for variational systems). In particular, multipliers are symmetries of a given PDE system only when the system is variational as written. There is a full discussion on connections between symmetries and conservation laws including the use of symmetries to find one or more additional conservation laws from a known conservation law.

Chapter 2 deals with the construction of local mappings relating a given PDE system to a target system of interest (or a member of a target class of PDE systems) from knowledge of the symmetries and/or conservation law multipliers of the given PDE system. In particular it is shown how to determine whether (1) a given nonlinear PDE system can be mapped invertibly to a linear PDE system and it is shown how to construct such a mapping when one exists; (2) a given linear PDE with variable coefficients can be mapped invertibly to a linear PDE with constant coefficients and it is shown how to construct such a mapping when one exists.

Chapter 3 considers perhaps the most important application of the material on conservation laws presented in Chapter 1. In particular, it is shown how to use local conservation laws and subsystems of a given PDE system to construct systematically a tree of equivalent nonlocally related systems. One of the many exhibited examples involves the planar gas dynamics equations, for which it is shown how the Euler and Lagrange systems are related systematically within such a tree of nonlocally related systems.

Chapter 4 considers the applications of such nonlocally related systems to find systematically nonlocal symmetries and nonlocal conservation laws of a given PDE system. In turn, it is shown how to use such nonlocal symmetries to construct nonlocal mappings of nonlinear PDE systems to equivalent linear PDE systems and to use conservation law multipliers of nonlocally related systems to construct nonlocal mappings of linear PDEs with variable coefficients to equivalent linear PDEs with constant coefficients.

The topics of Chapter 5 include how to use various kinds of symmetries to construct explicit solutions of PDEs, a discussion of the complexity involved in the construction of interesting nonlocally related systems in multi-dimensions, and a discussion of existing software to implement the procedures presented in this book.

If one is primarily interested in the material of Chapters 3–5, then Chapter 2 can be skipped. Chapter 1 is essential reading for all subsequent chapters.

Every topic is illustrated by examples. All sections have many exercises. It is essential to do some of the exercises to obtain a working knowledge of the presented material. Each chapter begins with a comprehensive Introduction section. The Discussion section at the end of each chapter discusses related work and puts the subject matter of the chapter in context for later chapters.

Within each section of a given chapter, definitions, theorems, corollaries, and remarks are numbered separately and consecutively. For example, Remark 4.2.1 refers to the first remark in Section 4.2. Exercises appear at the end of each section; Exercise 4.2.2 refers to the second problem of Exercises 4.2, i.e., the second problem at the end of Section 4.2.

There are separate Author and Subject indices as well as a References section. In addition there is a Theorem, Corollary and Lemma Index.

The authors are grateful to their many collaborators without whom this book would not have been possible. In particular we wish to thank Julian Cole (posthumously), Sukeyuki Kumei, Gregory Reid, Vladimir Shtelen, Zhenya Yan, Temuerchaolu, Oleg Bogoyavlenskij, Nataliya Ivanova, Dennis The, Sheng Liu, Thomas Wolf, and Juha Pohjanpelto.

Thanks also to Andy Wan, Chengzhong Wu, Raouf Dridi and Olivier Glorieux for valuable comments on preliminary versions of the manuscript.

Finally, we thank our respective families for their patience and understanding through the course of our writing including our children David, Benny, Vladimir, Tatiana, Maria, and Darren to whom this book is dedicated.

Vancouver, British Columbia, Canada — George W. Bluman
Saskatoon, Saskatchewan, Canada — Alexei F. Cheviakov
St. Catharines, Ontario, Canada — Stephen C. Anco

Introduction

This book is concerned with some modern developments related to symmetries and conservation laws for partial differential equations (PDEs). It is a sequel to *Symmetry and Integration Methods for Differential Equations* (2002) by George W. Bluman and Stephen C. Anco (2002), which focused on Lie groups of transformations and their applications to solving ordinary differential equations (ODEs) and finding invariant solutions of PDEs. The present volume primarily concentrates on recent research of the authors and their collaborators. Most important, we attempt to put this work in a form accessible to graduate students and researchers in applied mathematics, the physical sciences and engineering. Most of the material in this book did not appear in *Symmetries and Differential Equations* [(1989); reprinted with corrections (1996)], by George W. Bluman and Sukeyuki Kumei, and includes a significant updating of the final three chapters.

In the latter part of the 19th century, Sophus Lie initiated his studies on continuous groups (Lie groups) with the aim to put order to, and thereby extend systematically, the hodgepodge of heuristic techniques for solving ODEs. He showed that the problem of finding the Lie group of point transformations leaving invariant a DE (ordinary or partial), i.e., a point symmetry of a DE, reduced to solving related linear systems of determining equations for its infinitesimal generators. Lie also showed that a point symmetry of a DE leads, in the case of an ODE, to reducing the order of the DE (irrespective of any imposed initial conditions) and, in the case of a PDE, to finding special solutions called invariant (similarity) solutions of the DE. Moreover, he showed that a point symmetry of a DE generates a one-parameter family of solutions from any known solution of a DE that is not an invariant solution arising from the symmetry. Most importantly, Lie's work is applicable to nonlinear DEs. His work is discussed in the two above-mentioned books as well as many other excellent references therein. The direct applicability of Lie's work to PDEs, especially nonlinear PDEs, is rather limited, even when a given PDE has a point symmetry, since the resulting invariant solutions yield only a small subset of the solution set of the PDE and hence few posed boundary value problems can be solved.

The extensions of Lie's work to PDEs have focused on finding further applications of point symmetries to include linearization mappings and solutions of boundary value problems, extending the spaces of symmetries of a given PDE system to include local symmetries (higher-order symmetries) as well as nonlocal symmetries, extending the applications of symmetries to include variational symmetries that yield conservation laws for variational systems, extending variational symmetries to multipliers and resulting conservation laws for any given PDE system, finding further solutions that arise from the extension of Lie's method to the "nonclassical method" as well as other generalizations, and efficiently solving the (over-determined) linear system of symmetry and/or multiplier determining equations through the development of symbolic computation software as well as related calculations for solving the nonlinear system of determining equations for the nonclassical method.

A symmetry of a PDE system is any transformation of its solution manifold into itself, i.e., a symmetry transforms (maps) any solution of a PDE system to another solution of the same system. Consequently, continuous symmetries of PDE systems are defined topologically and hence are not restricted to just point symmetries. Thus, in principle, any nontrivial PDE system has symmetries. The problem is to find and use such symmetries. Practically, to find a symmetry of a PDE system, one must consider transformations, acting locally in some finite-dimensional space, whose variables include the dependent variables of the PDE system. However, as it will be seen, these transformation variables do not have to be restricted to the independent and dependent variables of a given PDE system.

One such extension is to consider *higher-order symmetries* (*local symmetries*) where the solutions of the linear determining equations for the components of infinitesimal generators of symmetries are allowed to depend on a finite number of derivatives of the given dependent variables of the PDE. [By comparison, components of infinitesimal generators of point symmetries allow dependence at most linearly on the first derivatives of the dependent variables whereas components of infinitesimal generators of contact symmetries allow arbitrary dependence on first derivatives of dependent variables.] In making this extension, it is essential to realize that the linear determining equations for local symmetries are the linearized system of the given PDE that holds for *all* of its solutions. *Globally*, point and contact symmetries act on finite-dimensional spaces whereas higher-order symmetries act on infinite-dimensional spaces consisting of the dependent and independent variables as well as *all* of their derivatives. Well-known integrable equations of mathematical physics such as the Korteweg–de Vries equation have an infinite number of higher-order local symmetries.

Another extension is to consider solutions of the determining equations that allow an ad-hoc dependence on nonlocal variables such as integrals of the dependent variables. Usually such symmetries are found formally through recursion operators that depend on inverse differentiation. Integrable equations such as the sine-Gordon and cubic Schrödinger equations have an infinite number of such nonlocal symmetries.

In her celebrated 1918 paper, Emmy Noether showed that if a system of DEs admits a variational principle, then any local transformation group leaving invariant the action integral for its Lagrangian density, i.e., an admitted *variational symmetry*, yields a local conservation law. Conversely, any local conservation law of a variational DE system arises from a variational symmetry, and hence there is a direct correspondence between local conservation laws and variational symmetries (Noether's theorem and its generalizations due to Bessel-Hagen (1921) and Boyer (1967)).

There are several limitations to Noether's theorem for finding the local conservation laws for a given DE system. First of all, it is restricted to variational systems. Consequently, for this theorem to be applicable to a given DE system *as written*, the system must have the same number of dependent variables as the number of equations in the given system, and have no dissipation. Moreover, if a given DE system consists of one scalar equation, it must be of even order. In particular, a given system of DEs, as written, is variational if and only if its linearized system is self-adjoint. There is also the difficulty of finding local symmetries of the action integral. In general, not all local symmetries of a variational DE system are variational symmetries. Moreover, the use of Noether's theorem to find local conservation laws is coordinate-dependent.

A conservation law of a given DE system is a divergence expression that vanishes on all solutions of the DE system. Conservation laws describe essential properties of the process modeled by a given DE system and are also used for existence, uniqueness and stability analysis and for the development of numerical methods. In general, all such divergences that yield local conservation laws arise from linear combinations of the DEs of a given system taken with sets of local multipliers in which each multiplier is an expression depending on the independent and dependent variables as well as derivatives (up to some finite order) of the dependent variables of a given DE system. It will be seen that a given DE system has a local conservation law if and only if there exists a set of local multipliers such that the corresponding linear combination of the DEs in the system is identically annihilated by the Euler operators associated with each of its dependent variables without restricting these variables to solutions of the DE system, i.e., the dependent variables are now treated as arbitrary functions. If a given DE system, as written, is variational then its local conservation law multipliers correspond to variational symmetries. In this case, it turns out that its local conservation law multipliers satisfy a system of determining equations that includes the linearizing system of the given DE system augmented by additional determining equations that taken together correspond to the action integral being invariant under the associated variational symmetry. More generally, for *any* given DE system, all local conservation law multipliers are the solutions of an easily found linear determining system that includes the adjoint system of the linearizing DE system. For any set of local conservation law multipliers, one can either directly find the fluxes and density of the corresponding local conservation law or, if this proves difficult, there is an integral formula that yields them without the need of a specific functional (Lagrangian) even in the case when the given DE system is variational.

Another important application of symmetries of PDEs is to determine whether a given PDE system can be mapped into an equivalent target PDE system of interest. This is especially significant if a target class of PDEs can be completely characterized in terms of its symmetries. Target classes with such complete characterizations include linear PDE systems and linear PDEs with constant coefficients. Consequently, from knowledge of the point or contact symmetries of a given PDE system, one can determine whether it can be mapped invertibly to a linear PDE system by a point or contact transformation and explicitly find such a mapping when one exists. Moreover, one can also see whether such a linearization is possible from knowledge of the local conservation law multipliers of a given PDE system. From knowledge of the point symmetries of a linear PDE with variable coefficients, one can determine whether it can be mapped by an invertible point transformation to a linear PDE with constant coefficients and find such an explicit mapping when one exists.

In order to effectively apply symmetry methods to PDE systems, one needs to work in some specific coordinate frame in order to perform calculations. A procedure to find symmetries that are nonlocal and yet are local in some related coordinate frame involves embedding a given PDE system in another PDE system obtained by adjoining nonlocal variables in such a way that the related PDE system is equivalent to the given system and the given system arises through projection. Consequently, any local symmetry of the related system yields a symmetry of the given system. If the local symmetry of the related system has an essential dependence on the nonlocal variables after projection, then it yields a nonlocal symmetry of the given PDE system.

A systematic way to find such an embedding is through local conservation laws of a given PDE system. For each local conservation law, one can introduce a potential variable(s). By adjoining the resulting potential equations to the given PDE system, one can construct an augmented system (*potential system*) of PDEs. By construction, such a potential system is nonlocally equivalent to the given PDE system since, through built in integrability conditions, any solution of the given PDE system yields a solution of the potential system and, conversely, through projection any solution of the potential system yields a solution of the given PDE system. But this relationship is nonlocal since there is no one-to-one correspondence between solutions of the given and potential systems. If a local symmetry of the potential system has an essential dependence on the potential variables when projected onto the space of variables of the given system, then it yields a nonlocal symmetry (*potential symmetry*) of the given PDE system. It turns out that many PDE systems have such potential symmetries. Moreover, one can find other nonlocal symmetries of a given PDE system through seeking local symmetries of an equivalent subsystem of the given system or one of its potential systems provided that such a subsystem is nonlocally related to the given PDE system. Invariant solutions of such potential systems and subsystems can yield further solutions of the given PDE system. A potential symmetry is a local symmetry of a potential system, thus it generates a one-parameter family of solutions from any known solution of the potential system that in turn yields a one-parameter family of solutions from a known solution of the given PDE system. Similarly, this will be the case for a nonlocal symmetry arising from a subsystem. Furthermore, local conservation laws of potential systems can yield nonlocal conservation laws of a given PDE system provided that their local conservation law multipliers have an essential dependence on the potential variables. Linearizations of such potential systems through local symmetry or local conservation law multiplier analysis can yield explicit nonlocal linearizations of a given PDE system. Moreover, through a potential system one can extend the mappings of linear systems with variable coefficients to linear systems with constant coefficients to include nonlocal mappings between such systems.

One can further extend embeddings through using local conservation laws to systematically construct trees of nonlocally related but equivalent systems of PDEs. If a given PDE system has n local conservation laws, then each conservation law yields potentials and corresponding potential systems. Most importantly, from the n local conservation laws, one can directly construct up to $2^n - 1$ independent nonlocally related systems of PDEs by considering the corresponding potential systems individually (n singlets), in pairs ($n(n-1)/2$ couplets), ..., and taken all together (one n-plet). In turn, any one of these $2^n - 1$ systems could lead to the discovery of new nonlocal symmetries and/or nonlocal conservation laws of the given PDE system or any of the other nonlocally related systems. Moreover, such nonlocal conservation laws could yield further nonlocally related systems, etc. Furthermore, subsystems of such nonlocally related systems could yield further nonlocally related systems. Correspondingly, a tree of nonlocally related systems is constructed. Through such constructions, one can systematically relate Eulerian and Lagrangian coordinate descriptions of gas dynamics and nonlinear elasticity. In both cases, for a corresponding PDE system written in Eulerian coordinates, there exists a nonlocally related system that yields a corresponding PDE system written in Lagrangian coordinates.

For a given class of PDEs with classifying (constitutive) functions, it is of interest to classify its trees of nonlocally related systems and corresponding symmetries and conservation laws with respect to various forms of its constitutive functions. When a system is variational, i.e., its linearized system is self-adjoint, then of course the local conservation laws arise from a subset of its local symmetries and, in particular, the number of linearly independent conservation laws cannot exceed the number of corresponding higher-order symmetries. But from the above, one can see that, in general, this will not be the case when a system is not variational. Here a specific constitutive function could yield more local conservation laws than local symmetries as well as vice versa.

For any given PDE system, a transformation group (continuous or discrete) that leaves it invariant yields a formula that maps a conservation law to a conservation law of the same system, whether or not the given system is variational. If the group is continuous, then in terms of a parameter expansion a given conservation law could map into more than one additional conservation law for the given PDE system.

Another important extension relates to Lie's work on finding invariant solutions for PDE systems. As mentioned previously, a point symmetry of a PDE system maps each of its solutions into a one-parameter family of solutions. But some solutions map into themselves, i.e., they are themselves invariant. Such solutions satisfy the characteristic PDE given by the invariant surface condition yielding the invariants of the point symmetry. The invariant solutions arising from the point symmetry are the solutions of the given PDE system that satisfy the augmented system consisting of this characteristic PDE with known coefficients (obtained from the point symmetry) and the given PDE system itself. The invariant solutions arise as solutions of a reduced system with one less independent variable. This method ("classical method") of Lie to find invariant solutions of a given PDE is generalized by the *nonclassical method* introduced in Bluman's 1967 PhD thesis where one seeks solutions of an augmented system consisting of the given PDE system and the characteristic PDE with unknown coefficients as well as differential consequences of the augmented system. Here the unknown coefficients are determined by substituting the characteristic equation, and its differential consequences, into the determining system for point symmetries of the augmented system. The resulting over-determined system is nonlinear (even if the given PDE system is linear) in these unknown coefficients, but less over-determined than is the case when finding point symmetries of the given PDE system. Each solution of the determining system for point symmetries is a solution of the determining system for the unknown coefficients of the characteristic PDE. Solving for the unknown coefficients, one then proceeds to find the corresponding "nonclassical" solutions of the augmented system that, by construction, include the classical invariant solutions.

The solutions of a PDE that can be obtained by the nonclassical method include all of its solutions that satisfy a particular functional form (ansatz) of some generality that allows an arbitrary dependence on a similarity variable (depending on the independent and dependent variables of the PDE) and an arbitrary dependence on a function of a similarity variable and the independent variables of the PDE. The solutions obtained by the nonclassical method include all solutions obtained "directly" from such an ansatz by the *direct method* introduced by Clarkson and Kruskal in 1988.

For many PDE systems arising in applications, the linear determining equations for local symmetry components or local conservation law multipliers split into over-determined linear PDE systems that can contain hundreds of equations. To generate, simplify and solve such PDE systems, symbolic software is used. Modern symbolic packages include routines for the automatic generation of determining equations, their subsequent simplification and solution (including classification with respect to constitutive functions and/or parameters of a given DE system), to yield local symmetries and conservation laws of a given DE system.

Chapter 1
Local Transformations and Conservation Laws

1.1 Introduction

A continuous symmetry of a system of partial differential equations (PDEs) is a transformation that leaves invariant the solution manifold of the system, i.e, it maps (deforms) any solution of the system into a solution of the same system. This definition is topological in nature. However, in practice, the direct calculation of the continuous symmetries of a given system of PDEs restricts one to consider symmetries that are local transformations acting on its space of independent variables, dependent variables and their derivatives. Lie's algorithm to determine Lie groups of point transformations (point symmetries) of differential equations was presented in Bluman & Anco (2002) [see also Ovsiannikov [(1962), (1982)]; Bluman & Cole (1974); Olver (1986); Bluman & Kumei (1989); Stephani (1989); Hydon (2000); Cantwell (2002)]. Point symmetries arise from solutions of linear systems of determining equations for components of infinitesimal generators for the independent and dependent variables of a given PDE system, where these components themselves depend only on the given PDE system's independent and dependent variables. Point transformations acting on the space of the given independent and dependent variables of a given PDE system can be extended (prolonged) to point transformations acting on the space of the given independent variables, dependent variables, and their derivatives to any finite order.

Lie's algorithm for finding point symmetries of a PDE system can be extended to find more general local symmetries admitted by PDEs. In the extension of Lie's algorithm, one uses differential consequences of the given PDE system, i.e., invariance of a given PDE system is understood to include its differential consequences. Here it is important to consider the infinitesimal generators for point symmetries in their evolutionary form where the independent variables are themselves invariant and the action of a group of point transformations is strictly an action on the dependent variables of the PDE

G.W. Bluman et al., *Applications of Symmetry Methods to Partial Differential Equations*, Applied Mathematical Sciences 168, DOI 10.1007/978-0-387-68028-6_1,

system, so that solutions are directly mapped into other solutions under the group action. In evolutionary form, the components of infinitesimal generators for dependent variables have at most a linear dependence on the first derivatives of the dependent variables (the coefficients of the first derivatives are the components of the independent variables when not in evolutionary form).

This allows one to readily extend Lie's algorithm to seek *contact symmetries*of PDEs (only existing for scalar PDEs) where now the components of infinitesimal generators for dependent variables can depend at most on the first derivatives of the dependent variable of a given scalar PDE. [If this dependence is at most linear on the first derivatives, then a contact symmetry is a point symmetry.] A contact transformation is equivalent to a point transformation acting on the space of the given independent variables, the dependent variable and its first derivatives, and, through this, can be naturally extended to point transformations acting on the space of the given independent variables, the dependent variable and its derivatives to any finite order greater than one.

Lie's algorithm can be still further extended by allowing the infinitesimal generators in evolutionary form to depend on derivatives of dependent variables to any finite order. This allows one to calculate symmetries that are called *higher-order symmetries*. In the scalar case, contact symmetries are first-order symmetries. It turns out that higher-order symmetries are not equivalent to point transformations acting on a finite-dimensional manifold including the independent variables, the dependent variables and their derivatives to some finite order. However, they are local symmetries in the sense that the components of the dependent variables in their infinitesimal generators depend at most on a finite number of derivatives of the given PDE system's dependent variables so that their calculation only depends on the local behaviour of solutions of the given PDE system. *Local symmetries* include point symmetries, contact symmetries and higher-order symmetries. Local symmetries are uniquely determined when infinitesimal generators are represented in evolutionary form.

Sophus Lie considered contact transformations and contact symmetries of PDEs. Emmy Noether (1918) introduced the notion of a one-parameter higher-order transformation in her celebrated paper on conservation laws. The well-known infinite sequences of conservation laws of the Korteweg–de Vries (KdV) and sine-Gordon equations are directly related to admitted infinite sequences of local symmetries (higher-order symmetries) obtained through the use of recursion operators [Olver (1977)].

A conservation law of a given PDE system is a divergence expression that vanishes on all solutions of the PDE system. In general, any such nontrivial divergence expression that yields a local conservation law of the given PDE system arises from a linear combination formed by *local multipliers* (*char-*

acteristics, *factors*), depending on the independent and dependent variables as well as at most a finite number of derivatives of the dependent variables of the given PDE system, with each PDE in the given system. It turns out that a divergence expression depending on independent variables, dependent variables and their derivatives to some finite order is annihilated by the Euler operators associated with each of its dependent variables; conversely, if the Euler operators, associated with each dependent variable in an expression involving independent variables, dependent variables and their derivatives to some finite order, annihilate the expression, then the expression is a divergence expression. From this it follows that a given PDE system has a local conservation law if and only if there exists a set of local multipliers whose linear combination with each of the PDEs in the system is identically annihilated by the Euler operators associated with each of its dependent variables *without restricting* these dependent variables in the linear combination to solutions of the PDE system, i.e, the dependent variables are treated as *arbitrary* functions.

Thus the problem of finding local conservation laws of a given PDE system reduces to the problem of finding sets of local multipliers whose linear combination with each PDE in the given system is annihilated by the Euler operators associated with each dependent variable with the dependent variables in the given PDE system replaced by arbitrary functions (and their derivatives). Each such set of multipliers yields a local conservation law of the given PDE system. Moreover, for any given set of local multipliers yielding a local conservation law, there is an integral formula to obtain the flux and densities of the conservation law [Anco & Bluman [(1997a), (2002a,b)], Anco (2003)]. Often it is straightforward to obtain the conservation law by direct calculations after its multipliers are known [Wolf (2002a)]. What has been outlined here is referred to as the *direct method* for obtaining local conservation laws.

For a given PDE system, Lie's algorithm yields a set of over-determined linear determining equations whose solutions yield admitted local symmetries. This set of linear PDEs arises from the linearization of the given PDE system (*Fréchet derivative*) about an *arbitrary* solution of the given PDE system, i.e., the resulting linear system must hold for each solution of the given PDE system. After the given PDE system and its differential consequences are substituted into the linearization of the given PDE system, the resulting linear system must hold with the dependent variables replaced by *arbitrary* functions (and their derivatives).

In contrast, for a given PDE system, sets of multipliers yielding local conservation laws are solutions of a set of over-determined linear determining equations arising from annihilations by Euler operators. It turns out that the set of linear multiplier determining equations for local conservation law multipliers includes the *adjoint* of the set of linear PDEs arising from the linearization of the given PDE system about an arbitrary solution of the given PDE system [Anco & Bluman (1997a)].

It follows that in the situation when the set of linearized equations of a given PDE system (Fréchet derivative) is self-adjoint, the set of multiplier determining equations includes the set of local symmetry determining equations. Consequently, here each set of local conservation law multipliers yields a local symmetry of the given PDE system. In particular, the conservation law multipliers are also components of the infinitesimal generators of local symmetries in evolutionary form. However, in the self-adjoint case, the set of linear determining equations for local conservation law multipliers is more over-determined than those for local symmetries since here the set of linear determining equations for local conservation law multipliers includes further linear PDEs in addition to the set of linear PDEs for local symmetries. Consequently, in the self-adjoint case, there can exist local symmetries that do not yield local conservation law multipliers.

Noether (1918) showed that if a system of differential equations admits a variational principle, then any one-parameter Lie group of point transformations that leaves invariant the action functional yields a local conservation law. In particular, she gave an explicit formula for the fluxes of the conservation law. Noether's theorem was extended by Bessel-Hagen (1921) to allow the one-parameter Lie group of point transformations to leave invariant the action functional to within a divergence term. As presented, their results depended on Lie groups of point transformations used in their "standard" form, i.e., not in evolutionary form. Boyer (1967) showed how all such conservation laws could be obtained from Lie groups of point transformations used in evolutionary form. From this point of view, it is straightforward to apply Noether's theorem to obtain a conservation law for any one-parameter higher-order transformation leaving invariant the action functional to within a divergence term. A one-parameter higher-order transformation that leaves invariant an action functional to within a divergence term is called a *variational symmetry*.

As might be expected, Noether's explicit formula for a local conservation law arises from local multipliers that yield components of local symmetries in evolutionary form. From this point of view, it follows that all local conservation laws arising from Noether's theorem are obtained by the direct method. Moreover, one can see that a variational symmetry must map an extremal of the action functional to another extremal. Since an extremal of an action functional is a solution of the system of differential equations arising from the variational principle, it follows that a variational symmetry must be a local symmetry of the given system of differential equations arising from the variational principle.

A system of differential equations (as written) has a variational principle if and only if its linearized system (Fréchet derivative) is self-adjoint [Volterra

(1913); Vainberg (1964); Olver (1986)]. From this point of view it also follows that all conservation laws obtained by Noether's theorem must arise from the direct method.

The direct method supersedes Noether's theorem. In particular, for Noether's theorem to be directly applicable to a given DE system, the following must hold.

- The linearized system of the given DE system is self-adjoint.
- One has an explicit action functional.
- One has a one-parameter local transformation that leaves the action functional invariant to within a divergence. In order to find such a symmetry systematically, one first finds local symmetries (solutions of the linearized system) and then checks whether such local symmetries leave the action functional invariant to within a divergence.

On the other hand, the direct method is applicable to any given DE system, whether or not its linearized system is self-adjoint. No functional needs to be determined. Moreover, a set of local conservation law multipliers is represented by *any* solution of an over-determined linear system of PDEs satisfied by the multipliers and this over-determined linear system is obtained directly from the given DE system. As mentioned above, in the case when the linearized system is self-adjoint, the symmetry determining equations are a subset of this over-determined linear system.

For any system of DEs, a contact transformation maps the given DE system into another system of DEs. Through the contact transformation one can give an explicit formula that transforms any conservation law of the given DE system into a conservation law of the transformed DE system. In the case when the contact transformation is a symmetry (not necessarily a continuous symmetry) of the given DE system, it follows that here the contact transformation maps any known conservation law of the given DE system into another conservation law of the same DE system. However, the resulting conservation law could be the same one! When the symmetry is a continuous symmetry (i.e., a point or contact symmetry), due to the parameter dependence of the contact transformation, through parameter expansion, one could obtain more than one new conservation law from a known conservation law.

1.2 Local Transformations

In Bluman & Anco (2002) [see also Bluman & Kumei (1989); Olver (1986); Stephani (1989); Hydon (2000)], point transformations and one-parameter Lie groups of point transformations were defined and it was shown how to find all the one-parameter Lie groups of point transformations (point symmetries) of a given system of partial differential equations. Moreover, it was shown how to use a point symmetry to obtain a one-parameter family of solutions from a known solution, except in the case when the known solution is itself invariant (i.e., maps into itself under the action of the point symmetry). In this case, it was shown how to use a point symmetry to find such invariant solutions, through a system of PDEs with one less independent variable. In this section, we first review some essential information about point transformations and do this in such a way that point transformations can be generalized to include the wider class of local transformations.

1.2.1 Point transformations

Consider the situation of n independent variables $x = (x^1, \ldots, x^n)$ and m dependent variables $u(x) = (u^1(x), \ldots, u^m(x))$. Partial derivatives are denoted by $u_i^\mu = \partial u^\mu(x)/\partial x^i$; the notation

$$\partial u \equiv \partial^1 u = \left(u_1^1(x), \ldots, u_n^1(x), \ldots, u_1^m(x), \ldots, u_n^m(x)\right)$$

denotes the set of all first-order partial derivatives;

$$\begin{aligned}\partial^p u &= \left\{u^\mu_{i_1 \ldots i_p} \mid \mu = 1, \ldots, m; \quad i_1, \ldots, i_p = 1, \ldots, n\right\}\\ &= \left\{\frac{\partial^p u^\mu(x)}{\partial x^{i_1} \ldots \partial x^{i_p}} \mid \mu = 1, \ldots, m; i_1, \ldots, i_p = 1, \ldots, n\right\}\end{aligned}$$

denotes the set of all partial derivatives of order p.

A point transformation is a one-to-one transformation acting on the $n+m$-dimensional space (x, u). In particular, a point transformation is of the form

$$x^* = f(x, u), \tag{1.1a}$$

$$u^* = g(x, u). \tag{1.1b}$$

Through invariance of contact conditions, a point transformation (assuming that (1.1) is differentiable as needed) naturally extends to a one-to-one transformation acting on $(x, u, \partial u, \ldots, \partial^p u)$-space for $p = 1, 2, \ldots$.

In particular the pth extended transformation of (1.1) is given by

$$
\begin{aligned}
(x^*)^i &= f^i(x,u), && (1.2a)\\
(u^*)^\mu &= g^\mu(x,u), && (1.2b)\\
(u^*)^\mu_i &= h^\mu_i(x,u,\partial u), && (1.2c)\\
&\vdots\\
(u^*)^\mu_{i_1\dots i_p} &= h^\mu_{i_1\dots i_p}(x,u,\partial u,\dots,\partial^p u), && (1.2d)
\end{aligned}
$$

where $i, i_1, \dots, i_p = 1, \dots, n$, $\mu = 1, \dots, m$; $(u^*)^\mu_i = \partial(u^*)^\mu/\partial(x^*)^i$, etc. In particular, the transformed components of first-order derivatives are determined by

$$
\begin{bmatrix} (u^*)^\mu_1 \\ \vdots \\ (u^*)^\mu_n \end{bmatrix} = \begin{bmatrix} h^\mu_1 \\ \vdots \\ h^\mu_n \end{bmatrix} = A^{-1} \begin{bmatrix} \mathrm{D}_1 g^\mu \\ \vdots \\ \mathrm{D}_n g^\mu \end{bmatrix}, \tag{1.3}
$$

where A^{-1} is the inverse of the Jacobian matrix

$$
A = \begin{bmatrix} \mathrm{D}_1 f^1 & \dots & \mathrm{D}_1 f^n \\ \vdots & & \vdots \\ \mathrm{D}_n f^1 & \dots & \mathrm{D}_n f^n \end{bmatrix}, \tag{1.4}
$$

in terms of total derivative operators

$$
\mathrm{D}_i = \frac{\partial}{\partial x^i} + u^\mu_i \frac{\partial}{\partial u^\mu} + u^\mu_{ii_1} \frac{\partial}{\partial u^\mu_{i_1}} + u^\mu_{ii_1i_2} \frac{\partial}{\partial u^\mu_{i_1i_2}} + \cdots, \tag{1.5}
$$

$i = 1, \dots, n$. The transformed components of higher-order derivatives are determined by

$$
\begin{bmatrix} (u^*)^\mu_{i_1\dots i_{p-1}1} \\ \vdots \\ (u^*)^\mu_{i_1\dots i_{p-1}n} \end{bmatrix} = \begin{bmatrix} h^\mu_{i_1\dots i_{p-1}1} \\ \vdots \\ h^\mu_{i_1\dots i_{p-1}n} \end{bmatrix} = A^{-1} \begin{bmatrix} \mathrm{D}_1 h^\mu_{i_1\dots i_{p-1}} \\ \vdots \\ \mathrm{D}_n h^\mu_{i_1\dots i_{p-1}} \end{bmatrix}. \tag{1.6}
$$

Now consider the situation where the point transformation (1.1) is a one-parameter Lie group of point transformations given by

$$
\begin{aligned}
(x^*)^i &= f^i(x,u;\varepsilon) = x^i + \varepsilon\xi^i(x,u) + O(\varepsilon^2), \quad i = 1,\dots,n, && (1.7a)\\
(u^*)^\mu &= g^\mu(x,u;\varepsilon) = u^\mu + \varepsilon\eta^\mu(x,u) + O(\varepsilon^2), \quad \mu = 1,\dots,m, && (1.7b)
\end{aligned}
$$

with the corresponding infinitesimal generator given by

$$
\mathrm{X} = \xi^i(x,u)\frac{\partial}{\partial x^i} + \eta^\mu(x,u)\frac{\partial}{\partial u^\mu}. \tag{1.8}
$$

A one-parameter Lie group of point transformations (1.7) induces one-parameter Lie groups of point transformations acting on $(x, u, \partial u)$-space, ..., $(x, u, \partial u, \ldots, \partial^k u)$-space, as follows.

$$(u^*)^\mu_i = u^\mu_i + \varepsilon \eta^{(1)\,\mu}_i(x, u, \partial u) + O(\varepsilon^2), \tag{1.9a}$$

$$\vdots$$

$$(u^*)^\mu_{i_1 \ldots i_k} = u^\mu_{i_1 \ldots i_k} + \varepsilon \eta^{(k)\,\mu}_{i_1 \ldots i_k}(x, u, \partial u, \ldots, \partial^k u) + O(\varepsilon^2), \tag{1.9b}$$

with the extended infinitesimals given by

$$\eta^{(1)\,\mu}_i = \mathrm{D}_i \eta^\mu - (\mathrm{D}_i \xi^j) u^\mu_j, \tag{1.10}$$

and

$$\eta^{(k)\,\mu}_{i_1 \ldots i_k} = \mathrm{D}_{i_k} \eta^{(k-1)\,\mu}_{i_1 \ldots i_{k-1}} - (\mathrm{D}_{i_k} \xi^j) u^\mu_{i_1 \ldots i_{k-1} j}, \tag{1.11}$$

$\mu = 1, \ldots, m$, $i, i_j = 1, \ldots, n$ for $j = 1, \ldots, k$ with $k = 2, 3, \ldots$.

The kth *extended infinitesimal generator* (kth *prolongation* of (1.8)) is given by

$$\begin{aligned} \mathrm{X}^{(k)} = {} & \xi^i(x,u)\frac{\partial}{\partial x^i} + \eta^\mu(x,u)\frac{\partial}{\partial u^\mu} + \eta^{(1)\,\mu}_i(x,u,\partial u)\frac{\partial}{\partial u^\mu_i} \\ & + \cdots + \eta^{(k)\,\mu}_{i_1 \ldots i_k}(x, u, \partial u, \ldots, \partial^k u)\frac{\partial}{\partial u^\mu_{i_1 \ldots i_k}}. \end{aligned} \tag{1.12}$$

1.2.2 Contact transformations

Consider the situation of n independent variables $x = (x^1, \ldots, x^n)$ and one dependent variable $u(x)$.

Definition 1.2.1. A *contact transformation* is a transformation of the form ($m = 1$)

$$(x^*)^i = f^i(x, u, \partial u), \tag{1.13a}$$
$$u^* = g(x, u, \partial u), \tag{1.13b}$$
$$u^*_i = h_i(x, u, \partial u), \tag{1.13c}$$

$i = 1, \ldots, n$, that is one-to-one on some domain D in $(x, u, \partial u)$-space and leaves invariant the contact condition, i.e.,

$$du^* = u^*_i dx^{*\,i}. \tag{1.14}$$

It is assumed that f^i, g have an essential dependence on the first derivatives of u. Otherwise a contact transformation is a point transformation.

Theorem 1.2.1 (Lie (1890), Mayer (1875)). *Equations* (1.13) *define a contact transformation if and only if the functions f^i, g, h_i satisfy*

$$\frac{\partial g}{\partial u_i} = h_j \frac{\partial f^j}{\partial u_i}, \tag{1.15a}$$

$$\frac{\partial g}{\partial x^i} + u_i \frac{\partial g}{\partial u} = h_j \left(\frac{\partial f^j}{\partial x^i} + u_i \frac{\partial f^j}{\partial u} \right), \tag{1.15b}$$

for $i = 1, \ldots, n$.

Proof. The proof is straightforward and is left to Exercise 1.2.1. □

Now consider the situation where a contact transformation (1.13) is a one-parameter (ε) Lie group of contact transformations given by

$$(x^*)^i = x^i + \varepsilon \xi^i(x, u, \partial u) + O(\varepsilon^2), \tag{1.16a}$$

$$u^* = u + \varepsilon \eta(x, u, \partial u) + O(\varepsilon^2), \tag{1.16b}$$

$$u_i^* = u_i + \eta_i^{(1)}(x, u, \partial u) + O(\varepsilon^2), \tag{1.16c}$$

$i = 1, \ldots, n$, with infinitesimal generator

$$\mathrm{X} = \xi^i(x, u, \partial u) \frac{\partial}{\partial x^i} + \eta(x, u, \partial u) \frac{\partial}{\partial u} + \eta_i^{(1)}(x, u, \partial u) \frac{\partial}{\partial u_i}. \tag{1.17}$$

Theorem 1.2.2. *Equations* (1.16) *define a one-parameter Lie group of contact transformations if and only if the functions ξ^i, η satisfy*

$$\frac{\partial \eta}{\partial u_i} - u_j \frac{\partial \xi^j}{\partial u_i} = 0, \quad i = 1, \ldots, n. \tag{1.18}$$

Proof. From (1.10), one has

$$\eta_j^{(1)} = \frac{\partial \eta}{\partial x^j} + \frac{\partial \eta}{\partial u} u_j + \frac{\partial \eta}{\partial u_i} u_{ij} - \left[\frac{\partial \xi^k}{\partial x^j} + \frac{\partial \xi^k}{\partial u} u_j + \frac{\partial \xi^k}{\partial u_i} u_{ij} \right] u_k, \quad j = 1, \ldots, n.$$

Equations (1.16) define a one-parameter Lie group of contact transformations if and only if $\partial \eta_j^{(1)} / \partial u_{ik} = 0$, $i, j, k = 1, \ldots, n$. This leads to (1.18). □

Let the *characteristic function* W of an infinitesimal generator (1.17) be defined by

$$W = \xi^i u_i - \eta. \tag{1.19}$$

Then it is straightforward to show that the following theorem holds [Exercise 1.2.2].

Theorem 1.2.3. *Let* (1.17) *be the infinitesimal generator of a one-parameter Lie group of contact transformations. In terms of the characteristic function* (1.19), *the infinitesimals are given by*

$$\begin{aligned}\xi^j &= \frac{\partial W}{\partial u_j},\\ \eta &= u_i \frac{\partial W}{\partial u_i} - W,\\ \eta_j^{(1)} &= -\frac{\partial W}{\partial x^j} - u_j \frac{\partial W}{\partial u},\end{aligned}$$

$j = 1, \ldots, n.$

1.2.3 Higher-order transformations

Now consider the situation of n independent variables $x = (x^1, \ldots, x^n)$ and m dependent variables $u(x) = (u^1(x), \ldots, u^m(x))$ and a transformation acting on some finite-dimensional $(x, u, \partial u, \ldots, \partial^k u)$-space such that the transformation leaves invariant all contact conditions. Such transformations are of the form

$$(x^*)^i = f^i(x, u, \partial u, \ldots, \partial^s u), \tag{1.20a}$$

$$(u^*)^\mu = g^\mu(x, u, \partial u, \ldots, \partial^s u). \tag{1.20b}$$

One can show [Exercise 1.2.3] that the only transformations of the form (1.20) that are one-to-one on $(x, u, \partial u, \ldots, \partial^k u)$-space for some finite k are point transformations of the form (1.1) or contact transformations of the form (1.13). In the proof, one can show that for all other transformations of the form (1.20), for any k, the components of kth-order partial derivatives in the transformations will have an essential dependence on components of partial derivatives of order at least $k+1$. Hence any one-to-one transformation of the form (1.20) must act on the space of all partial derivatives (an infinite-dimensional space) if it is not a point or contact transformation. Such a one-to-one transformation is called *a higher-order transformation.*

1.2.4 One-parameter higher-order transformations

Even though higher-order transformations do not have a one-to-one action on any finite-dimensional space $(x, u, \partial u, \ldots, \partial^k u)$, it turns out that one can characterize a class of one-parameter higher-order transformations in terms of infinitesimal generators (vector fields) $\mathrm{X} = \eta^i(x, u, \partial u, \ldots, \partial^k u)\partial/\partial u^i$. To do this effectively, it is important to first consider one-parameter Lie groups of point transformations in terms of mappings of surfaces to other surfaces and then consider an equivalent way of finding such mappings so that independent variables are not transformed, i.e., only dependent variables are transformed.

Mappings of surfaces

Consider a one-parameter Lie group of point transformations

$$(x^*)^i = f^i(x, u; \varepsilon) = e^{\varepsilon \mathrm{X}} x^i, \quad i = 1, \ldots, n, \tag{1.21a}$$
$$(u^*)^\mu = g^\mu(x, u; \varepsilon) = e^{\varepsilon \mathrm{X}} u^\mu, \quad \mu = 1, \ldots, m, \tag{1.21b}$$

with infinitesimal generator

$$\mathrm{X} = \xi^i(x, u)\frac{\partial}{\partial x^i} + \eta^\mu(x, u)\frac{\partial}{\partial u^\mu}. \tag{1.22}$$

Consider the family of surfaces $u^\mu = \Theta^\mu(x)$ that is not invariant under (1.21). For a specific value of ε, a transformation (1.21) maps a point (x, u) on the family of surfaces $u^\mu = \Theta^\mu(x)$ into the point (x^*, u^*) with

$$(x^*)^i = f(x, \Theta(x); \varepsilon), \quad i = 1, \ldots, n, \tag{1.23a}$$
$$(u^*)^\mu = g^\mu(x, \Theta(x); \varepsilon), \quad \mu = 1, \ldots, m. \tag{1.23b}$$

For this specified value of ε, one can eliminate x from (1.23) by substitution through the inverse transformation of (1.21a), i.e., by substitution of

$$x = f(x^*, u^*; -\varepsilon)$$

into (1.23b). Then

$$\begin{aligned} u^* &= g(f(x^*, u^*; -\varepsilon), \Theta(f(x^*, u^*; -\varepsilon)); \varepsilon) \\ &= g(e^{-\varepsilon \mathrm{X}} x^*, \Theta(e^{-\varepsilon \mathrm{X}} x^*); \varepsilon), \end{aligned} \tag{1.24}$$

with

$$\mathrm{X} = \xi^i(x, u)\frac{\partial}{\partial x^i} + \eta^\mu(x, u)\frac{\partial}{\partial u^\mu} = \xi^i(x^*, u^*)\frac{\partial}{\partial (x^*)^i} + \eta^\mu(x^*, u^*)\frac{\partial}{\partial (u^*)^\mu}.$$

Replacing $(x^*, u^*, -\varepsilon)$ by (x, u, ε) in (1.24), one then has

$$u = g(e^{\varepsilon \mathrm{X}} x, \Theta(e^{\varepsilon \mathrm{X}} x); -\varepsilon) = g(f(x, u; \varepsilon), \Theta(f(x, u; \varepsilon)); -\varepsilon). \tag{1.25}$$

Theorem 1.2.4. *Suppose the family of surfaces* $u = \Theta(x)$ *is not invariant under* (1.21). *Then* (1.25) *implicitly defines a mapping of the family of surfaces* $u^\mu = \Theta^\mu(x)$ *into a one-parameter family of surfaces* $u^\mu = \phi^\mu(x;\varepsilon)$.

In order to effectively generalize one-parameter Lie groups of point or contact transformations to one-parameter higher-order transformations, it is important to consider the mapping of surfaces from the point of view of transformations acting directly on the space of functions $u = u(x)$ instead of transformations acting on (x, u)-space (or $(x, u, \partial u)$-space in the case of contact transformations). In particular, this leads to an explicit formula for the family of surfaces $u^\mu = \phi^\mu(x;\varepsilon)$ defined by (1.25).

Consider again the mapping of a family of surfaces $u^\mu = \Theta^\mu(x)$ into the one-parameter family of surfaces $u^\mu = \phi^\mu(x;\varepsilon)$ under a one-parameter Lie group of point transformations (1.21). Geometrically, this transformation represents a mapping of points (x, u) into (x^*, u^*) as discussed above, leading to the implicit formula (1.25) for the family of surfaces $u^\mu = \Theta^\mu(x)$. For generalization to mappings under higher-order transformations, it is important to describe this mapping explicitly as a direct transformation of the family of surfaces $u^\mu = \Theta^\mu(x)$ to the one-parameter family of surfaces $u^\mu = \phi^\mu(x;\varepsilon)$. Formally, this mapping is given by

$$\begin{aligned} x^* &= x, \\ u^* &= \phi(x;\varepsilon) = (e^{\varepsilon \hat{\mathrm{X}}} u)\Big|_{u=\Theta(x)}, \end{aligned}$$

in terms of some infinitesimal generator $\hat{\mathrm{X}}$.

We now derive a formula for the infinitesimal generator $\hat{\mathrm{X}}$. Under a one-parameter Lie group of point transformations (1.21), one has

$$(x^*)^i = x^i + \varepsilon\xi^i(x, \Theta(x)) + O(\varepsilon^2), \quad i = 1, \dots, n, \tag{1.26a}$$

$$(u^*)^\mu = u^\mu + \varepsilon\eta^\mu(x, \Theta(x)) + O(\varepsilon^2), \quad \mu = 1, \dots, m. \tag{1.26b}$$

The dependence of u^* on x^* yields the image $u^\mu = \phi^\mu(x;\varepsilon)$ of the family of surfaces $u^\mu = \Theta^\mu(x)$. In order to obtain $\phi^\mu(x;\varepsilon)$, one eliminates x from (1.26). Solving (1.26a) for x yields

$$x^i = (x^*)^i - \varepsilon\xi^i(x^*, \Theta(x^*)) + O(\varepsilon^2), \quad i = 1, \dots, n. \tag{1.27}$$

After substituting (1.27) into (1.26b) and then expanding about $\varepsilon = 0$, one obtains

$$\phi^\mu(x^*, \varepsilon) = \Theta^\mu(x^*) + \varepsilon \left[\eta^\mu(x^*, \Theta(x^*)) - \frac{\partial\Theta^\mu(x^*)}{\partial(x^*)^i}\xi^i(x^*, \Theta(x^*))\right] + O(\varepsilon^2),$$
$$\mu = 1, \dots, m. \tag{1.28}$$

Then after replacing x^* by x in (1.28) through using (1.27), one obtains the direct image of the family of surfaces $u^\mu = \Theta^\mu(x)$ under the one-parameter Lie group of point transformations (1.21). In particular, the family of surfaces $u^\mu = \Theta^\mu(x)$ is mapped into the one-parameter family of surfaces given by

$$(u^*)^\mu = \phi(x;\varepsilon) = \Theta^\mu(x) + \varepsilon\left[\eta^\mu(x,\Theta(x)) - \tfrac{\partial \Theta^\mu(x)}{\partial x^i}\xi^i(x,\Theta(x))\right] + O(\varepsilon^2),$$
$$\mu = 1,\ldots,m. \tag{1.29}$$

Now observe that the same image of the family of surfaces $u^\mu = \Theta^\mu(x)$ can be obtained by a one-parameter family of transformations that leaves invariant the independent variables x:

$$\begin{aligned}(x^*)^i &= x^i, \quad i = 1,\ldots,n,\\ (u^*)^\mu &= u^\mu + \varepsilon\left[\eta^\mu(x,u) - u_i^\mu \xi^i(x,u)\right] + O(\varepsilon^2), \quad \mu = 1,\ldots,m.\end{aligned} \tag{1.30}$$

Consequently, the infinitesimal generator for the one-parameter family of transformations (1.30) is given by

$$\hat{\mathrm{X}} = \left[\eta^\mu(x,u) - u_i^\mu \xi^i(x,u)\right]\frac{\partial}{\partial u^\mu}. \tag{1.31}$$

Geometrically, one has moved from a transformation (1.21) acting on (x,u)-space to a transformation (1.30) acting on the space of functions $u = u(x)$. The infinitesimal generator (1.31) is the *characteristic form* (*evolutionary form*) of the infinitesimal generator (1.22).

As examples ($n = m = 1$), for the translation group

$$\begin{aligned}x^* &= x + \varepsilon,\\ u^* &= u,\end{aligned} \tag{1.32}$$

one has

$$\mathrm{X} = \frac{\partial}{\partial x}, \quad \hat{\mathrm{X}} = -u_x\frac{\partial}{\partial u}; \tag{1.33}$$

and for the scaling group

$$\begin{aligned}x^* &= e^\varepsilon x,\\ u^* &= e^{2\varepsilon} u,\end{aligned} \tag{1.34}$$

one has

$$\mathrm{X} = x\frac{\partial}{\partial x} + 2u\frac{\partial}{\partial u}, \quad \hat{\mathrm{X}} = [2u - xu_x]\frac{\partial}{\partial u}. \tag{1.35}$$

Local transformations

One can generalize one-parameter Lie groups of point transformations with infinitesimal generators in the characteristic form (1.31) to one-parameter higher-order local transformations with infinitesimal generators of the form

$$\hat{\mathrm{X}} = \hat{\eta}^{\mu}(x, u, \partial u, \ldots, \partial^{s} u)\frac{\partial}{\partial u^{\mu}}, \tag{1.36}$$

where the infinitesimal components depend on derivatives of u up to some finite order $s \geq 1$.

Formally, one can exponentiate (1.36) to obtain a corresponding one-parameter higher-order local transformation acting on the space of functions $u = u(x)$:

$$\begin{aligned}(x^*)^i &= x^i, \quad i = 1, \ldots, n,\\ (u^*)^{\mu} &= u^{\mu} + \varepsilon\hat{\eta}^{\mu}(x, u, \partial u, \ldots, \partial^s u) + O(\varepsilon^2), \quad \mu = 1, \ldots, m.\end{aligned} \tag{1.37}$$

To calculate the higher-order terms in (1.37), one extends (prolongs) the infinitesimal generator (1.36) to act on the components of derivatives of u by requiring that the contact conditions are invariant. Consequently, the extended infinitesimal generator (*the prolongation of* $\hat{\mathrm{X}}$) is given by

$$\hat{\mathrm{X}}^{\infty} = \hat{\eta}^{\mu}\frac{\partial}{\partial u^{\mu}} + \hat{\eta}_i^{(1)\,\mu}\frac{\partial}{\partial u_i^{\mu}} + \cdots + \hat{\eta}_{i_1\ldots i_p}^{(p)\,\mu}\frac{\partial}{\partial u_{i_1\ldots i_p}^{\mu}} + \cdots, \tag{1.38}$$

where, analogously to (1.10) and (1.11),

$$\hat{\eta}_i^{(1)\,\mu} = \mathrm{D}_i\hat{\eta}^{\mu}, \tag{1.39a}$$

$$\hat{\eta}_{i_1\ldots i_p}^{(p)\,\mu} = \mathrm{D}_{i_p}\hat{\eta}_{i_1\ldots i_{p-1}}^{(p-1)\,\mu}, \tag{1.39b}$$

$\mu = 1, \ldots, m$; $i, i_j = 1, \ldots, n$ for $p = 2, 3, \ldots$.

Hence the exponentiation of the infinitesimal generator (1.36) yields the following transformation.

Definition 1.2.2. *A one-parameter higher-order local transformation* is a transformation of the form

$$\begin{aligned}(x^*)^i &= x^i, \quad i = 1, \ldots, n,\\ (u^*)^{\mu} &= e^{\varepsilon\hat{\mathrm{X}}^{\infty}}u^{\mu} = u^{\mu} + \sum_{j=1}^{\infty}\frac{\varepsilon^j}{j!}(\hat{\mathrm{X}}^{\infty})^{j-1}\hat{\eta}^{\mu}, \quad \mu = 1, \ldots, m,\end{aligned} \tag{1.40}$$

where $\hat{\mathrm{X}}^{\infty}$ is given by (1.38).

Note that one can invert (1.40) through inverse exponentiation.

One can show that:

1. A one-parameter local transformation is equivalent to a one-parameter Lie group of point transformations if and only if all $\hat{\eta}^{\mu}$ are of the form $\hat{\eta}^{\mu} = \eta^{\mu}(x, u) - u_i^{\mu}\xi^i(x, u)$ for some $\eta^{\mu}(x, u), \xi^i(x, u)$ $(i = 1, \ldots, n;$

$\mu = 1, \ldots, m$), i.e, each $\hat{\eta}^\mu$ is linear in the first derivatives u_i^μ and has no dependence on higher-order derivatives of u.

2. A one-parameter local transformation is equivalent to a one-parameter Lie group of contact transformations if and only if $m = 1$ and $\hat{\eta}$ is of the form $\hat{\eta} = \hat{\eta}(x, u, \partial u)$ with a nonlinear dependence on first-order derivatives of u.

In particular, the following theorem holds.

Theorem 1.2.5. *A one-parameter local transformation with an infinitesimal generator of the form*

$$\mathrm{X} = W(x, u, \partial u) \frac{\partial}{\partial u} \tag{1.41}$$

is equivalent to a one-parameter Lie group of contact transformations with the infinitesimal generator

$$\mathrm{X} = \xi^j(x, u, \partial u) \frac{\partial}{\partial x^j} + \eta(x, u, \partial u) \frac{\partial}{\partial u} + \eta_j^{(1)}(x, u, \partial u) \frac{\partial}{\partial u_j}, \tag{1.42}$$

where

$$\xi^j(x, u, \partial u) = \frac{\partial W}{\partial u_j}, \tag{1.43a}$$

$$\eta(x, u, \partial u) = u_i \frac{\partial W}{\partial u_i} - W, \tag{1.43b}$$

$$\eta_j^{(1)}(x, u, \partial u) = -\frac{\partial W}{\partial x^j} - u_j \frac{\partial W}{\partial u}, \tag{1.43c}$$

$j = 1, \ldots, n.$

Proof. Let η, ξ^j satisfy (1.43a),(1.43b). Then

$$\frac{\partial \eta}{\partial u_j} = \frac{\partial W}{\partial u_j} + u_i \frac{\partial^2 W}{\partial u_i \partial u_j} - \frac{\partial W}{\partial u_j} = u_i \frac{\partial^2 W}{\partial u_i \partial u_j} = u_i \frac{\partial \xi^j}{\partial u_i}, \quad j = 1, \ldots, n.$$

Hence (1.18) is satisfied. Moreover,

$$\eta - \xi^j u_j = u_i \frac{\partial W}{\partial u_i} - W - u_j \frac{\partial W}{\partial u_j} = -W;$$

$$\eta_j^{(1)} = \frac{\partial \eta}{\partial x^j} + \frac{\partial \eta}{\partial u} u_j - \left[\frac{\partial \xi^k}{\partial x^j} + \frac{\partial \xi^k}{\partial u} u_j \right] u_k = -\frac{\partial W}{\partial x^j} - u_j \frac{\partial W}{\partial u}.$$

Hence (1.42), (1.43) defines a contact transformation group equivalent to the group generated by $W(x, u, \partial u)\partial/\partial u$. □

Consequently, any one-parameter local transformation with an infinitesimal generator of the form

$$\eta(x, u, \partial u)\frac{\partial}{\partial u} \tag{1.44}$$

is uniquely equivalent to an infinitesimal generator of a one-parameter Lie group of contact transformations with $\eta(x, u, \partial u)$ playing the role of a characteristic function.

The proof that an infinitesimal generator of the form (1.44) is uniquely equivalent to an infinitesimal generator of a one-parameter Lie group of point transformations when $\eta(x, u, \partial u)$ has a linear dependence on first derivatives is left to Exercise 1.2.5.

A *one-parameter higher-order local transformation* is a one-parameter local transformation that is neither a one-parameter Lie group of point transformations nor a one-parameter Lie group of contact transformations. In the literature, such transformations are also called *Lie–Bäcklund transformations* or *Noether transformations.* It should be pointed out that neither Lie nor Bäcklund considered such higher-order transformations whereas Noether (1918) implicitly considered such transformations in her famous paper on conservation laws. For further information on higher-order transformations, see Ibragimov (1985) and Anderson & Ibragimov (1979).

For example, the infinitesimal generator

$$\hat{\mathrm{Y}} = \left[\tfrac{1}{2}u^2 u_x + 2u_x u_{xx} + u u_{xxx} + \tfrac{3}{5}u_{xxxxx}\right]\frac{\partial}{\partial u} \tag{1.45}$$

corresponds to a higher-order transformation. The one-parameter transformation defined by (1.45) arises as a higher-order symmetry for the Korteweg–de Vries (KdV) equation

$$u_t + u u_x + u_{xxx} = 0, \tag{1.46}$$

which describes the amplitude $u(x,t)$ of long surface waves on shallow water.

1.2.5 Point symmetries

Consider a system $\mathbf{R}\{x\,; u\}$ of N PDEs of order k with n independent variables $x = (x^1, \ldots, x^n)$ and m dependent variables $u(x) = (u^1(x), \ldots, u^m(x))$, given by

$$R^\sigma(x, u, \partial u, \ldots, \partial^k u) = 0, \quad \sigma = 1, \ldots, N. \tag{1.47}$$

Consider a one-parameter Lie group of point transformations

$$(x^*)^i = f^i(x, u; \varepsilon), \tag{1.48a}$$
$$(u^*)^\mu = g^\mu(x, u; \varepsilon), \tag{1.48b}$$

with the corresponding infinitesimal generator

$$\mathrm{X} = \xi^i(x,u)\frac{\partial}{\partial x^i} + \eta^\mu(x,u)\frac{\partial}{\partial u^\mu}. \tag{1.49}$$

The kth extension (prolongation) of (1.49) is given by

$$\begin{aligned}\mathrm{X}^{(k)} &= \xi^i(x,u)\frac{\partial}{\partial x^i} + \eta^\mu(x,u)\frac{\partial}{\partial u^\mu} + \eta_i^{(1)\,\mu}(x,u,\partial u)\frac{\partial}{\partial u_i^\mu} \\ &\quad + \ldots + \eta_{i_1\ldots i_k}^{(k)\,\mu}(x,u,\partial u,\ldots,\partial^k u)\frac{\partial}{\partial u_{i_1\ldots i_k}^\mu},\end{aligned} \tag{1.50}$$

where $\eta_i^{(1)\,\mu}$, ..., $\eta_{i_1\ldots i_k}^{(k)\,\mu}$ are defined in terms of $\{\xi^i(x,u), \eta^\mu(x,u)\}$ by (1.10) and (1.11), for $\mu = 1,\ldots,m$, and $i, i_j = 1,\ldots,n$ for $j = 1,\ldots,k$.

Definition 1.2.3. A one-parameter Lie group of point transformations (1.48) *leaves the PDE system* $\mathbf{R}\{x\,;u\}$ (1.47) *invariant* if and only if its kth extension (1.50) leaves invariant the solution manifold of $\mathbf{R}\{x\,;u\}$ in $(x,u,\partial u,\ldots,\partial^k u)$-space, i.e., it maps any family of solution surfaces of the PDE system (1.47) into another family of solution surfaces of PDE system (1.47). In this case, the one-parameter Lie group of point transformations (1.48) is called a *point symmetry* of the PDE system $\mathbf{R}\{x\,;u\}$ (1.47).

Lie's algorithm to find the point symmetries of a given PDE system (1.47) is given by the following theorem.

Theorem 1.2.6 (Infinitesimal criterion of invariance under a one-parameter Lie group of point transformations). *Let* (1.49) *be the infinitesimal generator of a one-parameter Lie group of point transformations* (1.48). *Let* (1.50) *be its kth extension. Then the transformation* (1.48) *is a point symmetry of the PDE system* $\mathbf{R}\{x\,;u\}$ (1.47) *if and only if for each* $\alpha = 1,\ldots,N$,

$$\mathrm{X}^{(k)}R^\alpha(x,u,\partial u,\ldots,\partial^k u) = 0, \tag{1.51}$$

when

$$R^\sigma(x,u,\partial u,\ldots,\partial^k u) = 0, \quad \sigma = 1,\ldots,N. \tag{1.52}$$

The proof of this theorem appears in Olver (1986) with the restriction that $\mathbf{R}\{x\,;u\}$ (1.47) can be written in a solved form in terms of a set of leading derivatives.

Note that the invariance criterion (1.51), (1.52) involves substitutions of the N PDEs (1.52) and (possibly) their differential consequences into each of the N determining equations (1.51). In particular, the extended generator $\mathrm{X}^{(k)}$ (1.50) involves derivatives of u up to (and including) the order k. To carry out these substitutions, in practice one first expresses each PDE in the system (1.52) in a solved form with respect to some leading derivative. In particular, the resulting solved-form system of N PDEs must have the feature that all differential consequences of the left-hand sides of each PDE in the system (1.52) are linearly independent of the left-hand sides of the other $N-1$ PDEs in the system (1.52). Moreover, none of these N left-hand side leading derivatives or their differential consequences can appear in any of the right-hand sides of the N PDEs in the system (1.52). [In the case of a scalar PDE (1.52), a leading derivative is given by any kth-order derivative that appears linearly in (1.52). In the case of a dynamical PDE system (1.52), a time derivative yields a leading derivative.] Then one substitutes the established leading derivatives given by the N solved-form PDEs and (possibly) their differential consequences into each of the equations (1.51). [In particular, for any PDE $R^\sigma(x, u, \partial u, \ldots, \partial^{k'} u) = 0$ of order $k' < k$, its differential consequences up to order k, i.e., $\partial^l R^\sigma(x, u, \partial u, \ldots, \partial^{k'} u) = 0$, $l = 1, \ldots, k-k'$, must be computed and used in substitutions.] The resulting linear system to determine the components $\xi^i(x,u), \eta^\mu(x,u)$ of the infinitesimal generator (1.49) is called *the set of determining equations for the point symmetries of* $\mathbf{R}\{x\,;u\}$ (1.47).

In the resulting set of determining equations for the point symmetries of $\mathbf{R}\{x\,;u\}$, one can treat each u^μ and each of its derivatives u_i^μ, u_{ij}^μ, etc. as independent variables along with x^i. Consequently, the set of determining equations splits into an over-determined linear PDE system for $\{\xi^i(x,u), \eta^\mu(x,u)\}$. The number of arbitrary constants that arise in the solution of the set of determining equations is the number of point symmetry generators (1.49) when the number of generators is finite. [It can happen that the number of point symmetry generators is infinite – this is the case when the given PDE system $\mathbf{R}\{x\,;u\}$ (1.47) is a linear PDE system or a nonlinear PDE system that is linearizable by a point transformation.]

After the symmetry components $\{\xi^i(x,u), \eta^\mu(x,u)\}$ $(i = 1, \ldots, n;\ \mu = 1, \ldots, m)$ are found, one can find the global form of the Lie group of point transformations through either solving a corresponding system of first-order ODEs or exponentiation in terms of the infinitesimal point symmetry generators. For details, see any of Bluman & Anco (2002), Bluman & Kumei (1989), Olver (1986), Stephani (1989), or Hydon (2000).

As an example, consider the linear heat equation

$$u_t = u_{xx}. \tag{1.53}$$

One can show that X is a point symmetry of the heat equation (1.53) if and only if its second extension $\mathrm{X}^{(2)}$ satisfies

$$\left[\mathrm{X}^{(2)}(u_t - u_{xx})\right]\Big|_{u_{xx}=u_t} = (\eta_t^{(1)} - \eta_{xx}^{(2)})\Big|_{u_{xx}=u_t} = 0 \tag{1.54}$$

which leads to

$$X = \xi(x,t)\frac{\partial}{\partial x} + \tau(t)\frac{\partial}{\partial t} + [f(x,t)u + g(x,t)]\frac{\partial}{\partial u}$$

with the components of X satisfying the corresponding set of linear determining equations given by

$$\begin{aligned} \tau'(t) - 2\xi_x &= 0, \\ 2f_x - \xi_{xx} + \xi_t &= 0, \\ f_t - f_{xx} &= 0, \\ g_t - g_{xx} &= 0. \end{aligned} \tag{1.55}$$

After solving the linear determining system (1.55), one finds that the heat equation (1.53) has an infinite number of point symmetries given by the infinitesimal generators $X_\infty = g(x,t)\partial/\partial u$ with $g_t = g_{xx}$, corresponding to its linearity, and six nontrivial point symmetries given by the infinitesimal generators

$$\begin{aligned} &X_1 = \frac{\partial}{\partial t}, \quad X_2 = \frac{\partial}{\partial x}, \quad X_3 = 2t\frac{\partial}{\partial t} + x\frac{\partial}{\partial x}, \\ &X_4 = tx\frac{\partial}{\partial x} + t^2\frac{\partial}{\partial t} - \left(\tfrac{1}{2}t + \tfrac{1}{4}x^2\right)u\frac{\partial}{\partial u}, \\ &X_5 = t\frac{\partial}{\partial x} - \tfrac{1}{2}xu\frac{\partial}{\partial u}, \quad X_6 = \frac{\partial}{\partial u}. \end{aligned} \tag{1.56}$$

Now consider the point symmetry represented by the infinitesimal generator

$$X_4 = tx\frac{\partial}{\partial x} + t^2\frac{\partial}{\partial t} - \left(\tfrac{1}{2}t + \tfrac{1}{4}x^2\right)u\frac{\partial}{\partial u}. \tag{1.57}$$

The initial value problem to determine the corresponding global one-parameter Lie group of point transformations is given by

$$\begin{aligned} \frac{dx^*}{d\varepsilon} &= x^*t^*, \\ \frac{dt^*}{d\varepsilon} &= (t^*)^2, \\ \frac{du^*}{d\varepsilon} &= -\left(\tfrac{1}{2}(t^*) + \tfrac{1}{4}(x^*)^2\right)u^*, \end{aligned} \tag{1.58}$$

with $x^* = x$, $t^* = t$, $u^* = u$ when $\varepsilon = 0$. Solving (1.58), one obtains the one-parameter Lie group of point transformations

$$x^* = \frac{x}{1-\varepsilon t}, \quad t^* = \frac{t}{1-\varepsilon t}, \quad u^* = \left[\sqrt{1-\varepsilon t}\,\exp\left(-\frac{\varepsilon x^2}{4(1-\varepsilon t)}\right)\right]u, \tag{1.59}$$

admitted by the heat equation (1.53).

1.2.6 Contact and higher-order symmetries

Previously, we have seen that in order to generalize one-parameter Lie groups of point (or contact) transformations to one-parameter higher-order local transformations, it was natural to consider transformations from the point of view of directly mapping surfaces to other surfaces with independent variables fixed. This led to the consideration of one-parameter local transformations with infinitesimal generators of the form (1.36) with extended infinitesimal generators given by (1.38), (1.39).

A one-parameter local transformation (1.37) *leaves the PDE system* $\mathbf{R}\{x;u\}$ (1.47) *invariant* if and only if its kth extension

$$\mathrm{X}^{(k)} = \hat{\eta}^{\mu}\frac{\partial}{\partial u^{\mu}} + \hat{\eta}_i^{(1)\,\mu}\frac{\partial}{\partial u_i^{\mu}} + \cdots + \hat{\eta}_{i_1\ldots i_k}^{(k)\,\mu}\frac{\partial}{\partial u_{i_1\ldots i_k}^{\mu}} \tag{1.60}$$

leaves invariant the solution manifold of $\mathbf{R}\{x;u\}$ in $(x, u, \partial u, \ldots, \partial^k u)$-space, i.e., it maps any family of solution surfaces of the PDE system (1.47) into some (possibly the same) family of solution surfaces of PDE system (1.47). In this case, the one-parameter local transformation is called a *local* (*higher-order*, *contact*, or *point*) *symmetry* of the PDE system $\mathbf{R}\{x;u\}$ (1.47).

From this point of view, Lie's algorithm extends as follows in order to find the local symmetries of a given PDE system if $\mathbf{R}\{x;u\}$ (1.47) can be written in a solved form with respect to a set of leading derivatives.

Theorem 1.2.7 (Infinitesimal criterion of invariance under a one-parameter local transformation). *Let* (1.36) *be the infinitesimal generator of a one-parameter local transformation* (1.37) *of order* $s \geq 0$, *and let* $\mathrm{X}^{(k)}$ (1.60) *be its kth extension. Then the transformation* (1.37) *is a local (point, contact or higher-order) symmetry of the PDE system* $\mathbf{R}\{x;u\}$ (1.47) *if and only if for each* $\alpha = 1, \ldots, N$,

$$\mathrm{X}^{(k)} R^{\alpha}(x, u, \partial u, \ldots, \partial^k u) = 0, \tag{1.61}$$

when

$$\begin{aligned} R^{\sigma}(x, u, \partial u, \ldots, \partial^k u) &= 0, \quad \sigma = 1, \ldots, N, \\ \partial^l R^{\sigma}(x, u, \partial u, \ldots, \partial^k u) &= 0, \quad l = 1, \ldots, s, \quad \sigma = 1, \ldots, N. \end{aligned} \tag{1.62}$$

The invariance criterion (1.61), (1.62) involves substitutions of the N PDEs (1.62) and their differential consequences (up to order s) into each of the N equations (1.61). [Note that if a particular PDE $R^{\sigma}(x, u, \partial u, \ldots, \partial^{k'} u) = 0$ has order $k' < k$, then all its differential consequences $\partial^l R^{\sigma}(x, u, \partial u, \ldots, \partial^{k'} u) = 0$, $l = 1, \ldots, s + k - k'$, must be computed and used in substitutions.] After these substitutions, the resulting linear system of PDEs to determine the components of the infinitesimal generator (1.36) is called *the set of determining equations for the local symmetries of order s of* $\mathbf{R}\{x\,;u\}$ (1.47).

Note that the set of N PDEs (1.61) without substitutions arising from (1.62) is just the linearization of the given PDE system (1.62), i.e., the Fréchet derivative of the given PDE system (1.62). After the substitutions of (1.62) and their differential consequences, one sees that a local symmetry of a given system of PDEs is simply any solution of its linearized system that holds for every solution of the given PDE system. In particular, in terms of an *arbitrary* function $V(x) = (V^1(x), \ldots, V^m(x))$, the *linearizing operator* (Fréchet derivative) associated with the PDE system $\mathbf{R}\{x\,;u\}$ (1.47) is given by

$$\mathrm{L}^{\sigma}_{\rho}[U]V^{\rho} = \left[\frac{\partial R^{\sigma}[U]}{\partial U^{\rho}} + \frac{\partial R^{\sigma}[U]}{\partial U^{\rho}_{i}}\mathrm{D}_i + \cdots + \frac{\partial R^{\sigma}[U]}{\partial U^{\rho}_{i_1\ldots i_k}}\mathrm{D}_{i_1}\ldots\mathrm{D}_{i_k}\right]V^{\rho}, \quad \sigma = 1,\ldots,N. \tag{1.63}$$

It follows that a local transformation (1.37) is a local symmetry of the PDE system $\mathbf{R}\{x\,;u\}$ (1.47) if and only if

$$\mathrm{L}^{\sigma}_{\rho}[u]\hat{\eta}^{\rho}[u] = 0, \quad \sigma = 1,\ldots,N, \tag{1.64}$$

when $R^{\sigma}(x,u,\partial u,\ldots,\partial^k u) = 0$, $\partial^l R^{\sigma}(x,u,\partial u,\ldots,\partial^k u) = 0$, $l = 1,\ldots,s$, $\sigma = 1,\ldots,N$.

1.2.7 Equivalence transformations and symmetry classification

If a PDE system contains classifying (*constitutive*) functions and/or parameters, it is useful to consider *equivalence transformations* of the system, i.e., transformations that preserve the differential structure of the equations in the PDE system but may change the form of the constitutive functions and/or parameters. In particular, the consideration of equivalence transformations is useful in analyses that involve classifications with respect to constitutive functions and/or parameters, such as local symmetry and local conservation law analysis. Moreover, classification tables are usually presented modulo known equivalence transformations, i.e., only for forms of constitutive functions and/or parameters that are not related by an equivalence transformation.

Work on equivalence transformations was initiated by Ovsiannikov (1982). Multiple applications and extensions of the notion of equivalence transformations appear in the works of Akhatov, Gazizov & Ibragimov [(1987), (1991)],

Ibragimov, Torrisi & Valenti (1991), Lisle (1992), Popovych & Ivanova (2005a), and other authors.

Without loss of generality, consider the family $\mathcal{F}_K$ of PDE systems $\mathbf{R}\{x\,;u;K\}$:

$$R^\sigma(x,u,\partial u,\ldots,\partial^k u,K)=0,\quad \sigma=1,\ldots,N, \tag{1.65}$$

involving L constitutive functions and/or parameters $K=(K_1,\ldots,K_L)$. Such functions may depend on particular dependent and independent variables of the system, as well as derivatives of dependent variables.

Definition 1.2.4. A one-parameter Lie group of equivalence transformations of a family $\mathcal{F}_K$ of PDE systems is a one-parameter Lie group of transformations given by

$$\begin{aligned}\widetilde{x}^i&=f^i(x,u;\varepsilon), && i=1,\ldots,n,\\ \widetilde{u}^\mu&=g^\mu(x,u;\varepsilon), && \mu=1,\ldots,m,\\ \widetilde{K}_l&=G_l(x,u,K;\varepsilon), && l=1,\ldots,L,\end{aligned} \tag{1.66}$$

which maps a PDE system $\mathbf{R}\{x\,;u;K\}\in\mathcal{F}_K$ into another PDE system $\mathbf{R}\{\widetilde{x}\,;\widetilde{u};\widetilde{K}\}$ in the same family.

Note that if constitutive functions and/or parameters are not modified under the one-parameter Lie group of equivalence transformations (1.66), then the transformation (1.66) is simply a point symmetry of each PDE system in the family $\mathcal{F}_K$.

Simple one-parameter Lie groups of equivalence transformations can be often found by inspection. Consider the incompressible three-dimensional Navier–Stokes equations in Cartesian coordinates (x^1,x^2,x^3):

$$\begin{aligned}&\frac{\partial v^1}{\partial x^1}+\frac{\partial v^2}{\partial x^2}+\frac{\partial v^3}{\partial x^3}=0,\\ &\frac{\partial v^i}{\partial t}+v^j\frac{\partial v^i}{\partial x^j}+\frac{\partial p}{\partial x^i}=\nu\left(\frac{\partial^2 v^i}{\partial (x^1)^2}+\frac{\partial^2 v^i}{\partial (x^2)^2}+\frac{\partial^2 v^i}{\partial (x^3)^2}\right),\\ &i=1,2,3.\end{aligned} \tag{1.67}$$

Here the fluid viscosity $\nu=\text{const}>0$ is a constitutive parameter. One can check that the PDE system (1.67) admits the group of equivalence transformations

$$\widetilde{t}=t,\quad \widetilde{x}=ax,\quad \widetilde{v}^i=av^i,\quad \widetilde{p}=a^2p,\quad \widetilde{\nu}=a^2\nu,$$

where $a=e^\varepsilon$, which maps the PDE system (1.67) into the PDE system

$$
\begin{aligned}
&\frac{\partial \tilde{v}^1}{\partial \tilde{x}^1}+\frac{\partial \tilde{v}^2}{\partial \tilde{x}^2}+\frac{\partial \tilde{v}^3}{\partial \tilde{x}^3}=0,\\
&\frac{\partial \tilde{v}^i}{\partial t}+\tilde{v}^j\frac{\partial \tilde{v}^i}{\partial \tilde{x}^j}+\frac{\partial \tilde{p}}{\partial \tilde{x}^i}=\nu\left(\frac{\partial^2 \tilde{v}^i}{\partial (\tilde{x}^1)^2}+\frac{\partial^2 \tilde{v}^i}{\partial (\tilde{x}^2)^2}+\frac{\partial^2 \tilde{v}^i}{\partial (\tilde{x}^3)^2}\right),\\
&i=1,2,3.
\end{aligned}
$$

which coincides with (1.67) except for a different viscosity coefficient.

As a second example, consider the family of nonlinear PDE systems $\{\mathbf{R}\{x,t\,;u,v;K\}\}$

$$
\begin{aligned}
v_x &= u,\\
v_t &= K(u)u_x
\end{aligned}
\tag{1.68}
$$

related to the nonlinear diffusion equation where the conductivity $K(u)$ is an arbitrary constitutive function. To find one-parameter Lie groups of equivalence transformations of the family (1.68), one applies the standard Lie symmetry algorithm to the general PDE system (1.68), treating the constitutive function $K(u)$ as a new dependent variable. Additionally, one needs to assume that the symmetry components for t, x, u and v are independent of K, following the definition (1.66). The resulting symmetry generators are given by

$$
\begin{aligned}
&\mathrm{X}_1=\frac{\partial}{\partial t},\quad \mathrm{X}_2=\frac{\partial}{\partial x},\quad \mathrm{X}_3=\frac{\partial}{\partial v},\\
&\mathrm{X}_4=\frac{\partial}{\partial u}+x\frac{\partial}{\partial v},\quad \mathrm{X}_5=x\frac{\partial}{\partial x}+v\frac{\partial}{\partial v}+2t\frac{\partial}{\partial t},\\
&\mathrm{X}_6=u\frac{\partial}{\partial u}+v\frac{\partial}{\partial v},\quad \mathrm{X}_7=F(t)\frac{\partial}{\partial t}-KF'(t)\frac{\partial}{\partial K},\\
&\mathrm{X}_8=v\frac{\partial}{\partial x}-u^2\frac{\partial}{\partial u}+2Ku\frac{\partial}{\partial K},\\
&\mathrm{X}_9=x^2\frac{\partial}{\partial x}+(v-xu)\frac{\partial}{\partial u}+xv\frac{\partial}{\partial v}+4xK\frac{\partial}{\partial K},\\
&\mathrm{X}_{10}=xv\frac{\partial}{\partial x}+u(v-xu)\frac{\partial}{\partial u}+v^2\frac{\partial}{\partial v}+2(xu+v)K\frac{\partial}{\partial K}.
\end{aligned}
\tag{1.69}
$$

One can check that for the generators X_9 and X_{10}, if $K=K(u)$, then the transformed function $\widetilde{K}$ is not a function of $\widetilde{u}$ only. Therefore these generators do not correspond to equivalence transformations of the family (1.68). [The same is true for the generator X_7 with the arbitrary function $F(t)$, when $F''(t)\neq 0$.]

The finite form of the equivalence transformations of the family of PDEs (1.68) arises from the seven generators $\mathrm{X}_1,\ldots,\mathrm{X}_7$ ($F'(t)=\text{const}$) and correspondingly involves seven arbitrary parameters. It is given by

$$\begin{aligned}
\widetilde{t} &= a_5 t + a_1, \quad \widetilde{x} = a_6 x + a_2, \\
\widetilde{u} &= a_7 u + a_3, \quad \widetilde{v} = a_5 a_7 v + a_3 x + a_4, \\
\widetilde{K}(\widetilde{u}) &= \frac{a_5^2}{a_6} K(u) = \frac{a_5^2}{a_6} K\left(\frac{\widetilde{u} - a_3}{a_7}\right),
\end{aligned} \tag{1.70}$$

where $a_1, \ldots, a_7$ are arbitrary constants, with $a_5, a_6, a_7 \neq 0$. In particular, under an equivalence transformation (1.70), the PDE system $\mathbf{R}\{x, t\,; u, v; K\}$ (1.68) with a constitutive function $K(u)$ is mapped into a PDE system $\mathbf{R}\{\widetilde{x}, \widetilde{t}\,; \widetilde{u}, \widetilde{v}; \widetilde{K}\}$ of the form (1.68), with $\widetilde{K}(\widetilde{u}) = \frac{a_5^2}{a_6} K\left(\frac{\widetilde{u} - a_3}{a_7}\right)$.

The generator X_8 defines an additional one-parameter group of equivalence transformations given by

$$\begin{aligned}
&\widetilde{t} = t, \quad \widetilde{x} = x - a_8 v, \quad \widetilde{u} = \frac{u}{1 - a_8 u}, \quad \widetilde{v} = v, \\
&\widetilde{K}(\widetilde{u}) = (1 + a_8 \widetilde{u})^{-2} K\left(\frac{\widetilde{u}}{1 + a_8 \widetilde{u}}\right).
\end{aligned} \tag{1.71}$$

Here a_8 is an arbitrary constant.

It is worth making the following remark [Ovsiannikov (1982)]. Consider a family $\mathcal{F}_K$ of PDE systems with constitutive functions and/or parameters K. Consider a symmetry group that is common for all PDE systems in the family $\mathcal{F}_K$, i.e., an intersection of symmetry groups of PDE systems with all possible forms of constitutive functions and/or parameters K. Then such a group is always included in the group of equivalence transformations of the family $\mathcal{F}_K$.

1.2.8 Recursion operators for local symmetries

In Section 1.2.6, it was seen that the algorithm for finding contact or higher-order symmetries of PDEs is essentially the same as that for finding point symmetries. A difficulty in applying this algorithm is to determine a priori which derivatives could appear in an admitted infinitesimal (1.37). For most known scalar PDE examples, it happens that if a given scalar PDE has a higher-order symmetry then it has an infinite sequence of higher-order symmetries where successive terms of the sequence depend on higher-order derivatives of the dependent variable. Olver (1977) introduced the notion of *recursion operators* to generate such infinite sequences of higher-order symmetries. Firstly, we examine the situation for linear PDEs where local recursion operators and corresponding sequences of higher-order symmetries can arise naturally from a nontrivial local symmetry. Secondly, we consider the situation for nonlinear PDEs.

Recursion operators for linear PDEs

Consider a linear PDE system

$$\mathrm{L}u = 0, \tag{1.72}$$

where L is a linear differential operator. Suppose the linear PDE system (1.72) has a nontrivial local symmetry

$$\mathrm{X} = \eta^\mu(x, u, \partial u, \ldots, \partial^s u)\frac{\partial}{\partial u^\mu}. \tag{1.73}$$

["Trivial" local symmetries of any linear PDE system (1.72) include $\mathrm{X} = u^\mu \partial/\partial u^\mu$ and $\mathrm{X} = f^\mu(x)\partial/\partial u^\mu$ where $u = f(x)$ is any solution of (1.72).] From (1.64) and the linearity of (1.72), it follows that (1.72) has the local symmetry (1.73) if and only if η satisfies the local symmetry determining equations

$$\mathrm{L}\eta\Big|_{\mathrm{L}u=0} = 0. \tag{1.74}$$

Now suppose η is linear homogeneous in u, i.e.,

$$\eta = \mathcal{R}u \tag{1.75}$$

in terms of some linear differential operator $\mathcal{R}$. [This is the situation for any *point symmetry* of a linear scalar PDE of second or higher order [Bluman (1990)].] For a scalar u, equation (1.75) is of the form

$$\eta = r(x)u + r^i(x)u_i + \cdots + r^{i_1\cdots i_k}(x)u_{i_1\cdots i_k}, \tag{1.76}$$

with

$$\mathcal{R} = r(x) + r^i(x)\mathrm{D}_i + \cdots + r^{i_1\cdots i_k}(x)\mathrm{D}_{i_1}\ldots\mathrm{D}_{i_k}, \tag{1.77}$$

for some functions $r(x),\ r^i(x),\ \ldots, r^{i_1\cdots i_k}(x)$. For a linear PDE system, equations (1.75) are of the form

$$\eta^\alpha = r^\alpha_\beta(x)u^\beta + r^{\alpha\ i}_\beta(x)u^\beta_i + \cdots + r^{\alpha\ i_1\cdots i_k}_\beta(x)u^\beta_{i_1\cdots i_k}, \tag{1.78}$$

and $\mathcal{R}$ is a matrix differential operator with matrix elements

$$\mathcal{R}^\alpha_\beta = r^\alpha_\beta(x) + r^{\alpha\ i}_\beta(x)\mathrm{D}_i + \cdots + r^{\alpha\ i_1\cdots i_k}_\beta(x)\mathrm{D}_{i_1}\ldots\mathrm{D}_{i_k}, \tag{1.79}$$

for some functions $r^\alpha_\beta(x)$, $r^{\alpha\, i}_\beta(x)$, $\ldots, r^{\alpha\, i_1\cdots i_k}_\beta(x)$. When $\eta = \mathcal{R}u$, the local symmetry determining equations (1.74) become

$$\mathrm{L}\mathcal{R}u\Big|_{\mathrm{L}u=0} = 0, \tag{1.80}$$

i.e., if $u = \Theta(x)$ solves $\mathrm{L}u = 0$, then $u = \mathcal{R}\Theta(x)$ also solves $\mathrm{L}u = 0$. It immediately follows that if the linear PDE system (1.72) has a local symmetry (1.73) with an infinitesimal of the form (1.75), then it also has higher-order symmetries (1.73) with infinitesimals of the form

$$\eta = \mathcal{R}^j u, \quad j = 1, \ldots . \tag{1.81}$$

More generally, from the above it is easy to see that the following theorem holds.

Theorem 1.2.8. *If the determining equations* (1.74) *are satisfied by both* $\eta = \tilde{\eta}(x, u, \partial u, \ldots, \partial^p u)$ *and* $\eta = \mathcal{R}u$, *where* $\mathcal{R}$ *is given by* (1.77) *or* (1.79), *then*

$$\eta = \mathcal{R}^j \tilde{\eta}, \quad j = 1, \ldots \tag{1.82}$$

also satisfies the determining equations (1.74). *Hence the linear PDE system* (1.72) *has the infinite sequence of higher-order symmetries* $\mathcal{R}^j \tilde{\eta}^\mu \partial/\partial u^\mu$, $j = 1, \ldots$.

A linear differential operator $\mathcal{R}$ given by (1.77) or (1.79) is called a *recursion operator* for the linear PDE system (1.72) corresponding to an admitted nontrivial local symmetry (1.73) with an infinitesimal of the form (1.75).

In both quantum mechanics and the study of group properties of special functions [Miller (1968), (1977)], local symmetries of related linear PDE systems are considered in terms of corresponding recursion operators $\mathcal{R}$.

The following lemma is easily proved by direct calculation.

Lemma 1.2.1. *The extended operator* $\mathrm{X}^\infty = \eta^\gamma \partial/\partial u^\gamma + \mathrm{D}_j \eta^\gamma \partial/\partial u^\gamma_j + \cdots$ *commutes with any total derivative operator* D_i, *i.e.,* $[\mathrm{D}_i, \mathrm{X}^\infty] = 0$.

The proof of the following theorem follows from the above lemma.

Theorem 1.2.9. *Let* $\eta_i = \mathcal{R}_i u$ *and let* $\mathrm{X}_i = \eta_i \partial/\partial u$ *where* $\mathcal{R}_i$ *is a linear differential operator of the form* (1.77), $i = 1, \ldots$. *The commutation relation* $[\mathrm{X}_i, \mathrm{X}_j] = \mathrm{X}_k$ *holds if and only if the commutation relation* $[\mathcal{R}_i, \mathcal{R}_j] = -\mathcal{R}_k$ *is satisfied.*

Now consider two examples.

(1) Schrödinger equation for a harmonic oscillator

As a first example, consider the Schrödinger equation for a harmonic oscillator given by

$$\mathrm{L}u = (\mathrm{H} - i\mathrm{D}_t)u = (-\tfrac{1}{2}\mathrm{D}_x^2 + \tfrac{1}{2}x^2 - i\mathrm{D}_t)u = 0. \tag{1.83}$$

One can show that the PDE (1.83) has the recursion operators $\mathcal{R}_1 = e^{it}(x + \mathrm{D}_x)$ and $\mathcal{R}_2 = e^{-it}(x - \mathrm{D}_x)$ as well as the trivial operator $\mathcal{R}_3 = 1$, with $[\mathcal{R}_1, \mathcal{R}_2] = 2\mathcal{R}_3$. The corresponding local symmetries are

$$\mathrm{X}_1 = e^{it}(xu + u_x)\frac{\partial}{\partial u}, \quad \mathrm{X}_2 = e^{-it}(xu - u_x)\frac{\partial}{\partial u}, \quad \mathrm{X}_3 = u\frac{\partial}{\partial u}, \tag{1.84}$$

and satisfy the commutation relation $[\mathrm{X}_1, \mathrm{X}_2] = -2\mathrm{X}_3$. It is easy to see that the infinitesimal generators (1.84) are respectively equivalent to the point symmetries $e^{it}\,(-\partial/\partial x + xu\partial/\partial u)$, $e^{-it}\,(\partial/\partial x + xu\partial/\partial u)$, $u\partial/\partial u$. From Theorem 1.2.8, it follows that PDE (1.83) has the symmetries $[P(\mathcal{R}_1, \mathcal{R}_2)u]\partial/\partial u$ for any polynomial function $P(a, b)$ in a and b.

(2) Schrödinger equation for the hydrogen atom
As a second example, consider the time-independent Schrödinger equation for the hydrogen atom given by

$$\mathrm{L}u = \left(\frac{1}{2}\Delta + r^{-1} + E\right)u = 0, \tag{1.85}$$

where the Laplacian $\Delta = \mathrm{D}_x^2 + \mathrm{D}_y^2 + \mathrm{D}_z^2$, $r = \sqrt{x^2 + y^2 + z^2}$, $E = \text{const}$. It is well-known [Schiff (1968)] that PDE (1.84) has three recursion operators corresponding to the Runge–Lenz vector $\boldsymbol{\mathcal{R}} = \frac{1}{2}\left(\mathbf{p} \times \mathbf{L} - \mathbf{L} \times \mathbf{p}\right) - \mathbf{r}/r$ where $\mathbf{p}$ and $\mathbf{L}$ are, respectively, the linear and angular momentum operators. The x-component of $\boldsymbol{\mathcal{R}}$ is the recursion operator given by

$$\mathcal{R}_1 = -\left(x\mathrm{D}_z^2 - z\mathrm{D}_z\mathrm{D}_x - y\mathrm{D}_x\mathrm{D}_y + x\mathrm{D}_y^2 - \mathrm{D}_x + \frac{x}{r}\right).$$

Corresponding to $\mathcal{R}_1$, the PDE (1.84) has the local symmetry given by

$$\mathrm{X}_1 = (\mathcal{R}_1 u)\frac{\partial}{\partial u} = -\left(xu_{zz} - zu_{xz} - yu_{xy} + xu_{yy} - u_x + \frac{x}{r}u\right)\frac{\partial}{\partial u}. \tag{1.86}$$

Clearly, the local symmetry (1.86) is equivalent to neither a point symmetry nor a contact symmetry and hence is a genuine higher-order symmetry of the linear PDE (1.84).

Recursion operators for nonlinear PDEs

If a nonlinear scalar PDE

$$R(x, u, \partial u, \ldots, \partial^k u) = 0 \tag{1.87}$$

has a higher-order symmetry

$$\mathrm{X} = \eta(x, u, \partial u, \ldots, \partial^s u)\frac{\partial}{\partial u}, \quad s \geq 2, \tag{1.88}$$

then this indicates that it may have an infinite sequence of higher-order symmetries

$$\mathrm{X}_i = \mathcal{R}^i \eta \frac{\partial}{\partial u}, \quad i = 1, \ldots, \tag{1.89}$$

generated in terms of some *recursion operator* $\mathcal{R}$.

As one sees from Theorem 1.2.7, the PDE (1.87) has a higher-order symmetry (1.88) if and only if

$$\mathrm{X}^{(k)} R(x, u, \partial u, \ldots, \partial^k u) = \mathrm{L}[u]\eta = 0 \tag{1.90}$$

holds for every solution of PDE (1.87) where $\mathrm{L}[u]$ is the linearization operator (Fréchet derivative) associated with the nonlinear scalar PDE (1.87). In particular,

$$\mathrm{L}[u] = \frac{\partial R}{\partial u} + \frac{\partial R}{\partial u_j}\mathrm{D}_j + \cdots + \frac{\partial R}{\partial u_{j_1 \cdots j_k}}\mathrm{D}_{j_1} \ldots \mathrm{D}_{j_k}. \tag{1.91}$$

The *linearized equation* associated with PDE (1.87) is given by

$$\mathrm{L}[u]v = 0 \tag{1.92}$$

in terms of its linearization operator (1.91). For any solution $u = \Theta(x)$ of PDE (1.87), the linearized equation (1.92) is a linear PDE for v. The infinitesimal $\eta(x, u, \partial u, \ldots, \partial^s u)$ of a higher-order symmetry (1.88) of PDE (1.87) is obviously a solution $v = \eta(x, u, \partial u, \ldots, \partial^s u)$ of the linearized equation (1.92). Conversely, any solution $v = f(x, u, \partial u, \ldots, \partial^p u)$, $p \geq 2$, of (1.92), where $u = \Theta(x)$ is any solution of PDE (1.87), yields a higher-order symmetry $f(x, u, \partial u, \ldots, \partial^p u)\, \partial/\partial u$ of PDE (1.87).

Let $u = \Theta(x)$ be any solution of PDE (1.87). Now suppose the linearized equation (1.92) has a nontrivial local symmetry with its infinitesimal generator given by

$$(\mathcal{R}[u]v)\frac{\partial}{\partial v}, \tag{1.93}$$

where $\mathcal{R}[u]$ is of the local form (1.77), i.e.,

$$\begin{aligned} \mathcal{R}[u] = r(x, u, \partial u, \ldots, \partial^l u) + r^i(x, u, \partial u, \ldots, \partial^l u)\mathrm{D}_i + \cdots \\ + r^{i_1 \ldots i_m}(x, u, \partial u, \ldots, \partial^l u)\mathrm{D}_{i_1} \ldots \mathrm{D}_{i_m}, \end{aligned} \tag{1.94}$$

for some functions $r(x, u, \partial u, \ldots, \partial^l u)$, $r^i(x, u, \partial u, \ldots, \partial^l u)$, $\ldots$, $r^{i_1 \ldots i_k}(x, u, \partial u, \ldots, \partial^l u)$. The above and Theorem 1.2.8 lead to the following theorem.

Theorem 1.2.10. *If the nonlinear scalar PDE* (1.87) *has the higher-order symmetry* (1.88) *and its linearized equation* (1.92) *has the local symmetry* (1.93), *then* $v = (\mathcal{R}[u])^q\eta$ *solves the linearized equation* (1.92), *for each positive integer* q. *Hence PDE* (1.87) *has the infinite sequence of higher-order symmetries given by*

$$\mathrm{X}_q = (\mathcal{R}[u])^q\eta\frac{\partial}{\partial u}, \quad q = 1, \ldots . \tag{1.95}$$

Note that a recursion operator $\mathrm{R}[u]$ satisfies the determining equation

$$\mathrm{L}[u]\mathcal{R}[u]v = 0 \tag{1.96}$$

that holds for any (u, v) pair satisfying the given nonlinear scalar PDE (1.87) and its linearized equation (1.92).

In applying Theorem 1.2.10 to find a local recursion operator $\mathcal{R}[u]$ for some given nonlinear scalar PDE (1.87), it is important to note that one need not know a priori which derivatives of u enter in the coefficients of $\mathcal{R}[u]$. However, the highest derivative m appearing in $\mathcal{R}[u]$ must be chosen a priori in any calculation. Then $\mathcal{R}[u]$ is found through application of Lie's algorithm. The existence of a local recursion operator is often connected with an underlying point or contact transformation that invertibly maps the given nonlinear scalar PDE to some linear PDE.

As an example, consider the integrated Burgers' equation given by

$$u_{xx} - \tfrac{1}{2}u_x^2 - u_t = 0. \tag{1.97}$$

The associated linearized equation is given by

$$\mathrm{L}[u]v = 0, \tag{1.98}$$

in terms of the linearizing operator

$$\mathrm{L}[u] = \mathrm{D}_x^2 - u_x\mathrm{D}_x - \mathrm{D}_t. \tag{1.99}$$

Now suppose that the linearized equation (1.98), (1.99) has a recursion operator of the form

$$\mathcal{R}[u] = a + b\mathrm{D}_x + c\mathrm{D}_x^2, \tag{1.100}$$

whose coefficients a, b, and c depend on x, t, u, u_x, $u_{xx}, \ldots$. Without loss of generality, one need only consider a recursion operator given by a polynomial in D_x with coefficients depending at most on x-derivatives of u, since all t-derivatives of u can be expressed in terms of x-derivatives of u through substitutions for t-derivatives from the given PDE (1.97), and D_t, $\mathrm{D}_x\mathrm{D}_t$, etc. can be expressed in terms of D_x from the linearized equation (1.98), (1.99). Consequently, the determining equation for the coefficients of the recursion operator (1.100) is given by

$$\begin{aligned} \mathrm{L}[u]\mathcal{R}[u]v = {} & v\mathrm{D}_x^2 a + 2v_x \mathrm{D}_x a + v_{xx} a + v_x \mathrm{D}_x^2 b + 2v_{xx}\mathrm{D}_x b \\ & + v_{xxx} b + v_{xx}\mathrm{D}_x^2 c + 2v_{xxx}\mathrm{D}_x c + v_{xxxx} c - v u_x \mathrm{D}_x a \\ & - v_x u_x a - v_x u_x \mathrm{D}_x b - v_{xx} u_x b - v_{xx} u_x \mathrm{D}_x c - v_{xxx} u_x c \\ & - v\mathrm{D}_t a - v_t a - v_x \mathrm{D}_t b - v_{xt} b - v_{xx}\mathrm{D}_t c - v_{xxt} c = 0. \end{aligned} \tag{1.101}$$

Equation (1.101) must hold for *every* solution (u, v) of (1.97) and (1.98), (1.99). After substitution for the t-derivatives of (v, v_x, v_{xx}) in (1.101) through equation (1.98), and two differential consequences in terms of x-derivatives of v, the determining equation (1.101) becomes a linear homogeneous expression in terms of independent variables v_{xxx}, v_{xx}, v_x and v. Thus the determining equation (1.101) reduces to the following four equations for the coefficients a, b, and c:

$$\mathrm{D}_x c = 0, \tag{1.102a}$$

$$2\mathrm{D}_x b + 2u_{xx} c - \mathrm{D}_t c = 0, \tag{1.102b}$$

$$2\mathrm{D}_x a + \mathrm{D}_x^2 b - u_x \mathrm{D}_x b + u_{xx} b - \mathrm{D}_t b + u_{xxx} c = 0, \tag{1.102c}$$

$$\mathrm{D}_x^2 a - u_x \mathrm{D}_x a - \mathrm{D}_t a = 0. \tag{1.102d}$$

Equations (1.102) must hold for any solution $u = \Theta(x, t)$ of the integrated Burgers' equation (1.97). In solving (1.102), we use PDE (1.97) and its differential consequences to substitute for t-derivatives of $u, u_x, u_{xx}, \ldots$. From equation (1.102a), it immediately follows that

$$c = c(t). \tag{1.103}$$

Then equation (1.102b) yields

$$b = b(x, t, u_x) = -c(t)u_x + \tfrac{1}{2}xc'(t) + \alpha(t), \tag{1.104}$$

for arbitrary $\alpha(t)$. After substituting equations (1.103) and (1.104) into equation (1.102c), one finds that

$$\begin{aligned} a &= a(x, t, u_x, u_{xx}) \\ &= -\tfrac{1}{2}c(t)u_{xx} + \tfrac{1}{4}c(t)u_x^2 - \left[\tfrac{1}{4}xc'(t) + \tfrac{1}{2}\alpha(t)\right]u_x \\ &\quad + \tfrac{1}{8}x^2 c''(t) + \tfrac{1}{2}x\alpha'(t) + \beta(t), \end{aligned} \tag{1.105}$$

for arbitrary $\beta(t)$. Finally, substitution of equation (1.105) into equation (1.102d) leads to

$$c'''(t) = 0, \quad \alpha''(t) = 0, \quad \beta'(t) = \tfrac{1}{4}c''(t). \tag{1.106}$$

Consequently, one obtains six recursion operators for the integrated Burgers' equation (1.97) given by

$$\begin{aligned}
&\mathcal{R}_1[u] = 1, \quad \mathcal{R}_2[u] = \tfrac{1}{2}u_x - \mathrm{D}_x, \quad \mathcal{R}_3[u] = \tfrac{1}{2}(tu_x - x) - t\mathrm{D}_x,\\
&\mathcal{R}_4[u] = \tfrac{1}{4}[u_x^2 - 2u_{xx}] - u_x\mathrm{D}_x + \mathrm{D}_x^2,\\
&\mathcal{R}_5[u] = \tfrac{1}{4}[tu_x^2 - xu_x - 2tu_{xx}] + \left[\tfrac{1}{2}x - tu_x\right]\mathrm{D}_x + t\mathrm{D}_x^2,\\
&\mathcal{R}_6[u] = \tfrac{1}{4}[t^2u_x^2 - 2txu_x - 2t^2u_{xx} + x^2 + 2t]\\
&\qquad +[xt - t^2u_x]\mathrm{D}_x + t^2\mathrm{D}_x^2.
\end{aligned} \tag{1.107}$$

It is easy to see that if $\{\mathcal{R}_1[u], \ldots, \mathcal{R}_p[u]\}$ is a set of p recursion operators of a given nonlinear scalar PDE (1.87), then any polynomial operator $P(\mathcal{R}_1[u], \ldots, \mathcal{R}_p[u])$ is a recursion operator of PDE (1.87). Thus if a given nonlinear scalar PDE (1.87) has a local symmetry $\eta\,\partial/\partial u$ and a set of p recursion operators, then PDE (1.87) also has local symmetries $P(\mathcal{R}_1[u], \ldots, \mathcal{R}_p[u])\eta\,\partial/\partial u$ for any polynomial operator $P(\mathcal{R}_1[u], \ldots, \mathcal{R}_p[u])$.

Since for the recursion operators (1.107), one can show that $\mathcal{R}_4 = (\mathcal{R}_2)^2$, $\mathcal{R}_5 = \mathcal{R}_2\mathcal{R}_3$, $\mathcal{R}_6 = (\mathcal{R}_3)^2$, it follows that only $\mathcal{R}_2[u]$ and $\mathcal{R}_3[u]$ are independent recursion operators. From its invariance under translations in x, it follows that the integrated Burgers' equation (1.97) has the symmetry $u_x\,\partial/\partial u$. Consequently, PDE (1.97) has the local symmetry $P(\mathcal{R}_2[u], \mathcal{R}_3[u])u_x\,\partial/\partial u$ for any polynomial $P(\mathcal{R}_2[u], \mathcal{R}_3[u])$. Note that

$$\begin{aligned}
\mathcal{R}_2[u]u_x &= -u_{xx} + \tfrac{1}{2}u_x^2 = -u_t,\\
\mathcal{R}_3[u]u_x &= -tu_{xx} + \tfrac{1}{2}[tu_x^2 - xu_x] = -[tu_t + \tfrac{1}{2}xu_x]
\end{aligned}$$

correspond to the invariance of PDE (1.97) under translations in t and particular scalings of x and t, respectively. These two recursion operators applied to the point symmetry $u_x\,\partial/\partial u$ yield infinite sequences of higher-order symmetries of the integrated Burgers' equation (1.97). An example of such a higher-order symmetry is given by

$$(\mathcal{R}_2[u])^3 u_x \frac{\partial}{\partial u} = \left[-u_{xxt} + u_x u_{xt} + \tfrac{1}{4}(2u_{xx} - u_x^2)u_t\right]\frac{\partial}{\partial u}.$$

A higher-order symmetry classification problem

Now consider a higher-order symmetry classification problem for the nonlinear heat conduction equation given by

$$u_t - (K(u)u_x)_x = 0. \tag{1.108}$$

In particular, we find all conductivities $K(u)$ for which PDE (1.108) has higher-order symmetries [Bluman & Kumei (1980)].

For brevity of notation, in this example we denote x-derivatives by $u_x = u^{(1)}$, $u_{xx} = u^{(2)}$, $u_{xxx} = u^{(3)}$, etc.

Suppose PDE (1.108) has a higher-order symmetry

$$\mathrm{X} = \eta(x, t, u, u^{(1)}, \ldots, u^{(s)})\frac{\partial}{\partial u} \tag{1.109}$$

for some $s \geq 2$. For convenience, we introduce the notations

$$\eta_0 = \frac{\partial \eta}{\partial u}, \quad \eta_i = \frac{\partial \eta}{\partial u^{(i)}}, \quad \eta_{ij} = \frac{\partial^2 \eta}{\partial u^{(i)} \partial u^{(j)}}, \quad K' = \frac{dK}{du}, \quad K'' = \frac{d^2 K}{du^2}.$$

The PDE (1.108) has the higher-order symmetry (1.109) if and only if the determining equation (linearized equation)

$$\begin{aligned} &\mathrm{X}^{(2)}(u_t - (K(u)u_x)_x) \\ &\quad = \mathrm{L}[u]\eta \\ &\quad = [\mathrm{D}_t - K''(u^{(1)})^2 - 2K'u^{(1)}\mathrm{D}_x - K'u^{(2)} - K\mathrm{D}_x^2]\eta = 0 \end{aligned} \tag{1.110}$$

holds for every solution $u = \Theta(x, t)$ of the corresponding nonlinear heat conduction equation (1.108). In (1.110), one has

$$\mathrm{D}_t \eta = \frac{\partial \eta}{\partial t} + \eta_0 u_t + \sum_{i=1}^{s} \eta_i u_t^{(i)}, \quad \mathrm{D}_x \eta = \frac{\partial \eta}{\partial x} + \sum_{i=0}^{s} \eta_i u^{(i+1)},$$

$$\mathrm{D}_x^2 \eta = \frac{\partial^2 \eta}{\partial x^2} + 2\sum_{i=0}^{s} \frac{\partial \eta_i}{\partial x} u^{(i+1)} + \sum_{i,j=0}^{s} \eta_{ij} u^{(i+1)} u^{(j+1)} + \sum_{i=0}^{s} \eta_i u^{(i+2)}.$$

After substituting for u_t and $u_t^{(i)}$ in (1.110) through the given PDE (1.108) and its differential consequences with respect to x-derivatives of u, the determining equation (1.110) becomes a polynomial equation in terms of $u^{(s+1)}$. The coefficients of each power of $u^{(s+1)}$ must separately vanish. The vanishing of the coefficients of $(u^{(s+1)})^2$ and $u^{(s+1)}$, respectively, yields the equations

$$\eta_{ss} = 0, \quad sK'\eta_s u^{(1)} = 2K \sum_{i=0}^{s-1} \eta_{is} u^{(i+1)}. \tag{1.111}$$

The solution of (1.111) leads to

$$\eta = \alpha K^{s/2} u^{(s)} + f(u, u^{(1)}, \ldots, u^{(s-1)}), \tag{1.112}$$

where f is an arbitrary function of its arguments and $\alpha = \text{const} \neq 0$ if $s \geq 3$.

The substitution of (1.112) into the determining equation (1.110) reduces (1.110) to a polynomial equation in terms of $u^{(s)}$. The vanishing of its coefficients of $(u^{(s)})^2$ and $u^{(s)}$, respectively, yields the equations

$$\begin{aligned}
&\frac{\partial^2 f}{\partial (u^{(s-1)})^2} = 0, \\
&2K \sum_{i=0}^{s-2} u^{(i+1)} \frac{\partial^2 f}{\partial u^{(i)} \partial u^{(s-1)}} + (1-s) K' u^{(1)} \frac{\partial f}{\partial u^{(s-1)}} \\
&-\tfrac{1}{2}\alpha s(s+3) K' K^{s/2} u^{(2)} \\
&+\tfrac{1}{4}\left[s^2 (K')^2 K^{s/2-1} - 2s(s+2) K'' K^{s/2}\right] (u^{(1)})^2 = 0.
\end{aligned} \tag{1.113}$$

From (1.113), one can easily deduce that η is of the form

$$\begin{aligned}
\eta = \alpha &\left[K^{s/2} u^{(s)} + \tfrac{1}{4} s(s+3) K' K^{s/2-1} u^{(1)} u^{(s-1)}\right] \\
&+ g(u) u^{(s-1)} + h(u, u^{(1)}, \ldots, u^{(s-2)}),
\end{aligned} \tag{1.114}$$

where g and h are arbitrary functions of their respective arguments. More importantly, the substitution of (1.114) into (1.113) yields a nontrivial solution for η only if the conductivity $K(u)$ satisfies the ODE

$$2KK'' - 3K'^2 = 0. \tag{1.115}$$

The solution of ODE (1.115) leads to

$$K(u) = \frac{c}{(u+d)^2}, \tag{1.116}$$

for arbitrary constants c and d. Hence, modulo scalings and translations in u, the only nonlinear heat conduction equation (1.108) that could have a higher-order symmetry is given by

$$u_t - (u^{-2} u_x)_x = 0. \tag{1.117}$$

Let $K(u) = u^{-2}$. Then for $s = 3$, the substitution of (1.114) into the determining equation (1.110) yields a higher-order symmetry $\eta_1\, \partial/\partial u$ of PDE (1.117) with

$$\eta_1 = u^{-3} u^{(3)} - 9u^{-4} u^{(1)} u^{(2)} + 12 u^{-5} (u^{(1)})^3. \tag{1.118}$$

For $s = 4$, the substitution of (1.114) into the determining equation (1.110) yields a second higher-order symmetry $\eta_2\, \partial/\partial u$ of the PDE (1.117) with

$$\begin{aligned}
\eta_2 = u^{-4} u^{(4)} &- 14 u^{-5} u^{(1)} u^{(3)} - 10 u^{-5} (u^{(2)})^2 \\
&+ 95 u^{-6} (u^{(1)})^2 u^{(2)} - 90 u^{-7} (u^{(1)})^4.
\end{aligned} \tag{1.119}$$

The invariance of PDE (1.117) under translations in t leads to PDE (1.117) admitting $\eta_0\,\partial/\partial u$ with

$$\eta_0 = u_t = (u^{-2}u_x)_x = u^{-2}u^{(2)} - 2u^{-3}(u^{(1)})^2. \tag{1.120}$$

The form of η_0, η_1 and η_2 leads one to seek a recursion operator of the form

$$\mathcal{R}[u] = p\mathrm{D}_x + q + r\mathrm{D}_x^{-1} \tag{1.121}$$

so that $\mathcal{R}[u]\eta_i = \eta_{i+1}$, $i = 0,\ 1$, where $\mathrm{D}_x\mathrm{D}_x^{-1}$ is the identity operator and p, q, r are functions of $u, u^{(1)}, u^{(2)}$. Then one can show that $\mathcal{R}[u]\eta_0 = \eta_1$ if and only if $p = u^{-1}$ and $[u^{-2}u^{(2)} - 2u^{-3}(u^{(1)})^2]q + u^{-2}u^{(1)}r = -3u^{-4}u^{(1)}u^{(2)} + 6u^{-5}(u^{(1)})^3$, and, furthermore, that $\mathcal{R}[u]\eta_1 = \eta_2$ if and only if $q = -2u^{-2}u^{(1)}$ and $r = -u^{-2}u^{(2)} + 2u^{-3}(u^{(1)})^2$. Consequently, one can show that the operator (1.121) can be written more concisely as

$$\mathcal{R}[u] = \mathrm{D}_x^2 \circ u^{-1} \circ \mathrm{D}_x^{-1}. \tag{1.122}$$

It is left as an exercise to prove that the operator $\mathcal{R}[u]$ given by (1.122) is a recursion operator of the nonlinear heat conduction equation (1.117). Correspondingly, PDE (1.117) has the infinite sequence of higher-order symmetries given by $(\mathrm{R}[u])^s\eta_0\,\partial/\partial u$, $s = 1, \ldots$. Moreover, one can show that these are the only higher-order symmetries of PDE (1.117) [Bluman & Kumei (1980)].

Exercises 1.2

1.2.1. Prove Theorem 1.2.1 [Lie (1890); Mayer (1875)].

1.2.2. By direct calculation, show that Theorem 1.2.3 holds.

1.2.3. Prove that the only transformations of the form (1.20) that are one-to-one on $(x, u, \partial u, \ldots, \partial^k u)$-space for some finite k are point transformations of the form (1.1) or contact transformations of the form (1.13).

1.2.4. For $u = u(x,t)$, show that the transformation given by

$$\begin{aligned} x^* &= x + u_t, \\ t^* &= t + u_x, \\ u^* &= u + u_x u_t \end{aligned} \tag{1.123}$$

yields a contact transformation.

1.2.5. Show that an infinitesimal generator of the form (1.44) is uniquely equivalent to an infinitesimal generator of a one-parameter Lie group of point transformations when $\eta(x, u, \partial u)$ has a linear dependence on first derivatives.

1.2.6. For the linear heat equation (1.53), show that the one-parameter Lie group of transformations (1.59) maps any solution $u = \theta(x,t)$ that is not invariant under (1.57) into the one-parameter family of solutions of PDE (1.53) given by the expression

$$u = \phi(x,t;\varepsilon) = \frac{1}{\sqrt{1-\varepsilon t}} \exp\left[\frac{\varepsilon x^2}{4(1-\varepsilon t)}\right] \theta\left(\frac{x}{1-\varepsilon t}, \frac{t}{1-\varepsilon t}\right).$$

1.2.7. Find the contact symmetries of the Liouville equation $u_{xt} = e^u$.

1.2.8. Consider the PDE system given by

$$\begin{aligned} v_x &= u, \\ v_t &= F(x,t,u,u_x), \end{aligned} \tag{1.124}$$

and the related scalar PDE given by

$$v_t = F(x,t,v_x,v_{xx}). \tag{1.125}$$

(a) Show that any solution of the PDE system (1.124) yields a solution of the related scalar PDE (1.125) and, conversely, that any solution of the related scalar PDE (1.125) yields a solution of the PDE system (1.124).

(b) Show that a point symmetry of the PDE system (1.124) yields a contact symmetry (which could be a point symmetry) of the related scalar PDE (1.125) and, conversely, that a contact symmetry of the related scalar PDE (1.125) yields a point symmetry of the PDE system (1.124).

1.2.9. Consider Burgers' equation:

$$u_t + uu_x = u_{xx}. \tag{1.126}$$

(a) Show that the only admitted local symmetries of the form $\mathrm{X} = \eta(x,u,u_x,u_{xx},u_{xxx})\,\partial/\partial u$ are given by

$$\begin{aligned} &\mathrm{X}_1 = u_x\frac{\partial}{\partial u}, \quad \mathrm{X}_2 = (u_{xx} - uu_x)\frac{\partial}{\partial u}, \\ &\mathrm{X}_3 = \left(4u_{xxx} - 6uu_{xx} - 6u_x^2 + 3u^2u_x\right)\frac{\partial}{\partial u}. \end{aligned}$$

(b) Show that X_1 and X_2 correspond to point symmetries and that X_3 corresponds to a higher-order symmetry.

(c) Find the other three higher-order symmetries of the form $\mathrm{X} = \eta(x,t,u,u_x,u_{xx},u_{xxx})\,\partial/\partial u$ [Bluman & Kumei (1989)].

1.2.10. Find groups of equivalence transformations for the following families of PDE systems.

(a) Diffusion-convection equations given by

$$u_t = (A(u)u_x)_x + B(u)u_x,$$

involving constitutive functions $A(u)$ and $B(u)$ [Popovych & Ivanova (2004)].

(b) Nonlinear telegraph (NLT) equations given by

$$u_{tt} - (F(u)u_x)_x - (G(u))_x = 0,$$

involving constitutive functions $F(u)$ and $G(u)$ [Bluman & Temuerchaolu (2005a)].

(c) Lagrange PDE systems of planar gas dynamics equations given by

$$\begin{aligned} q_s - v_y &= 0, \\ v_s + p_y &= 0, \\ p_s + B(p,q)v_y &= 0, \end{aligned}$$

involving a constitutive function $B(p,q)$ [Akhatov, Gazizov & Ibragimov (1991)] .

(d) PDE systems for part (c) in the polytropic case, i.e., $B(p,q) = \gamma p/q$, where $\gamma = \text{const}$ is an arbitrary constitutive parameter.

(e) Linear PDE systems given by

$$\begin{aligned} p_v &= ut_u - t, \\ p_u &= uc^2(u)t_v, \end{aligned}$$

involving a constitutive function $c^2(u)$. [These systems arise in the analysis of nonlocally related systems for the nonlinear wave equations [Bluman & Cheviakov (2007)]; see Section 4.2.2.]

(f) Bragg–Hawthorne equations describing axially symmetric ideal fluid flow, given by

$$v_{rr} - \frac{v_r}{r} + v_{zz} + I(v)I'(v) = -r^2 P'(v),$$

involving constitutive functions $I(v), P(v)$.

1.2.11. Consider the linear PDE

$$x^2 u_{xx} + xu_x - u_{tt} + x^2 u = 0. \tag{1.127}$$

(a) Show that from its invariance under translations in t and scalings of u, it follows that PDE (1.127) has solutions of the form $u = e^{\nu t} y(x,\nu)$, $\nu = \text{const}$, where $y(x,\nu)$ satisfies a reduced ODE.

(b) Show that $y(x,\nu)$ satisfies Bessel's equation of order ν given by

$$y'' + x^{-1}y' + (1 - \nu^2 x^{-2})y = 0.$$

(c) Find the recursion operators of the form $\mathcal{R} = a(x,t) + b(x,t)\mathrm{D}_t + c(x,t)\mathrm{D}_x$ that are admitted by PDE (1.127). Find their commutation relations.

(d) Find two linear combinations of these recursion operators, labeled $\mathcal{R}_+$ and $\mathcal{R}_-$ with the property that they act, respectively, as raising and lowering operators in the sense that $\mathcal{R}_\pm e^{\nu t} y(x,\nu) = e^{(\nu\pm 1)t} y(x, \nu \pm 1)$ [Miller (1968)].

1.2.12. Show that if $\{\mathcal{R}_1[u], \ldots, \mathcal{R}_p[u]\}$ is a set of p recursion operators of a nonlinear scalar PDE (1.87), then any polynomial operator $P(\mathcal{R}_1[u], \ldots, \mathcal{R}_p[u])$ is a recursion operator of PDE (1.87).

1.2.13. Show that the operator $\mathcal{R}[u]$ given by (1.122) is a recursion operator of the nonlinear heat conduction equation (1.117).

1.2.14.

(a) Show that if $\mathcal{R}[u]$ is a recursion operator of the integrated Burgers' equation (1.97), then the operator $\mathrm{D}_x\mathcal{R}[\mathrm{D}_x^{-1}u]\mathrm{D}_x^{-1}$ is a recursion operator of Burgers' equation (1.126).

(b) For the recursion operators $\mathcal{R}_2[u] = \frac{1}{2}u_x - \mathrm{D}_x$, $\mathcal{R}_3[u] = \frac{1}{2}(tu_x - x) - t\mathrm{D}_x$, of PDE (1.97), find the corresponding recursion operators of PDE (1.126).

1.2.15. Consider the Korteweg–de Vries equation

$$u_t + uu_x + u_{xxx} = 0. \tag{1.128}$$

(a) Find the linearized equation associated with PDE (1.128).

(b) Show that

$$\mathcal{R}[u] = \mathrm{D}_x^2 + \tfrac{2}{3}u + u_x\mathrm{D}_x^{-1} \tag{1.129}$$

is a recursion operator of PDE (1.128).

(c) From its invariance under translations in x, clearly PDE (1.128) has the point symmetry $u_x\,\partial/\partial u$. Hence, apply the recursion operator (1.129) to $u_x\,\partial/\partial u$ to obtain the two higher-order symmetries of lowest order in the corresponding infinite sequence of higher-order symmetries of PDE (1.128).

(d) Find a scaling symmetry of PDE (1.128). Show that the application of the recursion operator (1.129) to this scaling symmetry does not yield local symmetries of PDE (1.128).

1.2.16. Consider the system of PDEs

$$u_{xx} + u^2 v + iu_t = 0, \tag{1.130a}$$

$$v_{xx} + uv^2 - iv_t = 0. \tag{1.130b}$$

Show that the matrix operator given by

$$\mathcal{R}[u,v] = \begin{bmatrix} i\mathrm{D}_x + iu\mathrm{D}_x^{-1} \circ v & iu\mathrm{D}_x^{-1} \circ u \\ -iv\mathrm{D}_x^{-1} \circ v & -i\mathrm{D}_x - iv\mathrm{D}_x^{-1} \circ u \end{bmatrix}$$

is a recursion operator of the system of PDEs (1.130). Note that if $v = \bar{u}$ is the complex conjugate of u, then PDE (1.130a) becomes the cubic Schrödinger equation given by $u_{xx} + \bar{u}u^2 + iu_t = 0$ and PDE (1.130b) is its complex conjugate equation [Ablowitz, Kaup, Newell & Segur (1974)].

1.2.17. Verify that $\mathcal{R}[u] = \mathrm{D}_x^2 + u_x\mathrm{D}_x^{-1} \circ u_x\mathrm{D}_x$ is a recursion operator for the sine-Gordon equation

$$u_{xt} - \sin u = 0. \tag{1.131}$$

Find two higher-order symmetries of PDE (1.131). [This recursion operator for PDE (1.131) was found by Olver (1977).]

1.3 Conservation Laws

In the study of DEs, conservation laws have many significant uses. They describe physical conserved quantities such as mass, energy, momentum and angular momentum, as well as charge and other constants of motion. They are important for investigating integrability and linearization mappings and for establishing existence and uniqueness of solutions. They are also used in the analysis of stability and global behavior of solutions. In addition, they play an essential role in the development of numerical methods and provide an essential starting point for finding nonlocally related systems and potential variables. In particular, a conservation law is fundamental in studying a given DE in the sense that it holds for any posed data (initial and/or boundary conditions). Moreover, the structure of conservation laws is coordinate-independent, as a point (contact) transformation maps a conservation law into a conservation law.

1.3.1 Local conservation laws

Consider a system $\mathbf{R}\{x\,;u\}$ of N partial differential equations of order k with n independent variables $x = (x^1, \ldots, x^n)$ and m dependent variables $u(x) = (u^1(x), \ldots, u^m(x))$, given by

$$R^\sigma[u] = R^\sigma(x, u, \partial u, \ldots, \partial^k u) = 0, \quad \sigma = 1, \ldots, N. \tag{1.132}$$

Definition 1.3.1. A *local conservation law* of PDE system (1.132) is a divergence expression

$$\mathrm{D}_i\Phi^i[u] = \mathrm{D}_1\Phi^1[u] + \cdots + \mathrm{D}_n\Phi^n[u] = 0 \tag{1.133}$$

holding for all solutions of PDE system (1.132). In (1.133), $\Phi^i[u] = \Phi^i(x, u, \partial u, \ldots, \partial^r u)$, $i = 1, \ldots, n$, are called the *fluxes* of the conservation law, and the highest-order derivative (r) present in the fluxes $\Phi^i[u]$ is called the (differential) *order of a conservation law.*

Remark 1.3.1. If one of the independent variables of $\mathbf{R}\{x\,;u\}$ is time t, the conservation law (1.133) takes the form

$$\mathrm{D}_t\Psi[u] + \operatorname{div}\mathbf{\Phi}[u] = 0, \tag{1.134}$$

where $\operatorname{div}\mathbf{\Phi}[u] = \mathrm{D}_i\Phi^i[u] = \mathrm{D}_1\Phi^1[u] + \cdots + \mathrm{D}_{n-1}\Phi^{n-1}[u]$ is a spatial divergence, and $x = (x^1, \ldots, x^{n-1})$ are $n-1$ spatial variables. Here $\Psi[u]$ is referred to as a *density*, and $\Phi^i[u]$ as *spatial fluxes* of the conservation law (1.134).

The following theorem is easily proved.

Theorem 1.3.1. *Suppose the conservation law* (1.134) *is defined on a domain $\mathcal{D}$ in x-space for $t > 0$, and its fluxes $\Phi^i[u]$ vanish on the boundary $\partial\mathcal{D}$ (or alternatively, $\mathcal{D}$ is unbounded and $\lim_{|x|\to\infty}\Phi^i[u] = 0$ for all i). Then $\int_{\mathcal{D}}\Psi[u]\,d^{n-1}x$ is time-independent.*

Proof. Integrating (1.134) over $\mathcal{D}$, one obtains

$$\int_{\mathcal{D}} \mathrm{D}_t\Psi[u]d^{n-1}x = -\int_{\mathcal{D}}(\operatorname{div}\mathbf{\Phi}[u])d^{n-1}x = -\oint_{\partial\mathcal{D}}(\mathbf{\Phi}[u]\cdot\mathbf{n})d^{n-2}x = 0.$$

□

The conserved quantity in Theorem 1.3.1 is called a *constant of motion.* It can be evaluated in terms of initial data for $u(x,t)$ at time $t = 0$, i.e.,

$$\int_{\mathcal{D}}\Psi[u]d^{n-1}x = \int_{\mathcal{D}}\Psi[f]d^{n-1}x,$$

where $f(x) = u(x,0)$.

As a first example, consider the adiabatic motion of an ideal gas on a bounded three-dimensional domain: $x = \mathbf{x} = (x^1, x^2, x^3) \in \mathcal{D} \subset \mathbb{R}^3$. At position x and time t, let $v(x,t) = \mathbf{v} = (v^1(x,t), v^2(x,t), v^3(x,t))$, $\rho = \rho(x,t)$ and $p = p(x,t)$ be the velocity, the density, and the pressure of the gas, respectively. For adiabatic processes in an ideal gas, the entropy density is given by $S = c_v \ln p\rho^{-\gamma}$, where $c_v = \text{const}$ is the specific heat of the gas at constant volume, and γ = const is the adiabatic exponent. The Euler equations describing such a gas motion are given by

$$\mathrm{D}_t\rho + \mathrm{D}_j(\rho v^j) = 0, \tag{1.135a}$$
$$\rho(\mathrm{D}_t + v^j\mathrm{D}_j)v^i + \mathrm{D}_i p = 0, \quad i = 1, 2, 3, \tag{1.135b}$$
$$\rho(\mathrm{D}_t + v^j\mathrm{D}_j)p + \gamma\rho p\mathrm{D}_j v^j = 0. \tag{1.135c}$$

In particular, the equation (1.135c) results from the conservation of entropy along the streamlines of the flow: $(\mathrm{D}_t + v^j\mathrm{D}_j)S = 0$.

Equation (1.135a) is a divergence expression (1.134) and thus a conservation law of the PDE system (1.135). It expresses the local *conservation of mass*. In particular, the integral version of (1.135a) is given by

$$\mathrm{D}_t \int_{\mathcal{D}} \rho d^3x = -\int_{\mathcal{D}} \mathrm{div}\,(\rho\mathbf{v})d^3x = -\int_{\partial\mathcal{D}} \rho(\mathbf{v}\cdot\mathbf{n})dS, \tag{1.136}$$

showing that the rate of change of the total gas mass $M(t) = \int_{\mathcal{D}} \rho d^3x$ in $\mathcal{D}$ is due to the flux of gas proportional to the component of velocity $\mathbf{v}\cdot\mathbf{n}$ normal to the domain boundary $\partial\mathcal{D}$. Hence, if the velocity is tangent to the domain boundary ($\mathbf{v}\cdot\mathbf{n} = 0$), the total mass is a constant of motion.

Other conservation laws of the Euler PDE system (1.135) arise from linear combinations of the equations of the system. For example, multiplying equation (1.135a) by v^i and adding equation (1.135b), for $i = 1, 2, 3$, one obtains the three components of the vector equation of *conservation of momentum* (three scalar conservation laws):

$$\begin{aligned} &v^i\left[\mathrm{D}_t\rho + \mathrm{D}_i(\rho v^j)\right] + \left[\rho(\mathrm{D}_t + v^j\mathrm{D}_j)v^i + \mathrm{D}_i p\right] \\ &\quad = \mathrm{D}_t(\rho v^i) + \mathrm{D}_j(\rho v^i v^j + p\delta^{ij}) = 0, \quad i = 1, 2, 3. \end{aligned} \tag{1.137}$$

In integral form, this yields

$$\mathrm{D}_t \int_{\mathcal{D}} \rho\mathbf{v} d^3x = -\int_{\mathcal{D}} \mathrm{div}(\rho\mathbf{v}\otimes\mathbf{v} + p\mathbb{I})d^3x = -\int_{\partial\mathcal{D}} \left[\rho\mathbf{v}(\mathbf{v}\cdot\mathbf{n}) + p\mathbf{n}\right] dS,$$

where $\mathbf{P}(t) = \int_{\mathcal{D}} \rho\mathbf{v}d^3x$ is the total momentum of the gas in $\mathcal{D}$ and $\rho\mathbf{v}(\mathbf{v}\cdot\mathbf{n}) + p\mathbf{n} = \mathbf{n}\cdot(\rho\mathbf{v}\otimes\mathbf{v} + p\mathbb{I})$ is the flux of the momentum of the gas normal to the boundary of $\mathcal{D}$. [Here $\mathbf{v}\otimes\mathbf{v}$ is a 3×3 tensor with components $v^i v^j$, and $\mathbb{I}$ is the identity tensor with components δ^{ij}.]

As another example, taking a linear combination of equations (1.135a), (1.135b), and (1.135c) with respective multipliers $(v^i)^2/2$, ρv^i, and $1/\rho(\gamma - 1)$, followed by summation in i from 1 to 3, one obtains the equation of *conservation of energy*:

$$\sum_{i=1}^{3}\Big(\tfrac{1}{2}(v^i)^2\left[\mathrm{D}_t\rho+\mathrm{D}_j(\rho v^j)\right]+\rho v^i\left[\rho(\mathrm{D}_t+v^j\mathrm{D}_j)v^i+\mathrm{D}_ip\right]$$
$$+\frac{1}{\rho(\gamma-1)}\left[\rho(\mathrm{D}_t+v^j\mathrm{D}_j)p+\gamma\rho p\mathrm{D}_jv^j\right]\Big) \tag{1.138}$$
$$=\mathrm{D}_t(E)+\mathrm{D}_j\left(v^j(E+p)\right)=0,$$

where $E=\frac{1}{2}\rho\mathbf{v}^2+p/(\gamma-1)$ is the total energy density. Thus, one has

$$\mathrm{D}_t\int_{\mathcal{D}}\left[\tfrac{1}{2}\rho\mathbf{v}^2+\frac{p}{\gamma-1}\right]d^3x=-\int_{\mathcal{D}}\operatorname{div}\left((E+p)\mathbf{v}\right)d^3x=-\int_{\partial\mathcal{D}}(E+p)(\mathbf{v}\cdot\mathbf{n})dS,$$

relating the rate of change of the total energy in the gas in $\mathcal{D}$ to the normal component of the flux of the energy $(E+p)\mathbf{v}$ through the boundary of $\mathcal{D}$.

Similarly, the linear combination arising after multiplying equation (1.135a) by the vector $\mathbf{x}\times\mathbf{v}$, and cross-multiplying the vector equation (1.135a) by the vector $\mathbf{v}$, yields the three components of the *angular momentum conservation law*:

$$\mathrm{D}_t(m^i)+\mathrm{D}_j\left(v^jm^i\right)-\left(\operatorname{curl}(p\mathbf{x})\right)^i=0,\quad i=1,2,3, \tag{1.139}$$

where $\mathbf{m}=\mathbf{x}\times\rho\mathbf{v}=(m^1,m^2,m^3)$ is the angular momentum density [Exercise 1.3.1].

As an example of a time-independent conservation law, consider Maxwell's equation

$$\operatorname{div}\mathbf{B}\equiv\mathrm{D}_1B^1+\mathrm{D}_2B^2+\mathrm{D}_3B^3=0 \tag{1.140}$$

in $\mathbb{R}^3$, for the magnetic field $\mathbf{B}=B(x,t)=(B^1(x,t),B^2(x,t),B^3(x,t))$. The integral form of the conservation law (1.140) for any closed domain $\mathcal{D}$ with boundary surface $\partial\mathcal{D}$ in $\mathbb{R}^3$ yields

$$\int_{\mathcal{D}}\operatorname{div}\,\mathbf{B}d^3x=\oint_{\partial\mathcal{D}}\mathbf{B}\cdot\mathbf{n}dS=0,$$

which shows that there is no magnetic flux through the boundary of $\mathcal{D}$. This expresses the fact that magnetic charges do not exist.

Certain PDE systems have an infinite number of local conservation laws. A well-known example of this situation is given by the Korteweg–de Vries equation

$$u_t+uu_x+u_{xxx}=0, \tag{1.141}$$

where $u(x,t)$ is the amplitude of long surface waves on shallow water. In addition to conservation laws for mass, momentum and energy, given, respectively, by

$$\mathrm{D}_t(u) + \mathrm{D}_x\left(\tfrac{1}{2}u^2 + u_{xx}\right) = 0, \tag{1.142a}$$

$$\mathrm{D}_t\left(\tfrac{1}{2}u^2\right) + \mathrm{D}_x\left(\tfrac{1}{3}u^3 + uu_{xx} - \tfrac{1}{2}u_x^2\right) = 0, \tag{1.142b}$$

$$\begin{aligned}&\mathrm{D}_t\left(\tfrac{1}{6}u^3 - \tfrac{1}{2}u_x^2\right)\\&+\mathrm{D}_x\left(\tfrac{1}{8}u^4 - uu_x^2 + \tfrac{1}{2}(u^2u_{xx} + u_{xx}^2) - u_xu_{xxx}\right) = 0,\end{aligned} \tag{1.142c}$$

the KdV equation (1.141) has an infinite sequence of local conservation laws of increasing order in which the conserved densities are polynomials in u and its x-derivatives [Miura, Gardner & Kruskal (1968)]. In particular, the next higher-order local conservation law in this sequence is given by

$$\begin{aligned}&\mathrm{D}_t\left(\tfrac{5}{72}u^4 - \tfrac{5}{6}uu_x^2 + \tfrac{1}{2}u_{xx}^2\right)\\&+\mathrm{D}_x\left(\tfrac{1}{18}u^5 - \tfrac{5}{12}u^2u_x^2 + \tfrac{5}{6}u_x^2u_{xx} + \tfrac{4}{3}uu_{xx}^2\right.\\&\left.-\tfrac{5}{3}uu_xu_{xxx} - \tfrac{1}{2}u_{xxx}^2 + u_{xx}u_{xxxx}\right) = 0.\end{aligned} \tag{1.143}$$

The KdV equation (1.141) also has local conservation laws with explicit dependence on t and x. An example of such a conservation law is given by

$$\mathrm{D}_t\left(\tfrac{1}{2}tu^2 - xu\right) + \mathrm{D}_x\left(-\tfrac{1}{2}xu^2 + tuu_{xx} - \tfrac{1}{2}tu_x^2 - xu_{xx} + u_x\right) = 0 \tag{1.144}$$

which can be shown to describe the motion of the center of mass of a surface wave.

1.3.2 Equivalent conservation laws

Definition 1.3.2. A local conservation law (1.133) of the PDE system $\mathbf{R}\{x\,;u\}$ (1.132) is *trivial* if its fluxes are of the form $\Phi^i[u] = M^i[u] + H^i[u]$, where $M^i[u]$ and $H^i[u]$ are functions of x, u and derivatives of u such that $M^i[u]$ vanishes on the solutions of the system (1.132), and $\mathrm{D}_iH^i[u] \equiv 0$ is identically divergence-free.

In particular, a trivial conservation law contains no information about a given PDE system $\mathbf{R}\{x\,;u\}$ (1.132) and arises in two cases:

1. Each of its fluxes vanishes identically on the solutions of the given PDE system.
2. The conservation law vanishes identically as a differential identity. In particular, this second type of trivial conservation law is simply an identity holding for arbitrary fluxes.

As an example, consider the PDE system

$$v_x = u, \quad v_t = K(u)u_x. \tag{1.145}$$

The conservation law

$$\mathrm{D}_t(u(u - v_x)) + \mathrm{D}_x(2(v_t - K(u)u_x)) = 0$$

is a trivial conservation law of the first type for PDE system (1.145), and

$$\mathrm{D}_t(u_{xx}) - \mathrm{D}_x(u_{tx}) = 0$$

is a trivial conservation law of the second type.

An important general example of a trivial conservation law of the second type is given by $\mathrm{div}(\mathrm{curl}\,\mathbf{f}) = 0$ for any vector function $\mathbf{f}(x)$ in $\mathbb{R}^3$, and its multi-dimensional generalizations.

The notion of a trivial conservation law leads to the following definitions of equivalence and linear dependence of conservation laws.

Definition 1.3.3. Two conservation laws $\mathrm{D}_i\Phi^i[u] = 0$ and $\mathrm{D}_i\Psi^i[u] = 0$ are *equivalent* if $\mathrm{D}_i(\Phi^i[u] - \Psi^i[u]) = 0$ is a trivial conservation law. An *equivalence class* of conservation laws consists of all conservation laws equivalent to some given nontrivial conservation law.

Definition 1.3.4. A set of l conservation laws $\{\mathrm{D}_i\Phi^i_{(j)}[u] = 0\}_{j=1}^{l}$ is *linearly dependent* if there exists a set of constants $\{a^{(j)}\}_{j=1}^{l}$, not all zero, such that the linear combination

$$\mathrm{D}_i(a^{(j)}\Phi^i_{(j)}[u]) = 0 \tag{1.146}$$

is a trivial conservation law. In this case, up to equivalence, one of the conservation laws in the set can be expressed as a linear combination of the others.

In practice, one is interested in finding linearly independent sets of conservation laws of a given PDE system.

1.3.3 *Multipliers for conservation laws. Euler operators*

In general, for a given PDE system (1.132), nontrivial local conservation laws arise from linear combinations of the equations of the PDE system (1.132) with *multipliers* (*factors*, *characteristics*) that yield nontrivial divergence expressions. In seeking such expressions, the dependent variables (and each of their derivatives) that arise in the PDE system (1.132), or appear in the multipliers, are replaced by arbitrary functions (and their derivatives). By their construction, such divergence expressions vanish on all solutions of the PDE system (1.132).

In particular, a set of multipliers $\{\Lambda_\sigma[U]\}_{\sigma=1}^N = \{\Lambda_\sigma(x, U, \partial U, \ldots, \partial^l U)\}_{\sigma=1}^N$ yields a divergence expression for the PDE system $\mathbf{R}\{x\,; u\}$ (1.132) if the identity

$$\Lambda_\sigma[U]R^\sigma[U] \equiv \mathrm{D}_i\Phi^i[U] \tag{1.147}$$

holds for *arbitrary* functions $U(x)$. Then on the solutions $U(x) = u(x)$ of the PDE system (1.132), if $\Lambda_\sigma[U]$ is non-singular, one has a local conservation law

$$\Lambda_\sigma[u]R^\sigma[u] = \mathrm{D}_i\Phi^i[u] = 0. \tag{1.148}$$

A multiplier $\Lambda_\sigma[U]$ is *singular* if it is a singular function when evaluated on solutions $U(x) = u(x)$ of the given PDE system (1.132). [In practice, one is only interested in non-singular sets of multipliers, since considering singular multipliers can lead to arbitrary divergence expressions that are not conservation laws of the given system. For example, $\Lambda_\sigma[U] = \mathrm{D}_i\Phi^i[U]/R^\sigma[U]$ yields $\Lambda_\sigma[U]R^\sigma[U] \equiv \mathrm{D}_i(N\Phi^i[U])$, in terms of arbitrary functions $\Phi^1[U], \ldots, \Phi^n[U]$.]

Through this approach, the determination of local conservation laws for a given PDE system (1.132) reduces to finding sets of local multipliers. The following essential questions arise.

1. How can one formulate the determining equations to find all sets of local multipliers of a given PDE system (1.132) that only yield its nontrivial local conservation laws?
2. Under what conditions do all nontrivial local conservation laws arise from sets of local multipliers? Conversely, under what conditions does a set of local multipliers yield only nontrivial local conservation laws?
3. How can one construct the fluxes of a local conservation law arising from a given set of local multipliers?

The first question is answered through the use of Euler operators that are introduced below. The second question involves writing a given PDE system in a solved form with respect to some leading derivatives, as shown below and discussed further in Section 1.3.4. The third question is considered in Section 1.3.7.

Definition 1.3.5. The *Euler operator* with respect to U^j is the operator defined by

$$\mathrm{E}_{U^j} = \frac{\partial}{\partial U^j} - \mathrm{D}_i\frac{\partial}{\partial U^j_i} + \cdots + (-1)^s \mathrm{D}_{i_1}\ldots\mathrm{D}_{i_s}\frac{\partial}{\partial U^j_{i_1\ldots i_s}} + \cdots \tag{1.149}$$

for each $j = 1, \ldots, m$.

By direct calculation, one can show that the Euler operators (1.149) annihilate *any* divergence expression $\mathrm{D}_i\Phi^i[U]$. In particular, the following identities hold for arbitrary $U(x)$:

$$\mathrm{E}_{U^j}(\mathrm{D}_i\Phi^i(x,U,\partial U,\ldots,\partial^r U))\equiv 0,\quad j=1,\ldots,m. \tag{1.150}$$

The converse also holds. Specifically, the only scalar expressions annihilated by Euler operators are divergence expressions. This establishes the following theorem.

Theorem 1.3.2. *The equations* $\mathrm{E}_{U^j}F(x,U,\partial U,\ldots,\partial^s U)\equiv 0$, $j=1,\ldots,m$ *hold for arbitrary* $U(x)$ *if and only if* $F(x,U,\partial U,\ldots,\partial^s U)\equiv \mathrm{D}_i\Psi^i(x,U,\partial U,\ldots,\partial^{s-1}U)$ *holds for some functions* $\Psi^i(x,U,\partial U,\ldots,\partial^{s-1}U)$, $i=1,\ldots,n$.

Proof. See Exercises 1.3.2 and 1.3.3. □

From Theorem 1.3.2, the proof of the following theorem connecting local multipliers and local conservation laws is immediate.

Theorem 1.3.3. *A set of non-singular local multipliers* $\{\Lambda_\sigma(x,U,\partial U,\ldots,\partial^l U)\}_{\sigma=1}^N$ *yields a local conservation law for the PDE system* $\mathbf{R}\{x\,;u\}$ (1.132) *if and only if the set of identities*

$$\begin{aligned}&\mathrm{E}_{U^j}(\Lambda_\sigma(x,U,\partial U,\ldots,\partial^l U)R^\sigma(x,U,\partial U,\ldots,\partial^k U))\equiv 0,\\ &j=1,\ldots,m,\end{aligned} \tag{1.151}$$

holds for arbitrary functions $U(x)$.

The set of equations (1.151) yields the set of *linear determining equations* to find *all* sets of local conservation law multipliers of the PDE system $\mathbf{R}\{x\,;u\}$ (1.132) by considering multipliers of all orders $l=1,2,\ldots$. Since equations (1.151) hold for arbitrary $U(x)$, it follows that one can treat each U^μ and each of its derivatives U^μ_i, U^μ_{ij}, etc. as independent variables along with x^i, and consequently the linear PDE system (1.151) splits into an over-determined linear system of determining equations whose solutions are the sets of local multipliers $\{\Lambda_\sigma(x,U,\partial U,\ldots,\partial^l U)\}_{\sigma=1}^N$ of the PDE system $\mathbf{R}\{x\,;u\}$ (1.132).

It is important to note that for PDE systems in solved form with respect to a set of leading derivatives, there is a simple converse to Theorem 1.3.3. In particular, suppose each PDE of a given kth order PDE system $\mathbf{R}\{x\,;u\}$ (1.132) can be written in a solved form

$$R^\sigma[u]=u^{j_\sigma}_{i_{\sigma,1}\cdots i_{\sigma,s}}-G^\sigma(x,u,\partial u,\ldots,\partial^k u)=0,\quad \sigma=1,\ldots,N, \tag{1.152}$$

where $s\le k$, $1\le j_\sigma\le m$, $1\le i_{\sigma,1},\ldots,i_{\sigma,s}\le n$ for all $\sigma=1,\ldots,N$. In (1.152), $\{u^{j_\sigma}_{i_{\sigma,1}\cdots i_{\sigma,s}}\}$ is a set of N linearly independent sth order leading partial derivatives, with the property that none of them or their differential consequences appears in $\{G^\sigma[u]\}_{\sigma=1}^N$. Then, without loss of generality, all leading derivatives and their differential consequences can be eliminated in the fluxes $\Phi^i[u]$ of any given local conservation law (1.148) through the PDEs (1.152) and their differential consequences. This leads to the following theorem.

Theorem 1.3.4. *For each local conservation law* $\mathrm{D}_i\Phi^i[u] = 0$ *of a PDE system* $\mathbf{R}\{x\,;u\}$ (1.132) *written in solved form* (1.152), *there exists an equivalent local conservation law* $\mathrm{D}_i\widetilde{\Phi}^i[u] = 0$ *that can be expressed in a* characteristic form

$$\mathrm{D}_i\widetilde{\Phi}^i[U] = \widetilde{\Lambda}_\sigma[U]\left(U^{j_\sigma}_{i_{\sigma,1}\ldots i_{\sigma,s}} - G^\sigma[U]\right) \tag{1.153}$$

in terms of a set of non-singular local multipliers $\{\widetilde{\Lambda}_\sigma[U]\}_{\sigma=1}^N$, *with fluxes that contain no leading derivatives* $U^{j_\sigma}_{i_{\sigma,1}\ldots i_{\sigma,s}}$ *nor their differential consequences.*

Proof. See Exercise 1.3.4. □

Most importantly, Theorem 1.3.4 establishes that essentially, to within equivalence, all local conservation laws of a PDE system $\mathbf{R}\{x\,;u\}$ (1.132), written in a solved form (1.152), arise from local multipliers that are the solutions of the determining equations (1.151).

Remark 1.3.2. The assumption that $\mathbf{R}\{x\,;u\}$ (1.132) can be written in a solved form is the same assumption that is required when finding the local symmetries of $\mathbf{R}\{x\,;u\}$ (1.132). Even in the situation when a given PDE system $\mathbf{R}\{x\,;u\}$ (1.132) cannot be written in a solved form (1.152), the multiplier approach still can be used to seek local conservation laws of $\mathbf{R}\{x\,;u\}$ (1.132).

1.3.4 The direct method for construction of conservation laws. Cauchy–Kovalevskaya form

Following from Theorems 1.3.3 and 1.3.4, a systematic procedure for the construction of local conservation laws, referred to as the *direct method*, is now outlined.

- For a given kth-order PDE system $\mathbf{R}\{x\,;u\}$ (1.132), seek sets of multipliers of the form $\{\Lambda_\sigma(x, U, \partial U, \ldots, \partial^l U)\}_{\sigma=1}^N$ to some specified order l. Choose the dependence of multipliers on their arguments so that singular multipliers do not arise.
- Solve the set of determining equations (1.151) for arbitrary $U(x)$ to find all such sets of multipliers.
- Find the corresponding fluxes $\Phi^i(x, U, \partial U, \ldots, \partial^r U)$ satisfying the identity

$$\begin{aligned}&\Lambda_\sigma(x,U,\partial U,\ldots,\partial^l U)R^\sigma(x,U,\partial U,\ldots,\partial^k U)\\&\qquad\equiv \mathrm{D}_i\Phi^i(x,U,\partial U,\ldots,\partial^r U).\end{aligned}\tag{1.154}$$

- Each set of fluxes and multipliers yields a local conservation law

$$\mathrm{D}_i\Phi^i(x,u,\partial u,\ldots,\partial^r u)=0,$$

holding for all solutions $u(x)$ of the given PDE system $\mathbf{R}\{x\,;u\}$ (1.132).

In practice, in setting up and then solving the set of multiplier determining equations (1.151), it is convenient to write the given PDE system $\mathbf{R}\{x\,;u\}$ (1.132) in a solved form (1.152) with respect to some leading derivatives. In particular, when a given PDE system is written in a solved form, Theorem 1.3.4 ensures that all local conservation laws to a specified order are found. Moreover, it is preferable to have a PDE system written in a solved form since this leads to a straightforward elimination of singular multipliers (which do not lead to conservation laws) and trivial multipliers (which lead to trivial conservation laws).

The fluxes $\Phi^i[U]$ are found either by directly matching the two sides of equation (1.154), or in the case of complicated forms of multipliers and/or PDEs, by an integral formula that is presented in Section 1.3.7. In the situation when the given PDE system has a scaling symmetry, the fluxes can often be found by an algebraic formula in terms of the corresponding sets of multipliers without integration [Section 1.5.2].

Cauchy–Kovalevskaya form

It is important to note that in general, for a given PDE system, whether or not it is in a solved form (1.152), there need not be a one-to-one correspondence between sets of conservation law multipliers $\{\Lambda_\sigma[U]\}_{\sigma=1}^N$ and equivalence classes of conservation laws. In particular, this relation can be many-to-one, as illustrated by the following example.

Consider the PDE system for Maxwell's equations in a vacuum in three space dimensions, given by

$$\begin{aligned}&\mathbf{E}_t-\operatorname{curl}\mathbf{B}=0,\quad \mathbf{B}_t+\operatorname{curl}\mathbf{E}=0,\\&\operatorname{div}\mathbf{E}=0,\quad \operatorname{div}\mathbf{B}=0,\end{aligned}\tag{1.155}$$

for time-dependent electric and magnetic fields given, respectively, by $\mathbf{E}=(E^1(\mathbf{x},t),\ E^2(\mathbf{x},t),\ E^3(\mathbf{x},t))$, $\mathbf{B}=(B^1(\mathbf{x},t),\ B^2(\mathbf{x},t),\ B^3(\mathbf{x},t))$. The PDE system (1.155) can be written in a solved form (1.152) in many ways. For instance, a straightforward choice of eight leading derivatives for the eight PDEs in system (1.155) is given by E^1_x and B^1_x and the components of $\mathbf{E}_t$

and $\mathbf{B}_t$. It is easy to verify that the set of eight respective multipliers given by $\{\boldsymbol{\Lambda}^E = \operatorname{grad} F,\ \boldsymbol{\Lambda}^B = \operatorname{grad} G,\ \Lambda_4^E = -\mathrm{D}_t F,\ \Lambda_4^B = -\mathrm{D}_t G\}$ of the PDE system (1.155) yields the local conservation law

$$\begin{aligned} &\mathrm{D}_t(-F \operatorname{div} \mathbf{E} - G \operatorname{div} \mathbf{B}) \\ &+ \operatorname{div}\left(F(\mathbf{E}_t - \operatorname{curl} \mathbf{B}) + G(\mathbf{B}_t + \operatorname{curl} \mathbf{E})\right) = 0, \end{aligned} \tag{1.156}$$

where $F(\mathbf{x}, t)$ and $G(\mathbf{x}, t)$ are arbitrary functions. Since the density and the fluxes of the conservation law (1.156) vanish on all solutions of the PDE system (1.155), it follows that the conservation law (1.156) is trivial even though its multipliers are nontrivial. The existence of the conservation law (1.156) arises from Maxwell's equations (1.155) satisfying the differential identities

$$\operatorname{div}(\mathbf{E}_t - \operatorname{curl} \mathbf{B}) = \mathrm{D}_t \operatorname{div} \mathbf{E}, \quad \operatorname{div}(\mathbf{B}_t + \operatorname{curl} \mathbf{E}) = \mathrm{D}_t \operatorname{div} \mathbf{B},$$

that hold for *arbitrary* functions $\mathbf{E}(\mathbf{x}, t), \mathbf{B}(\mathbf{x}, t)$. A similar situation will arise for any PDE system that satisfies a differential identity. For such PDE systems, the correspondence between sets of nontrivial multipliers and nontrivial fluxes is many-to-one.

In general, a one-to-one correspondence between sets of nontrivial local multipliers and nontrivial fluxes will hold only for PDE systems that admit a *Cauchy–Kovalevskaya* form.

Definition 1.3.6. A PDE system $\mathbf{R}\{x\,;u\}$ (1.132) *is in Cauchy–Kovalevskaya form* with respect to an independent variable x^j, if the system is in a solved form for the highest derivative of each dependent variable with respect to x^j, i.e.,

$$\frac{\partial^{s_\sigma}}{\partial (x^j)^{s_\sigma}} u^\sigma = G^\sigma(x, u, \partial u, \ldots, \partial^k u), \quad 1 \leq s_\sigma \leq k, \quad \sigma = 1, \ldots, m, \tag{1.157}$$

where all derivatives with respect to x^j appearing in the right-hand side of each PDE of (1.157) are of lower order than those appearing on the left-hand side.

Definition 1.3.7. A PDE system $\mathbf{R}\{x\,;u\}$ (1.132) *admits a Cauchy - Kovalevskaya form* if it can be written in Cauchy–Kovalevskaya form with respect to some independent variable (after a point (contact) transformation if necessary).

Note that a Cauchy–Kovalevskaya form (1.157) is a special case of a solved-form PDE system (1.152) with respect to the same leading derivatives for all dependent variables. Consequently, a PDE system can admit a Cauchy–Kovalevskaya form only if its number of dependent variables equals the number of PDEs in the system, i.e., $N = m$.

As simple examples, the KdV equation admits the Cauchy–Kovalevskaya forms $u_t = -uu_x - u_{xxx}$ and $u_{xxx} = -u_t - uu_x$ in terms of leading t- and x-derivatives, respectively; the wave equation $u_{tx} = 0$ is not in Cauchy–Kovalevskaya form as written, but admits the Cauchy–Kovalevskaya form $u_{TT} = u_{XX}$ after the point transformation $T = t - x$, $X = t + x$.

As a less obvious example, consider the two-dimensional Euler equations for an incompressible fluid given by

$$\begin{aligned} u_t + uu_x + vu_y + \frac{1}{\rho}p_x &= 0, \\ v_t + uv_x + vv_y + \frac{1}{\rho}p_y &= 0, \\ u_x + v_y &= 0, \end{aligned} \tag{1.158}$$

where (u, v) is the fluid velocity, p is the fluid pressure, and $\rho = \text{const}$ is the density of the fluid. The PDE system (1.158) cannot be written in Cauchy–Kovalevskaya form with respect to leading t-derivatives but can be written in Cauchy–Kovalevskaya form with respect to leading x-derivatives:

$$u_x = -v_y, \quad p_x = -\rho(u_t + uu_x + vu_y), \quad v_x = -\frac{1}{u}\left(\frac{1}{\rho}p_y + v_t + vv_y\right).$$

An example of a PDE system that does not admit a Cauchy–Kovalevskaya form is given by Maxwell's equations (1.155) since in this system there are $N = 8$ PDEs and only $m = 6$ dependent variables, i.e., $N \neq m$.

Now one can show that the following theorem holds.

Theorem 1.3.5. *Suppose a PDE system admits a Cauchy–Kovalevskaya form* (1.157)*. Then all of its nontrivial (up to equivalence) local conservation laws arise from multipliers. Moreover, there is a one-to-one correspondence between equivalence classes of conservation laws and sets of conservation law multipliers with no dependence on derivatives of u^σ with respect to x^j.*

For details of the proof, see Anco & Bluman (2002b).

Remark 1.3.3. For a PDE system $\mathbf{R}\{x\,;u\}$ (1.132) that has a Cauchy–Kovalevskaya PDE form (1.157), there is a particularly effective formulation of the direct method to find local conservation laws. Let t denote the independent variable in the derivative appearing in solved form in each PDE of the system and let $x = (x^1, \ldots, x^{n-1})$ denote the remaining independent variables. It is convenient to express the given PDE system in its equivalent first-order (evolutionary) form with respect to t:

$$R^\sigma[u] = \frac{\partial u^\sigma}{\partial t} - g^\sigma(x, t, u, \partial_x u, \ldots, \partial_x^k u) = 0, \quad \sigma = 1, \ldots, m. \tag{1.159}$$

Since t-derivatives $\partial u^j/\partial t$ and their differential consequences can be expressed through the equations (1.159), one can show from Theorems 1.3.4 and 1.3.5 that all nontrivial local conservation laws (up to equivalence) of the system (1.159) arise from non-singular sets of multipliers of the form $\Lambda_\sigma(x,t,U,\partial_x U,\ldots,\partial_x^l U)$, $\sigma=1,\ldots,m$.

1.3.5 Examples

To illustrate the direct method to obtain local conservation laws of a given system of PDEs, we consider two examples.

(1) A nonlinear telegraph system
As a first example, consider a nonlinear telegraph system ($u^1=u,\ u^2=v$) given by

$$\begin{aligned} R^1[u,v] &= v_t-(u^2+1)u_x-u=0,\\ R^2[u,v] &= u_t-v_x=0. \end{aligned} \tag{1.160}$$

This is a first-order Cauchy–Kovalevskaya PDE system with leading derivatives v_t and u_t.

We seek all local conservation law multipliers of the form

$$\Lambda_1=\xi(x,t,U,V),\quad \Lambda_2=\phi(x,t,U,V) \tag{1.161}$$

of the PDE system (1.160). In terms of the Euler operators

$$\mathrm{E}_U=\frac{\partial}{\partial U}-\mathrm{D}_x\frac{\partial}{\partial U_x}-\mathrm{D}_t\frac{\partial}{\partial U_t},\quad \mathrm{E}_V=\frac{\partial}{\partial V}-\mathrm{D}_x\frac{\partial}{\partial V_x}-\mathrm{D}_t\frac{\partial}{\partial V_t},$$

the determining equations (1.151) for the multipliers (1.161) become

$$\begin{aligned} &\mathrm{E}_U\left[\xi(x,t,U,V)(V_t-(U^2+1)U_x-U)\right.\\ &\qquad\left.+\,\phi(x,t,U,V)(U_t-V_x)\right]\equiv 0,\\ &\mathrm{E}_V\left[\xi(x,t,U,V)(V_t-(U^2+1)U_x-U)\right.\\ &\qquad\left.+\,\phi(x,t,U,V)(U_t-V_x)\right]\equiv 0, \end{aligned} \tag{1.162}$$

where $U(x,t)$ and $V(x,t)$ are arbitrary functions. Equations (1.162) split with respect to U_t,V_t,U_x,V_x to yield the over-determined linear PDE system given by

$$\begin{aligned} \phi_V-\xi_U&=0, &\quad \phi_U-(U^2+1)\xi_V&=0,\\ \phi_x-\xi_t-U\xi_V&=0, &\quad (U^2+1)\xi_x-\phi_t-U\xi_U-\xi&=0, \end{aligned} \tag{1.163}$$

whose solutions $(\xi(x,t,U,V), \phi(x,t,U,V))$ are the sets of local multipliers of all nontrivial local conservation laws of zeroth order of the NLT system (1.160).

The solutions of the determining system (1.163) are the five sets of local multipliers given by

$$\begin{aligned}
&(\xi_1, \phi_1) = (0, 1), \quad (\xi_2, \phi_2) = (t, x - \tfrac{1}{2}t^2),\\
&(\xi_3, \phi_3) = (1, -t), \quad (\xi_4, \phi_4) = (e^{x+\frac{1}{2}U^2+V}, Ue^{x+\frac{1}{2}U^2+V}),\\
&(\xi_5, \phi_5) = (e^{x+\frac{1}{2}U^2-V}, -Ue^{x+\frac{1}{2}U^2-V}).
\end{aligned} \tag{1.164}$$

Each set (ξ, ϕ) determines a nontrivial zeroth-order local conservation law $\mathrm{D}_t\Psi(x,t,u,v) + \mathrm{D}_x\Phi(x,t,u,v) = 0$ with the characteristic form

$$\begin{aligned}
&\mathrm{D}_t\Psi(x,t,U,V) + \mathrm{D}_x\Phi(x,t,U,V)\\
&\quad \equiv \xi(x,t,U,V)R^1[U,V] + \phi(x,t,U,V)R^2[U,V].
\end{aligned} \tag{1.165}$$

In particular, after equating like derivative terms of (1.165), one has the relations

$$\begin{aligned}
\Psi_U &= \mathrm{E}_U\Psi = \phi, \quad \Psi_V = \mathrm{E}_V\Psi = \xi,\\
\Phi_U &= \xi\frac{\partial R^1}{\partial U_x} + \phi\frac{\partial R^2}{\partial U_x}, \quad \Phi_V = \xi\frac{\partial R^1}{\partial V_x} + \phi\frac{\partial R^2}{\partial V_x} = -\phi,\\
\Psi_t + \Phi_x &= -U\xi.
\end{aligned} \tag{1.166}$$

The integration of the equations (1.166) for each set of multipliers yields the following five linearly independent zeroth order local conservation laws of the PDE system (1.160):

$$\begin{aligned}
&\mathrm{D}_t u + \mathrm{D}_x[-v] = 0,\\
&\mathrm{D}_t[(x - \tfrac{1}{2}t^2)u + tv] + \mathrm{D}_x[(\tfrac{1}{2}t^2 - x)v - t(\tfrac{1}{3}u^3 + u)] = 0,\\
&\mathrm{D}_t[v - tu] + \mathrm{D}_x[tv - (\tfrac{1}{3}u^3 + u)] = 0,\\
&\mathrm{D}_t[e^{x+\frac{1}{2}u^2+v}] + \mathrm{D}_x[-ue^{x+\frac{1}{2}u^2+v}] = 0,\\
&\mathrm{D}_t[e^{x+\frac{1}{2}u^2-v}] + \mathrm{D}_x[ue^{x+\frac{1}{2}u^2-v}] = 0.
\end{aligned} \tag{1.167}$$

For further details, see Bluman & Temuerchaolu (2005a), where conservation laws are found for wide classes of nonlinear telegraph systems.

(2) Korteweg–de Vries equation
As a second example, consider the KdV equation

$$R[u] = u_t + uu_x + u_{xxx} = 0. \tag{1.168}$$

Since the PDE (1.168) can be directly expressed in the solved form $u_t = g[u] = -(uu_x + u_{xxx})$, it follows from Remark 1.3.3 that local multipliers yielding local conservation laws of PDE (1.168) are of the form $\Lambda = \Lambda(x, t, U, \partial_x U, \ldots, \partial_x^l U)$, $l = 1, 2, \ldots$ i.e., multipliers can be assumed to depend at most on x-derivatives of U. Consequently, $\Lambda(x, t, U, \partial_x U, \ldots, \partial_x^l U)$ is a local conservation law multiplier for the PDE (1.168) if and only if it satisfies the determining equation (1.151) given by

$$\begin{aligned}
&\mathrm{E}_U\left(\Lambda(x, t, U, \partial_x U, \ldots, \partial_x^l U)(U_t + UU_x + U_{xxx})\right) \\
&= -\mathrm{D}_t\Lambda - U\mathrm{D}_x\Lambda - \mathrm{D}_x^3\Lambda + (U_t + UU_x + U_{xxx})\Lambda_U \\
&\quad -\mathrm{D}_x((U_t + UU_x + U_{xxx})\Lambda_{\partial_x U}) \\
&\quad + \cdots + (-1)^l\mathrm{D}_x^l((U_t + UU_x + U_{xxx})\Lambda_{\partial_x^l U}) \equiv 0
\end{aligned} \tag{1.169}$$

for an arbitrary function $U(x, t)$ where here the Euler operator

$$\mathrm{E}_U = \frac{\partial}{\partial U} - (\mathrm{D}_t\frac{\partial}{\partial U_t} + \mathrm{D}_x\frac{\partial}{\partial U_x}) + \mathrm{D}_x^2\frac{\partial}{\partial U_{xx}} + \cdots$$

truncates after $\max(3, l)$ x-derivatives of U. Note that the linear determining equation (1.169) is of the form

$$\alpha_1[U] + \alpha_2[U]U_t + \alpha_3[U]\partial_x U_t + \cdots + \alpha_{l+2}[U]\partial_x^l U_t \equiv 0 \tag{1.170}$$

where each $\alpha_i[U]$ depends at most on x, t, U and x-derivatives of U. Since $U(x, t)$ is an arbitrary function, in equation (1.170) each of $U_t, \partial_x U_t, \ldots, \partial_x^l U_t$ can be treated as independent variables, and hence $\alpha_i[U] = 0$, $i = 1, \ldots, l{+}2$. Furthermore, there is a further splitting of each of these $l + 2$ determining equations with respect to each of the x-derivatives of U. This yields an overdetermined linear PDE system for the multipliers. One can show that there is a one-to-one correspondence between multipliers of order at most l and conserved densities of order at most $l/2$ (to within total derivatives with respect to x) for the KdV equation (1.168).

Now consider zeroth-order multipliers, i.e., $\Lambda = \Lambda(x, t, U)$. Then from equations (1.169) and (1.170), it follows that

$$\begin{aligned}
&(\Lambda_t + U\Lambda_x + \Lambda_{xxx}) + 3\Lambda_{xxU}U_x + 3\Lambda_{xUU}U_x^2 \\
&+\Lambda_{UUU}U_x^3 + 3\Lambda_{xU}U_{xx} + 3\Lambda_{UU}U_xU_{xx} \equiv 0.
\end{aligned} \tag{1.171}$$

Equation (1.171) is a polynomial identity in the variables U_x, U_{xx}. Hence equation (1.171) splits into the three equations (the other three equations are differential consequences)

$$\Lambda_t + U\Lambda_x + \Lambda_{xxx} = 0, \quad \Lambda_{xU} = 0, \quad \Lambda_{UU} = 0,$$

whose solution yields the three local multipliers

$$\Lambda_1 = 1, \quad \Lambda_2 = U, \quad \Lambda_3 = tU - x. \tag{1.172}$$

It is easy to check that the three multipliers (1.172) yield the conservation laws for mass, momentum and center of mass motion, given, respectively, by equations (1.142a)–(1.142c), of the KdV equation (1.168).

Next consider first order multipliers, i.e., $\Lambda = \Lambda(x, t, U, U_x)$. From equations (1.169) and (1.170), one gets $-\alpha_3 = \Lambda_{U_x} = 0$. Thus the KdV equation has no first order multipliers. It is left as an exercise to show that there is only one second order multiplier $\Lambda = \Lambda(x, t, U, U_x, U_{xx})$, given by

$$\Lambda_4 = U_{xx} + \tfrac{1}{2}U^2. \tag{1.173}$$

1.3.6 Linearizing operators and adjoint equations

Consider a system of N PDEs $\mathbf{R}\{x\,;u\}$ given by

$$R^\sigma[u] = R^\sigma(x, u, \partial u, \ldots, \partial^k u) = 0, \quad \sigma = 1, \ldots, N, \tag{1.174}$$

with n independent variables $x = (x^1, \ldots, x^n)$ and m dependent variables $u = (u^1, \ldots, u^m)$. Let

$$R^\sigma[U] = R^\sigma(x, U, \partial U, \ldots, \partial^k U), \quad \sigma = 1, \ldots, N, \tag{1.175}$$

for an *arbitrary* function $U(x) = (U^1(x), \ldots, U^m(x))$. The *linearizing operator* $\mathrm{L}[U]$ associated with the PDE system (1.174) is given by

$$\begin{aligned} \mathrm{L}^\sigma_\rho[U]V^\rho = \Big[&\frac{\partial R^\sigma[U]}{\partial U^\rho} + \frac{\partial R^\sigma[U]}{\partial U^\rho_i}\mathrm{D}_i + \cdots \\ &+ \frac{\partial R^\sigma[U]}{\partial U^\rho_{i_1\ldots i_k}}\mathrm{D}_{i_1}\ldots\mathrm{D}_{i_k}\Big]V^\rho, \qquad \sigma = 1, \ldots, N, \end{aligned} \tag{1.176}$$

in terms of an *arbitrary* function $V(x) = (V^1(x), \ldots, V^m(x))$. The *adjoint operator* $\mathrm{L}^*[U]$ associated with the PDE system (1.174) is obtained formally through integration by parts and is given by

$$\begin{aligned}
\mathrm{L}_\rho^{*\,\sigma}[U]W_\sigma &= \frac{\partial R^\sigma[U]}{\partial U^\rho}W_\sigma - \mathrm{D}_i\left(\frac{\partial R^\sigma[U]}{\partial U_i^\rho}W_\sigma\right) + \cdots \\
&= +(-1)^k\mathrm{D}_{i_1}\ldots\mathrm{D}_{i_k}\left(\frac{\partial R^\sigma[U]}{\partial U_{i_1\ldots i_k}^\rho}W_\sigma\right), \qquad (1.177)\\
\rho &= 1,\ldots,m,
\end{aligned}$$

in terms of an *arbitrary* function $W(x) = (W_1(x),\ldots,W_N(x))$.

In particular, the operators (1.176) and (1.177) satisfy the divergence relation

$$W_\sigma\mathrm{L}_\rho^\sigma[U]V^\rho - V^\rho\mathrm{L}_\rho^{*\,\sigma}[U]W_\sigma \equiv \mathrm{D}_i\Psi^i[U] \qquad (1.178)$$

with

$$\begin{aligned}
\Psi^i[U] = \sum_{p=0}^{k-1}\sum_{q=0}^{k-p-1} &(-1)^q\left(\mathrm{D}_{i_1}\ldots\mathrm{D}_{i_p}V^\rho\right) \\
&\times\mathrm{D}_{j_1}\ldots\mathrm{D}_{j_q}\left(W_\sigma\frac{\partial R^\sigma[U]}{\partial U^\rho_{j_1\ldots j_q i i_1\ldots i_p}}\right), \qquad (1.179)
\end{aligned}$$

where $j_1\ldots j_q$ and $i_1\ldots i_p$ are ordered combinations of indices such that $1 \le j_1 \le \ldots \le j_q \le i \le i_1 \le \ldots \le i_p \le n$. [Exercise 1.3.16].

Now let $W_\sigma = \Lambda_\sigma[U] = \Lambda_\sigma(x, U, \partial U, \ldots, \partial^l U)$, $\sigma = 1,\ldots,N$. By direct calculation, in terms of the Euler operators defined by (1.149), one can show that

$$\mathrm{E}_{U^\rho}(\Lambda_\sigma[U]R^\sigma[U]) \equiv \mathrm{L}_\rho^{*\,\sigma}[U]\Lambda_\sigma[U] + F_\rho(R[U]) \qquad (1.180)$$

with

$$\begin{aligned}
F_\rho(R[U]) &= \frac{\partial\Lambda_\sigma[U]}{\partial U^\rho}R^\sigma[U] - \mathrm{D}_i\left(\frac{\partial\Lambda_\sigma[U]}{\partial U_i^\rho}R^\sigma[U]\right) \\
&= +\cdots+(-1)^l\mathrm{D}_{i_1}\ldots\mathrm{D}_{i_l}\left(\frac{\partial\Lambda_\sigma[U]}{\partial U^\rho_{i_1\ldots i_l}}R^\sigma[U]\right), \qquad (1.181)\\
\rho &= 1,\ldots,m.
\end{aligned}$$

From expression (1.180), it immediately follows that $\{\Lambda_\sigma[U]\}_{\sigma=1}^N$ is a set of local multipliers of the PDE system $\mathbf{R}\{x\,;u\}$ (1.174) yielding a divergence expression if and only if the right hand side of (1.180) vanishes for arbitrary $U(x)$. Now suppose each multiplier is *non-singular* for each solution $U(x) = u(x)$ of the PDE system (1.174). Since the expression (1.181) then vanishes for each solution $U(x) = u(x)$ of the PDE system $\mathbf{R}\{x\,;u\}$ (1.174), it follows that every set of non-singular multipliers $\{\Lambda_\sigma[U]\}_{\sigma=1}^N$ of $\mathbf{R}\{x\,;u\}$ is a solution of the adjoint linearizing system of PDEs when $U(x) = u(x)$ is a solution of the PDE system $\mathbf{R}\{x\,;u\}$, i.e.,

$$\mathrm{L}_\rho^{*\,\sigma}[u]\Lambda_\sigma[u] = 0, \quad \rho = 1,\ldots,m. \qquad (1.182)$$

In particular, the following two results have been proved.

Theorem 1.3.6. *For a given PDE system* $\mathbf{R}\{x\,;u\}$ (1.174), *every set of local conservation law multipliers* $\{\Lambda_\sigma[U] = \Lambda_\sigma(x, U, \partial U, \ldots, \partial^l U)\}_{\sigma=1}^N$ *satisfies the identity*

$$\begin{aligned} &\mathrm{L}_\rho^{*\,\sigma}[U]\Lambda_\sigma[U] + \frac{\partial \Lambda_\sigma[U]}{\partial U^\rho} R^\sigma[U] - \mathrm{D}_i\left(\frac{\partial \Lambda_\sigma[U]}{\partial U_i^\rho} R^\sigma[U]\right) \\ &+ \cdots + (-1)^l \mathrm{D}_{i_1}\ldots\mathrm{D}_{i_l}\left(\frac{\partial \Lambda_\sigma[U]}{\partial U_{i_1\ldots i_l}^\rho} R^\sigma[U]\right) \equiv 0, \quad \rho = 1, \ldots, m, \end{aligned} \tag{1.183}$$

holding for arbitrary functions $U(x) = (U^1(x), \ldots, U^m(x))$, *where the components* $\{\mathrm{L}_\rho^{*\,\sigma}[U]\}$ *of the adjoint operator of the linearizing operator (Fréchet derivative) for the PDE system* (1.174) *are given by expressions* (1.177).

Corollary 1.3.1. *For any solution* $U(x) = u(x) = (u^1(x), \ldots, u^m(x))$ *of a given PDE system* $\mathbf{R}\{x\,;u\}$ (1.174), *each set of local multipliers* $\{\Lambda_\sigma[U]\}_{\sigma=1}^N$ *satisfies the adjoint linearizing system*

$$\mathrm{L}_\rho^{*\,\sigma}[u]\Lambda_\sigma[u] = 0, \tag{1.184}$$

where $\{\mathrm{L}_\rho^{*\,\sigma}[U]\}$ *is given by the components of the adjoint operator* (1.183).

The identity (1.183) provides the explicit general form of the multiplier determining system (1.151) in Theorem 1.3.3. In general, the adjoint system (1.184) is strictly a subset of system (1.151) when one takes into account the splitting of (1.183) with respect to a set of leading derivatives for $R^\sigma[U]$, $\sigma = 1\ldots, N$.

As an example, consider the KdV equation (1.168). Its linearizing operator is given by

$$\mathrm{L}[U] = \mathrm{D}_x^3 + U\mathrm{D}_x + U_x + \mathrm{D}_t, \tag{1.185}$$

in terms of an arbitrary function $U(x,t)$. The formal *adjoint* operator of the linearizing operator (1.185) is given by

$$\mathrm{L}^*[U] = -\mathrm{D}_x^3 - U\mathrm{D}_x - \mathrm{D}_t, \tag{1.186}$$

since for any two functions $V(x,t)$ and $W(x,t)$, the expression $W\mathrm{L}[U]V - V\mathrm{L}^*[U]W$ is a divergence. In particular, one has

$$\begin{aligned} &W\mathrm{L}[U]V - V\mathrm{L}^*[U]W \\ &\quad\equiv \mathrm{D}_t(WV) + \mathrm{D}_x(WV_{xx} - W_xV_x + W_{xx}V + WVU). \end{aligned} \tag{1.187}$$

Then one can show that the set of determining equations (1.151) for local multipliers $\Lambda(x, t, U, \partial_x U, \ldots, \partial_x^l U)$ of the KdV equation splits into an overdetermined linear system consisting of the $l+2$ equations

$$-\widetilde{\mathrm{D}}_t\Lambda - U\mathrm{D}_x\Lambda - \mathrm{D}_x^3\Lambda = 0, \tag{1.188a}$$

$$\sum_{k=1}^{l}(-\mathrm{D}_x)^k\Lambda_{\partial_x^k U} = 0, \tag{1.188b}$$

$$(1-(-1)^q)\Lambda_{\partial_x^q U} + \sum_{k=q+1}^{l}\frac{k!}{q!(k-q)!}(-\mathrm{D}_x)^{k-q}\Lambda_{\partial_x^k U} = 0, \quad q = 1,\dots,l-1, \tag{1.188c}$$

$$(1-(-1)^l)\Lambda_{\partial_x^l U} = 0, \tag{1.188d}$$

where $\hat{\mathrm{D}}_t = \partial/\partial t + g[U]\,\partial/\partial U + (g[U])_x\,\partial/\partial U_x + \dots$ is the total time derivative operator restricted to the KdV equation, with $g[U] = -(UU_x + U_{xxx})$. In particular, $\Lambda(x,t,u,\partial_x u,\dots,\partial_x^l u)$ satisfies

$$\mathrm{L}^*[u]\Lambda[u]|_{R[u]=u_t-g[u]=0} = 0, \tag{1.189}$$

and after substitution for u_t in (1.189), one obtains the first determining equation (1.188a) holding for *arbitrary* $u = U(x,t)$.

As one can see, the determining equation (1.188a) for multipliers $\Lambda(x,t,U,\partial_x U,\dots,\partial_x^l U)$ is the adjoint of the linearizing equation of the KdV equation given by

$$\mathrm{L}[u]\eta[u]|_{R[u]=u_t-g[u]=0} = 0, \tag{1.190}$$

in terms of the infinitesimals $\eta[u]$ of local symmetries $\eta[u]\,\partial/\partial u$ of the KdV equation. In particular, the adjoint determining equation (1.188a) is just one of the set of $l+2$ linear determining equations (1.188) that are the necessary and sufficient conditions for $\Lambda(x,t,U,\partial_x U,\dots,\partial_x^l U)$ to yield a local conservation law of the KdV equation (1.168).

1.3.7 Determination of fluxes of conservation laws from multipliers

Consider a set of non-singular multipliers $\{\Lambda_\sigma[U] = \Lambda_\sigma(x,U,\partial U,\dots,\partial^l U)\}_{\sigma=1}^N$ that yields a divergence expression for a PDE system $\mathbf{R}\{x\,;u\}$ (1.174) with some set of fluxes $\{\Phi^i[U] = \Phi^i(x,U,\partial U,\dots,\partial^r U)\}_{i=1}^n$, i.e.

$$\Lambda_\sigma[U]R^\sigma[U] \equiv \mathrm{D}_i\Phi^i[U], \tag{1.191}$$

which thus gives a local conservation law

$$\operatorname{div}\mathbf{\Phi}[u] = 0 \tag{1.192}$$

holding on the solutions $U(x) = u(x)$ of the PDE system (1.132).

Given a set of conservation law multipliers, the problem of finding the fluxes $\{\Phi^i[U]\}_{i=1}^n$ is formally a problem of inversion of the divergence differential operator. Modulo trivial conservation laws of the first type [Section 1.3.2], a set of multipliers defines an *equivalence class* of conservation laws, up to free constants and arbitrary functions. In particular, for any fluxes $\{\Phi^i[U]\}_{i=1}^n$ of a local conservation law in characteristic form

$$\mathrm{D}_i\Phi^i[U] \equiv \Lambda_\sigma[U]R^\sigma[U],$$

the addition of curl-type expressions $\mathrm{D}_jH^{ij}[U]$ given in terms of arbitrary functions $H^{ij}(x, U, \partial U, \ldots, \partial^{r-1}U) = -H^{ji}(x, U, \partial U, \ldots, \partial^{r-1}U)$, $i, j = 1, \ldots, n$, yields an equivalent local conservation law having the same characteristic form

$$\mathrm{D}_i\widetilde{\Phi}^i[U] \equiv \Lambda_\sigma[U]R^\sigma[U]$$

with fluxes

$$\widetilde{\Phi}^i[U] = \tilde{\Phi}^i[U] + \mathrm{D}_jH^{ij}[U]$$

since $\mathrm{D}_i\mathrm{D}_jH^{ij}[U] \equiv 0$ holds identically. Thus, the fluxes arising from a set of local conservation law multipliers are always arbitrary to within curl-type expressions, which are described by an n-dimensional generalization of the vector operator identity $\operatorname{div}\operatorname{curl} = 0$ in three dimensions. From the practical point of view, for a given set of multipliers, it is sufficient to find just one corresponding set of fluxes.

There are several ways of finding the fluxes of local conservation laws from a known set of multipliers. Each method has its own advantages in different situations depending on the precise form of the multipliers and PDE system. First, we present a direct method that converts (1.191) directly into the set of determining equations to be solved for the fluxes $\Phi^i[U]$. This method is easy to implement for simple types of conservation laws, including those involving arbitrary functions. The second method sets up a one-dimensional integral (homotopy) formula that yields $\Phi^i[U]$. This formula is systematic and hence more generally applicable, but it can be awkward for conservation laws involving arbitrary functions or if singularities arise at the endpoints of integration.

It is important to note that in the situation when a PDE system has a scaling symmetry, one can often find fluxes of conservation laws by an algebraic formula in terms of the corresponding sets of multipliers without integration. This is considered in Section 1.5.2.

Direct computation of fluxes of conservation laws

For conservation laws arising from simple forms of multipliers, the fluxes are most easily found by direct matching of the two sides of equation (1.191). Often, to find the divergence expression in (1.191), one can use integration by parts on the terms in the expression $\Lambda_\sigma[U]R^\sigma[U]$.

If integration by parts is not obvious, one may alternatively assume a general form for the fluxes $\Phi^i[U]$ with an appropriate choice of dependence on derivatives of U, and directly solve the resulting determining equations that arise from equating both sides of (1.191). In particular, if the maximal order of derivatives of $U(x)$ present in the multipliers is l, and the maximal order in the equations $R^\sigma[U]$ appearing in the linear combination (1.191) is k, then, without loss of generality (i.e., up to a trivial conservation law) one may assume that the fluxes are of the form $\Phi^i(x, U, \partial U, \ldots, \partial^r U)$, where $r = \max(l, k)$ [Olver (1983)]. Moreover, in the common situation when the left-hand side of (1.191) is linear in the partial derivatives of highest order (i.e., $\max(l, k)$), one may assume that the fluxes $\Phi^i[U]$ depend only on the derivatives of $U(x)$ up to order $r = \max(l, k) - 1$. After determining the order r of derivatives appearing in the fluxes $\Phi^i[U]$, one proceeds directly by first equating terms in (1.191) that contain the highest derivatives of $U(x)$, second solving for the dependence of $\Phi^i[U]$ on these derivatives, then repeating these steps on the remaining terms of successively lower orders.

As an example, consider the nonlinear wave equation

$$R[u] = u_{tt} - (c^2(u)u_x)_x = 0, \tag{1.193}$$

with an arbitrary wave speed $c(u)$. For simplicity, consider multipliers of the form $\Lambda[U] = \Lambda(x, t, U)$. The determining equations (1.150) yield the solution $\Lambda(x, t) = C_1 + C_2 x + C_3 t + C_4 tx$, where $C_1, \ldots, C_4$ are arbitrary constants. As a result, one obtains four linearly independent conservation laws, arising from the multipliers $\Lambda^{(1)} = 1$, $\Lambda^{(2)} = x$, $\Lambda^{(3)} = t$, $\Lambda^{(4)} = tx$. We now determine the corresponding density-flux pairs.

For the multiplier $\Lambda^{(1)} = 1$, one has

$$\Lambda^{(1)}[U]R[U] \equiv \mathrm{D}_t(U_t) - \mathrm{D}_x(c^2(U)U_x), \tag{1.194}$$

since PDE (1.193) is in divergence form as it stands:

$$\mathrm{D}_t(u_t) - \mathrm{D}_x(c^2(u)u_x) = 0. \tag{1.195}$$

For the multiplier $\Lambda^{(2)} = x$, one can determine the flux and density using integration by parts:

$$\begin{aligned}\Lambda^{(2)}[U]R[U] &\equiv x(\mathrm{D}_t(U_t) - \mathrm{D}_x(c^2(U)U_x)) \\ &\equiv \mathrm{D}_t(xU_t) - \mathrm{D}_x(xc^2(U)U_x) + c^2(U)U_x \\ &\equiv \mathrm{D}_t(xU_t) - \mathrm{D}_x\left(xc^2(U)U_x - \int c^2(U)dU\right),\end{aligned} \tag{1.196}$$

and thus the corresponding conservation law is given by

$$\mathrm{D}_t(xu_t) - \mathrm{D}_x\left(xc^2(u)u_x - \int c^2(u)du\right) = 0. \tag{1.197}$$

Similarly, for the multiplier $\Lambda^{(3)} = t$, one finds

$$\Lambda^{(3)}[U]R[U] \equiv \mathrm{D}_t(tU_t - U) - \mathrm{D}_x(tc^2(U)U_x), \tag{1.198}$$

giving the corresponding conservation law

$$\mathrm{D}_t(tu_t - u) - \mathrm{D}_x(tc^2(u)u_x) = 0. \tag{1.199}$$

To find the flux and density of the somewhat more complicated fourth conservation law of PDE (1.193) arising from the multiplier $\Lambda^{(4)} = tx$, it is straightforward to solve the flux-density determining equation

$$\Lambda^{(4)}[U]R[U] = tx(\mathrm{D}_t(U_t) - \mathrm{D}_x(c^2(U)U_x)) \equiv \mathrm{D}_tT[U] + \mathrm{D}_xX[U]. \tag{1.200}$$

Since the left-hand side of (1.200) is linear in the highest derivatives U_{tt} and U_{xx}, one can assume that $T[U] = T(x,t,U,U_t,U_x)$ and $X[U] = X(x,t,U,U_t,U_x)$. Expanding both sides of (1.200), one obtains

$$\begin{aligned}&tx(U_{tt} - 2c(U)c'(U)(U_x)^2 - c^2(U)U_{xx}) \\ &\quad = (T_t + T_UU_t + T_{U_t}U_{tt} + T_{U_x}U_{tx}) \\ &\qquad + (X_x + X_UU_x + X_{U_t}U_{tx} + X_{U_x}U_{xx}).\end{aligned} \tag{1.201}$$

Matching the terms of the highest order derivatives U_{tt}, U_{xx} and U_{tx}, one finds that

$$tx = T_{U_t}, \quad -txc^2(U) = X_{U_x}, \quad T_{U_x} = -X_{U_t}. \tag{1.202}$$

The third equation in (1.202) can be replaced by $T_{U_x} = X_{U_t} = 0$, since one can show that the general solution of $T_{U_x} = -X_{U_t}$ just leads to equivalent conservation laws. The first two equations in (1.202) yield

$$T[U] = txU_t + \alpha(x,t,U), \quad X[U] = -txc^2(U)U_x + \beta(x,t,U), \tag{1.203}$$

for arbitrary $\alpha(x,t,U)$, $\beta(x,t,U)$. Substituting (1.203) into the determining equations (1.201) and setting to zero coefficients of first-order partial derivatives of U, one finds

$$x = -\alpha_U, \quad tc^2(U) = \beta_U, \quad \alpha_t = -\beta_x.$$

Therefore

$$\alpha(x,t,U) = -xU + \widetilde{\alpha}(x,t), \quad \beta(x,t,U) = t\int c^2(U)dU + \widetilde{\beta}(x,t), \quad \widetilde{\alpha}_t = -\widetilde{\beta}_x.$$

It is evident that any choice of $\widetilde{\alpha}$ and $\widetilde{\beta}$ satisfying $\widetilde{\alpha}_t = -\widetilde{\beta}_x$ yields an equivalent conservation law, with the simplest one having $\widetilde{\alpha} = \widetilde{\beta} = 0$. Thus, the fluxes of the fourth conservation law of PDE (1.193) are given by

$$T[U] = txU_t - xU, \quad X[U] = -txc^2(U)U_x + t\int c^2(U)dU,$$

and the corresponding conservation law is given by

$$\mathrm{D}_t(txu_t - xu) - \mathrm{D}_x\left(txc^2(u)u_x - t\int c^2(u)du\right) = 0. \tag{1.204}$$

Integral formula for fluxes of a conservation law

In the case of complicated forms of multipliers and/or PDE systems, the problem of finding fluxes can be reduced to an integral (homotopy) formula.

Consider a set of local conservation law multipliers $\{\Lambda_\sigma[U]\}_{\sigma=1}^N$ of a given PDE system $\mathbf{R}\{x\,;u\}$ (1.174), i.e.,

$$\Lambda_\sigma[U]R^\sigma[U] \equiv \mathrm{D}_i\Phi^i[U]. \tag{1.205}$$

Denote the linearizing operator (1.176) associated with the PDE system $\mathbf{R}\{x\,;u\}$ (1.174) by

$$\begin{aligned}(\mathrm{L}_R)^\sigma_\rho[U]V^\rho = &\Big[\frac{\partial R^\sigma[U]}{\partial U^\rho} + \frac{\partial R^\sigma[U]}{\partial U^\rho_i}\mathrm{D}_i \\ &+ \cdots + \frac{\partial R^\sigma[U]}{\partial U^\rho_{i_1\ldots i_k}}\mathrm{D}_{i_1}\ldots\mathrm{D}_{i_k}\Big]V^\rho, \\ \sigma = &\ 1,\ldots,N,\end{aligned} \tag{1.206}$$

and its adjoint (1.177) by

$$\begin{aligned}(\mathrm{L}_R^*)^\sigma_\rho[U]W_\sigma &= \frac{\partial R^\sigma[U]}{\partial U^\rho}W_\sigma - \mathrm{D}_i\left(\frac{\partial R^\sigma[U]}{\partial U^\rho_i}W_\sigma\right) \\ &\quad + \cdots + (-1)^k\mathrm{D}_{i_1}\ldots\mathrm{D}_{i_k}\left(\frac{\partial R^\sigma[U]}{\partial U^\rho_{i_1\ldots i_k}}W_\sigma\right), \\ \rho &= 1,\ldots,m,\end{aligned} \tag{1.207}$$

acting, respectively, on arbitrary functions $V = (V^1(x),\ldots,V^m(x))$ and $W = (W_1(x),\ldots,W_N(x))$.

For each multiplier $\Lambda_\sigma[U] = \Lambda_\sigma(x,U,\partial U,\ldots,\partial^l U)$, introduce the corresponding linearizing operator

$$\begin{aligned}(\mathrm{L}_\Lambda)_{\sigma\rho}[U]\widetilde{V}^\rho &= \left[\frac{\partial \Lambda_\sigma[U]}{\partial U^\rho} + \frac{\partial \Lambda_\sigma[U]}{\partial U^\rho_i}\mathrm{D}_i\right. \\ &\quad \left. + \cdots + \frac{\partial \Lambda_\sigma[U]}{\partial U^\rho_{i_1\ldots i_l}}\mathrm{D}_{i_1}\ldots\mathrm{D}_{i_l}\right]\widetilde{V}^\rho, \\ \sigma &= 1,\ldots,N,\end{aligned} \tag{1.208}$$

and its adjoint

$$\begin{aligned}(\mathrm{L}_\Lambda^*)_{\sigma\rho}[U]\widetilde{W}^\sigma &= \frac{\partial \Lambda_\sigma[U]}{\partial U^\rho}\widetilde{W}^\sigma - \mathrm{D}_i\left(\frac{\partial \Lambda_\sigma[U]}{\partial U^\rho_i}\widetilde{W}^\sigma\right) \\ &\quad + \cdots + (-1)^k\mathrm{D}_{i_1}\ldots\mathrm{D}_{i_l}\left(\frac{\partial \Lambda_\sigma[U]}{\partial U^\rho_{i_1\ldots i_l}}\widetilde{W}^\sigma\right), \\ \rho &= 1,\ldots,m,\end{aligned} \tag{1.209}$$

acting, respectively, on arbitrary functions $\widetilde{V} = (\widetilde{V}^1(x),\ldots,\widetilde{V}^m(x))$ and $\widetilde{W} = (\widetilde{W}_1(x),\ldots,\widetilde{W}_N(x))$.

It is straightforward to show that the operators (1.206)–(1.209) satisfy the following divergence identities [Exercise 1.3.16]:

$$W_\sigma(\mathrm{L}_R)^\sigma_\rho[U]V^\rho - V^\rho(\mathrm{L}_R^*)^\sigma_\rho[U]W_\sigma \equiv \mathrm{D}_i S^i[V,W;R[U]], \tag{1.210}$$

$$\widetilde{W}^\sigma(\mathrm{L}_\Lambda)_{\sigma\rho}[U]\widetilde{V}^\rho - \widetilde{V}^\rho(\mathrm{L}_\Lambda^*)_{\sigma\rho}[U]\widetilde{W}^\sigma \equiv \mathrm{D}_i\widetilde{S}^i[\widetilde{V},\widetilde{W};\Lambda[U]], \tag{1.211}$$

with

$$\begin{aligned}S^i[V,W;R[U]] &= \sum_{p=0}^{k-1}\sum_{q=0}^{k-p-1}(-1)^q\left(\mathrm{D}_{i_1}\ldots\mathrm{D}_{i_p}V^\rho\right) \\ &\quad \times\mathrm{D}_{j_1}\ldots\mathrm{D}_{j_q}\left(W_\sigma\frac{\partial R^\sigma[U]}{\partial U^\rho_{j_1\ldots j_q i i_1\ldots i_p}}\right),\end{aligned} \tag{1.212}$$

and

$$\begin{aligned}\widetilde{S}^i[\widetilde{V},\widetilde{W};\Lambda[U]] = \sum_{p=0}^{l-1}\sum_{q=0}^{l-p-1}(-1)^q\left(\mathrm{D}_{i_1}\ldots\mathrm{D}_{i_p}\widetilde{V}^\rho\right)\\ \times\mathrm{D}_{j_1}\ldots\mathrm{D}_{j_q}\left(\widetilde{W}^\sigma\frac{\partial\Lambda_\sigma[U]}{\partial U^\rho_{j_1\ldots j_q i i_1\ldots i_p}}\right),\end{aligned} \tag{1.213}$$

where k is the order of the PDE system $\mathbf{R}\{x\,;u\}$ (1.174), l is the maximal order of the derivatives appearing in the multipliers $\{\Lambda_\sigma[U]\}_{\sigma=1}^N$, and $j_1\ldots j_q$, $i_1\ldots i_p$ are ordered combinations of indices such that $1\le j_1\le\ldots\le j_q\le i\le i_1\le\ldots\le i_p\le n$.

Now introduce a one-parameter family of functions

$$U_{(\lambda)}\equiv U+(\lambda-1)V, \tag{1.214}$$

where $U=(U^1(x),\ldots,U^m(x))$ and $V=(V^1(x),\ldots,V^m(x))$ are arbitrary functions, and λ is a scalar parameter. Replacing U by $U_{(\lambda)}$ in the conservation law identity (1.205), one has

$$\frac{\partial}{\partial\lambda}(\Lambda_\sigma[U_{(\lambda)}]R^\sigma[U_{(\lambda)}])\equiv\frac{\partial}{\partial\lambda}\mathrm{D}_i\Phi^i[U_{(\lambda)}]=\mathrm{D}_i\left(\frac{\partial}{\partial\lambda}\Phi^i[U_{(\lambda)}]\right). \tag{1.215}$$

[The last identity holds since λ can be viewed as an additional independent variable.] The left-hand side of (1.215) can then be expressed in terms of the linearizing operators (1.206) and (1.208) as follows:

$$\frac{\partial}{\partial\lambda}(\Lambda_\sigma[U_{(\lambda)}]R^\sigma[U_{(\lambda)}])=\Lambda_\sigma[U_{(\lambda)}](\mathrm{L}_R)^\sigma_\rho[U_{(\lambda)}]V^\rho+R^\sigma[U_{(\lambda)}](\mathrm{L}_\Lambda)_{\sigma\rho}[U_{(\lambda)}]V^\rho.$$

From (1.210) and (1.211) with $W_\sigma=\Lambda_\sigma[U_{(\lambda)}]$ and $\widetilde{W}^\sigma=R^\sigma[U_{(\lambda)}]$, respectively, one obtains

$$\begin{aligned}&\frac{\partial}{\partial\lambda}(\Lambda_\sigma[U_{(\lambda)}]R^\sigma[U_{(\lambda)}])\\ &\quad=V^\rho(\mathrm{L}^*_R)^\sigma_\rho[U_{(\lambda)}]\Lambda_\sigma[U_{(\lambda)}]+\mathrm{D}_iS^i[V,\Lambda[U_{(\lambda)}];R[U_{(\lambda)}]]\\ &\qquad+V^\rho(\mathrm{L}^*_\Lambda)_{\sigma\rho}[U_{(\lambda)}]R^\sigma[U_{(\lambda)}]+\mathrm{D}_i\widetilde{S}^i[V,R[U_{(\lambda)}];\Lambda[U_{(\lambda)}]]\\ &\quad=\mathrm{D}_i\left(S^i[V,\Lambda[U_{(\lambda)}];R[U_{(\lambda)}]]+\widetilde{S}^i[V,R[U_{(\lambda)}]\Lambda[U_{(\lambda)}]]\right),\end{aligned} \tag{1.216}$$

where the last equality follows from the identity (1.183) holding for local conservation law multipliers in Theorem 1.3.6 [Exercise 1.3.17].

Comparing (1.215) and (1.216), one finds that

$$\mathrm{D}_i\left(\frac{\partial}{\partial\lambda}\Phi^i[U_{(\lambda)}]\right)=\mathrm{D}_i\left(S^i[V,\Lambda[U_{(\lambda)}];R[U_{(\lambda)}]]+\widetilde{S}^i[V,R[U_{(\lambda)}];\Lambda[U_{(\lambda)}]]\right),$$

which implies

$$\frac{\partial}{\partial \lambda}\Phi^i[U_{(\lambda)}] = S^i[V, \Lambda[U_{(\lambda)}]; R[U_{(\lambda)}]] + \widetilde{S}^i[V, R[U_{(\lambda)}]; \Lambda[U_{(\lambda)}]], \tag{1.217}$$

up to fluxes of a trivial conservation law. Now let $V = U - \widetilde{U}$, for an arbitrary function $\widetilde{U} = (\widetilde{U}^1(x), \ldots, \widetilde{U}^m(x))$, so that

$$U_{(\lambda)} = \lambda U + (1-\lambda)\widetilde{U}$$

with $U_{(0)} = \widetilde{U}$ and $U_{(1)} = U$. Then, integrating (1.217) with respect to λ from 0 to 1, one obtains

$$\begin{aligned}\Phi^i[U] &= \Phi^i[\widetilde{U}] \\ &+ \int_0^1 \Big(S^i\left[U-\widetilde{U}, \Lambda[\lambda U + (1-\lambda)\widetilde{U}]; R[\lambda U + (1-\lambda)\widetilde{U}]\right] \\ &\qquad + \widetilde{S}^i\left[U-\widetilde{U}, R[\lambda U + (1-\lambda)\widetilde{U}]; \Lambda[\lambda U + (1-\lambda)\widetilde{U}]\right]\Big)\, d\lambda,\end{aligned} \tag{1.218}$$

$i = 1, \ldots, n$.

The following theorem has been proven.

Theorem 1.3.7. *For a given set of local conservation law multipliers* $\{\Lambda_\sigma[U]\}_{\sigma=1}^N$ *of a PDE system* $\mathbf{R}\{x\,; u\}$ (1.174), *the corresponding fluxes are given by the integral formula* (1.218).

In the formula (1.218), $\widetilde{U}$ is an arbitrary function of x, chosen so that the integral converges. Different choices of $\widetilde{U}$ yield fluxes of equivalent conservation laws. One normally chooses $\widetilde{U} = 0$ (provided that the integral converges). Once $\widetilde{U} = \widetilde{U}(x)$ has been chosen, the fluxes $\{\Phi^i[\widetilde{U}]\}_{i=1}^N$ can be found by direct integration through the divergence relation

$$\mathrm{D}_i \Phi^i[\widetilde{U}] = \Lambda_\sigma[\widetilde{U}] R^\sigma[\widetilde{U}] \equiv F(x).$$

For example, one may choose

$$\Phi^1[\widetilde{U}] = \int F(x)dx^1, \quad \Phi^2[\widetilde{U}] = \cdots = \Phi^n[\widetilde{U}] = 0.$$

As an example, consider the modified Korteweg–de Vries equation

$$u_t + u^2 u_x + u_{xxx} = 0. \tag{1.219}$$

One can show that PDE (1.141) has a local conservation law arising from the multiplier $\Lambda[U] = U$, i.e., $f[U] = U(U_t + U^2 U_x + U_{xxx}) \equiv \mathrm{D}_t\Phi^1[U] + \mathrm{D}_x\Phi^2[U]$ is a divergence expression. Using $\widetilde{U} = 0$, one finds $\Phi^1[0] = \Phi^2[0] = 0$, and hence

$$\Phi^1[U] = \tfrac{1}{2}U^2, \quad \Phi^2[U] = \tfrac{1}{4}U^4 + UU_{xx} - \tfrac{1}{2}U_x^2.$$

Alternatively, using $\widetilde{U} = x$, one finds $\Phi^1[\widetilde{U}] = tx^3, \Phi^2[\widetilde{U}] = 0$, and hence

$$\Phi^1[U] = \tfrac{1}{2}(U-x)^2 + x(U-x) + tx^3,$$

$$\Phi^2[U] = \tfrac{1}{4}(U-x)^4 + x(U-x)^3 - \tfrac{1}{2}(U_x-1)^2 + (U-x)U_{xx} + \tfrac{3}{2}x^2(U-x)^2 + x^3(U-x) - U_x + xU_{xx} + 1,$$

which are a density and a flux of an equivalent conservation law [Exercise 1.3.12].

1.3.8 Self-adjoint PDE systems

An especially interesting situation arises when the linearizing operator (Fréchet derivative) $\mathrm{L}[U]$ of a given PDE system (1.174) is self-adjoint.

Definition 1.3.8. Let $\mathrm{L}[U]$, with its components $\mathrm{L}_\rho^\sigma[U]$ given by (1.176), be the linearizing operator associated with a PDE system (1.174). The adjoint operator of $\mathrm{L}[U]$ is $\mathrm{L}^*[U]$, with its components $\mathrm{L}_\rho^{*\sigma}[U]$ given by (1.177). $\mathrm{L}[U]$ is a *self-adjoint* operator if and only if $\mathrm{L}[U] \equiv \mathrm{L}^*[U]$, i.e., $\mathrm{L}_\rho^\sigma[U] \equiv \mathrm{L}_\rho^{*\sigma}[U]$, $\sigma, \rho = 1, \ldots, m$.

It is straightforward to see that if a PDE system, as written, has a self-adjoint linearizing operator, then

- the number of dependent variables appearing in the system must equal the number of equations in the system, i.e., $N = m$;
- if the PDE system is a scalar PDE, the highest-order partial derivative appearing in it must be of even order.

The converse of the latter statement is false. For example, consider the linear heat equation

$$u_t - u_{xx} = 0. \tag{1.220}$$

The linearizing operator of PDE (1.220) is obviously given by $\mathrm{L} = \mathrm{D}_t - \mathrm{D}_x^2$, with adjoint operator $\mathrm{L}^* = -\mathrm{D}_t - \mathrm{D}_x^2 \not\equiv \mathrm{L}$.

One can show that a given PDE system, as written, has a variational (Lagrangian) formulation if and only if its associated linearizing operator is self-adjoint [Volterra (1913); Vainberg (1964); Olver (1986)]. Lagrangians and variational formulations are considered in Section 1.4.

Most importantly, if the linearizing operator associated with a given PDE system is self-adjoint, then any set of local multipliers yields the components, in evolutionary form, of a local symmetry of the given PDE system. In particular, one has the following theorem.

Theorem 1.3.8. *Consider a given PDE system* $\mathbf{R}\{x\,;u\}$ (1.174) *with* $N = m$. *Suppose its associated linearizing operator* $\mathrm{L}[U]$, *with its components given by* (1.176), *is self-adjoint. Suppose* $\{\Lambda_\sigma(x, U, \partial U, \ldots, \partial^l U)\}_{\sigma=1}^m$ *is a set of local conservation law multipliers of the PDE system* (1.174). *Let* $\eta^\sigma(x, u, \partial u, \ldots, \partial^l u) = \Lambda_\sigma(x, u, \partial u, \ldots, \partial^l u)$, $\sigma = 1, \ldots, m$, *where* $U(x) = u(x)$ *is any solution of the PDE system* $\mathbf{R}\{x\,;u\}$ (1.174). *Then*

$$\eta^\sigma(x, u, \partial u, \ldots, \partial^l u)\frac{\partial}{\partial u^\sigma} \tag{1.221}$$

is a local symmetry of the PDE system $\mathbf{R}\{x\,;u\}$ (1.174).

Proof. From the equations (1.184) of Corollary 1.3.1 with $\mathrm{L}[U] = \mathrm{L}^*[U]$, it follows that in terms of the components (1.176) of the associated linearizing operator $\mathrm{L}[U]$, one has

$$\mathrm{L}_\rho^\sigma[u]\Lambda_\sigma(x, u, \partial u, \ldots, \partial^l u) = 0, \quad \rho = 1, \ldots, m, \tag{1.222}$$

where $u = \Theta(x)$ is any solution of the PDE system $\mathbf{R}\{x\,;u\}$ (1.174). But the set of equations (1.222) is the set of determining equations for a local symmetry $\Lambda_\sigma(x, u, \partial u, \ldots, \partial^l u)\,\partial/\partial u^\sigma$ of the PDE system $\mathbf{R}\{x\,;u\}$ (1.174). Hence, it follows that (1.221) is a local symmetry of the PDE system $\mathbf{R}\{x\,;u\}$ (1.174). □

The converse of Theorem 1.3.8 is false. In particular, suppose $\eta^\sigma(x, u, \partial u, \ldots, \partial^l u)\,\partial/\partial u^\sigma$ is a local symmetry of a given PDE system $\mathbf{R}\{x\,;u\}$ (1.174) with a self-adjoint linearizing operator $\mathrm{L}[U]$. Let $\Lambda_\sigma(x, U, \partial U, \ldots, \partial^l U) = \eta^\sigma(x, U, \partial U, \ldots, \partial^l U)$, $\sigma = 1, \ldots, m$, where $U(x) = (U^1(x), \ldots, U^m(x))$ is an arbitrary function. Then it does not necessarily follow that $\{\Lambda_\sigma(x, U, \partial U, \ldots, \partial^l U)\}_{\sigma=1}^m$ is a set of local multipliers of a local conservation law of the PDE system (1.174). This can be seen as follows: In the self-adjoint case, the set of local symmetry determining equations is a subset of the set of local multiplier determining equations, so consequently *each* local symmetry yields a set of local multipliers if and only if *each* solution of the set of local symmetry determining equations also solves the remaining set of local multiplier determining equations.

For example, it can happen that a nonlinear PDE whose linearizing operator is self-adjoint has a point symmetry that does not yield a local conservation law multiplier. In particular, consider the nonlinear wave-speed equation [Anco & Bluman (2002a)] given by

$$u_{tt} = u^2 u_{xx} + u u_x^2. \tag{1.223}$$

It is easy to see that PDE (1.223) is invariant under the scaling point symmetry $x \to \alpha x$, $u \to \alpha u$, corresponding to the infinitesimal generator

$$\mathrm{X} = (u - xu_x)\frac{\partial}{\partial u}. \tag{1.224}$$

The linearizing operator associated with PDE (1.223) is given by

$$\mathrm{L}[U] = \mathrm{D}_t^2 - U^2\mathrm{D}_x^2 - 2UU_x\mathrm{D}_x - 2UU_{xx} - U_x^2. \tag{1.225}$$

It is easy to check that $\mathrm{L}[U]$ is a self-adjoint operator, i.e., $\mathrm{L}^*[U] = \mathrm{L}[U]$. Let $R[U] = U_{tt} - U^2U_{xx} - UU_x^2$. Then $\Lambda(x,t,U,U_t,U_x)$ is a multiplier of a local conservation law of the PDE (1.223) if and only if

$$\begin{aligned}\mathrm{E}_U(R[U]\Lambda) = \mathrm{D}_t^2\Lambda - U^2\mathrm{D}_x^2\Lambda - 2UU_x\mathrm{D}_x\Lambda - (2UU_{xx} + U_x^2)\Lambda \\ +R[U]\Lambda_U - \mathrm{D}_x(R[U]\Lambda_{U_x}) - \mathrm{D}_t(R[U]\Lambda_{U_t}) \equiv 0,\end{aligned} \tag{1.226}$$

in terms of the Euler operator $\mathrm{E}_U = \partial/\partial U - \mathrm{D}_x\,\partial/\partial U_x - \mathrm{D}_t\,\partial/\partial U_t + \mathrm{D}_x^2\,\partial/\partial U_{xx} + \mathrm{D}_t^2\,\partial/\partial U_{tt}$. Equation (1.226) is an identity holding for all values of $x, t, U, U_x, U_t, U_{xx}, U_{xt}, U_t, U_{xxx}, U_{xxt}, U_{xtt}, U_{ttt}$. It is left as an exercise to show that equation (1.226) splits into a system of two determining equations for $\Lambda(x,t,U,U_x,U_t)$, consisting of

$$\hat{\mathrm{D}}_t^2\Lambda - U^2\mathrm{D}_x^2\Lambda - 2UU_x\mathrm{D}_x\Lambda - (2UU_{xx} + U_x^2)\Lambda = 0, \tag{1.227}$$

and

$$2\Lambda_U + \hat{\mathrm{D}}_t\Lambda_{U_t} - \mathrm{D}_x\Lambda_{U_t} = 0, \tag{1.228}$$

in terms of the total derivative $\hat{\mathrm{D}}_t = \partial/\partial t + U_t\,\partial/\partial U + U_{tx}\,\partial/\partial U_x + g[U]\,\partial/\partial U_t + U_{txx}\,\partial/\partial U_{xx} + \mathrm{D}_t(g[U])\,\partial/\partial U_{tt}$ associated with the PDE (1.223) where $g[U] = U(UU_x)_x$.

The first determining equation (1.227) is the determining equation for $\Lambda(x,t,u,u_t,u_x)\,\partial/\partial u$ to be a contact symmetry of the given PDE (1.223). If a contact symmetry satisfies the second determining equation then it yields a local multiplier $\Lambda(x,t,U,U_t,U_x)$ for the PDE (1.223). It is easy to check that the scaling symmetry (1.224) obviously satisfies the symmetry determining equation (1.227) but does not satisfy the second determining equation (1.228) when $u(x,t)$ is replaced by an arbitrary function $U(x,t)$. Hence the scaling symmetry (1.224) does not yield a local conservation law of PDE (1.223).

Exercises 1.3

1.3.1. Derive the conservation law (1.139) of the Euler equations (1.135) for the adiabatic motion of an ideal gas. Write down the integral form of the conservation law (1.139).

1.3.2. Show that the Euler operator (1.149) annihilates any divergence expression, i.e.,

$$\mathrm{E}_{U^j}(\mathrm{D}_i\Phi^i(x,U,\partial U,\ldots,\partial^r U))\equiv 0$$

holds for any set of functions $\{\Phi^i(x,U,\partial U,\ldots,\partial^r U)\}_{i=1}^n$, $j=1,\ldots,m$ where $U=U(x)=(U^1(x),\ldots,U^m(x))$ is an arbitrary function.

1.3.3. Show that in terms of the Euler operators (1.149), the expressions $\mathrm{E}_{U^j}F(x,U,\partial U,\ldots,\partial^s U)\equiv 0$, $j=1,\ldots,m$, hold if and only if $F(x,U,\partial U,\ldots,\partial^s U)\equiv \mathrm{D}_i\Psi^i(x,U,\partial U,\ldots,\partial^{s-1}U)$ holds for some set of functions $\{\Psi^i(x,U,\partial U,\ldots,\partial^r U)\}_{i=1}^n$, where $U=U(x)=(U^1(x),\ldots,U^m(x))$ is an arbitrary function.

1.3.4. Prove Theorem 1.3.4. [Hint: by adding trivial fluxes (of the first type) to fluxes $\Phi^i[U]$ if necessary, one can assume that each $\Phi^i[U]$ contains no leading derivatives nor their differential consequences.]

1.3.5. Discuss whether each of the following two equations are in Cauchy–Kovalevskaya form with respect to the given independent variables. If not, find a point transformation that yields one or more Cauchy–Kovalevskaya forms.

(a) Benjamin–Bona–Mahoney equation [Benjamin, Bona & Mahoney (1972)]: $u_t+(1+u^2)u_x-u_{xxt}=0$.
(b) Symmetric regularized long wave equation [Seyler & Fenstermacher (1984)]: $u_{tt}+u_{xx}+uu_{tx}+u_xu_t+u_{ttxx}=0$.

1.3.6. Consider a PDE system $\mathbf{R}\{x\,;u\}$ (1.132) written in Cauchy–Kovalevskaya form (1.157) with respect to the independent variable x^1. Denote the leading and the subleading derivatives by

$$u^\sigma_{\mathrm{L}}=\frac{\partial^{s_\sigma}}{\partial(x^1)^{s_\sigma}}u^\sigma,\quad u^\sigma_{\mathrm{S}}=\frac{\partial^{s_\sigma-1}}{\partial(x^1)^{s_\sigma-1}}u^\sigma,\quad \sigma=1,\ldots,m.$$

Let $\mathrm{D}_i\Phi^i[u]=0$ be a nontrivial conservation law of $\mathbf{R}\{x\,;u\}$ (1.132).

(a) Show that the corresponding multipliers are given by

$$\Lambda_\sigma[U]=\mathrm{E}_{U^\sigma_{\mathrm{S}}}(\Phi^1[U]),\quad \sigma=1,\ldots,m,$$

where $\mathrm{E}_{U^\sigma_{\mathrm{S}}}$ is the Euler operator with respect to a subleading derivative u^σ_{S}, given by

$$\mathrm{E}_{U^\sigma_{\mathrm{S}}}=\frac{\partial}{\partial U^\sigma_{\mathrm{S}}}-\sum_{i=2}^n \mathrm{D}_i\frac{\partial}{\partial(U^\sigma_{\mathrm{S}})_i}+\sum_{i,j=2}^n \mathrm{D}_i\mathrm{D}_j\frac{\partial}{\partial(U^\sigma_{\mathrm{S}})_{ij}}+\cdots.$$

(b) Show that the multipliers $\Lambda_\sigma[U]$ do not depend on leading derivatives u^σ_{L} and their differential consequences $\partial u^\sigma_{\mathrm{L}},\ldots$.

(c) Show that there exists an equivalent conservation law $\mathrm{D}_i\widetilde{\Phi}^i[u] = 0$, such that the fluxes $\{\widetilde{\Phi}^i[U]\}_{i=1}^m$ do not depend on leading derivatives u_{L}^σ and their differential consequences $\partial u_{\mathrm{L}}^\sigma, \partial^2 u_{\mathrm{L}}^\sigma$, etc.

1.3.7. Show that a set of functions $\{\xi(x,t,U,V),\ \phi(x,t,U,V)\}$ solves equations (1.162) if and only if $\{\xi(x,t,U,V),\ \phi(x,t,U,V)\}$ solves equations (1.163).

1.3.8. Show that (1.164) yields all solutions of the equations (1.163).

1.3.9. Show that the only conservation law multipliers of the form $\Lambda(x,t,U,U_x,U_{xx})$ of the KdV equation (1.168) are given by (1.172) and (1.173).

1.3.10. Show that the expression (1.180), (1.181) holds.

1.3.11. Consider the generalized KdV equation

$$u_t + u^n u_x + u_{xxx} = 0, \tag{1.229}$$

with parameter $n > 0$.

(a) For all $n > 0$, show that the only multipliers of the form $\Lambda(x,t,U,U_x,U_{xx})$ of PDE (1.229) are given by $\Lambda_1 = 1$, $\Lambda_2 = U$, $\Lambda_3 = U_{xx} + U^{n+1}/(n+1)$.
(b) Show that the only additional multipliers of the form $\Lambda(x,t,U,U_x,U_{xx})$ of PDE (1.229) are given by $\Lambda_4 = tU - x$ if $n = 1$; $\Lambda_5 = t\left(U_{xx} + \frac{1}{3}U^3\right) - \frac{1}{3}xU$ if $n = 2$. For details, see Anco & Bluman (2002a).

1.3.12. Fill in the details for the calculations of fluxes of local conservation laws of the modified Korteweg–de Vries equation (1.219).

1.3.13. Consider the class of Klein–Gordon wave equations of the form

$$u_{tx} - g(u) = 0 \tag{1.230}$$

with a nonlinear interaction term $g(u)$.

(a) Show that the linearizing operator of PDE (1.230) is self-adjoint.
(b) Derive the set of determining equations for local conservation law multipliers of the form $\Lambda(x,U,U_x,\ldots,\partial_x^p U)$ of PDE (1.230) where $\partial_x^j U = \partial^j U/\partial x^j$, $j = 1,\ldots,p$. Isolate the determining equation that yields the local symmetries of the form $\mathrm{X} = \Lambda(x,u,u_x,\ldots,\partial_x^p u)\partial/\partial u$ of PDE (1.230).
(c) A conservation law is said to be of order q if its fluxes depend at most on derivatives of order q. For multipliers of the form $\Lambda(x,U,U_x,\ldots,\partial_x^p U)$, show that the only PDEs of the form (1.230), that have a conservation law of order $q = 2$ are given by

 (i) Liouville equation where $g(u) = e^u$;

(ii) sine-Gordon equation where $g(u) = \sin u$;
(iii) sinh-Gordon equation where $g(u) = e^u \pm e^{-u}$,

modulo scalings and translations in u [Anco & Bluman (2002a)].

(d) Find the multipliers of the form $\Lambda(x, U, U_x, U_{xx}, U_{xxx})$ and corresponding second-order conservation laws for each of the Liouville, sine-Gordon and sinh-Gordon equations.

1.3.14. Consider the nonlinear wave equation

$$u_{tt} - u_{xx} + u^p = 0, \quad p > 1. \tag{1.231}$$

(a) Show that the linearizing operator of PDE (1.231) is self-adjoint.
(b) Find all point symmetries of PDE (1.231).
(c) Find which point symmetries yield local conservation law multipliers of PDE (1.231) and find the fluxes of corresponding conservation laws [Anco & Bluman (1997a)].

1.3.15. Consider the nonlinear telegraph system of PDEs given by

$$\begin{aligned} R^1 &= v_t - F(u)u_x - G(u) = 0, \\ R^2 &= u_t - v_x = 0. \end{aligned} \tag{1.232}$$

(a) Assuming that $F(u)$ is *arbitrary*, find all local multipliers of the form $\Lambda^1 = \xi(x, t, U, V), \Lambda^2 = \phi(x, t, U, V)$ of the PDE system (1.232), i.e., do a local multiplier classification with respect to $G(u)$ [Bluman & Temuerchaolu (2005a)].
(b) Find the fluxes of the corresponding conservation laws.

1.3.16.

(a) By a direct computation, show that the Fréchet derivative (linearization) of an expression $R^\sigma[U] = R^\sigma(x, U, \partial U, \ldots, \partial^k U)$ can be expressed as the action of the linearization operator $\mathrm{L}_\rho^\sigma[U]$ (1.176):

$$\mathcal{F}(R^\sigma[U]) \equiv \frac{d}{dh} R^\sigma[U + Vh]\Big|_{h=0} = \mathrm{L}_\rho^\sigma[U]V^\rho.$$

(b) Using integration by parts, find the adjoint of the linearizing operator $\mathrm{L}^{*}{}_\rho^\sigma[U]$ and fluxes $\Phi^i[U]$, such that

$$W_\sigma \mathrm{L}_\rho^\sigma[U]V^\rho \equiv V^\rho \mathrm{L}^{*}{}_\rho^\sigma[U]W_\sigma + \mathrm{D}_i\Phi^i[U]$$

for arbitrary functions W_σ and V^ρ. Thus derive formulas (1.177) and (1.179).

1.3.17. Consider a system of N PDEs $R^\sigma[u] = 0$, $\sigma = 1, \ldots, N$, given by (1.174), and a set of multipliers $\{\Lambda_\sigma[U]\}_{\sigma=1}^N$ for these equations.

The components of the adjoint linearization operator for PDE system (1.174) are given by (1.177)

$$(\mathrm{L}_R^*)_\rho^\sigma[U]W_\sigma = \frac{\partial R^\sigma[U]}{\partial U^\rho}W_\sigma - \mathrm{D}_i\left(\frac{\partial R^\sigma[U]}{\partial U_i^\rho}W_\sigma\right) + \cdots + (-1)^k \mathrm{D}_{i_1}\ldots\mathrm{D}_{i_k}\left(\frac{\partial R^\sigma[U]}{\partial U_{i_1\ldots i_k}^\rho}W_\sigma\right),$$

$$\rho = 1,\ldots,m,$$

acting on arbitrary functions $W = (W_1(x),\ldots,W_N(x))$.

Define the adjoint linearization operator of a PDE system $\Lambda_\sigma[u] = 0$, $\sigma = 1,\ldots,N$, by

$$(\mathrm{L}_\Lambda^*)_{\sigma\,\rho}[U]\widetilde{W}^\sigma = \frac{\partial \Lambda_\sigma[U]}{\partial U^\rho}\widetilde{W}^\sigma - \mathrm{D}_i\left(\frac{\partial \Lambda_\sigma[U]}{\partial U_i^\rho}\widetilde{W}^\sigma\right) + \cdots + (-1)^k \mathrm{D}_{i_1}\ldots\mathrm{D}_{i_k}\left(\frac{\partial \Lambda_\sigma[U]}{\partial U_{i_1\ldots i_k}^\rho}\widetilde{W}^\sigma\right),$$

$$\rho = 1,\ldots,m,$$

acting on arbitrary functions $\widetilde{W} = (\widetilde{W}^1(x),\ldots,\widetilde{W}^N(x))$.

Show that the condition (1.183) for the set of multipliers $\{\Lambda_\sigma[U]\}_{\sigma=1}^N$ to yield a divergence expression can be written as a symmetric expression

$$(\mathrm{L}_R^*)_\rho^\sigma[U]\Lambda_\sigma[U] + (\mathrm{L}_\Lambda^*)_{\sigma\,\rho}R^\sigma[U] = 0. \tag{1.233}$$

1.4 Noether's Theorem

In 1918, Noether (1918) presented her celebrated procedure (Noether's theorem) to find local conservation laws for systems of DEs that admit a variational principle (action functional). When a given DE system admits a variational principle, then the extremals of its action functional yield the given DE system (the *Euler–Lagrange equations*). In this case, Noether showed that if one has a point symmetry of the action functional (action integral), then one obtains the fluxes of a local conservation law through an explicit formula that involves the infinitesimals of the point symmetry and the Lagrangian (Lagrangian density) of the action functional.

In this section, we present Noether's theorem and its generalizations due to Bessel-Hagen (1921) and Boyer (1967).

1.4.1 Euler–Lagrange equations

Consider a functional $J[U]$ in terms of n independent variables $x = (x^1, \ldots, x^n)$ and m arbitrary functions $U = (U^1(x), \ldots, U^m(x))$ and their partial derivatives to order k, defined on a domain Ω,

$$J[U] = \int_\Omega L[U]dx = \int_\Omega L(x, U, \partial U, \ldots, \partial^k U)dx. \tag{1.234}$$

The function $L[U] = L(x, U, \partial U, \ldots, \partial^k U)$ is called a *Lagrangian* and the functional $J[U]$ is called an *action integral*. Consider an infinitesimal change of U given by $U(x) \to U(x) + \varepsilon v(x)$ where $v(x)$ is any function such that $v(x)$ and its derivatives to order $k - 1$ vanish on the boundary $\partial\Omega$ of the domain Ω. The corresponding change (variation) in the Lagrangian $L[U]$ is given by

$$\begin{aligned}\delta L &= L(x, U + \varepsilon v, \partial U + \varepsilon \partial v, \ldots, \partial^k U + \varepsilon \partial^k v)\\ &\quad -L(x, U, \partial U, \ldots, \partial^k U)\\ &= \varepsilon \left(\frac{\partial L[U]}{\partial U^\sigma} v^\sigma + \frac{\partial L[U]}{\partial U^\sigma_j} v^\sigma_j + \cdots + \frac{\partial L[U]}{\partial U^\sigma_{j_1 \cdots j_k}} v^\sigma_{j_1 \cdots j_k} \right) + O(\varepsilon^2).\end{aligned} \tag{1.235}$$

Then after repeatedly using integration by parts, one can show that

$$\delta L = \varepsilon(v^\sigma \mathrm{E}_{U^\sigma}(L[U]) + \mathrm{D}_i W^i[U, v]) + O(\varepsilon^2), \tag{1.236}$$

where E_{U^σ} is the Euler operator with respect to U^σ and

$$\begin{aligned}W^i[U, v] &= v^\sigma \left(\frac{\partial L[U]}{\partial U^\sigma_i} + \cdots + (-1)^{k-1} \mathrm{D}_{j_1} \cdots \mathrm{D}_{j_{k-1}} \frac{\partial L[U]}{\partial U^\sigma_{i j_1 \cdots j_{k-1}}} \right)\\ &\quad + v^\sigma_{j_1} \left(\frac{\partial L[U]}{\partial U^\sigma_{i j_1}} + \cdots + (-1)^{k-2} \mathrm{D}_{j_2} \cdots \mathrm{D}_{j_{k-1}} \frac{\partial L[U]}{\partial U^\sigma_{i j_1 j_2 \cdots j_{k-1}}} \right)\\ &\quad + \cdots + v^\sigma_{j_1 \cdots j_{k-1}} \frac{\partial L[U]}{\partial U^\sigma_{i j_1 j_2 \cdots j_{k-1}}}.\end{aligned} \tag{1.237}$$

From expression (1.236) and the divergence theorem, the corresponding variation in the action integral $J[U]$ is given by

$$\begin{aligned}\delta J &= J[U + \varepsilon v] - J[U] = \textstyle\int_\Omega \delta L dx\\ &= \varepsilon \textstyle\int_\Omega (v^\sigma \mathrm{E}_{U^\sigma}(L[U]) + \mathrm{D}_l W^l[U, v])dx + O(\varepsilon^2)\\ &= \varepsilon(\textstyle\int_\Omega v^\sigma \mathrm{E}_{U^\sigma}(L[U])dx + \int_{\partial\Omega} W^l[U, v] n^l dS) + O(\varepsilon^2)\end{aligned} \tag{1.238}$$

where $\int_{\partial\Omega}$ represents the surface integral over the boundary $\partial\Omega$ of the domain Ω with $n = (n^1, \ldots n^n)$ being the unit outward normal vector to $\partial\Omega$. It follows that if $U = u(x)$ extremizes the action integral $J[U]$, then the $O(\varepsilon)$ term of δJ must vanish and hence

$$\int_\Omega v^\sigma \mathrm{E}_{u^\sigma}(L[u])dx = 0 \tag{1.239}$$

for an *arbitrary* $v(x)$ defined on the domain Ω. Hence, it follows that if $U = u(x)$ extremizes the action integral $J[U]$ given by (1.234), then $u(x)$ must satisfy the PDE system

$$\begin{aligned} \mathrm{E}_{u^\sigma}(L[u]) &= \frac{\partial L[u]}{\partial u^\sigma} + \cdots + (-1)^k \mathrm{D}_{j_1} \cdots \mathrm{D}_{j_k} \frac{\partial L[u]}{\partial u^\sigma_{j_1 \cdots j_k}} = 0, \\ \sigma &= 1, \ldots, m. \end{aligned} \tag{1.240}$$

Equations (1.240) are called the *Euler–Lagrange equations* satisfied by an extremum $U = u(x)$ of the action integral $J[U]$. The following theorem has been proved.

Theorem 1.4.1. *If a smooth function* $U(x) = u(x)$ *is an extremum of an action integral* $J[U] = \int_\Omega L[U]dx$ *with* $L[U] = L(x, U, \partial U, \ldots, \partial^k U)$*, then* $u(x)$ *satisfies the Euler–Lagrange equations* (1.240).

1.4.2 Noether's formulation of Noether's theorem

We now present Noether's formulation of her theorem. In this formulation, the action integral $J[U]$ (1.234) is required to be invariant under a one-parameter Lie group of point transformations

$$\begin{aligned} (x^*)^i &= x^i + \varepsilon \xi^i(x, U) + O(\varepsilon^2), \quad i = 1, \ldots, n, \\ (U^*)^\mu &= U^\mu + \varepsilon \eta^\mu(x, U) + O(\varepsilon^2), \quad \mu = 1, \ldots, m, \end{aligned} \tag{1.241}$$

with corresponding infinitesimal generator given by

$$\mathrm{X} = \xi^i(x, U)\frac{\partial}{\partial x^i} + \eta^\nu(x, U)\frac{\partial}{\partial U^\nu}. \tag{1.242}$$

Invariance holds if and only if $\int_{\Omega^*} L[U^*]dx^* = \int_\Omega L[U]dx$ where Ω^* is the image of Ω under the point transformation (1.241). The Jacobian of the transformation (1.241) is given by

$$\mathrm{J} = \det(\mathrm{D}_i(x^*)^j) = 1 + \varepsilon \mathrm{D}_i \xi^i(x, U) + O(\varepsilon^2). \tag{1.243}$$

Then $dx^* = \mathrm{J}dx$. Moreover, since (1.241) is a Lie group of transformations, it follows that $L[U^*] = e^{\varepsilon \mathrm{X}^{(k)}} L[U]$ in terms of the kth extension of the infinitesimal generator (1.242). Consequently, in Noether's formulation, the one-parameter Lie group of point transformations (1.241) is a point symmetry of $J[U]$ (1.234) if and only if

$$\begin{aligned} &\textstyle\int_\Omega (\mathrm{J}e^{\varepsilon \mathrm{X}^{(k)}} - 1)L[U]dx \\ &\quad = \varepsilon \textstyle\int_\Omega \Big(L[U](\mathrm{D}_i\xi^i(x,U)) + \mathrm{X}^{(k)} L[U] \Big) dx + O(\varepsilon^2) \end{aligned} \tag{1.244}$$

holds for arbitrary $U(x)$ where $\mathrm{X}^{(k)}$ is the kth extended infinitesimal generator given by expression (1.12) with U replacing u. Hence, if $J[U]$ (1.234) has the point symmetry (1.241), then the $O(\varepsilon)$ term in (1.244) vanishes, and thus one obtains the identity

$$L[U]\mathrm{D}_i\xi^i(x,U) + \mathrm{X}^{(k)} L[U] \equiv 0. \tag{1.245}$$

In Section 1.2.4, it was shown that the one-parameter Lie group of point transformations (1.241) is equivalent to the one-parameter family of local transformations

$$\begin{aligned} &(x^*)^i = x^i, \quad i = 1, \dots, n, \\ &(U^*)^\mu = U^\mu + \varepsilon[\eta^\mu(x,U) - U_i^\mu \xi^i(x,U)] + O(\varepsilon^2), \quad \mu = 1, \dots, m, \end{aligned} \tag{1.246}$$

with the corresponding kth extended infinitesimal generator $\hat{\mathrm{X}}^{(k)}$ given through the appropriate truncation of expressions (1.30), (1.31) with U replacing u.

Under the transformation (1.246), the corresponding infinitesimal change $U(x) \to U(x) + \varepsilon v(x)$ has components $v^\mu(x) = \hat{\eta}^\mu[U] = \eta^\mu(x,U) - U_i^\mu \xi^i(x,U)$ in terms of the transformations (1.246). Moreover, from the group property of (1.246), it follows that

$$\delta L = \varepsilon \hat{\mathrm{X}}^{(k)} L[U] + O(\varepsilon^2). \tag{1.247}$$

Thus

$$\int_\Omega \delta L dx = \varepsilon \int_\Omega \hat{\mathrm{X}}^{(k)} L[U] dx + O(\varepsilon^2). \tag{1.248}$$

Consequently, after comparing expression (1.248) to expression (1.238) with $v^\mu(x) = \hat{\eta}^\mu[U] = \eta^\mu(x,U) - U_i^\mu \xi^i(x,U)$, it follows that

$$\hat{\mathrm{X}}^{(k)} L[U] \equiv \hat{\eta}^\mu[U]\mathrm{E}_{U^\mu}(L[U]) + \mathrm{D}_i W^i[U, \hat{\eta}[U]], \tag{1.249}$$

where $W^i[U, \hat{\eta}[U]]$ is given by expression (1.237) with the obvious substitutions.

The proof of the following theorem is straightforward and is left to Exercise 1.4.3.

Theorem 1.4.2. *Let* $\mathrm{X}^{(k)}$ *be the kth extended infinitesimal generator of the one-parameter Lie group of point transformations* (1.241) *and let* $\hat{\mathrm{X}}^{(k)}$ *be the kth extended infinitesimal generator of the equivalent one-parameter family of transformations* (1.246). *Let* $F[U] = F(x, U, \partial U, \ldots, \partial^k U)$ *be an arbitrary function of its arguments. Then the following identity holds:*

$$\mathrm{X}^{(k)} F[U] + F[U]\mathrm{D}_i \xi^i(x, U) \equiv \hat{\mathrm{X}}^{(k)} F[U] + \mathrm{D}_i(F[U]\xi^i(x, U)). \qquad (1.250)$$

Putting all of the above together, one obtains the following theorem.

Theorem 1.4.3 (Noether's formulation of Noether's theorem). *Suppose a given PDE system* $\mathbf{R}\{x\,; u\}$ (1.174), *as written, arises from a variational principle, i.e., the given PDE system is a set of Euler–Lagrange equations* (1.239) *whose solutions* $u(x)$ *are extrema* $U(x) = u(x)$ *of an action integral* $J[U]$ (1.234) *with Lagrangian* $L[U]$. *Suppose the one-parameter Lie group of point transformations* (1.241) *is a point symmetry of* $J[U]$. *Let* $W^l[U, v]$ *be defined by* (1.237) *for arbitrary functions* $U(x)$, $v(x)$. *Then*

(1) The identity

$$\hat{\eta}^{\mu}[U]\mathrm{E}_{U^{\mu}}(L[U]) \equiv -\mathrm{D}_i(\xi^i(x, U)L[U] + W^i[U, \hat{\eta}[U]]) \qquad (1.251)$$

holds for arbitrary functions $U(x)$, *i.e.,* $\{\hat{\eta}^{\mu}[U]\}_{\mu=1}^{m}$ *is a set of local multipliers of the Euler–Lagrange system* (1.239);

(2) The local conservation law

$$\mathrm{D}_i(\xi^i(x, u)L[u] + W^i[u, \hat{\eta}[u]]) = 0 \qquad (1.252)$$

holds for any solution $u = \Theta(x)$ *of the Euler–Lagrange system* (1.239).

Proof. Let $F[U] = L[U]$ in the identity (1.250). Then from the identity (1.245), one obtains

$$\hat{\mathrm{X}}^{(k)} L[U] + \mathrm{D}_i(L[U]\xi^i(x, U)) \equiv 0 \qquad (1.253)$$

holding for arbitrary functions $U(x)$. Substitution for $\hat{\mathrm{X}}^{(k)} L[U]$ in (1.253) through (1.249) yields (1.251). If $U(x) = u(x)$ solves the Euler–Lagrange system (1.239), then the left-hand side of the identity (1.251) vanishes. This yields the conservation law (1.252). □

1.4.3 Boyer's formulation of Noether's theorem

Boyer (1967) extended Noether's theorem to enable one to conveniently find conservation laws arising from invariance under higher-order transformations by generalizing Noether's definition of invariance of an action integral $J[U]$ (1.234). In particular, under the following definition, an action integral $J[U]$ (1.234) is invariant under a one-parameter higher-order transformation if its integrand $L[U]$ is invariant to within a divergence under such a transformation.

Definition 1.4.1. Let

$$\hat{\mathrm{X}} = \hat{\eta}^{\mu}(x, U, \partial U, \ldots, \partial^{s} U)\frac{\partial}{\partial U^{\mu}} \tag{1.254}$$

be the infinitesimal generator of a one-parameter higher-order local transformation (1.40) with its extension $\hat{\mathrm{X}}^{\infty}$ given by (1.38), (1.39). Let $\hat{\eta}^{\mu}[U] = \hat{\eta}^{\mu}(x, U, \partial U, \ldots, \partial^{s} U)$. The transformation is a *local symmetry* of $J[U]$ (1.234) if and only if

$$\hat{\mathrm{X}}^{\infty} L[U] \equiv \mathrm{D}_i A^i[U] \tag{1.255}$$

holds for some set of functions $A^i[U] = A^i(x, U, \partial U, \ldots, \partial^r U)$, $i = 1, \ldots, n$.

Definition 1.4.2. A local transformation with infinitesimal generator (1.40) that is a local symmetry of $J[U]$ (1.234) is called a *variational symmetry* of $J[U]$ (1.234).

The proof of the following theorem follows from the property of Euler operators annihilating divergences.

Theorem 1.4.4. *A variational symmetry with infinitesimal generator* (1.254) *of the action integral* $J[U]$ (1.234) *yields a local symmetry with infinitesimal generator* $\hat{\mathrm{X}} = \hat{\eta}^{\mu}(x, u, \partial u, \ldots, \partial^s u)\, \partial/\partial u^{\mu}$ *of the corresponding Euler–Lagrange system* (1.239).

The following theorem generalizes Noether's formulation of her theorem.

Theorem 1.4.5 (Boyer's generalization of Noether's theorem)**.** *Suppose a given PDE system* $\mathbf{R}\{x\,; u\}$ (1.174)*, as written, arises from a variational principle, i.e., the given PDE system is a set of Euler–Lagrange equations* (1.239) *whose solutions* $u(x)$ *are extrema* $U(x) = u(x)$ *of an action integral* $J[U]$ (1.234) *with Lagrangian* $L[U]$*. Suppose a local transformation with infinitesimal generator* (1.254) *yields a variational symmetry of* $J[U]$*. Let* $W^l[U, v]$ *be defined by* (1.237) *for arbitrary functions* $U(x)$, $v(x)$*. Then*

(1) The identity

$$\hat{\eta}^{\mu}[U]\mathrm{E}_{U^{\mu}}(L[U]) \equiv \mathrm{D}_i(A^i[U] - W^i[U, \hat{\eta}[U]]) \tag{1.256}$$

holds for arbitrary functions $U(x)$, *i.e.*, $\{\hat{\eta}^\mu[U]\}_{\mu=1}^m$ *is a set of local multipliers of the Euler–Lagrange system* (1.239);

(2) The local conservation law

$$\mathrm{D}_i(W^i[u,\hat{\eta}[u]] - A^i[u]) = 0 \tag{1.257}$$

holds for any solution $u = \Theta(x)$ *of the Euler–Lagrange system* (1.239).

Proof. For a local transformation with infinitesimal generator (1.254), it follows that the corresponding infinitesimal change $U(x) \to U(x) + \varepsilon v(x)$ has components $v^\mu(x) = \hat{\eta}^\mu[U]$. Consequently, equation (1.247) becomes

$$\delta L = \varepsilon \hat{\mathrm{X}}^\infty L[U] + O(\varepsilon^2). \tag{1.258}$$

But from (1.236) it follows that

$$\delta L = \varepsilon(\hat{\eta}^\mu[U]\mathrm{E}_{U^\mu}(L[U]) + \mathrm{D}_i(W^i[U,\hat{\eta}[U]])) + O(\varepsilon^2). \tag{1.259}$$

Hence it immediately follows that

$$\hat{\mathrm{X}}^\infty L[U] = \hat{\eta}^\mu[U]\mathrm{E}_{U^\mu}(L[U]) + \mathrm{D}_i(W^i[U,\hat{\eta}[U]]) \tag{1.260}$$

holds for arbitrary functions $U(x)$. Since the local transformation with infinitesimal generator (1.254) is a variational symmetry of $J[U]$ (1.234), it follows that equation (1.255) holds. Substitution for $\hat{\mathrm{X}}^\infty L[U]$ in (1.260) through (1.255) yields the identity (1.256). If $U(x) = u(x)$ solves the Euler–Lagrange system (1.239), then the left-hand side of the identity (1.256) vanishes. This yields the conservation law (1.257). □

Theorem 1.4.6. *If a conservation law is obtained through Noether's formulation (Theorem 1.4.3), then the conservation law can be obtained through Boyer's formulation (Theorem 1.4.5).*

Proof. Suppose the one-parameter Lie group of point transformations (1.241) yields a conservation law. Then the identity (1.253) holds. Consequently,

$$\hat{\mathrm{X}}^{(k)} L[U] = \hat{\mathrm{X}}^\infty L[U] = \mathrm{D}_i A^i[U], \tag{1.261}$$

where $A^i[U] = -\mathrm{D}_i(L[U]\xi^i(x,U))$. But equation (1.261) is just the condition for the one-parameter Lie group of point transformations (1.241) to be a variational symmetry of $J[U]$ (1.234). Consequently, one obtains the same conservation law from Boyer's formulation. □

1.4.4 Limitations of Noether's theorem

There are several limitations inherent in using Noether's theorem to find local conservation laws for a given PDE system $\mathbf{R}\{x\,;u\}$. First of all, it is restricted to variational systems. Consequently, the linearizing operator (Fréchet derivative) for $\mathbf{R}\{x\,;u\}$, *as written*, must be self-adjoint, which implies that the number of PDEs must be the same as the number of dependent variables appearing in $\mathbf{R}\{x\,;u\}$. [In particular, this can be seen from comparing expressions (1.176) and (1.177).] Moreover, if $\mathbf{R}\{x\,;u\}$ is a scalar equation, it must be of even order. In addition, one must find an explicit Lagrangian $L[U]$ whose Euler–Lagrange equations yield $\mathbf{R}\{x\,;u\}$.

There is also the difficulty of finding the variational symmetries of a given variational PDE system $\mathbf{R}\{x\,;u\}$. First, for the given PDE system, one must determine local symmetries depending on derivatives of dependent variables up to some chosen order. Second, one must find an explicit Lagrangian $L[U]$ and check if each symmetry of the given PDE system leaves invariant the Lagrangian $L[U]$ to within a divergence, i.e., if a symmetry is indeed a variational symmetry.

Finally, the use of Noether's theorem to find local conservation laws is coordinate dependent since the action of a point (contact) transformation can transform a DE having a variational principle to one that does not have one. On the other hand, in Section 1.5, it is shown that conservation laws are *coordinate-independent* in the sense that a point (contact) transformation maps a conservation law into a conservation law, and therefore it follows that an ideal method for finding conservation laws should be coordinate-independent.

Artifices may make a given PDE system variational. Such artifices include:

1. *The use of multipliers.* As an example, the PDE

$$u_{tt} + H'(u_x)u_{xx} + H(u_x) = 0, \tag{1.262}$$

as written, does not admit a variational principle since its linearized equation $v_{tt} + H'(u_x)v_{xx} + (H''(u_x)u_{xx} + H'(u_x))v_x = 0$ is not self-adjoint. However, the equivalent PDE

$$e^x[u_{tt} + H'(u_x)u_{xx} + H(u_x)] = 0, \tag{1.263}$$

as written, is self-adjoint!

2. *The use of a contact transformation of the variables.* As an example, the PDE

$$e^x u_{tt} - e^{3x}(u + u_x)^2(u + 2u_x + u_{xx}) = 0, \tag{1.264}$$

as written, does not admit a variational principle, since its linearized PDE and the adjoint PDE are different. But the point transformation

$x^* = x$, $t^* = t$, $u^*(x^*, t^*) = y(x,t) = e^x u(x,t)$, maps the PDE (1.264) into the self-adjoint PDE

$$y_{tt} - (y_x)^2 y_{xx} = 0, \tag{1.265}$$

which is the Euler–Lagrange equation for an extremum $Y = y$ of the action integral with Lagrangian $L[Y] = Y_t^2/2 - Y_x^4/12$.

3. *The use of a differential substitution.* As an example, the Korteweg–de Vries equation

$$u_t + uu_x + u_{xxx} = 0, \tag{1.266}$$

as written, obviously does not admit a variational principle since it is of odd order. But the well-known differential substitution $u = v_x$ yields the related transformed KdV equation

$$v_{xt} + v_x v_{xx} + v_{xxxx} = 0, \tag{1.267}$$

which arises from the Lagrangian $L[V] = (V_{xx})^2/2 - (V_x)^3/6 - V_x V_t/2$.

4. *The use of an artificial additional equation.* For example, the linear heat equation $u_t - u_{xx} = 0$ is not self-adjoint since its adjoint equation is given by $w_t + w_{xx} = 0$. However the decoupled PDE system

$$u_t - u_{xx} = 0, \quad \tilde{u}_t + \tilde{u}_{xx} = 0$$

is evidently self-adjoint! [In general, the formal system, obtained through appending any given PDE system by the adjoint of its linearized system, is self-adjoint.]

The direct method for finding conservation laws [Section 1.3.4] is free of all of the above problems. It is directly applicable to any PDE system, whether or not it is variational. Moreover, it does not require the knowledge of a Lagrangian, whether or not one exists. Indeed, under the direct method, variational and non-variational PDE systems are treated in the same manner.

The direct method is naturally coordinate-independent. This follows from the fact that a point (contact) transformation maps a conservation law into a conservation law, and hence either form of a conservation law (in original or transformed variables) will arise from corresponding sets of multipliers, which can be found by the direct method in either coordinate system.

Finding conservation laws through the direct method is computationally more straightforward than through Noether's theorem even when a given PDE system is variational. One simply writes down the set of linear determining equations (1.151) holding for arbitrary functions $U(x)$, which in the case of a variational system, includes the symmetry determining equations as a subset of the multiplier determining equations. Hence, the resulting set of linear determining equations for local multipliers is usually not as difficult to solve as the set of linear determining equations for local symmetries since the determining system is more over-determined in the variational case.

On the other hand, if a given PDE system is variational and one has obtained the Lagrangian for the PDE system, then it is worthwhile to combine the direct method with Noether's theorem as follows. First, use the direct method to find the multipliers of local conservation laws and hence the corresponding variational symmetries. Second, for each variational symmetry, find the corresponding divergence term $\mathrm{D}_i A^i[U]$ that arises from the use of Boyer's formulation of the extended Noether's theorem. Third, use the expression (1.237) in conjunction with Boyer's formula (1.257) to find the resulting conservation law.

1.4.5 Examples

Now examples are considered, including two that compare the use of Noether's theorem and the direct method for finding conservation laws (for PDE systems that admit a variational formulation), and another that compares the local symmetry and conservation law structure of PDE systems that are not variational [Bluman & Temuerchaolu (2005b)].

Klein–Gordon wave equation

Consider the class of Klein–Gordon wave equations

$$R[u] = u_{tx} + g(u) = 0 \tag{1.268}$$

with a general nonlinear interaction term $g(u)$. This class has a variational principle given by the action functional

$$J[U] = \int L[U]\, dtdx, \tag{1.269}$$

with Lagrangian

$$L[U] = -\tfrac{1}{2}U_t U_x + h(U), \quad h'(U) = g(U). \tag{1.270}$$

For a general $g(u)$, the point symmetries of PDE (1.268) are translations in t and x and a scaling, respectively, given by the generators (in characteristic form)

$$\mathrm{X}_1 = u_t \frac{\partial}{\partial u}, \quad \mathrm{X}_2 = u_x \frac{\partial}{\partial u}, \quad \mathrm{X}_3 = (tu_t - xu_x)\frac{\partial}{\partial u}. \tag{1.271}$$

It is easy to check that all three symmetries (1.271) are variational symmetries of the action functional $J[U]$ (1.269). In particular, the actions of the extensions (1.38), (1.39) of the generators (1.271) on the Lagrangian (1.270) yield the divergence expressions

$$\hat{\mathrm{X}}_1^\infty L[U] = -\mathrm{D}_t\left(\tfrac{1}{2}U_t U_x + h(U)\right), \tag{1.272a}$$

$$\hat{\mathrm{X}}_2^\infty L[U] = -\mathrm{D}_x\left(\tfrac{1}{2}U_t U_x + h(U)\right), \tag{1.272b}$$

$$\hat{\mathrm{X}}_3^\infty L[U] = -\mathrm{D}_t\left(\tfrac{1}{2}tU_t U_x + t\,h(U)\right) + \mathrm{D}_x\left(\tfrac{1}{2}xU_t U_x + xh(U)\right). \tag{1.272c}$$

Hence through Boyer's formulation of Noether's theorem [Theorem 1.4.5], the three symmetries (1.271) yield three conservation laws with multipliers given by

$$\Lambda_1[U] = \hat{\eta}_1[U] = U_t, \quad \Lambda_2[U] = \hat{\eta}_2[U] = U_x, \quad \Lambda_3[U] = \hat{\eta}_3[U] = tU_t - xU_x.$$

In this example, due to the simplicity of the form of multipliers and the given PDE, fluxes of the three conservation laws are readily found through integration by parts. However, in more complicated practical situations when Noether's theorem is used, one would normally compute fluxes using the formula (1.257) which involves no integration. Here we illustrate the use of this formula.

Denoting $x^1 = t$, $x^2 = x$, we first compute the quantities $W^i[U,v]$ (1.237), $i = 1, 2$, using the Lagrangian (1.270):

$$W^1[U,v] = -\tfrac{1}{2}vU_x, \quad W^2[U,v] = -\tfrac{1}{2}vU_t.$$

Then for the three conservation laws, from (1.272), we identify

$$(A_1^1, A_1^2) = \left(-\tfrac{1}{2}U_t U_x - h(U), 0\right),$$

$$(A_2^1, A_2^2) = \left(0, -\tfrac{1}{2}U_t U_x - h(U)\right),$$

$$(A_3^1, A_3^2) = \left(-\tfrac{1}{2}tU_t U_x - th(U), \tfrac{1}{2}xU_t U_x + xh(U)\right).$$

Therefore from (1.257), the three conservation laws of PDE (1.268) corresponding to variational symmetries (1.271) have the form

$$\mathrm{D}_t(h[u]) - \mathrm{D}_x\left(\tfrac{1}{2}u_t^2\right) = 0, \tag{1.273a}$$

$$\mathrm{D}_t\left(\tfrac{1}{2}u_x^2\right) - \mathrm{D}_x(h[u]) = 0, \tag{1.273b}$$

$$\mathrm{D}_t\left(\tfrac{1}{2}xu_x^2 + t\,h[u]\right) - \mathrm{D}_x\left(\tfrac{1}{2}tu_t^2 + x\,h[u]\right) = 0. \tag{1.273c}$$

Now consider the Klein–Gordon equation with a power nonlinearity, i.e., $g(u) = u^n$, for $n \neq 0, 1$. In this case, the corresponding Klein–Gordon equation (1.268) has an extra scaling symmetry given by the infinitesimal generator (in evolutionary form)

$$\hat{\mathrm{X}}_4 = (u - (1-n)xu_x)\frac{\partial}{\partial u}. \tag{1.274}$$

One can check that the symmetry (1.274) does not yield a variational symmetry of the action functional $J[U]$ (1.269). This is considered from three points of view: Noether's formulation of Noether's theorem, Boyer's formulation of Noether's theorem and, finally, the direct method.

(1) Noether's formulation of Noether's theorem
First of all, in terms of using Noether's formulation, the additional infinitesimal generator (1.274) corresponds to the scaling symmetry $x^* = \alpha^{1-n}x$, $t^* = t$, $u^* = \alpha u$ of the Klein–Gordon PDE $u_{tx} - u^n = 0$. Now one checks whether the scaling transformation $x^* = \alpha^{1-n}x$, $t^* = t$, $U^* = \alpha U$ is a symmetry of the action functional $J[U]$. In particular,

$$J[U^*] = J[\alpha U] = \int L[U^*]dt^*dx^* = \alpha^{1-n}\int L[\alpha U]dtdx.$$

But $L[\alpha U] = \alpha^{1+n}L[U]$. Hence $J[U^*] = \alpha^2 J[U] \neq J[U]$. Thus, using Noether's formulation of Noether's theorem, the scaling symmetry (1.274) does not yield an additional conservation law of the Klein–Gordon equation $u_{tx} - u^n = 0$.

(2) Boyer's formulation of Noether's theorem
Secondly, in terms of the more general Boyer's formulation of Noether's theorem, using the extension of the infinitesimal generator (1.274) with $u(x)$ replaced by an arbitrary function $U(x)$, one obtains the expression

$$\begin{aligned}\hat{\mathrm{X}}_4^\infty L[U] &= U^n(U - xU_x(1-n)) \\ &\quad -\tfrac{1}{2}\Big[U_x(U_t - xU_{xt}(1-n)) + U_t(U_x - xU_{xx}(1-n))\Big].\end{aligned} \tag{1.275}$$

The right-hand side of (1.275) cannot be expressed as a divergence expression. To show this, it is best to directly apply the Euler operator (1.149) with respect to U to this expression. In particular, one obtains

$$\mathrm{E}_U\left(\hat{\mathrm{X}}_4^\infty L[U]\right) = 2(U_{xt} + U^n) \neq 0,$$

which means that $\hat{\mathrm{X}}_4^\infty L[U]$ is not a divergence expression, and hence X_4 does not yield a variational symmetry of the action (1.269). Hence, in these power law cases, the scaling symmetry (1.274) of the Klein–Gordon equation (1.268) does not yield a variational symmetry of the corresponding action functional $J[U]$ (1.269). Thus this scaling symmetry does not yield a conservation law multiplier in terms of using Boyer's formulation of Noether's theorem.

(3) Direct method
Finally, it is easiest to show that the scaling symmetry (1.274) does not yield a variational symmetry through using the direct method. Here one checks to see whether $(U - (1-n)xU_x)$ is a multiplier for a conservation law. In particular, one merely applies the Euler operator (1.149) with respect to U, i.e., E_U given by (1.149), to the expression $(U - (1-n)xU_x)(U_{tx} - U^n)$, to show that $\mathrm{E}_U\left[(U - (1-n)xU_x)(U_{tx} - U^n)\right] \not\equiv 0$ for an arbitrary function $U(x,t)$.

Generalized Korteweg–de Vries equation

Consider the generalized Korteweg–de Vries equation

$$R[u] = u_t + u^n u_x + u_{xxx} = 0 \tag{1.276}$$

with parameter $n > 0$. For $n = 1, 2$, the evolution PDE (1.276) reduces to the KdV and modified KdV equations, respectively.

The Fréchet derivative (linearized equation) of PDE (1.276) is given by

$$\mathrm{L}[u]v = \mathrm{D}_t v + u^n \mathrm{D}_x v + nu^{n-1}u_x v + \mathrm{D}_x^3 v = 0 \text{ when } R[u] = 0, \tag{1.277}$$

and the adjoint linearized equation by

$$\mathrm{L}^*[u]w = -\mathrm{D}_t w - u^n \mathrm{D}_x w - \mathrm{D}_x^3 w = 0 \text{ when } R[u] = 0. \tag{1.278}$$

Since (1.277) and (1.278) are different, the generalized Korteweg–de Vries equation (1.276), as written, has no variational principle and thus Noether's theorem does not hold. Hence the symmetry and conservation law structure of the equation are not directly related. A comparison is now made between the local symmetries and local conservation law multipliers of PDE (1.276), in terms of seeking symmetry components and local multipliers of the respective comparable forms: $\hat{\eta}[u] = \hat{\eta}(x,t,u,u_x,u_{xx},u_{xxx})$, $\Lambda[U] = \Lambda(x,t,U,U_x,U_{xx},U_{xxx})$. In particular, such symmetry components $\hat{\eta}[u]$ are solutions of the linearized PDE (1.277), whereas such conservation law multipliers $\Lambda[U]$ are solutions of the determining equations (1.151), following the direct method. The resulting symmetries and conservation law multipliers are summarized in Table 1.1.

Table 1.1 Comparison of local symmetries and conservation law multipliers of the generalized Korteweg–de Vries equation (1.276)

n	Symmetries $\hat{\eta}[u]=\hat{\eta}(x,t,u,u_x,u_{xx},u_{xxx})$	Conservation Law Multipliers $\Lambda[U]=\Lambda(x,t,U,U_x,U_{xx},U_{xxx})$
Arbitrary	$\hat{\eta}_1[u]=u_x$, $\hat{\eta}_2[u]=u^n u_x+u_{xxx}$, $\hat{\eta}_2[u]=u^n u_x+u_{xxx}$, $\hat{\eta}_3[u]=t(u^n u_x+u_{xxx})$ $-\frac{1}{3n}(2u+nxu_x)$.	$\Lambda_1[U]=1$, $\Lambda_2[U]=U$, $\Lambda_3[U]=U^{n+1}+(n+1)U_{xx}$.
1	$\hat{\eta}_1[u]$, $\hat{\eta}_2[u]$, $\hat{\eta}_3[u]\,(n=1)$, $\hat{\eta}_4[u]=tu_x-1$.	$\Lambda_1[U]$, $\Lambda_2[U]$, $\Lambda_3[U]\,(n=1)$, $\Lambda_4[U]=\frac{1}{2}U^2+U_{xx}$.
2	$\hat{\eta}_1[u]$, $\hat{\eta}_2[u]$, $\hat{\eta}_3[u]\,(n=2)$.	$\Lambda_1[U]$, $\Lambda_2[U]$, $\Lambda_3[U]\,(n=2)$, $\Lambda_5[U]=t\left(\frac{1}{3}U^3+U_{xx}\right)-\frac{1}{3}xU$.

Note that in the case $n=2$, the generalized Korteweg–de Vries equation (1.276) has one additional conservation law and no additional symmetry of the same order.

However the non-self-adjoint PDE (1.276) is special in the sense that, similarly to the KdV equation (1.266), it can be transformed into the self-adjoint PDE

$$v_{xt}+(v_x)^n v_{xx}+v_{xxxx}=0 \tag{1.279}$$

by the differential substitution $u=v_x$. One can show [Exercise 1.4.2] that equation (1.279) is the Euler–Lagrange equation for an extremum of the action integral with Lagrangian

$$L[V]=\tfrac{1}{2}(V_{xx})^2-\frac{1}{(n+1)(n+2)}(V_x)^{n+3}-\tfrac{1}{2}V_xV_t. \tag{1.280}$$

Now a comparison is made of the local symmetries and conservation law multipliers of the transformed equation (1.279), using the same ansatz for symmetry components and conservation law multipliers, respectively: $\hat{\eta}[v]=\hat{\eta}(x,t,v,v_x,v_{xx},v_{xxx},v_{xxxx})$, $\Lambda[V]=\Lambda(x,t,V,V_x,V_{xx},V_{xxx},V_{xxxx})$.

From Table 1.2, one sees that:

- For a general n, symmetries $\hat{\eta}_i[v]$, $i=1,2,3$, are variational, since $\Lambda_1[V]=\hat{\eta}_i[V]$. Symmetry $\hat{\eta}_4[v]$ is variational only for $n=2$.

Table 1.2 Comparison of local symmetries and conservation law multipliers of the transformed generalized Korteweg–de Vries equation (1.279)

n	Symmetries $\hat{\eta}(x,t,v,v_x,v_{xx},v_{xxx},v_{xxxx})$	Conservation Law Multipliers $\Lambda(x,t,V,V_x,V_{xx},V_{xxx},V_{xxxx})$
Arbitrary	$\hat{\eta}_1[v]=f(t)$, $\hat{\eta}_2[v]=v_x$, $\hat{\eta}_3[v]=\frac{1}{n+1}v_x^{n+1}+v_{xxx}$, $\hat{\eta}_4[v]=t\left(\frac{1}{n+1}v_x^{n+1}+v_{xxx}\right)-\frac{1}{3n}(nxv_x-(n-2)v)$.	$\Lambda_1[V]=f(t)$, $\Lambda_2[V]=V_x$, $\Lambda_3[V]=\frac{1}{n+1}V_x^{n+1}+V_{xxx}$.
1	$\hat{\eta}_1[v]$, $\hat{\eta}_2[v]$, $\hat{\eta}_3[v]\ (n=1)$, $\hat{\eta}_5[v]=x-tv_x$.	$\Lambda_1[V]$, $\Lambda_2[V]$, $\Lambda_3[V]\ (n=1)$, $\Lambda_5[V]=x-tV_x$.
2	$\hat{\eta}_1[v]$, $\hat{\eta}_2[v]$, $\hat{\eta}_3[v]\ (n=2)$, $\hat{\eta}_4[v]\ (n=2)$.	$\Lambda_1[V]$, $\Lambda_2[V]$, $\Lambda_3[V]\ (n=2)$, $\Lambda_4[V]=t\left(\frac{1}{3}V_x^3+V_{xxx}\right)-\frac{1}{3}xV_x$.

- For $n=1$, an additional variational symmetry $\hat{\eta}_5[v]$ arises, with a corresponding conservation law multiplier given by $\Lambda_5[V]=\hat{\eta}_5[V]$.

Nonlinear telegraph system

Consider the nonlinear telegraph (NLT) PDE system

$$\begin{aligned} u_t &= v_x, \\ v_t &= F(u)u_x + G(u). \end{aligned} \tag{1.281}$$

One can show that for all forms of the functions $F(u)$ and $G(u)$, the PDE system (1.281) is not variational. In particular, the linearized PDE system for (1.281) is given by

$$\begin{aligned} \mathrm{D}_t\tilde{v}^1 &= \mathrm{D}_x\tilde{v}^2, \\ \mathrm{D}_t\tilde{v}^2 &= F(u)\mathrm{D}_x\tilde{v}^1 + (F'(u)u_x + G'(u))\tilde{v}^1, \end{aligned} \tag{1.282}$$

holding for any solution of PDE system (1.281), whereas the adjoint linearized system is given by

$$\begin{aligned} \mathrm{D}_t\tilde{w}_1 &= F(u)\mathrm{D}_x\tilde{w}_2 - G'(u)\tilde{w}_2, \\ \mathrm{D}_t\tilde{w}_2 &= \mathrm{D}_x\tilde{w}_1, \end{aligned} \tag{1.283}$$

holding for any solution of PDE system (1.281). Since in general, for any $F(u)$ and $G(u)$, $F'(u)u_x + G'(u) \neq -G'(u)$ for the solutions $u=u(x,t)$ of the NLT system (1.281), the PDE system (1.281) is not self-adjoint.

For a variational PDE system, every conservation law multiplier yields a local symmetry, and so the number of local conservation laws never exceeds the number of local symmetries. This situation does not hold for the non-variational PDE system (1.281) as shown by a comparison [Bluman & Temuerchaolu (2005b)] between the point symmetries and zeroth order conservation law multipliers of the NLT PDE system (1.281) in terms of the classifying functions $F(u)$ and $G(u)$. Point symmetries

$$\mathrm{X} = \xi(x,t,u,v)\frac{\partial}{\partial x} + \tau(x,t,u,v)\frac{\partial}{\partial t} + \eta(x,t,u,v)\frac{\partial}{\partial u} + \kappa(x,t,u,v)\frac{\partial}{\partial v}, \quad (1.284)$$

are considered in characteristic form given by

$$\hat{\mathrm{X}} = \hat{\eta}\frac{\partial}{\partial u} + \hat{\omega}\frac{\partial}{\partial v} \quad (1.285)$$

with $\hat{\eta} = \eta - u_x\xi - u_t\tau$, $\hat{\omega} = \kappa - v_x\xi - v_t\tau$. A comparison is made with conservation law multipliers of zeroth order

$$\Lambda_i = \Lambda_i(x,t,U,V), \quad i = 1,2. \quad (1.286)$$

First note that the PDE system (1.281) has the group of equivalence transformations

$$\widetilde{x} = a_1x + a_4, \quad \widetilde{t} = a_2t + a_5, \quad \widetilde{u} = a_3u + a_6, \quad \widetilde{v} = a_3v + a_2a_7t + a_8,$$
$$\widetilde{F}(\widetilde{u}) = a_1^2a_2^{-2}F(u), \quad \widetilde{G}(\widetilde{u}) = a_1a_2^{-2}a_3G(u) + a_7, \quad (1.287)$$

where $a_1, \ldots, a_8$ are arbitrary constants, with $a_1a_2a_3 \neq 0$. Symmetries and conservation laws of PDE system (1.281) should therefore only be classified up to equivalence transformations (1.287), i.e., pairs of constitutive functions $(F(u), G(u))$ and $(\tilde{F}(u), \tilde{G}(u))$ are equivalent if

$$\tilde{F}(u) = \alpha F(\beta u + \gamma), \quad \tilde{G}(u) = \delta G(\beta u + \gamma) + \lambda,$$

for arbitrary constants $\alpha \neq 0, \beta \neq 0, \gamma, \delta \neq 0, \lambda$. Classification results in Table 1.3 are presented modulo these transformations.

For arbitrary $F(u)$ and $G(u)$, the NLT system (1.281) admits three point symmetries

$$\mathrm{X}_1 = \frac{\partial}{\partial x}, \quad \mathrm{X}_2 = \frac{\partial}{\partial t}, \quad \mathrm{X}_3 = \frac{\partial}{\partial v}, \quad (1.288)$$

and one conservation law with multipliers $\Lambda_1 = 1, \Lambda_2 = 0$ (i.e., the first equation of (1.281)). Table 1.3 lists only numbers of additional symmetries and conservation laws for each classification case.

For the sake of brevity, Table 1.3 includes only cases where $G(u)$ is a power, logarithmic or exponential function. Additional classification cases arise for

$G(u) = (u^m \pm 1)/(u^m \mp 1)$, $G(u) = \tan(\alpha \ln u)$, $G(u) = \tan u$, $G(u) = \tanh u$, and $G(u) = \coth u$. For details and the complete classification, see Bluman & Temuerchaolu (2005b).

Table 1.3 Numbers of additional point symmetries and conservation laws of the NLT system (1.281) arising for some particular forms of constitutive functions $F(u)$ and $G(u)$

$G(u)$	$F(u)$	Additional Symmetries	Additional Conservation Laws
const	Arbitrary	∞	∞
u	const	∞	∞
	u	1	4
	u^α $(\alpha \neq 0, 1)$	1	2
	$e^{\alpha u}$ $(\alpha \neq 0)$	1	2
	$u^2 + \alpha u + \beta$ $(\alpha^2 \neq 4\beta)$	0	4
	All other $F(u)$	0	2
u^{-1}	u^{-2}	∞	∞
	u^{-1}	1	4
	$(u+1)/u^2$	1	4
	$(u \pm 1)^\alpha/u^2$ $(\alpha \neq 0, 1)$	1	2
	u^α $(\alpha \neq -1, -2)$	1	2
	$u^{-2}e^{\alpha u}$ $(\alpha \neq 0)$	1	2
	$u^{-2} + \alpha u^{-1} + \beta$ $(\alpha^2 \neq 4\beta,\ \beta \neq 0)$	0	4
	All other $F(u)$	0	2
u^α	$u^{\alpha-1}$	1	3
$(\alpha \neq 0,$	u^β $(\beta \neq \alpha - 1)$	1	0
$\pm 1)$	$u^{\alpha-1} + \beta$ $(\beta \neq 0)$	0	2
	$u^{\alpha-1} + (u^\alpha + \beta)^2$	0	2
$\ln u$	u^{-1}	1	3
	u^β $(\beta \neq -1)$	1	0
	$u^{-1} + \alpha$ $(\alpha \neq 0)$	0	2
	$u^{-1} + \alpha(\ln u)^2$ $(\alpha \neq 0)$	0	2
$1/\ln u$	$1/[u(\ln u)^2]$	1	3
	$u^\beta/\ln u$ $(\beta \neq -1)$	1	0
	$1/[u(\ln u)^2] + \alpha$ $(\alpha \neq 0)$	0	2
	$1/[u(\ln u)^2] + \alpha(1/\ln u + \beta)^2$ $(\alpha \neq 0)$	0	2
e^u	e^u	1	4
	$e^{\alpha u}$, $(\alpha \neq 0, 1)$	1	0
	$e^u + \alpha$, $(\alpha \neq 0)$	0	4
	$e^{2u} + \alpha e^u + \beta$ $(\alpha^2 \neq 4\beta)$	0	4
	$(e^u + \alpha)^2$ $(\alpha \neq 0)$	0	2

In Table 1.3, α and β are arbitrary constants.

The case $G = u, F = \text{const}$ is a linear case; in cases $G(u) = u^{-1}, F(u) = u^{-2}$ and $G(u) = \text{const}, F(u)$ arbitrary, the NLT system (1.281) is linearizable by a point transformation [see Section 3.4.3 and Exercise 3.4.6]. It follows that in these three cases, PDE system (1.281) has an infinite number of point symmetries and conservation laws.

Note that from the point of view of Noether's theorem, a correspondence between symmetries of the form (1.285) and conservation law multipliers of the form (1.286) might seem invalid. Indeed, the ansatz for conservation law multipliers is more restrictive, since the form of the considered multipliers has no dependence on derivatives and, in particular, involves two arbitrary functions instead of four arbitrary functions in the case of point symmetries. However even with this restriction, the complete classification with respect to constitutive functions $F(u)$ and $G(u)$ shows that for most specific cases, the number of conservation laws exceeds the number of point symmetries! [This is never the case for variational PDE systems where the number of conservation laws is at most equal to the number of symmetries for any particular form since conservation laws arise from local multipliers that must be local symmetries.]

Exercises 1.4

1.4.1. Find the linearized equations (Fréchet derivatives) and adjoint linearized equations for the following PDEs.

(a) PDE (1.262) and equivalent PDE (1.263).
(b) PDE (1.264) and equivalent PDE (1.265).
(c) Korteweg–de Vries equation (1.266) and the transformed Korteweg-de Vries equation (1.267).

1.4.2. Consider the generalized Korteweg–de Vries equation (1.276).

(a) Show that equation (1.276) is not self-adjoint.
(b) Show that the transformed generalized KdV equation (1.279) is self-adjoint.
(c) Show that the transformed generalized KdV equation (1.279) corresponds to an extremum of the action integral with Lagrangian (1.280).
(d) Find fluxes of the conservation laws arising from multipliers in Tables 1.1 and 1.2. [Hint: For conservation laws in Table 1.2, use formula (1.257).]

1.4.3. By direct calculation, derive the identity (1.250).

1.4.4. Consider the PDE

$$u_{tt} - (u_x)^2 u_{xx} = 0. \tag{1.289}$$

(a) Show that the linearized PDE of (1.289) is self-adjoint.
(b) Show that PDE (1.289) is the Euler–Lagrange equation that arises for an extremum $U(x,t) = u(x,t)$ of the action integral with Lagrangian $L[U] = U_t^2/2 - U_x^4/12$.
(c) Show that if the action integral has the local symmetry $\mathrm{X} = \eta[U]\,\partial/\partial U$, then correspondingly $W^1[U,\eta[U]] = -U_x^3\eta[U]/3$, $W^2[U,\eta[U]] = U_t\eta[U]$.
(d) Find the most general scaling symmetry of
   (i) the PDE (1.289);
   (ii) the action integral.
(e) Use both Noether's formulation and Boyer's formulation to find all conservation laws of (1.289) that result from scaling symmetries of (1.289).
(f) Show that the PDE (1.289) has the point symmetry $t\,\partial/\partial u$. Check whether the point symmetry $t\,\partial/\partial u$ is a variational symmetry and find the corresponding conservation law if one exists. In doing so, compare Boyer's formulation and Noether's formulation.

1.4.5. Show that the indicated generators for one-parameter groups of local transformations yield variational symmetries of the corresponding action integrals for the given Lagrangians. Obtain the resulting conservation laws. In each case, derive the determining equations for conservation law multipliers and check that the variational symmetries yield solutions of the determining equations.

(a) $L[U] = \frac{1}{2}U_xU_t - \cos U, \quad \mathrm{X} = \left[\frac{1}{2}U_x^3 + U_{xxx}\right]\partial/\partial U$;
(b) $L[U,V] = 2U_xV_x + i(UV_t - VU_t) - U^2V^2, \quad \mathrm{X} = [3UVU_x + U_{xxx}]\,\partial/\partial U + [3UVV_x + V_{xxx}]\,\partial/\partial V$.

1.4.6. Consider the Boussinesq system of PDEs

$$\begin{aligned} u_t - v_{xx} &= 0, \\ v_t - u_{xx} + u + u^2 &= 0. \end{aligned} \tag{1.290}$$

(a) Show that the linearized system of the PDE system (1.290) is self-adjoint.
(b) By inspection, find three point symmetries of the PDE system (1.290). Show that each of these symmetries yields a set of conservation law multipliers and find the corresponding conservation laws through
   (i) using the direct method;
   (ii) Noether's theorem.

(c) Use the direct method and then Boyer's formulation of Noether's theorem to find the conservation law of the PDE system (1.290) that results from the set of conservation law multipliers given by

$$\Lambda_1[U,V] = U_{xxxx} - (3U+1)\,U_{xx} - \tfrac{3}{2}(U_x)^2 - \tfrac{1}{2}(V_x)^2 + \tfrac{2}{3}U^3 + \tfrac{1}{2}U^2,$$

$$\Lambda_2[U,V] = -V_{xxxx} + UV_{xx} + U_x V_x.$$

1.4.7. Consider the PDE

$$u_{tt} - c(u)[c(u)u_x]_x = 0. \tag{1.291}$$

(a) Show that the linearized equation of PDE (1.291) is self-adjoint and find a corresponding Lagrangian.
(b) For arbitrary $c(u)$, show that PDE (1.291) only has conservation law multipliers of the form $\Lambda[U] = \Lambda(x,t,U,U_t,U_x)$ given by $\Lambda[U] = U_t, U_x, tU_t + xU_x$. Find the fluxes for the corresponding three linearly independent conservation laws separately through the direct method and Noether's theorem.
(c) Show that the PDE (1.291) admits additional conservation laws resulting from multipliers of the form $\Lambda[U] = \Lambda(x,t,U,U_t,U_x)$ if and only if $c(u) = a(u+b)^2$, for arbitrary constants $a \neq 0$ and b. Find the resulting three additional conservation laws [Anco & Bluman (2002a)].

1.5 Some Connections Between Symmetries and Conservation Laws

So far it has been seen that if a PDE system is variational, i.e., its linearization operator (Fréchet derivative) is self-adjoint, then a set of local conservation law multipliers directly corresponds to a local symmetry of the PDE system. In general, the converse does not hold: a local symmetry of a variational PDE system may not necessarily yield a set of local conservation law multipliers for a conservation law.

If a PDE system is not variational, then local conservation law multipliers do not in general correspond to local symmetries of the PDE system.

In this section, it is shown that if a given PDE system $\mathbf{R}\{x\,;u\}$ is mapped into another PDE system $\mathbf{S}\{z\,;w\}$ by an invertible transformation (point or contact transformation) then each conservation law of $\mathbf{R}\{x\,;u\}$ is transformed to a corresponding conservation law of $\mathbf{S}\{z\,;w\}$. When the invertible transformation is a symmetry (discrete or continuous) of $\mathbf{R}\{x\,;u\}$, then the corresponding conservation law is a conservation law of $\mathbf{R}\{x\,;u\}$. Related to this, two formulas are presented [Bluman, Temuerchaolu & Anco (2006)].

The first formula yields the transformed conservation law. The second formula checks a priori whether the action of a symmetry on a conservation law of a given PDE system $\mathbf{R}\{x\,;u\}$ can yield one or more new conservation laws of $\mathbf{R}\{x\,;u\}$.

Furthermore, for any given PDE system $\mathbf{R}\{x\,;u\}$, it is shown that any paired set of functions consisting of any solution of its linearizing system (1.64) (i.e., components of a local symmetry in characteristic form) and a solution of the adjoint system (1.184) of the linearizing system (1.64) can yield directly a conservation law of $\mathbf{R}\{x\,;u\}$ through a simple algebraic formula [Anco & Bluman (1997a)]. However, it could happen that the resulting conservation law is trivial. In the important special case when the local symmetry is a scaling symmetry and the solution of the adjoint system is a local multiplier for a conservation law, it is shown that this formula directly yields the conservation law obtained from the local multiplier [Anco (2003)] (provided that the conservation law and the given PDE system are homogeneous under the scaling symmetry, and the conservation law has non-zero scaling weight).

In the variational (self-adjoint) case, it immediately follows that any pair of local symmetries of $\mathbf{R}\{x\,;u\}$ could yield directly a conservation law of $\mathbf{R}\{x\,;u\}$ [Anco & Bluman (1996)] through this simple formula. Furthermore, all local conservation laws of $\mathbf{R}\{x\,;u\}$ that have non-zero scaling weight are obtained directly through this simple formula if one of the symmetries of $\mathbf{R}\{x\,;u\}$ is a scaling symmetry and one determines all local multipliers (equivalent to determining all variational symmetries). Note that no Lagrangian is needed.

In the next chapter, it is shown how one can use either the local symmetries or local conservation law multipliers of a nonlinear PDE system $\mathbf{R}\{x\,;u\}$ to determine whether it can be invertibly mapped into some linear PDE system $\mathbf{S}\{z\,;w\}$ as well as obtain a specific mapping when one exists.

In subsequent chapters, it is shown that local conservation laws of a given PDE system $\mathbf{R}\{x\,;u\}$ yield nonlocally related PDE systems that in turn can yield nonlocal symmetries and nonlocal conservation laws of $\mathbf{R}\{x\,;u\}$.

1.5.1 Use of symmetries to find new conservation laws from known conservation laws

We derive two formulas related to obtaining new conservation laws from known conservation laws under the action of an invertible (point or contact) transformation. The first formula shows how to use an invertible transformation, including a discrete one, that maps any given PDE system $\mathbf{R}\{x\,;u\}$ to another PDE system $\mathbf{S}\{z\,;w\}$ to obtain directly a conservation law of $\mathbf{S}\{z\,;w\}$

from any known conservation law of $\mathbf{R}\{x\,;u\}$. The situation is particularly interesting when the invertible transformation is a symmetry of the given PDE system $\mathbf{R}\{x;u\}$ since here one could obtain new conservation laws from a known conservation law of $\mathbf{R}\{x\,;u\}$. No differential consequences of the given PDE system $\mathbf{R}\{x\,;u\}$ are used in these formulas.

The second formula uses the action of a symmetry of $\mathbf{R}\{x\,;u\}$ on the set of multipliers of a known conservation law of $\mathbf{R}\{x\,;u\}$ to construct the sets of multipliers of conservation laws of $\mathbf{R}\{x\,;u\}$. This allows one to check a priori whether a new conservation law of $\mathbf{R}\{x\,;u\}$ is obtained under the action of the symmetry of $\mathbf{R}\{x\,;u\}$. It is also shown that if the symmetry is a point or contact symmetry (i.e., a one-parameter Lie group of point or contact transformations), then one could obtain more than one new conservation law.

For the rest of this section, we restrict an invertible transformation to a point transformation (which must be the situation when $\mathbf{R}\{x\,;u\}$ has two or more dependent variables). The extension to contact transformations in the case when $\mathbf{R}\{x\,;u\}$ is a scalar PDE is straightforward [Bluman, Temuerchaolu & Anco (2006)].

Consider a system of N PDEs $\mathbf{R}\{x\,;u\}$ given by

$$R^\sigma[u] = R^\sigma(x, u, \partial u, \ldots, \partial^k u) = 0, \quad \sigma = 1, \ldots, N, \tag{1.292}$$

with n independent variables $x = (x^1, \ldots, x^n)$ and m dependent variables $u = (u^1, \ldots, u^m)$. Let

$$R^\sigma[U] = R^\sigma(x, U, \partial U, \ldots, \partial^k U), \quad \sigma = 1, \ldots, N, \tag{1.293}$$

where $U(x) = (U^1(x), \ldots, U^m(x))$ is an arbitrary function with $U(x) = u(x)$ solving the system of PDEs (1.292).

Consider an invertible point transformation

$$\begin{aligned} x^i &= x^i(z, W), \quad i = 1, \ldots, n, \\ U^\mu &= U^\mu(z, W), \quad \mu = 1, \ldots, m, \end{aligned} \tag{1.294}$$

where $U(x) = (U^1(x), \ldots, U^m(x))$, $z = (z^1, \ldots, z^m)$, $W(z) = (W^1(z), \ldots, W^m(z))$.

Under a point transformation (1.294) and its natural extensions (prolongations) to actions on derivatives, each function $R^\sigma[U]$ is mapped to some function $S^\sigma[W] = S^\sigma(z, W, \partial W, \ldots, \partial^k W)$. In particular,

$$S^\sigma[W] = R^\sigma[U], \tag{1.295}$$

where the components of $x, U, \partial U, \ldots, \partial^k U$ are expressed in terms of the components of $z, W, \partial W, \ldots, \partial^k W$ through (1.294). If $U(x) = u(x)$ solves PDE system $\mathbf{R}\{x\,;u\}$ (1.292), then correspondingly $W(z) = w(z)$ solves PDE system $\mathbf{S}\{z\,;w\}$ given by

$$S^\sigma[w] = S^\sigma(z, w, \partial w, \ldots, \partial^k w) = 0, \quad \sigma = 1, \ldots, N, \tag{1.296}$$

with n independent variables $z = (z^1, \ldots, z^m)$, and m dependent variables $w(z) = (w^1(z), \ldots, w^m(z))$.

Theorem 1.5.1. *Suppose* $\mathrm{D}_i\Phi^i[u] = 0$ *is a conservation law of PDE system* $\mathbf{R}\{x\,;u\}$ (1.292). *Under the point transformation* (1.294), *there exist functions* $\{\Psi^i[W]\}_{i=1}^n$ *such that the formula*

$$\mathrm{J}[W]\mathrm{D}_i\Phi^i[U] = \tilde{\mathrm{D}}_i\Psi^i[W] \tag{1.297}$$

holds, where $\Psi^i[W]$ *is given explicitly in terms of the determinant obtained by replacing the* i*th column of the Jacobian determinant*

$$\mathrm{J}[W] = \frac{\mathrm{D}(x^1, \ldots, x^n)}{\mathrm{D}(z^1, \ldots, z^n)} \tag{1.298}$$

by $\begin{bmatrix} \Phi^1[U] \\ \vdots \\ \Phi^n[U] \end{bmatrix}$, *and where* $\mathrm{D}_i, \tilde{\mathrm{D}}_i$ *are total derivative operators, respectively, given by*

$$\begin{aligned}
\mathrm{D}_i &= \frac{\partial}{\partial x^i} + U_i^\mu \frac{\partial}{\partial U^\mu} + U_{ii_1}^\mu \frac{\partial}{\partial U_{i_1}^\mu} + \cdots, \\
\tilde{\mathrm{D}}_i &= \frac{\partial}{\partial z^i} + W_i^\mu \frac{\partial}{\partial W^\mu} + W_{ii_1}^\mu \frac{\partial}{\partial W_{i_1}^\mu} + \cdots, \qquad i = 1, \ldots, n
\end{aligned}$$

with $U_i^\mu = \frac{\partial U^\mu}{\partial x^i}$, $W_i^\mu = \frac{\partial W^\mu}{\partial z^i}$, *etc.*

Proof. Consider the determinants

$$\Psi^1[W] = \begin{vmatrix} \Phi^1[U] & \tilde{\mathrm{D}}_2 x^1 & \cdots & \tilde{\mathrm{D}}_n x^1 \\ \Phi^2[U] & \tilde{\mathrm{D}}_2 x^2 & \cdots & \tilde{\mathrm{D}}_n x^2 \\ \vdots & \vdots & & \vdots \\ \Phi^n[U] & \tilde{\mathrm{D}}_2 x^n & \cdots & \tilde{\mathrm{D}}_n x^n \end{vmatrix}, \quad \Psi^2[W] = \begin{vmatrix} \tilde{\mathrm{D}}_1 x^1 & \Phi^1[U] & \cdots & \tilde{\mathrm{D}}_n x^1 \\ \tilde{\mathrm{D}}_1 x^2 & \Phi^2[U] & \cdots & \tilde{\mathrm{D}}_n x^2 \\ \vdots & \vdots & & \vdots \\ \tilde{\mathrm{D}}_1 x^n & \Phi^n[U] & \cdots & \tilde{\mathrm{D}}_n x^n \end{vmatrix},$$

$$\ldots, \quad \Psi^n[W] = \begin{vmatrix} \tilde{\mathrm{D}}_1 x^1 & \cdots & \tilde{\mathrm{D}}_{n-1} x^1 & \Phi^1[U] \\ \tilde{\mathrm{D}}_1 x^2 & \cdots & \tilde{\mathrm{D}}_{n-1} x^2 & \Phi^2[U] \\ \vdots & & \vdots & \vdots \\ \tilde{\mathrm{D}}_1 x^n & \cdots & \tilde{\mathrm{D}}_{n-1} x^n & \Phi^n[U] \end{vmatrix}. \tag{1.299}$$

Then

$$
\tilde{\mathrm{D}}_i \varPsi^i[W] = \left\{ \begin{vmatrix} \tilde{\mathrm{D}}_1 \varPhi^1[U] & \tilde{\mathrm{D}}_2 x^1 & \cdots & \tilde{\mathrm{D}}_n x^1 \\ \tilde{\mathrm{D}}_1 \varPhi^2[U] & \tilde{\mathrm{D}}_2 x^2 & \cdots & \tilde{\mathrm{D}}_n x^2 \\ \vdots & \vdots & & \vdots \\ \tilde{\mathrm{D}}_1 \varPhi^n[U] & \tilde{\mathrm{D}}_2 x^n & \cdots & \tilde{\mathrm{D}}_n x^n \end{vmatrix} + \begin{vmatrix} \varPhi^1[U] & \tilde{\mathrm{D}}_1 \tilde{\mathrm{D}}_2 x^1 & \tilde{\mathrm{D}}_3 x^1 & \cdots & \tilde{\mathrm{D}}_n x^1 \\ \varPhi^2[U] & \tilde{\mathrm{D}}_1 \tilde{\mathrm{D}}_2 x^2 & \tilde{\mathrm{D}}_3 x^2 & \cdots & \tilde{\mathrm{D}}_n x^2 \\ \vdots & \vdots & \vdots & & \vdots \\ \varPhi^n[U] & \tilde{\mathrm{D}}_1 \tilde{\mathrm{D}}_2 x^n & \tilde{\mathrm{D}}_3 x^n & \cdots & \tilde{\mathrm{D}}_n x^n \end{vmatrix} \right.
$$

$$
\left. + \cdots + \begin{vmatrix} \varPhi^1[U] & \tilde{\mathrm{D}}_2 x^1 & \cdots & \tilde{\mathrm{D}}_{n-1} x^1 & \tilde{\mathrm{D}}_1 \tilde{\mathrm{D}}_n x^1 \\ \varPhi^2[U] & \tilde{\mathrm{D}}_2 x^2 & \cdots & \tilde{\mathrm{D}}_{n-1} x^2 & \tilde{\mathrm{D}}_1 \tilde{\mathrm{D}}_n x^2 \\ \vdots & \vdots & & \vdots & \vdots \\ \varPhi^n[U] & \tilde{\mathrm{D}}_2 x^n & \cdots & \tilde{\mathrm{D}}_{n-1} x^n & \tilde{\mathrm{D}}_1 \tilde{\mathrm{D}}_n x^n \end{vmatrix} \right\}
$$

$$
+ \left\{ \begin{vmatrix} \tilde{\mathrm{D}}_1 \tilde{\mathrm{D}}_2 x^1 & \varPhi^1[U] & \tilde{\mathrm{D}}_3 x^1 & \cdots & \tilde{\mathrm{D}}_n x^1 \\ \tilde{\mathrm{D}}_1 \tilde{\mathrm{D}}_2 x^2 & \varPhi^2[U] & \tilde{\mathrm{D}}_3 x^2 & \cdots & \tilde{\mathrm{D}}_n x^2 \\ \vdots & \vdots & \vdots & & \vdots \\ \tilde{\mathrm{D}}_1 \tilde{\mathrm{D}}_2 x^n & \varPhi^n[U] & \tilde{\mathrm{D}}_3 x^n & \cdots & \tilde{\mathrm{D}}_n x^n \end{vmatrix} \right.
$$

$$
+ \begin{vmatrix} \tilde{\mathrm{D}}_1 x^1 & \tilde{\mathrm{D}}_2 \varPhi^1[U] & \tilde{\mathrm{D}}_3 x^1 & \cdots & \tilde{\mathrm{D}}_n x^1 \\ \tilde{\mathrm{D}}_1 x^2 & \tilde{\mathrm{D}}_2 \varPhi^2[U] & \tilde{\mathrm{D}}_3 x^2 & \cdots & \tilde{\mathrm{D}}_n x^2 \\ \vdots & \vdots & \vdots & & \vdots \\ \tilde{\mathrm{D}}_1 x^n & \tilde{\mathrm{D}}_2 \varPhi^n[U] & \tilde{\mathrm{D}}_3 x^n & \cdots & \tilde{\mathrm{D}}_n x^n \end{vmatrix} + \begin{vmatrix} \tilde{\mathrm{D}}_1 x^1 & \varPhi^1[U] & \tilde{\mathrm{D}}_2 \tilde{\mathrm{D}}_3 x^1 & \tilde{\mathrm{D}}_4 x^1 & \cdots & \tilde{\mathrm{D}}_n x^1 \\ \tilde{\mathrm{D}}_1 x^2 & \varPhi^2[U] & \tilde{\mathrm{D}}_2 \tilde{\mathrm{D}}_3 x^2 & \tilde{\mathrm{D}}_4 x^2 & \cdots & \tilde{\mathrm{D}}_n x^2 \\ \vdots & \vdots & \vdots & \vdots & & \vdots \\ \tilde{\mathrm{D}}_1 x^n & \varPhi^n[U] & \tilde{\mathrm{D}}_2 \tilde{\mathrm{D}}_3 x^n & \tilde{\mathrm{D}}_4 x^n & \cdots & \tilde{\mathrm{D}}_n x^n \end{vmatrix}
$$

$$
\left. + \cdots + \begin{vmatrix} \tilde{\mathrm{D}}_1 x^1 & \varPhi^1[U] & \tilde{\mathrm{D}}_3 x^1 & \cdots & \tilde{\mathrm{D}}_{n-1} x^1 & \tilde{\mathrm{D}}_2 \tilde{\mathrm{D}}_n x^1 \\ \tilde{\mathrm{D}}_1 x^2 & \varPhi^2[U] & \tilde{\mathrm{D}}_3 x^2 & \cdots & \tilde{\mathrm{D}}_{n-1} x^2 & \tilde{\mathrm{D}}_2 \tilde{\mathrm{D}}_n x^2 \\ \vdots & \vdots & \vdots & & \vdots & \vdots \\ \tilde{\mathrm{D}}_1 x^n & \varPhi^n[U] & \tilde{\mathrm{D}}_3 x^n & \cdots & \tilde{\mathrm{D}}_{n-1} x^n & \tilde{\mathrm{D}}_2 \tilde{\mathrm{D}}_n x^n \end{vmatrix} \right\} + \cdots
$$

$$
+ \left\{ \begin{vmatrix} \tilde{\mathrm{D}}_1 \tilde{\mathrm{D}}_n x^1 & \tilde{\mathrm{D}}_2 x^1 & \cdots & \tilde{\mathrm{D}}_{n-1} x^1 & \varPhi^1[U] \\ \tilde{\mathrm{D}}_1 \tilde{\mathrm{D}}_n x^2 & \tilde{\mathrm{D}}_2 x^2 & \cdots & \tilde{\mathrm{D}}_{n-1} x^2 & \varPhi^2[U] \\ \vdots & \vdots & & \vdots & \vdots \\ \tilde{\mathrm{D}}_1 \tilde{\mathrm{D}}_n x^n & \tilde{\mathrm{D}}_2 x^n & \cdots & \tilde{\mathrm{D}}_{n-1} x^n & \varPhi^n[U] \end{vmatrix} \right.
$$

$$
+ \begin{vmatrix} \tilde{\mathrm{D}}_1 x^1 & \tilde{\mathrm{D}}_2 \tilde{\mathrm{D}}_n x^1 & \tilde{\mathrm{D}}_3 x^1 & \cdots & \tilde{\mathrm{D}}_{n-1} x^1 & \varPhi^1[U] \\ \tilde{\mathrm{D}}_1 x^2 & \tilde{\mathrm{D}}_2 \tilde{\mathrm{D}}_n x^2 & \tilde{\mathrm{D}}_3 x^2 & \cdots & \tilde{\mathrm{D}}_{n-1} x^2 & \varPhi^2[U] \\ \vdots & \vdots & \vdots & & \vdots & \vdots \\ \tilde{\mathrm{D}}_1 x^n & \tilde{\mathrm{D}}_2 \tilde{\mathrm{D}}_n x^n & \tilde{\mathrm{D}}_3 x^n & \cdots & \tilde{\mathrm{D}}_{n-1} x^n & \varPhi^n[U] \end{vmatrix}
$$

$$
+ \cdots + \begin{vmatrix} \tilde{\mathrm{D}}_1 x^1 & \tilde{\mathrm{D}}_2 x^1 & \cdots & \tilde{\mathrm{D}}_{n-2} x^1 & \tilde{\mathrm{D}}_{n-1} \tilde{\mathrm{D}}_n x^1 & \varPhi^1[U] \\ \tilde{\mathrm{D}}_1 x^2 & \tilde{\mathrm{D}}_2 x^2 & \cdots & \tilde{\mathrm{D}}_{n-2} x^2 & \tilde{\mathrm{D}}_{n-1} \tilde{\mathrm{D}}_n x^2 & \varPhi^2[U] \\ \vdots & \vdots & & \vdots & \vdots & \vdots \\ \tilde{\mathrm{D}}_1 x^n & \tilde{\mathrm{D}}_2 x^n & \cdots & \tilde{\mathrm{D}}_{n-2} x^n & \tilde{\mathrm{D}}_{n-1} \tilde{\mathrm{D}}_n x^n & \varPhi^n[U] \end{vmatrix}
$$

$$+\left|\begin{matrix} \tilde{\mathrm{D}}_1 x^1 & \tilde{\mathrm{D}}_2 x^1 & \cdots & \tilde{\mathrm{D}}_{n-1} x^1 & \tilde{\mathrm{D}}_n \Phi^1[U] \\ \tilde{\mathrm{D}}_1 x^2 & \tilde{\mathrm{D}}_2 x^2 & \cdots & \tilde{\mathrm{D}}_{n-1} x^2 & \tilde{\mathrm{D}}_n \Phi^2[U] \\ \vdots & \vdots & & \vdots & \vdots \\ \tilde{\mathrm{D}}_1 x^n & \tilde{\mathrm{D}}_2 x^n & \cdots & \tilde{\mathrm{D}}_{n-1} x^n & \tilde{\mathrm{D}}_n \Phi^n[U] \end{matrix}\right| \Bigg\}. \tag{1.300}$$

In expression (1.300), the pairs of determinants with entries $\tilde{\mathrm{D}}_j\tilde{\mathrm{D}}_k x^i$, $j \neq k$, $i = 1, \ldots, n$, cancel each other out since their respective columns are odd permutations of each other. Hence expression (1.300) reduces to

$$\tilde{\mathrm{D}}_i \Psi^i[W] = \left|\begin{matrix} \tilde{\mathrm{D}}_1 \Phi^1[U] & \tilde{\mathrm{D}}_2 x^1 & \cdots & \tilde{\mathrm{D}}_n x^1 \\ \tilde{\mathrm{D}}_1 \Phi^2[U] & \tilde{\mathrm{D}}_2 x^2 & \cdots & \tilde{\mathrm{D}}_n x^2 \\ \vdots & \vdots & & \vdots \\ \tilde{\mathrm{D}}_1 \Phi^n[U] & \tilde{\mathrm{D}}_2 x^n & \cdots & \tilde{\mathrm{D}}_n x^n \end{matrix}\right| + \left|\begin{matrix} \tilde{\mathrm{D}}_1 x^1 & \tilde{\mathrm{D}}_2 \Phi^1[U] & \tilde{\mathrm{D}}_3 x^1 & \cdots & \tilde{\mathrm{D}}_n x^1 \\ \tilde{\mathrm{D}}_1 x^2 & \tilde{\mathrm{D}}_2 \Phi^2[U] & \tilde{\mathrm{D}}_3 x^2 & \cdots & \tilde{\mathrm{D}}_n x^2 \\ \vdots & \vdots & \vdots & & \vdots \\ \tilde{\mathrm{D}}_1 x^n & \tilde{\mathrm{D}}_2 \Phi^n[U] & \tilde{\mathrm{D}}_3 x^n & \cdots & \tilde{\mathrm{D}}_n x^n \end{matrix}\right|$$

$$+ \cdots + \left|\begin{matrix} \tilde{\mathrm{D}}_1 x^1 & \tilde{\mathrm{D}}_2 x^1 & \cdots & \tilde{\mathrm{D}}_{n-1} x^1 & \tilde{\mathrm{D}}_n \Phi^1[U] \\ \tilde{\mathrm{D}}_1 x^2 & \tilde{\mathrm{D}}_2 x^2 & \cdots & \tilde{\mathrm{D}}_{n-1} x^2 & \tilde{\mathrm{D}}_n \Phi^2[U] \\ \vdots & \vdots & & \vdots & \vdots \\ \tilde{\mathrm{D}}_1 x^n & \tilde{\mathrm{D}}_2 x^n & \cdots & \tilde{\mathrm{D}}_{n-1} x^n & \tilde{\mathrm{D}}_n \Phi^n[U] \end{matrix}\right|. \tag{1.301}$$

Let ξ_i^j denote the cofactor of $\tilde{\mathrm{D}}_j x^i$ for the Jacobian matrix given by

$$\begin{bmatrix} \tilde{\mathrm{D}}_1 x^1 & \cdots & \tilde{\mathrm{D}}_n x^1 \\ \vdots & & \vdots \\ \tilde{\mathrm{D}}_1 x^n & \cdots & \tilde{\mathrm{D}}_n x^n \end{bmatrix}. \tag{1.302}$$

Then expression (1.301) becomes

$$\tilde{\mathrm{D}}_i \Psi^i[W] = (\tilde{\mathrm{D}}_j \Phi^i[U])\xi_i^j. \tag{1.303}$$

Using the chain rule, one has $\tilde{\mathrm{D}}_j \Phi^i[U] = (\mathrm{D}_k \Phi^i[U])(\tilde{\mathrm{D}}_j x^k)$. Thus (1.303) becomes

$$\tilde{\mathrm{D}}_i \Psi^i[W] = (\mathrm{D}_k \Phi^i[U])(\tilde{\mathrm{D}}_j x^k)\xi_i^j. \tag{1.304}$$

But $(\tilde{\mathrm{D}}_j x^k)\xi_i^j = \delta_i^k \mathrm{J}[W]$, where $\mathrm{J}[W]$ is the Jacobian determinant given by (1.298) and δ_i^k is the Kronecker symbol: $\delta_i^i = 1$, $\delta_i^k = 0$ for $i \neq k$.

Consequently, equation (1.304) yields (1.297). □

From Theorem 1.5.1, one immediately obtains the following important result.

Corollary 1.5.1. *Under a point transformation* (1.294), *a conservation law* $\mathrm{D}_i \Phi^i[u] = 0$ *of PDE system* $\mathbf{R}\{x\,; u\}$ (1.292) *is transformed to the conservation law*

$$\tilde{\mathrm{D}}_i \Psi^i[w] = 0 \tag{1.305}$$

of PDE system $\mathbf{S}\{z\,;w\}$ (1.296) *with fluxes* $\Psi^i[w]$ *given by* (1.299) *in terms of the solutions* $W(z)=w(z)$ *of PDE system* $\mathbf{S}\{z\,;w\}$.

Symmetry action on a conservation law to yield new conservation laws

We now restrict our attention to the most important situation when the invertible point transformation (1.294) is a *symmetry* of PDE system $\mathbf{R}\{x\,;u\}$ (1.292). We show that the action of a symmetry on a conservation law of $\mathbf{R}\{x\,;u\}$ could yield a new conservation law of $\mathbf{R}\{x\,;u\}$. Since any point transformation that is a symmetry of PDE system $\mathbf{R}\{x\,;u\}$ leaves invariant the solution manifold of $\mathbf{R}\{x\,;u\}$, it follows that there exist specific functions $A^\sigma_\tau[W]$ so that (1.295) is of the form

$$R^\sigma[U]=S^\sigma[W]=A^\sigma_\tau[W]R^\tau[W]. \tag{1.306}$$

Hence through formulas (1.297) and (1.299) one obtains a symmetry mapping formula for conservation laws.

Corollary 1.5.2. *If the invertible point transformation* $(x,u)\rightarrow(\tilde{x}(x,u),\tilde{u}(x,u))$ *is a symmetry of the PDE system* $\mathbf{R}\{x\,;u\}$ (1.292), *then a conservation law* $\mathrm{D}_i\Phi^i[u]=0$ *of* $\mathbf{R}\{x\,;u\}$ *yields the conservation law*

$$\mathrm{D}_i\Psi^i[u]=0 \tag{1.307}$$

of $\mathbf{R}\{x\,;u\}$ *with fluxes given by*

$$\Psi^1[u]=\begin{vmatrix}\Phi^1[\tilde{u}] & \mathrm{D}_2\tilde{x}^1 & \cdots & \mathrm{D}_n\tilde{x}^1\\ \Phi^2[\tilde{u}] & \mathrm{D}_2\tilde{x}^2 & \cdots & \mathrm{D}_n\tilde{x}^2\\ \vdots & \vdots & & \vdots\\ \Phi^n[\tilde{u}] & \mathrm{D}_2\tilde{x}^n & \cdots & \mathrm{D}_n\tilde{x}^n\end{vmatrix},\ldots,\Psi^n[u]=\begin{vmatrix}\mathrm{D}_1\tilde{x}^1 & \cdots & \mathrm{D}_{n-1}\tilde{x}^1 & \Phi^1[\tilde{u}]\\ \mathrm{D}_1\tilde{x}^2 & \cdots & \mathrm{D}_{n-1}\tilde{x}^2 & \Phi^2[\tilde{u}]\\ \vdots & & \vdots & \vdots\\ \mathrm{D}_1\tilde{x}^n & \cdots & \mathrm{D}_{n-1}\tilde{x}^n & \Phi^n[\tilde{u}]\end{vmatrix}.$$

Proof. From (1.306), it follows that $S^\sigma[U]=A^\sigma_\tau[U]R^\tau[U]$ holds for arbitrary functions $U(x)$. Hence $S^\sigma[u]=0$ for any solution $U(x)=u(x)$ of the PDE system $\mathbf{R}\{x\,;u\}$ (1.292). Consequently, from Corollary 1.5.1, one obtains the conservation law (1.307) with fluxes $\Psi^i[u]$ given by formula (1.299) after first replacing x^i by $\tilde{x}^i$, u^μ by $\tilde{u}^\mu$, etc. and then z^i by x^i, $W^\mu(z)$ by $u^\mu(x)$, W^μ_i by u^μ_i, etc. □

Corollary 1.5.2 shows that the action of a symmetry of the PDE system $\mathbf{R}\{x\,;u\}$ (1.292) on a known conservation law $\mathrm{D}_i\Phi^i[u]=0$ of $\mathbf{R}\{x\,;u\}$ yields the conservation law (1.307) of $\mathbf{R}\{x\,;u\}$ through use of the formula (1.297). Clearly, any symmetry of $\mathbf{R}\{x\,;u\}$ must yield a symmetry of the determining equations for conservation law multipliers. The problem is to find how the multipliers themselves change under symmetries of $\mathbf{R}\{x\,;u\}$.

Now another theorem and immediate corollary are presented that together yield a second formula that directly allows one to check *a priori* whether the action of a symmetry (1.294) on a known conservation law $\mathrm{D}_i\Phi^i[u]=0$ of $\mathbf{R}\{x\,;u\}$ yields *new* conservation laws (1.307) of $\mathbf{R}\{x\,;u\}$. In particular, for each symmetry of $\mathbf{R}\{x\,;u\}$ and any given set of conservation law multipliers $\{\Lambda_\sigma[U]\}_{\sigma=1}^N$ of $\mathbf{R}\{x\,;u\}$, this formula will yield a transformed set of conservation multipliers $\{\hat{\Lambda}_\sigma[U]\}_{\sigma=1}^N$. If the set of multipliers $\{\hat{\Lambda}_\sigma[U]\}_{\sigma=1}^N$ is independent of the given set of multipliers $\{\Lambda_\sigma[U]\}_{\sigma=1}^N$, then one obtains a new conservation law of $\mathbf{R}\{x\,;u\}$.

Theorem 1.5.2. *Suppose the point transformation* (1.294) *is a symmetry of PDE system* $\mathbf{R}\{x\,;u\}$ (1.292)*. If* $\{\Lambda_\sigma[U]\}_{\sigma=1}^N$ *is a set of multipliers for a conservation law of* $\mathbf{R}\{x\,;u\}$ *with fluxes* $\Phi^i[u]$*, then*

$$\hat{\Lambda}_\tau[W]R^\tau[W]=\tilde{\mathrm{D}}_i\Psi^i[W], \tag{1.308}$$

where

$$\hat{\Lambda}_\tau[W]=\mathrm{J}[W]A_\tau^\sigma[W]\Lambda_\sigma[U(z,W)],\quad \tau=1,\dots,N, \tag{1.309}$$

with $U(z,W)$ *(and its derivatives) given by the transformation* (1.294) *(and its natural extensions). In* (1.308)*,* $\Psi^i[W]$ *is given by* (1.299) *and, in* (1.309)*, the Jacobian determinant* $\mathrm{J}[W]$ *and* $A_\tau^\sigma[W]$ *are given by* (1.298) *and* (1.306)*, respectively.*

Proof. Since the point transformation (1.294) is a symmetry of PDE system $\mathbf{R}\{x\,;u\}$ (1.292), it follows that equation (1.306) holds for arbitrary functions $W(z)$. Since $\{\Lambda_\sigma[U]\}_{\sigma=1}^N$ is a set of multipliers for a conservation law of $\mathbf{R}\{x\,;u\}$ with fluxes $\Phi^i[u]$, the identity

$$\Lambda_\sigma[U]R^\sigma[U]\equiv\mathrm{D}_i\Phi^i[U] \tag{1.310}$$

holds for arbitrary functions $U(x)$. After substituting (1.306) into (1.310), one obtains

$$\mathrm{D}_i\Phi^i[U]=\Lambda_\sigma[U]R^\sigma[U]=\Lambda_\sigma[U]A_\tau^\sigma[W]R^\tau[W]. \tag{1.311}$$

After multiplying (1.311) by $J[W]$ and using formula (1.297), one finds that

$$\mathrm{J}[W](\mathrm{D}_i\Phi^i[U])=\mathrm{J}[W]\Lambda_\sigma[U]A_\tau^\sigma[W]R^\tau[W]=\tilde{\mathrm{D}}_i\Psi^i[W].$$

Hence

$$\hat{\Lambda}_\tau[W]R^\tau[W]=\tilde{\mathrm{D}}_i\Psi^i[W],$$

where $\hat{\Lambda}_\tau[W]$ is given by (1.309). □

After replacing the coordinates z^i by x^i, $W^\mu(z)$ by $U^\mu(x)$, W_i^μ by U_i^μ, etc., in (1.308), the following important corollary is immediately obvious.

Corollary 1.5.3. *If $\{\Lambda_\sigma[U]\}_{\sigma=1}^N$ is a set of multipliers for a conservation law of PDE system $\mathbf{R}\{x\,;u\}$ (1.292) and PDE system $\mathbf{R}\{x\,;u\}$ is invariant under the point transformation $(x,u)\to(\tilde{x}(x,u),\tilde{u}(x,u))$, then $\{\hat{\Lambda}_\tau[U]\}_{\tau=1}^N$ yields a set of multipliers for a conservation law of $\mathbf{R}\{x\,;u\}$ where $\hat{\Lambda}_\tau[U]=$ $\mathrm{J}[\tilde{U}]A_\tau^\sigma[\tilde{U}]\Lambda_\sigma[U]$.*

The following proposition indicates whether a set of multipliers $\{\hat{\Lambda}_\tau[U]\}_{\tau=1}^N$ yields a *new* conservation law of PDE system $\mathbf{R}\{x\,;u\}$ (1.292).

Proposition 1.5.1. *A set of multipliers $\{\hat{\Lambda}_\tau[U]\}_{\tau=1}^N$ yields a* new *conservation law of PDE system $\mathbf{R}\{x\,;u\}$ (1.292) if and only if this set is independent of $\{\Lambda_\sigma[U]\}_{\sigma=1}^N$ on all solutions $U(x)=u(x)$ of $\mathbf{R}\{x\,;u\}$, i.e., $\hat{\Lambda}_\tau[U]\not\equiv c\Lambda_\tau[U]$, $\tau=1,\dots,N$, for any constant c.*

Proof. Two conservation laws of a PDE system $\mathbf{R}\{x\,;u\}$ are equivalent if and only if their corresponding fluxes differ by a curl term $\mathrm{D}_jH^{ij}[u]$ on all solutions $U(x)=u(x)$ of $\mathbf{R}\{x\,;u\}$ (where $H^{ij}[u]=-H^{ji}[u]$). For a PDE system $\mathbf{R}\{x\,;u\}$ in Cauchy–Kovalevskaya form, all equivalent conservation laws have the same set of multipliers when the multipliers are restricted to solutions $U(x)=u(x)$ of $\mathbf{R}\{x\,;u\}$ [Anco & Bluman (2002b)]. Hence two sets of multipliers are equivalent when they agree on all solutions $u(x)$ of $\mathbf{R}\{x\,;u\}$. In particular, there is a one-to-one correspondence between nontrivial conservation laws (up to equivalence) and sets of nontrivial multipliers. This establishes the proposition. □

Formulas (1.297) and (1.308), after the appropriate coordinate substitutions, use finite transformations to yield a new conservation law from a known conservation law of $\mathbf{R}\{x\,;u\}$ and hence are applicable for any group of transformations admitted by $\mathbf{R}\{x\,;u\}$, including discrete symmetries. Now consider the important situation when the point transformation (1.294) is a one-parameter Lie group of point transformations of PDE system $\mathbf{R}\{x\,;u\}$ (1.292). Using the infinitesimal form of the group, it is now shown that here it is possible to obtain more than one new conservation law from a known conservation law.

Suppose the point transformation (1.294) is a one-parameter Lie group of point transformations. Then in terms of its infinitesimals $\xi^i[W]=\xi^i(z,W)$, $\eta^\mu[W]=\eta^\mu(z,W)$, one has

$$\begin{aligned} x^i &= z^i+\varepsilon\xi^i[W]+O(\varepsilon^2), \quad i=1,\dots,n,\\ U^\mu &= W^\mu+\varepsilon\eta^\mu[W]+O(\varepsilon^2), \quad \mu=1,\dots,m. \end{aligned} \tag{1.312}$$

The corresponding extended infinitesimal generator is given by

$$\tilde{\mathrm{X}} = \xi^j[W]\frac{\partial}{\partial z^j} + \eta^\mu[W]\frac{\partial}{\partial W^\mu} + \cdots + \eta^\mu_{i_1\cdots i_k}[W]\frac{\partial}{\partial W^\mu_{i_1\cdots i_k}} + \cdots \tag{1.313}$$

with

$$\eta^\mu_i[W] = \tilde{\mathrm{D}}_i\eta^\mu[W] - (\tilde{\mathrm{D}}_i\xi^j[W])W^\mu_j, \quad i = 1,\ldots,n,$$

$$\eta^\mu_{i_1\cdots i_k}[W] = \tilde{\mathrm{D}}_{i_k}\eta^\mu_{i_1\cdots i_{k-1}}[W] - (\tilde{\mathrm{D}}_{i_k}\xi^j[W])W^\mu_{i_1\cdots i_{k-1}j}, \quad \mu = 1,\ldots,m,$$

where $i_l = 1,\ldots,n$ for $l = 1,\ldots,k,\ k \geq 2$.

In terms of the infinitesimal generator (1.313), one has

$$x^i = e^{\varepsilon\tilde{\mathrm{X}}}z^i, \quad U^\mu = e^{\varepsilon\tilde{\mathrm{X}}}W^\mu, \quad U^\mu_i = e^{\varepsilon\tilde{\mathrm{X}}}W^\mu_i, \ldots . \tag{1.314}$$

Let

$$\mathrm{X} = \xi^j[U]\frac{\partial}{\partial x^j} + \eta^\mu[U]\frac{\partial}{\partial U^\mu} + \cdots + \eta^\mu_{i_1\cdots i_k}[U]\frac{\partial}{\partial U^\mu_{i_1\cdots i_k}} + \cdots . \tag{1.315}$$

Now suppose the one-parameter Lie group of point transformations (1.312) is a point symmetry of PDE system $\mathbf{R}\{x\,;u\}$ (1.292). Then

$$\mathrm{X}R^\sigma[U] = a^\sigma_\tau[U]R^\tau[U] \tag{1.316}$$

for some specific functions $a^\sigma_\tau[U]$. Consequently, it is easy to see that

$$\mathrm{X}^k R^\sigma[U] = a^\sigma_\tau[U]_k\, R^\tau[U] \tag{1.317}$$

where $a^\sigma_\tau[U]_1 = a^\sigma_\tau[U]$ and, for $k \geq 1$, $a^\sigma_\tau[U]_{k+1} = \mathrm{X}a^\sigma_\tau[U]_k + a^\lambda_\tau[U]_k a^\sigma_\lambda[U]$.

The Jacobian determinant of the point transformations (1.312) is given by

$$\mathrm{J}[U;\varepsilon] = \begin{vmatrix} \mathrm{D}_1(e^{\varepsilon\mathrm{X}}x^1) & \cdots & \mathrm{D}_1(e^{\varepsilon\mathrm{X}}x^n) \\ \vdots & & \vdots \\ \mathrm{D}_n(e^{\varepsilon\mathrm{X}}x^1) & \cdots & \mathrm{D}_n(e^{\varepsilon\mathrm{X}}x^n) \end{vmatrix}. \tag{1.318}$$

Now let

$$\Psi^1[U;\varepsilon] = \begin{vmatrix} e^{\varepsilon\mathrm{X}}\Phi^1[U] & \mathrm{D}_2(e^{\varepsilon\mathrm{X}}x^1) & \cdots & \mathrm{D}_n(e^{\varepsilon\mathrm{X}}x^1) \\ e^{\varepsilon\mathrm{X}}\Phi^2[U] & \mathrm{D}_2(e^{\varepsilon\mathrm{X}}x^2) & \cdots & \mathrm{D}_n(e^{\varepsilon\mathrm{X}}x^2) \\ \vdots & \vdots & & \vdots \\ e^{\varepsilon\mathrm{X}}\Phi^n[U] & \mathrm{D}_2(e^{\varepsilon\mathrm{X}}x^n) & \cdots & \mathrm{D}_n(e^{\varepsilon\mathrm{X}}x^n) \end{vmatrix},$$

$$\Psi^2[U;\varepsilon] = \begin{vmatrix} \mathrm{D}_1(e^{\varepsilon \mathrm{X}}x^1) & e^{\varepsilon \mathrm{X}}\Phi^1[U] & \cdots & \mathrm{D}_n(e^{\varepsilon \mathrm{X}}x^1) \\ \mathrm{D}_1(e^{\varepsilon \mathrm{X}}x^2) & e^{\varepsilon \mathrm{X}}\Phi^2[U] & \cdots & \mathrm{D}_n(e^{\varepsilon \mathrm{X}}x^2) \\ \vdots & \vdots & & \vdots \\ \mathrm{D}_1(e^{\varepsilon \mathrm{X}}x^n) & e^{\varepsilon \mathrm{X}}\Phi^n[U] & \cdots & \mathrm{D}_n(e^{\varepsilon \mathrm{X}}x^n) \end{vmatrix},$$

$$\vdots$$

$$\Psi^n[U;\varepsilon] = \begin{vmatrix} \mathrm{D}_1(e^{\varepsilon \mathrm{X}}x^1) & \cdots & \mathrm{D}_{n-1}(e^{\varepsilon \mathrm{X}}x^1) & e^{\varepsilon \mathrm{X}}\Phi^1[U] \\ \mathrm{D}_1(e^{\varepsilon \mathrm{X}}x^2) & \cdots & \mathrm{D}_{n-1}(e^{\varepsilon \mathrm{X}}x^2) & e^{\varepsilon \mathrm{X}}\Phi^2[U] \\ \vdots & & \vdots & \vdots \\ \mathrm{D}_1(e^{\varepsilon \mathrm{X}}x^n) & \cdots & \mathrm{D}_{n-1}(e^{\varepsilon \mathrm{X}}x^n) & e^{\varepsilon \mathrm{X}}\Phi^n[U] \end{vmatrix}.$$

This leads to the following theorem.

Theorem 1.5.3. *Suppose the one-parameter Lie group of point transformations* (1.312) *is a point symmetry of PDE system* $\mathbf{R}\{x\,;u\}$ (1.292) *and* $\{\Lambda_\sigma[U]\}_{\sigma=1}^N$ *is a set of multipliers for a conservation law of* $\mathbf{R}\{x\,;u\}$ *with fluxes* $\{\Phi^j[u]\}_{j=1}^n$. *Let the determinants* $\mathrm{J}[U;\varepsilon]$ *and* $\Psi^i[U;\varepsilon]$ *be defined by expressions* (1.318) *and* (1.319), *respectively. Then*

$$\hat{\Lambda}_\tau[U]_p = \sum_{n+k+l=p} \frac{1}{k!l!n!}(a_\tau^\sigma[U]_l)(\mathrm{X}^k\Lambda_\sigma[U])\left.\frac{d^n}{d\varepsilon^n}\mathrm{J}[U;\varepsilon]\right|_{\varepsilon=0}, \tag{1.319}$$

$$\tau = 1,\dots,N,$$

defines sets of multipliers for conservation laws of PDE system $\mathbf{R}\{x\,;u\}$ (1.292) *with fluxes given by*

$$\Psi^i[u]_p = \left.\frac{1}{p!}\frac{d^p}{d\varepsilon^p}\Psi^i[u;\varepsilon]\right|_{\varepsilon=0}, \quad i=1,\dots,n, \tag{1.320}$$

for $p=0,1,2,\dots$; $\hat{\Lambda}_\tau[U]_o = \Lambda_\tau[U]$, $\Psi^i[u]_0 = \Phi^i[u]$, $a_\tau^\sigma[U]_0 = \delta_\tau^\sigma$.

Proof. Since $\{\Lambda_\sigma[U]\}_{\sigma=1}^N$ is a set of multipliers for a conservation law of PDE system $\mathbf{R}\{x\,;u\}$ (1.292) with fluxes $\{\Phi^j[u]\}_{j=1}^n$, the identity (1.310) holds for arbitrary functions $U(x)$. Under the one-parameter Lie group of point transformations (1.312), the relation (1.297) holds for arbitrary functions $U(x)$ and $W(z)$ related through (1.312). Substituting (1.310) into (1.297), one obtains

$$\mathrm{J}[W;\varepsilon]\Lambda_\sigma[U]R^\sigma[U] = \tilde{\mathrm{D}}_i\Psi^i[W;\varepsilon], \tag{1.321}$$

where $\mathrm{J}[W;\varepsilon]$ and $\Psi^i[W;\varepsilon]$ are given by (1.318) and (1.319), respectively, with D_i replaced by $\tilde{\mathrm{D}}_i$, X by $\tilde{\mathrm{X}}$, x^i by $\tilde{x}^i$, $U^\mu(x)$ by $W^\mu(z)$, U_i^μ by W_i^μ, etc. After substituting (1.314) into (1.321), one obtains

$$\mathrm{J}[W;\varepsilon]\Lambda_\sigma[e^{\varepsilon\tilde{\mathrm{X}}}W]R^\sigma[e^{\varepsilon\tilde{\mathrm{X}}}W] = \tilde{\mathrm{D}}_i\Psi^i[W;\varepsilon]. \tag{1.322}$$

After replacing $\tilde{\mathrm{D}}_i$ by D_i, $\tilde{\mathrm{X}}$ by X, $\tilde{x}^i$ by x^i, $W^\mu(z)$ by $U^\mu(x)$, W_i^μ by U_i^μ, etc. in (1.322), and using the group property $F[e^{\varepsilon \mathrm{X}}U] = e^{\varepsilon \mathrm{X}}F[U]$ for any function $F[U]$, one obtains

$$\mathrm{J}[U;\varepsilon](e^{\varepsilon \mathrm{X}}\Lambda_\sigma[U])(e^{\varepsilon \mathrm{X}}R^\sigma[U]) = \mathrm{D}_i\Psi^i[U;\varepsilon], \tag{1.323}$$

where $\mathrm{J}[U;\varepsilon]$ and $\Psi^i[U;\varepsilon]$ are given by (1.318) and (1.319), respectively. Each term in (1.323) can be expressed by a power series in ε. In particular,

$$\mathrm{J}[U;\varepsilon] = \sum_{n=0}^{\infty}\left(\frac{1}{n!}\left(\frac{d^n}{d\varepsilon^n}\mathrm{J}[U;\varepsilon]\bigg|_{\varepsilon=0}\right)\right)\varepsilon^n, \tag{1.324}$$

$$e^{\varepsilon \mathrm{X}}\Lambda_\sigma[U] = \sum_{k=0}^{\infty}\left(\frac{1}{k!}\mathrm{X}^k\Lambda_\sigma[U]\right)\varepsilon^k, \tag{1.325}$$

$$e^{\varepsilon \mathrm{X}}R^\sigma[U] = \sum_{l=0}^{\infty}\left(\frac{1}{l!}\mathrm{X}^l R^\sigma[U]\right)\varepsilon^l, \tag{1.326}$$

$$\Psi^i[U;\varepsilon] = \sum_{p=0}^{\infty}\left(\frac{1}{p!}\left(\frac{d^p}{d\varepsilon^p}\Psi^i[U;\varepsilon]\bigg|_{\varepsilon=0}\right)\right)\varepsilon^p. \tag{1.327}$$

Then using (1.316) and (1.317), one sees that (1.326) becomes

$$\sum_{l=0}^{\infty}\left(\frac{1}{l!}\mathrm{X}^l R^\sigma[U]\right)\varepsilon^l = \sum_{l=0}^{\infty}\left(\frac{1}{l!}\mathrm{X}^l a_\tau^\sigma[U]_l\right)\varepsilon^l R^\tau[U]. \tag{1.328}$$

After substituting expressions (1.324)–(1.327) into (1.323), one obtains the power series identity

$$\begin{aligned}&\sum_{p=0}^{\infty}\varepsilon^p\left(\sum_{n+k+l=p}\frac{1}{k!l!n!}(a_\tau^\sigma[U]_l)(\mathrm{X}^k\Lambda_\sigma[U])\,\frac{d^n}{d\varepsilon^n}\mathrm{J}[U;\varepsilon]\bigg|_{\varepsilon=0}\right)R^\tau[U]\\ &\quad=\sum_{p=0}^{\infty}\varepsilon^p\mathrm{D}_i\left(\frac{1}{p!}\left(\frac{d^p}{d\varepsilon^p}\Psi^i[U;\varepsilon]\right)\bigg|_{\varepsilon=0}\right)\end{aligned} \tag{1.329}$$

that holds for arbitrary functions $U(x)$. Consequently, after comparing coefficients in the power series representation (1.329), one obtains the sequence of identities

$$\hat{\Lambda}_\tau[U]_p R^\tau[U] = \mathrm{D}_i\left(\frac{1}{p!}\left(\frac{d^p}{d\varepsilon^p}\Psi^i[U;\varepsilon]\right)\bigg|_{\varepsilon=0}\right), \tag{1.330}$$

where $\hat{\Lambda}_\tau[U]_p$ is given by (1.319) for each power $p = 0, 1, 2, \ldots$. Thus, for the PDE system $\mathbf{R}\{x\,;u\}$ (1.292), from a given conservation law $\mathrm{D}_i\Phi^i[u] = 0$

arising from a set of multipliers $\{\Lambda_\sigma[U]\}$, and a point symmetry (1.312), one obtains the sequence of conservation laws

$$\mathrm{D}_i\left(\frac{1}{p!}\left(\frac{d^p}{d\varepsilon^p}\Psi^i[u;\varepsilon]\right)\bigg|_{\varepsilon=0}\right)=0,\quad p=1,2,\ldots, \tag{1.331}$$

where $\{\Psi^i[u;\varepsilon]\}$ is defined by (1.319) for any solution $U(x)=u(x)$ of $\mathbf{R}\{x\,;u\}$. □

It is important to note that it can happen that none of the terms of a sequence (1.331) yields a conservation law that is different from the given conservation law $\mathrm{D}_i\Phi^i[u]=0$.

In the case of two independent variables $(x^1,x^2)=(t,x)$ and two dependent variables $(U^1,U^2)=(U,V)$, one has

$$\begin{aligned}
\mathrm{X}= {}&\tau(x,t,U,V)\frac{\partial}{\partial t}+\xi(x,t,U,V)\frac{\partial}{\partial x}\\
&+\eta(x,t,U,V)\frac{\partial}{\partial U}+\phi(x,t,U,V)\frac{\partial}{\partial V}+\cdots,\\
\mathrm{D}_1=\mathrm{D}_t={}&\frac{\partial}{\partial t}+U_t\frac{\partial}{\partial U}+V_t\frac{\partial}{\partial V}+\cdots,\\
\mathrm{D}_2=\mathrm{D}_x={}&\frac{\partial}{\partial x}+U_x\frac{\partial}{\partial U}+V_x\frac{\partial}{\partial V}+\cdots.
\end{aligned} \tag{1.332}$$

It is left as an exercise to show that here for $p=1$, one has

$$\hat{\Lambda}_\tau[U]_1=(\mathrm{D}_t\tau+\mathrm{D}_x\xi)\Lambda_\tau[U]+\mathrm{X}\Lambda_\tau[U]+a_\tau^\sigma[U]\Lambda_\sigma[U], \tag{1.333}$$

$$\begin{aligned}
\Psi^1[U]_1&=\mathrm{X}\Phi^1[U]+\Phi^1[U]\mathrm{D}_x\xi-\Phi^2[U]\mathrm{D}_x\tau,\\
\Psi^2[U]_1&=\mathrm{X}\Phi^2[U]+\Phi^2[U]\mathrm{D}_t\tau-\Phi^1[U]\mathrm{D}_t\xi,
\end{aligned} \tag{1.334}$$

and for $p=2$, one has

$$\begin{aligned}
\hat{\Lambda}_\tau[U]_2={}&\tfrac{1}{2}\{[\mathrm{D}_t\mathrm{X}\tau+\mathrm{D}_x\mathrm{X}\xi+2(\mathrm{D}_t\tau\mathrm{D}_x\xi-\mathrm{D}_t\xi\mathrm{D}_x\tau)]\Lambda_\tau[U]\\
&+2(\mathrm{D}_t\tau+\mathrm{D}_x\xi)[\mathrm{X}\Lambda_\tau[U]+a_\tau^\sigma[U]\Lambda_\sigma[U]]+2a_\tau^\sigma[U]\mathrm{X}\Lambda_\sigma[U]\\
&+\mathrm{X}^2\Lambda_\tau[U]+\Lambda_\lambda[U](\mathrm{X}a_\tau^\lambda[U]+a_\sigma^\lambda[U]a_\tau^\sigma[U])\},\\
\Psi^1[U]_2={}&\tfrac{1}{2}\{\mathrm{X}^2\Phi^1[U]+2\mathrm{X}\Phi^1[U]\mathrm{D}_x\xi+\Phi^1[U]\mathrm{D}_x\mathrm{X}\xi\\
&-\Phi^2[U]\mathrm{D}_x\mathrm{X}\tau-2\mathrm{X}\Phi^2[U]\mathrm{D}_x\tau\},\\
\Psi^2[U]_2={}&\tfrac{1}{2}\{\mathrm{X}^2\Phi^2[U]+2\mathrm{X}\Phi^2[U]\mathrm{D}_t\tau+\Phi^2[U]\mathrm{D}_t\mathrm{X}\tau\\
&-\Phi^1[U]\mathrm{D}_t\mathrm{X}\xi-2\mathrm{X}\Phi^1[U]\mathrm{D}_t\xi\},
\end{aligned}$$

in terms of the set of multipliers $\{\Lambda_1[U], \Lambda_2[U]\}$ for a known conservation law with fluxes $\Phi^1[u]$ and $\Phi^2[u]$.

Examples

Now consider two examples involving nonlinear telegraph (NLT) systems that illustrate how one can obtain further conservation laws from symmetry action on a known conservation law. Examples include actions of discrete symmetries as well as point symmetries and, in the case of a point symmetry, it is shown that one can obtain more than one further conservation law. The examples involve two independent and two dependent variables. The following notation is used. For the two independent variables $x^1 = t$, $x^2 = x$ with $z^1 = \tilde{t}$, $z^2 = \tilde{x}$ and for the two dependent variables $u^1 = u$, $u^2 = v$ with $U^1 = U$, $U^2 = V$, $w^1 = \tilde{u}$, $w^2 = \tilde{v}$, $W^1 = \tilde{U}$, $W^2 = \tilde{V}$.

(1) First NLT system
One can show that the NLT system

$$\begin{aligned} R^1[u] &= v_t - (2e^{2u} - 1)u_x - e^u = 0, \\ R^2[u] &= u_t - v_x = 0, \end{aligned} \tag{1.335}$$

has a conservation law with fluxes given by

$$\begin{aligned} \Phi^1[u] &= -2e^{-\frac{1}{2}(u+t/\sqrt{2})} \cos\left[\tfrac{1}{2}\left(v + \tfrac{1}{\sqrt{2}}(x + 2e^u)\right)\right], \\ \Phi^2[u] &= 2e^{-\frac{1}{2}(u+t/\sqrt{2})}\left(\sqrt{2}e^u \cos\left[\tfrac{1}{2}\left(v + \tfrac{1}{\sqrt{2}}(x + 2e^u)\right)\right]\right. \\ &\quad \left. - \sin\left[\tfrac{1}{2}\left(v + \tfrac{1}{\sqrt{2}}(x + 2e^u)\right)\right]\right). \end{aligned} \tag{1.336}$$

By direction calculation, it is easy to see that the conservation law with fluxes (1.336) results from the set of multipliers

$$\begin{aligned} \Lambda_1[U] &= e^{-\frac{1}{2}(U+t/\sqrt{2})} \sin\left[\tfrac{1}{2}\left(V + \tfrac{1}{\sqrt{2}}(x + 2e^U)\right)\right], \\ \Lambda_2[U] &= e^{-\frac{1}{2}(U+t/\sqrt{2})}\left(\sqrt{2}e^U \sin\left[\tfrac{1}{2}\left(V + \tfrac{1}{\sqrt{2}}(x + 2e^U)\right)\right]\right. \\ &\quad \left. + \cos\left[\tfrac{1}{2}\left(V + \tfrac{1}{\sqrt{2}}(x + 2e^U)\right)\right]\right). \end{aligned} \tag{1.337}$$

Clearly, the NLT system (1.335) has the discrete symmetry (reflection)

$$t = -\tilde{t}, \quad x = \tilde{x}, \quad u = \tilde{u}, \quad v = -\tilde{v}, \tag{1.338}$$

and the point symmetry (translations)

$$t = \tilde{t}, \quad x = \tilde{x}, \quad u = \tilde{u}, \quad v = \tilde{v} + \varepsilon. \tag{1.339}$$

It is now shown that the symmetries (1.338) and (1.339) acting on the known conservation law with fluxes (1.336) that arises from the set of multipliers (1.337), lead to three further conservation laws of the NLT system (1.335). Through the second formula (1.308), it is shown that the actions of the reflection symmetry (1.338) and the translation symmetry (1.339), respectively, on this known conservation law, yield two further conservation laws. Furthermore, it is shown that further action of the reflection symmetry (1.338) on the conservation law obtained from the action of the translation symmetry (1.339) on the known conservation law, yields a fourth conservation law of the NLT system (1.335).

(I) Under the action of the point transformation corresponding to the reflection symmetry (1.338), one has $\mathrm{J}[W] = -1$, $S^1[W] = R^1[U] = R^1[W]$, $S^2[W] = R^2[U] = -R^2[W]$. Consequently, in (1.306), one obtains $A_1^1[W] = 1$, $A_2^2[W] = -1$, $A_1^2[W] = A_2^1[W] = 0$. After applying the point transformation corresponding to the reflection symmetry (1.338) to the set of multipliers (1.337), from formula (1.309) and Corollary 1.5.3 one gets a new set of multipliers

$$\begin{aligned}\hat{\Lambda}_1[U] &= e^{-\frac{1}{2}(U - t/\sqrt{2})} \sin\left[\tfrac{1}{2}\left(V - \tfrac{1}{\sqrt{2}}(x + 2e^U)\right)\right],\\ \hat{\Lambda}_2[U] &= e^{-\frac{1}{2}(U - t/\sqrt{2})}\left(\cos\left[\tfrac{1}{2}\left(V - \tfrac{1}{\sqrt{2}}(x + 2e^U)\right)\right]\right.\\ &\qquad \left. -\sqrt{2}e^U \sin\left[\tfrac{1}{2}\left(V - \tfrac{1}{\sqrt{2}}(x + 2e^U)\right)\right]\right).\end{aligned}$$

Then the first formula (1.299) leads to a second conservation law of the NLT system (1.335) with its fluxes given by

$$\begin{aligned}\Psi^1[u] &= -2e^{-\frac{1}{2}(u - t/\sqrt{2})} \cos\left[\tfrac{1}{2}\left(v - \tfrac{1}{\sqrt{2}}(x + 2e^u)\right)\right],\\ \Psi^2[u] &= -2e^{-\frac{1}{2}(u - t/\sqrt{2})}\left(\sqrt{2}e^u \cos\left[\tfrac{1}{2}\left(v - \tfrac{1}{\sqrt{2}}(x + 2e^u)\right)\right]\right.\\ &\qquad \left. + \sin\left[\tfrac{1}{2}\left(v - \tfrac{1}{\sqrt{2}}(x + 2e^u)\right)\right]\right).\end{aligned}$$

(II) Under the action of the point transformation corresponding to the translation symmetry (1.339), one has $\mathrm{J}[W] = 1$, $S^1[W] = R^1[U] = R^1[W]$, $S^2[W] = R^2[U] = R^2[W]$. Thus, in (1.306), one obtains $A_1^1[W] = A_2^2[W] = 1$, $A_1^2[W] = A_2^1[W] = 0$. After applying the point transformation corresponding to the translation symmetry (1.339) to the set of multipliers (1.337), from formula (1.309) and Corollary 1.5.3 one has a new set of multipliers

$$\begin{aligned}
\hat{\Lambda}_1[U] &= \lambda_1[U] = e^{-\frac{1}{2}(U+t/\sqrt{2})} \cos\left[\tfrac{1}{2}\left(V + \tfrac{1}{\sqrt{2}}(x + 2e^U)\right)\right],\\
\hat{\Lambda}_2[U] &= \lambda_2[U] = e^{-\frac{1}{2}(U+t/\sqrt{2})} \left(\sqrt{2}e^U \cos\left[\tfrac{1}{2}\left(V + \tfrac{1}{\sqrt{2}}(x + 2e^U)\right)\right]\right. \\
&\qquad \left. - \sin\left[\tfrac{1}{2}\left(V + \tfrac{1}{\sqrt{2}}(x + 2e^U)\right)\right]\right).
\end{aligned} \tag{1.340}$$

Hence, formula (1.299) leads to a third conservation law of the NLT system (1.335) with fluxes

$$\begin{aligned}
\Psi^1[u] &= \phi^1[u] = 2e^{-\frac{1}{2}(u+t/\sqrt{2})} \sin\left[\tfrac{1}{2}\left(v + \tfrac{1}{\sqrt{2}}(x + 2e^u)\right)\right],\\
\Psi^2[u] &= \phi^2[u] = -2e^{-\frac{1}{2}(u+t/\sqrt{2})} \left(\sqrt{2}e^u \sin\left[\tfrac{1}{2}\left(v + \tfrac{1}{\sqrt{2}}(x + 2e^u)\right)\right]\right. \\
&\qquad \left. + \cos\left[\tfrac{1}{2}\left(v + \tfrac{1}{\sqrt{2}}(x + 2e^u)\right)\right]\right).
\end{aligned} \tag{1.341}$$

(III) Under the action of the point transformation, corresponding to the reflection symmetry (1.338), to the set of multipliers (1.340) that leads to the third conservation law with fluxes (1.341), formula (1.309) and Corollary 1.5.3 yield a third new set of multipliers given by

$$\begin{aligned}
\hat{\Lambda}_1[U] &= \hat{\lambda}_1[U] = -e^{-\frac{1}{2}(U-t/\sqrt{2})} \cos\left[\tfrac{1}{2}\left(V - \tfrac{1}{\sqrt{2}}(x + 2e^U)\right)\right],\\
\hat{\Lambda}_2[U] &= \hat{\lambda}_2[U] = e^{-\frac{1}{2}(U-t/\sqrt{2})} \left(\sqrt{2}e^U \cos\left[\tfrac{1}{2}\left(V - \tfrac{1}{\sqrt{2}}(x + 2e^U)\right)\right]\right. \\
&\qquad \left. + \sin\left[\tfrac{1}{2}\left(V - \tfrac{1}{\sqrt{2}}(x + 2e^U)\right)\right]\right).
\end{aligned}$$

Thus formula (1.299) leads to a fourth conservation law of the NLT system (1.335) with fluxes

$$\begin{aligned}
\Psi^1[u] &= -2e^{-\frac{1}{2}(u-t/\sqrt{2})} \sin\left[\tfrac{1}{2}\left(v - \tfrac{1}{\sqrt{2}}(x + 2e^u)\right)\right],\\
\Psi^2[u] &= 2e^{-\frac{1}{2}(u-t/\sqrt{2})} \left(\cos\left[\tfrac{1}{2}\left(v - \tfrac{1}{\sqrt{2}}(x + 2e^u)\right)\right]\right. \\
&\qquad \left. - \sqrt{2}e^u \sin\left[\tfrac{1}{2}\left(v - \tfrac{1}{\sqrt{2}}(x + 2e^u)\right)\right]\right).
\end{aligned}$$

(2) Second NLT system

One can show that the NLT system

$$\begin{aligned}
R^1[u] &= v_t - (\operatorname{sech}^2 u)u_x - \tanh u = 0,\\
R^2[u] &= u_t - v_x = 0,
\end{aligned} \tag{1.342}$$

has a conservation law with fluxes given by

$$\begin{aligned}\Phi^1[u] &= e^x[2tu - \tfrac{1}{3}v^3 + v(t^2 + 2x - 2\log(\cosh u))],\\ \Phi^2[u] &= e^x[(v^2 - t^2 - 2x + 2(1 + \log(\cosh u)))\tanh u - 2(vt + u)].\end{aligned} \tag{1.343}$$

By direction calculation, it is easy to see that the conservation law with fluxes (1.343) results from the set of multipliers

$$\begin{aligned}\Lambda_1[U] &= e^x[2x + t^2 - V^2 - 2\log(\cosh U)],\\ \Lambda_2[U] &= 2e^x[t - V\tanh U].\end{aligned} \tag{1.344}$$

The NLT system (1.342) has the point symmetry (translations)

$$t = \tilde{t} + \varepsilon, \quad x = \tilde{x}, \quad u = \tilde{u}, \quad v = \tilde{v}, \tag{1.345}$$

and the point symmetry [Bluman, Temuerchaolu & Sahadevan (2005)] with infinitesimal generator

$$\mathrm{X} = v\frac{\partial}{\partial t} + \tanh u\frac{\partial}{\partial x} + \frac{\partial}{\partial u} + t\frac{\partial}{\partial v}. \tag{1.346}$$

It is now shown that the actions of the symmetries (1.345) and (1.346) on the known conservation law with fluxes (1.343) that arises from the set of multipliers (1.344), lead to three further conservation laws of the NLT system (1.342). Through the second formula (1.308), it is shown that the action of the translation symmetry (1.345) on this known conservation law yields two additional conservation laws resulting from the $O(\varepsilon)$ and $O(\varepsilon^2)$ terms, respectively. Moreover, it is shown that the action of the point symmetry (1.346) on the additional conservation law, resulting from the $O(\varepsilon)$ term, yields a fourth conservation law of the NLT system (1.342).

(I) Under the action of the point transformation corresponding to the translation symmetry (1.345), one has $\mathrm{J}[W] = 1$, $S^1[W] = R^1[U] = R^1[W]$, $S^2[W] = R^2[U] = R^2[W]$. Thus, in (1.306), one obtains $A_1^1[W] = A_2^2[W] = 1$, $A_1^2[W] = A_2^1[W] = 0$. After applying the point transformation corresponding to the translation symmetry (1.345) to the set of multipliers (1.344), from formula (1.309) and Corollary 1.5.3 one obtains

$$\begin{aligned}\hat{\Lambda}_1[W] &= \Lambda_1[W] + 2\tilde{t}e^{\tilde{x}}\varepsilon + e^{\tilde{x}}\varepsilon^2,\\ \hat{\Lambda}_2[W] &= \Lambda_2[W] + 2e^{\tilde{x}}\varepsilon.\end{aligned} \tag{1.347}$$

Then the first formula (1.299) leads to

$$\begin{aligned} \Psi^1[W] &= \Phi^1[W] + 2e^{\tilde{x}}(\tilde{t}\tilde{V} + \tilde{U})\varepsilon + e^{\tilde{x}}\tilde{V}\varepsilon^2, \\ \Psi^2[W] &= \Phi^2[W] - 2e^{\tilde{x}}(\tilde{t}\tanh\tilde{U} + \tilde{V})\varepsilon - e^{\tilde{x}}(\tanh\tilde{U})\varepsilon^2. \end{aligned} \tag{1.348}$$

Comparing (1.344) and (1.347), one immediately sees that the $O(\varepsilon)$ and $O(\varepsilon^2)$ terms in (1.347) yield two additional conservation laws of the NLT system (1.342). In particular, the $O(\varepsilon)$ terms in (1.347) and (1.348), respectively, yield a new set of multipliers

$$\hat{\Lambda}_1[U]_1 = \lambda_1[U] = te^x, \quad \hat{\Lambda}_2[U]_1 = \lambda_2[U] = e^x, \tag{1.349}$$

and a corresponding conservation law with fluxes

$$\Psi^1[u]_1 = \phi_1[u] = e^x(tv + u), \quad \Psi^2[u]_1 = \phi_2[u] = -e^x(t\tanh u + v); \tag{1.350}$$

the $O(\varepsilon^2)$ terms in (1.347) and (1.348), respectively, yield a new set of multipliers

$$\hat{\Lambda}_1[U]_2 = e^x, \quad \hat{\Lambda}_2[U]_2 = 0,$$

and a corresponding conservation law with fluxes

$$\Psi^1[u]_2 = e^x v, \quad \Psi^2[u]_2 = -e^x\tanh u.$$

(II) Under the action of the point transformation corresponding to the point symmetry (1.346), one has

$$\begin{aligned} \mathrm{X}R^1[U] &= -[V_t + (\operatorname{sech}^2 U)U_x + \tanh U]R^1[U], \\ sgXR^2[U] &= -[V_t + (\operatorname{sech}^2 U)U_x]R^2[U]. \end{aligned}$$

Thus in (1.316), one has $a_1^1[U] = -[V_t + (\operatorname{sech}^2 U)U_x + \tanh U]$, $a_2^2[U] = -[V_t + (\operatorname{sech}^2 U)U_x]$, $a_2^1[U] = a_1^2[U] = 0$. The point symmetry (1.346) is now applied to the $O(\varepsilon)$ term conservation law with fluxes (1.350). Then in formula (1.333) with $\Lambda_1[U] = \lambda_1[U]$ and $\Lambda_2[U] = \lambda_2[U]$ given by (1.349), one has $\tau = V$, $\xi = \tanh U$, $\mathrm{X}\lambda_1[U] = e^x(V + \tanh U)$, and $\mathrm{X}\lambda_2[U] = e^x\tanh U$. Consequently, here formula (1.333) yields a third set of multipliers given by

$$\hat{\lambda}_1[U]_1 = e^x V, \quad \hat{\lambda}_2[U]_1 = e^x\tanh U.$$

Hence formula (1.334) with $\Phi^1[U] = \phi^1[U]$ and $\Phi^2[U] = \phi^2[U]$ given by (1.350), yields a fourth conservation law of the NLT system (1.342) with fluxes

$$\phi^1[u]_1 = e^x[\tfrac{1}{2}v^2 + \log(\cosh u)], \quad \phi^2[u]_1 = -e^x v\tanh u.$$

1.5.2 Relationships among symmetries, solutions of adjoint equations, and conservation laws

Consider a system of PDEs (1.292). The linearizing operator $\mathrm{L}[U]$ associated with the PDE system (1.292) is given by (1.176), and the corresponding adjoint operator $\mathrm{L}^*[U]$ is given by (1.177). From integration by parts, it follows that for arbitrary functions $V(x) = (V^1(x), \ldots, V^m(x))$ and $W(x) = (W_1(x), \ldots, W_N(x))$, one has the relation [Exercise 1.3.16]

$$W_\sigma \mathrm{L}_\rho^\sigma[U]V^\rho - V^\rho \mathrm{L}_\rho^{*\,\sigma}[U]W_\sigma \equiv \mathrm{D}_i \Psi^i[U], \tag{1.351}$$

where $\Psi^i[U]$ is given by formula (1.179).

Suppose also that the PDE system $\mathbf{R}\{x\,;u\}$ (1.292) has a local symmetry with infinitesimal generator, in characteristic form,

$$\hat{\mathrm{X}} = \hat{\eta}^\rho[u]\frac{\partial}{\partial u^\rho}. \tag{1.352}$$

It follows that the symmetry components $\hat{\eta}^\rho[u]$ are solutions of the symmetry determining equations, i.e., the linearizing system

$$\mathrm{L}_\rho^\sigma[u]\hat{\eta}^\rho[u] = 0. \tag{1.353}$$

Now let $\{\omega_\sigma[u]\}_{\sigma=1}^N$ be some solution of the adjoint linearizing system

$$\mathrm{L}_\rho^{*\,\sigma}[u]\omega_\sigma[u] = 0 \tag{1.354}$$

for any solution $u(x)$ of $\mathbf{R}\{x\,;u\}$ (1.292). In the literature [Gordon (1986); Sarlet, Cantrijn & Crampin (1987); Sarlet, Prince & Crampin (1990); Anco & Bluman [(1998), (2002a,b)]; Bluman & Anco (2002); Anco & Pohjanpelto [(2001), (2003), (2004)]; Anco & The (2005)], such solutions are often called *adjoint symmetries*.

In the divergence identity (1.351), let $V^\rho = \hat{\eta}^\rho[U]$ and $W_\sigma = \omega_\sigma[U]$. It follows that one has the conservation law

$$\mathrm{D}_i \Psi^i[u] = 0 \tag{1.355}$$

on solutions $U(x) = u(x)$ of the PDE system $\mathbf{R}\{x\,;u\}$ (1.292).

Thus the following theorem holds [Anco & Bluman (1997a)].

Theorem 1.5.4. *For a PDE system* (1.292), *any pair consisting of a symmetry* (1.352) *and a solution* $\{\omega_\sigma[u]\}_{\sigma=1}^N$ *of the corresponding adjoint linearizing system* (1.354) *yields a conservation law* (1.355), *with fluxes* $\Psi^i[u]$ *given by* (1.179).

As an example, consider the NLT system (1.281) for power nonlinearities $G(u) = u^{\alpha+1}$, $F(u) = G'(u)$, $\alpha \neq 0, -1, -2$, i.e.,

$$u_t = v_x, \quad v_t = (\alpha+1)u^\alpha u_x + u^{\alpha+1}.$$

One can show that this system, for a general power α, has four point symmetries given by

$$\mathrm{X}_1 = \frac{\partial}{\partial t}, \quad \mathrm{X}_2 = \frac{\partial}{\partial x}, \quad \mathrm{X}_3 = \frac{\partial}{\partial v}, \quad \mathrm{X}_4 = -\tfrac{1}{2}\alpha t u\frac{\partial}{\partial t} + u\frac{\partial}{\partial u} + \tfrac{1}{2}(\alpha+2)v\frac{\partial}{\partial v}.$$

The infinitesimal generators in characteristic form are given by

$$\hat{\mathrm{X}}_j = \hat{\eta}_j^1[u,v]\frac{\partial}{\partial u} + \hat{\eta}_j^2[u,v]\frac{\partial}{\partial v}, \quad j = 1, \ldots, 4,$$

with components

$$\begin{aligned} &(\hat{\eta}_1^1, \hat{\eta}_1^2) = (-u_t, -v_t), \quad (\hat{\eta}_2^1, \hat{\eta}_2^2) = (-u_x, -v_x), \\ &(\hat{\eta}_3^1, \hat{\eta}_3^2) = (0, 1), \quad (\hat{\eta}_4^1, \hat{\eta}_4^2) = (u + \tfrac{1}{2}\alpha t u_t, \tfrac{1}{2}(\alpha+2)v + \tfrac{1}{2}\alpha t v_t). \end{aligned} \tag{1.356}$$

The adjoint linearizing system (1.283) also has four solutions $(\hat{w}_1^j[u,v], \hat{w}_2^j[u,v])$, $j = 1, \ldots, 4$, given by

$$\begin{aligned} &(\hat{w}_1^1, \hat{w}_2^1) = (1, 0), \quad (\hat{w}_1^2, \hat{w}_2^2) = (0, e^x), \\ &(\hat{w}_1^3, \hat{w}_2^3) = (e^x, te^x), \quad (\hat{w}_1^4, \hat{w}_2^4) = (e^x u^{\alpha+1}, e^x v). \end{aligned} \tag{1.357}$$

Now Theorem 1.5.4 is used to generate corresponding conservation laws of the NLT system (1.281). In particular, the simplified version (1.378) of formula (1.179) is used to obtain the fluxes of conservation laws

$$\mathrm{D}_t\Psi[u,v] + \mathrm{D}_x\Phi[u,v] = 0. \tag{1.358}$$

One can show that using the symmetry $(\hat{\eta}_4^1, \hat{\eta}_4^2)$ and the solution $(\hat{w}_1^3, \hat{w}_2^3)$ of the adjoint system (1.283), one obtains the fluxes

$$\begin{aligned} \Psi[u,v] &= e^x\left[u + \tfrac{1}{2}\alpha t u_t + \tfrac{1}{2}t\left((\alpha+2)v + \alpha t v_t\right)\right] \\ &= e^x\left[u + tv\right] + D_x\left(\tfrac{1}{2}\alpha e^x t[v + tu^{\alpha+1}]\right), \end{aligned}$$

$$\begin{aligned} \Phi[u,v] &= -e^x\left[(\alpha+1)tu^\alpha\left(u + \tfrac{1}{2}\alpha t u_t\right) + \tfrac{1}{2}\left((\alpha+2)v + \alpha t v_t\right)\right] \\ &= -e^x[v + tu^{\alpha+1}] - D_t\left(\tfrac{1}{2}\alpha e^x t[v + tu^{\alpha+1}]\right), \end{aligned}$$

corresponding to a non-obvious conservation law (1.358) of the NLT system (1.281).

Using the same symmetry $(\hat{\eta}_4^1, \hat{\eta}_4^2)$ and another solution $(\hat{w}_1^2, \hat{w}_2^2)$ of the adjoint system (1.283), one obtains the conservation law with the fluxes

$$\Psi[u,v] = u + \tfrac{1}{2}\alpha u_t = u + \mathrm{D}_x(\tfrac{1}{2}\alpha t v),$$

$$\Phi[u,v] = -\tfrac{1}{2}\left[(\alpha+2)v + \alpha t v_t\right] = -v - \mathrm{D}_t(\tfrac{1}{2}\alpha t v),$$

which is equivalent to the obvious known conservation law $u_t - v_x = 0$ (the first equation of the NLT system (1.281)).

The conservation law obtained from Theorem 1.5.4 can be trivial. For example, for the symmetry $(\hat{\eta}_1^1, \hat{\eta}_1^2)$ and the solution $(\hat{w}_1^4, \hat{w}_2^4)$ of the adjoint system, one obtains fluxes

$$\Psi[u,v] = -e^x\left[u^{\alpha+1}u_t + v v_t\right] = \mathrm{D}_x(e^x u^{\alpha+1} v),$$

$$\Phi[u,v] = e^x u^\alpha\left[(\alpha+1)v u_t + u v_t\right] = \mathrm{D}_t(e^x u^{\alpha+1} v).$$

Theorem 1.5.4 and formula (1.179) are not commonly used for the computation of fluxes of conservation laws, since the resulting conservation laws are sometimes trivial or already known; the resulting fluxes are often unnecessarily complicated; and there is the lack of completeness in knowing whether one has obtained all local conservation laws for sets of multipliers of fixed form.

However, as shown in the next section, formula (1.179) can be rather useful to generate fluxes of conservation laws for complicated forms of multipliers and/or PDE systems, when other methods fail.

Computation of fluxes of a conservation law of scaling-invariant PDE systems

Suppose the PDE system $\mathbf{R}\{x\,;u\}$ (1.292) has a conservation law

$$\Lambda_\sigma[u]R^\sigma[u] = \mathrm{D}_i\Phi^i[u] = 0, \tag{1.359}$$

with a set of multipliers $\{\Lambda_\sigma[U] = \Lambda_\sigma(x, U, \partial U, \ldots, \partial^l U)\}_{\sigma=1}^N$. From Corollary 1.3.1, it follows that the multipliers $\{\Lambda_\sigma[U]\}_{\sigma=1}^N$ are solutions of the adjoint system

$$\mathrm{L}_\rho^{*\,\sigma}[u]\Lambda_\sigma[u] = 0 \tag{1.360}$$

for every solution $U(x) = u(x)$ of PDE system $\mathbf{R}\{x\,;u\}$. Hence a set of multipliers $\{\Lambda_\sigma[U]\}_{\sigma=1}^N$ can be used in conjunction with components of local symmetries in Theorem 1.5.4 to obtain a conservation law (1.355).

It is important to note that in general, for any local symmetry (1.352) used in Theorem 1.5.4, the conservation law (1.359) corresponding to the set of multipliers $\{\Lambda_\sigma[U]\}_{\sigma=1}^N$ and the conservation law (1.355) arising from Theorem 1.5.4 are *different*, i.e., the conservation law arising from Theorem 1.5.4 has a different set of multipliers.

However if the symmetry (1.352) is a *scaling symmetry*

$$\begin{aligned} &\mathrm{X}_s[u] = p^{(i)} x^i \frac{\partial}{\partial x^i} + q^{(\rho)} u^\rho \frac{\partial}{\partial u^\rho} \\ &p^{(i)} = \text{const}, \quad i = 1, \dots, n, \quad q^{(\rho)} = \text{const}, \quad \rho = 1, \dots, m, \end{aligned} \tag{1.361}$$

i.e.,

$$x^i \to \tilde{x}^i = e^{\varepsilon \mathrm{X}_s[u]} x^i = e^{\varepsilon p^{(i)}} x^i,$$

$$u^\rho \to \tilde{u}^\rho = e^{\varepsilon \mathrm{X}_s[u]} u^\rho = e^{\varepsilon q^{(\rho)}} u^\rho,$$

then it often happens that conservation laws (1.355) and (1.359) coincide (to within fluxes of a trivial conservation law). In such cases, for complicated forms of PDEs $\mathbf{R}\{x\,;u\}$ and sets of multipliers $\Lambda_\sigma[U]$, formula (1.179) can be the most efficient way to compute fluxes as follows.

Lemma 1.5.1. *Suppose the PDE system* $\mathbf{R}\{x\,;u\}$ (1.292) *has a scaling symmetry* (1.361), *and a conservation law* (1.359). *Let* $r^{(\sigma)} = \text{const}$ *be the scaling weight of each PDE* $R^\sigma[U]$ *under the scaling symmetry* (1.361), *i.e.,* $\mathrm{X}_s^{(k)}[U]R^\sigma[U] = r^{(\sigma)}R^\sigma[U]$. *If the conservation law* (1.359) *is homogeneous under the scaling symmetry* (1.361), *i.e.,* $\mathrm{X}_s^{(l)}[U]\mathrm{D}_i\Phi^i[U] = P\mathrm{D}_i\Phi^i[U]$, $P =$ const, *then each of the multipliers* $\Lambda_\sigma[U]$ *appearing in* (1.359) *is homogeneous under the scaling symmetry* (1.361). *In particular,* $\mathrm{X}_s^{(l)}[U]\Lambda_\sigma[U] = s_{(\sigma)}\Lambda_\sigma[U]$, *where* $s_{(\sigma)} = P - r^{(\sigma)}$ *is the scaling weight of each* $\Lambda_\sigma[U] \neq 0$.

Proof. Assuming that the given PDE system $\mathbf{R}\{x\,;u\}$ (1.292) can be written in a solved form (1.152), one notes that the scaling homogeneity of each PDE of the given system $\mathbf{R}\{x\,;u\}$ follows from considering the action of the scaling symmetry (1.361) on the leading derivative in each $R^\sigma[U]$. Then applying the operator $\mathrm{X}_s^{(l)}[U]$ (1.361) to both sides of the equality $\Lambda_\sigma[U]R^\sigma[U] = \mathrm{D}_i\Phi^i[U]$, for an arbitrary function $U(x)$, yields

$$\left(\mathrm{X}_s^{(l)}[U]\Lambda_\sigma[U]\right) R^\sigma[U] + r^{(\sigma)}\Lambda_\sigma[U]R^\sigma[U] = P\mathrm{D}_i\Phi^i[U] = P\Lambda_\sigma[U]R^\sigma[U].$$

Hence

$$\left[\mathrm{X}_s^{(l)}[U]\Lambda_\sigma[U] - (P - r^{(\sigma)})\Lambda_\sigma[U]\right] R^\sigma[U] = 0.$$

Since without loss of generality, functions $\{R^\sigma[U]\}_{\sigma=1}^N$ can be assumed to be linearly independent, it follows that

$$\mathrm{X}_s^{(l)}[U]\Lambda_\sigma[U] - (P - r^{(\sigma)})\Lambda_\sigma[U] = 0$$

for each σ such that $\Lambda_\sigma[U] \not\equiv 0$. □

Definition 1.5.1. Suppose the conditions of Lemma 1.5.1 are satisfied. Then the conservation law (1.359) is called *noncritical with respect to the scaling symmetry* (1.361) if

$$\chi = s_{(\sigma)} + r^{(\sigma)} + \sum_{i=1}^{n} p^{(i)} \neq 0 \tag{1.362}$$

for each σ such that $\Lambda_\sigma[U] \neq 0$.

The scaling symmetry generator (1.361) in characteristic form is given by

$$\hat{\mathrm{X}}_s[u] = \hat{\eta}^\rho[u]\frac{\partial}{\partial u^\rho} = \left(q^{(\rho)}u^\rho - p^{(i)}x^i u_i^\rho\right)\frac{\partial}{\partial u^\rho}. \tag{1.363}$$

The following theorem holds [Anco (2003)].

Theorem 1.5.5. *Suppose the PDE system* $\mathbf{R}\{x\,;u\}$ (1.292) *has a scaling symmetry* (1.363), *and a conservation law* (1.359) *that is homogeneous under the scaling symmetry* (1.361), *i.e.,* $\mathrm{X}_s^{(l)}[U]\mathrm{D}_i\Phi^i[U] = P\mathrm{D}_i\Phi^i[U]$, $P =$ const. *For each* σ *such that* $\Lambda_\sigma[U] \neq 0$, *let* $r^{(\sigma)}$ *be the scaling weight of the corresponding function* $R^\sigma[U]$ *under the scaling symmetry* (1.361), *i.e.,* $\mathrm{X}_s^{(k)}[U]R^\sigma[U] = r^{(\sigma)}R^\sigma[U]$. *If the conservation law* (1.359) *is noncritical,* $\chi \neq 0$, *then it is equivalent to the conservation law* (1.355) *with fluxes* $\Psi^i[U]$ *given by* (1.179) *with* $V^\rho = \hat{\eta}^\rho[U]$ *and* $W_\sigma = \Lambda_\sigma[U]$. *In particular,*

$$\Psi^i[U] = (P - p^i)\Phi^i[U], \quad (P - p^i) = \text{const}, \tag{1.364}$$

holds modulo fluxes of a trivial conservation law.

Proof. From the homogeneity of $\Lambda_\sigma[U]$ and $R^\sigma[U]$ under the scaling symmetry [Lemma 1.5.1], it follows that for the corresponding extensions of $\hat{\mathrm{X}}_s[U]$, one has

$$\hat{\mathrm{X}}_s^{(l)}[U]\Lambda_\sigma[U] = s_{(\sigma)}\Lambda_\sigma[U] - p^{(i)}x^i\mathrm{D}_i\Lambda_\sigma[U], \tag{1.365}$$

$$\hat{\mathrm{X}}_s^{(k)}[U]R^\sigma[U] = r^{(\sigma)}R^\sigma[U] - p^{(i)}x^i\mathrm{D}_iR^\sigma[U]. \tag{1.366}$$

Let

$$(\mathrm{L}_R)^\sigma_\rho[U] \equiv \mathrm{L}^\sigma_\rho[U] \tag{1.367}$$

be the linearizing operator associated with the PDE system $\mathbf{R}\{x\,;u\}$ (1.292), with components given by (1.176). Let

$$(\mathrm{L}^*_R)^\sigma_\rho[U] \equiv (\mathrm{L}^*)^\sigma_\rho[U] \tag{1.368}$$

be the adjoint operator, with components given by (1.177). Now formally define the linear operators $(\mathrm{L}_\Lambda)_{\sigma\,\rho}[U]$ and $(\mathrm{L}^*_\Lambda)_{\sigma\,\rho}[U]$ by replacing $R^\sigma[U]$ by

$\Lambda_\sigma[U]$ in the operators (1.367) and (1.368). For these two pairs of linear operators, following from the identity (1.178), one obtains the relations

$$W_\sigma(\mathrm{L}_R)^\sigma_\rho[U]V^\rho - V^\rho(\mathrm{L}^*_R)^\sigma_\rho[U]W_\sigma \equiv \mathrm{D}_i\Psi^i[U],$$
$$\widetilde{W}^\sigma(\mathrm{L}_\Lambda)_{\sigma\,\rho}[U]\widetilde{V}^\rho - V^\rho(\mathrm{L}^*_\Lambda)_{\sigma\,\rho}[U]\widetilde{W}^\sigma \equiv \mathrm{D}_iH^i[U],$$

where $V(x)$, $W(x)$, $\widetilde{V}(x)$, and $\widetilde{W}(x)$ are arbitrary functions; the functions $\Psi^i[U]$ are given by (1.179), and the functions $H^i[U]$ are given by (1.179) with $R^\sigma[U]$ replaced by $\Lambda_\sigma[U]$.

Moreover, the necessary and sufficient conditions (1.183) for the set of multipliers $\{\Lambda_\sigma[U]\}_{\sigma=1}^N$ to yield a divergence expression can be written as a symmetric expression [Exercise 1.3.17]

$$(\mathrm{L}^*_R)^\sigma_\rho[U]\Lambda_\sigma[U] + (\mathrm{L}^*_\Lambda)_{\sigma\,\rho}R^\sigma[U] = 0. \tag{1.369}$$

Writing equations (1.369) and (1.369) with $V^\rho = \widetilde{V}^\rho = \hat{\eta}^\rho_s[U]$, $W_\sigma = \Lambda_\sigma[U]$, and $\widetilde{W}^\sigma = R^\sigma[U]$, and using (1.369), one obtains

$$\begin{aligned}\Lambda_\sigma[U](\mathrm{L}_R)^\sigma_\rho[U]\hat{\eta}^\rho_s[U] - \hat{\eta}^\rho_s[U](\mathrm{L}^*_R)^\sigma_\rho[U]\Lambda_\sigma[U] &\equiv \mathrm{D}_i\Psi^i[U],\\ R^\sigma[U](\mathrm{L}_\Lambda)_{\sigma\,\rho}[U]\hat{\eta}^\rho_s[U] + \hat{\eta}^\rho_s[U](\mathrm{L}^*_R)^\sigma_\rho[U]\Lambda_\sigma[U] &\equiv \mathrm{D}_iH^i[U].\end{aligned} \tag{1.370}$$

Hence

$$\Lambda_\sigma[U](\mathrm{L}_R)^\sigma_\rho[U]\hat{\eta}^\rho_s[U] + R^\sigma[U](\mathrm{L}_\Lambda)_{\sigma\,\rho}[U]\hat{\eta}^\rho_s[U] = \mathrm{D}_i(\Psi^i[U] + H^i[U]). \tag{1.371}$$

From the definition of a linearizing operator, it follows that

$$(\mathrm{L}_\Lambda)_{\sigma\,\rho}[U]\hat{\eta}^\rho_s[U] = \hat{\mathrm{X}}^{(k)}_s[U]\Lambda_\sigma[U],$$
$$(\mathrm{L}_R)^\sigma_\rho[U]\hat{\eta}^\rho_s[U] = \hat{\mathrm{X}}^{(k)}_s[U]R^\sigma[U].$$

Then using the scaling relations (1.365) and (1.366), one obtains the equations

$$(\mathrm{L}_\Lambda)_{\sigma\,\rho}[U]\hat{\eta}^\rho_s[U] = s_{(\sigma)}\Lambda_\sigma[U] - p^{(i)}x^i\mathrm{D}_i\Lambda_\sigma[U],$$
$$(\mathrm{L}_R)^\sigma_\rho[U]\hat{\eta}^\rho_s[U] = r^{(\sigma)}R^\sigma[U] - p^{(i)}x^i\mathrm{D}_iR^\sigma[U].$$

Therefore after an integration by parts, the expression (1.371) becomes

$$\begin{aligned}&\left(s_{(\sigma)} + r^{(\sigma)} + \textstyle\sum_{i=1}^n p^{(i)}\right) R^\sigma[U]\Lambda_\sigma[U]\\ &\quad = \mathrm{D}_i(\Psi^i[U] + H^i[U] + p^{(i)}x^iR^\sigma[U]\Lambda_\sigma[U]).\end{aligned} \tag{1.372}$$

From Lemma 1.5.1 it follows that the quantity $\chi = s_{(\sigma)} + r^{(\sigma)} + \sum_{i=1}^n p^{(i)} = P + \sum_{i=1}^n p^{(i)}$ is independent of σ. Since the conservation law (1.359) is non-critical, $\chi \neq 0$. Hence the left-hand side of (1.372) has the form $\chi\mathrm{D}_i\Phi^i[U]$.

On the right-hand side of (1.372), the flux parts $H^i[U]$ and $p^{(i)}x^iR^\sigma[U]\Lambda_\sigma[U]$ vanish on solutions $U(x) = u(x)$ of PDE system $\mathbf{R}\{x\,;u\}$ (1.292). [In particular, functions $H^i[U]$ vanish on solutions of the PDE system $\mathbf{R}\{x\,;u\}$, since they are given by (1.179) with $R^\sigma[U]$ replaced by $\Lambda_\sigma[U]$, and $\widetilde{W}^\sigma = R^\sigma[U]$, with $R^\sigma[U] = 0$ on solutions $U(x) = u(x)$ of $\mathbf{R}\{x\,;u\}$.] Hence the flux parts $H^i[U]$ and $p^{(i)}x^iR^\sigma[U]\Lambda_\sigma[U]$ correspond to fluxes of a trivial conservation law of the first kind [Section 1.3.2]. □

As an example, consider the two-dimensional nonlinear G-equation

$$R[g] = g_t - |\,\mathrm{grad}\, g| \equiv g_t - \sqrt{g_x^2 + g_y^2} = 0, \tag{1.373}$$

which describes flame propagation in a static gas [Oberlack, Wenzel & Peters (2001)]. Here $g(t,x,y) = 0$ implicitly defines the position of the flame surface at time t, and the surface advances at a constant speed in the normal direction. Let $(x^1, x^2) = (x, y)$.

Assuming the dependence $\Lambda[G] = \Lambda(t,x,y,G,G_x,G_y,G_{xx},G_{xy},G_{yy})$, one can show that the G-equation (1.373) has the following conservation law multipliers:

$$\begin{aligned}
\Lambda_{(1)}[G] &= \frac{1}{G_y^3}(G_xG_{yy} - G_yG_{xy}),\\
\Lambda_{(2)}[G] &= \frac{1}{G_x^3}(G_yG_{xx} - G_xG_{xy}),\\
\Lambda_{(3)}[G] &= F(G_x,G_y)(G_{xx}G_{yy} - G_{xy}^2),
\end{aligned} \tag{1.374}$$

where $F(G_x, G_y)$ is an arbitrary function of its arguments. One now seeks the corresponding densities and fluxes of conservation laws

$$\mathrm{D}_t\Phi^0_{(k)}[g] + \mathrm{D}_x\Phi^1_{(k)}[g] + \mathrm{D}_y\Phi^2_{(k)}[g] = 0, \quad k = 1,2,3. \tag{1.375}$$

The direct method for finding densities and fluxes [Section 1.3.7] leads to a complicated PDE system for the densities and fluxes $\Phi^0_{(k)}, \Phi^1_{(k)}, \Phi^2_{(k)}$, for each of the three conservation laws. The integral formulas (1.218) cannot be readily used for the multiplier $\Lambda_{(3)}[G]$ since it contains an arbitrary function. Moreover, one can show that for the multipliers $\Lambda_{(1)}[G]$ and $\Lambda_{(2)}[G]$, formulas (1.218) yield a divergent integral when $\widetilde{U} = 0$, and highly complicated integrals when $\widetilde{U} \neq 0$ [Exercise 1.5.2].

It is evident that the PDE (1.373) has a scaling symmetry

$$\mathrm{X}_1 = t\frac{\partial}{\partial t} + x\frac{\partial}{\partial x} + y\frac{\partial}{\partial y}. \tag{1.376}$$

Writing (1.376) in characteristic form, one finds $\hat{\eta}[g] = -tg_t - xg_x - yg_y$. For functions $F(G_x, G_y)$ that are homogeneous under scalings in G and that yield noncritical conservation laws, the conditions of Theorem 1.5.5 are satisfied for all three multipliers (1.374). Using a simplified version (1.378) of formula (1.178) [Exercise 1.5.1], one readily finds three corresponding conservation laws:

$$\mathrm{D}_t\left(\hat{\eta}[g]\Lambda_{(k)}[g]\right)-\mathrm{D}_x\left(\frac{\hat{\eta}[g]g_x\Lambda_{(k)}[g]}{\sqrt{g_x^2+g_y^2}}\right)-\mathrm{D}_y\left(\frac{\hat{\eta}[g]g_y\Lambda_{(k)}[g]}{\sqrt{g_x^2+g_y^2}}\right)=0,$$
$$k=1,2,3. \tag{1.377}$$

Exercises 1.5

1.5.1. Many PDE systems arising in applications are of first or second order. Suppose a given PDE system $\mathbf{R}\{x\,;u\}$ (1.292) is of order $k\leq 2$. Suppose it has a local symmetry $\mathrm{X}=\hat{\eta}^\rho[u]\,\partial/\partial u^\rho$, and $\{\omega_\sigma[u]\}_{\sigma=1}^N$ is a solution of the corresponding adjoint linearizing system (1.354). Show that the formulas (1.179) for the fluxes $\Psi^i[u]$ of the resulting conservation law simplify to

$$\begin{aligned}\Psi^i[u]=\hat{\eta}^\rho[u]\omega_\sigma[u]\frac{\partial R^\sigma[u]}{\partial u_i^\rho}+(\mathrm{D}_s\hat{\eta}^\rho[u])\omega_\sigma[u]\frac{\partial R^\sigma[u]}{\partial u_{is}^\rho}\\-\hat{\eta}^\rho[u]\mathrm{D}_j\left(\omega_\sigma[u]\frac{\partial R^\sigma[u]}{\partial u_{ji}^\rho}\right),\end{aligned} \tag{1.378}$$

where $i=1,\ldots,n$, $s\geq i$ and $j\leq i$. For details, see Anco (2003).

1.5.2. Consider the G-equation (1.373) and its local conservation law multipliers (1.374). Study the applicability of the integral formulas (1.218) to the computation of the corresponding fluxes of the conservation laws.

1.5.3. Show that in addition to the scaling symmetry (1.376), the G-equation (1.373) also has the scaling symmetry

$$\mathrm{X}_2=g\frac{\partial}{\partial g},$$

the translation symmetries

$$\mathrm{X}_3=\frac{\partial}{\partial t},\quad \mathrm{X}_4=\frac{\partial}{\partial x},\quad \mathrm{X}_5=\frac{\partial}{\partial y},\quad \mathrm{X}_6=\frac{\partial}{\partial g},$$

and a "relabeling symmetry"

$$\mathrm{X}_7=H(g)\frac{\partial}{\partial g}$$

for arbitrary $H(g)$ (which geometrically corresponds to the fact that the flame surface is, in general, implicitly defined by the equation $g(t,x,y) = \text{const}$, and may be equivalently defined by $K(g(t,x,y)) = \text{const}$, where $K(z)$ is some smooth function of its argument).

Use the symmetries $\mathrm{X}_2, \ldots, \mathrm{X}_7$ and multipliers (1.374) in formula (1.378) to generate fluxes of other local conservation laws of the G-equation (1.373). Check whether such conservation laws are linearly independent and nontrivial (up to equivalence).

1.5.4. Derive the expressions (1.333) and (1.334).

1.5.5. Prove that if the linearized system of a given PDE system is self-adjoint, then any two symmetries (local or nonlocal) can yield a conservation law (not necessarily a local conservation law) through an appropriate modification of Theorem 1.5.4 [Anco & Bluman (1996)].

1.5.6. Consider the linear wave equation

$$u_{xx} - [c(x)]^{-2} u_{tt} = 0. \tag{1.379}$$

(a) Show that the PDE (1.379) is self-adjoint for an arbitrary wave speed $c(x)$.

(b) Consider the nonlocally related PDE system given by

$$\begin{aligned} v_t - u_x &= 0, \\ v_x - [c(x)]^{-2} u_t &= 0. \end{aligned} \tag{1.380}$$

Show that the PDE system (1.380) is not self-adjoint for any wave speed $c(x)$.

(c) Show that if $c(x)$ satisfies the ODE $cc'(c/c')'' = 1$, then the PDE system (1.380) has two point symmetries given by the infinitesimal generators [Bluman & Kumei (1987)]

$$\begin{aligned} \mathrm{X}^{\pm} = {} & \pm 2e^{\pm t} \frac{c}{c'} \frac{\partial}{\partial x} + 2e^{\pm t} \left[\left(\frac{c}{c'} \right)' - 1 \right] \frac{\partial}{\partial t} \\ & \pm e^{\pm t} \left[2 - \left(\frac{c}{c'} \right)' \right] u \frac{\partial}{\partial u} - e^{\pm t} \frac{c}{c'} v \frac{\partial}{\partial v}. \end{aligned} \tag{1.381}$$

It turns out that the point symmetries (1.381) of the PDE system (1.380) are nonlocal symmetries of the PDE (1.379). For further details about this relationship between (1.379) and (1.380), see Chapters 3 and 4.

(d) Clearly, the linear PDE (1.379) has the scaling symmetry $\mathrm{X} = u\,\partial/\partial u$. Use this scaling symmetry in conjunction with the nonlocal symmetries (1.381) to obtain (nonlocal) conservation laws of the PDE (1.379) [Anco & Bluman (1996)].

(e) Check that these conservation laws are local conservation laws of the PDE system (1.380). Find the corresponding sets of local multipliers that yield these conservation laws.

1.5.7. Consider the NLT system

$$\begin{aligned} v_t - F(u)u_x - G(u) &= 0, \\ u_t - v_x &= 0 \end{aligned} \tag{1.382}$$

for arbitrary $F(u)$ and $G(u)$.

(a) Find a discrete symmetry of (1.382).
(b) Suppose $\{\Lambda_1(x,t,u,v,u_x,u_t), \Lambda_2(x,t,u,v,u_x,u_t)\}$ is a set of conservation law multipliers of the PDE system (1.382). Use the discrete symmetry and the translation symmetries of the PDE system (1.382) to obtain other sets of conservation law multipliers. Show that it does not necessarily follow that the resulting conservation laws are linearly independent. As examples, consider the PDE systems (1.335) and (1.342). For further examples, see Bluman & Temuerchaolu (2005b).

1.5.8. Consider the cylindrical KdV PDE given by

$$u_t + \tfrac{1}{2}t^{-1}u + uu_x + u_{xxx} = 0 \tag{1.383}$$

(a) Show that the point transformation

$$T = -2t^{-1/2}, \quad X = xt^{-1/2}, \quad U = tu - \tfrac{1}{2}x \tag{1.384}$$

invertibly maps the PDE (1.383) to the KdV equation

$$U_T + UU_X + U_{XXX} = 0. \tag{1.385}$$

(b) The KdV equation has an obvious conservation law resulting from the multiplier $\Lambda = 1$. Find the corresponding conservation law of the cylindrical KdV equation (1.383) under the action of the point transformation (1.384). For further examples of conservation laws obtained for PDE (1.383) from those known for PDE (1.385) through the action of the point transformation (1.384), see Bluman, Temuerchaolu & Anco (2006).
(c) Find the multiplier for this conservation law of the cylindrical KdV equation (1.383). Use obvious symmetries of PDE (1.383) to generate further conservation law multipliers of PDE (1.383).
(d) Check to see if further conservation laws are obtained from these multipliers.
(e) Check to see if further conservation laws are obtained from the constructed multiplier/symmetry pairs, using Theorem 1.5.4.

1.6 Discussion

Recursion operators for linear PDEs include, as a special case, ladder operators (raising and lowering operators) which are used in quantum mechanics [cf. Wybourne (1974)] and the theory of special functions [Talman (1968); Vilenkin (1968); Miller (1968)]. For linear PDEs, the existence of recursion operators is closely related to separation of variables [Miller (1977)]. Winternitz et al. (1967) give a method for finding recursion operators for time-independent Schrödinger equations. Anderson, Kumei & Wulfman (1972) appear to have been the first to exhibit nontrivial higher-order symmetries for PDEs and, in particular, found such symmetries and constructed corresponding recursion operators for linear PDEs.

For a PDE system that can be written in Hamiltonian form in two distinct ways, i.e., a bi-Hamiltonian PDE system, one can directly obtain recursion operators [Magri (1978), Olver (1986)].

The origin of much of the significant research activities in the study of nonlinear time evolution equations during the last four decades can be traced to the discovery of an infinite sequence of higher-order conservation laws for the Korteweg–de Vries equation by Miura, Gardner & Kruskal (1968) – see also Miura (1976) for a comprehensive review. These conservation laws were constructed with the aid of the Miura transformation [Miura (1968)]. In turn, this led to the important discovery of the inverse scattering method for initial value problems of integrable nonlinear evolution equations. The relationship between conservation laws and symmetries of integrable nonlinear evolution equations such as the KdV, nonlinear Schrödinger, and sine-Gordon equations was studied by Steudel (1975a,b) and Kumei [(1975), (1977)]. A list of known integrable nonlinear equations in 1+1 dimensions and their recursion operators is presented in Wang (2002). Sanders & Wang [(1998), (2000)] give general results to determine when a nonlinear evolutionary equation of the form

$$u_t = u_k + f(u, u_1, \ldots, u_{k-1}) \quad \text{with} \quad u_i \equiv \frac{\partial^i u}{\partial x^i},$$

has an infinite number of higher-order symmetries, assuming knowledge of one higher-order symmetry of the PDE. Moreover, for such PDEs, the higher-order symmetries are generated through recursion operators.

A very extensive study of the relationship among conservation laws, symmetries and multipliers appears in Olver (1986), especially in terms of Lagrangian and Hamiltonian formulations for Cauchy–Kovalevskaya PDE systems. Olver's book includes references to his many important papers on this subject. A discussion restricted to the action of a local symmetry on a conservation law to yield an additional conservation law appears in Kara & Mohamed (2002) and Olver (1986). The extension to the action of any symmetry (continuous or discrete) on a conservation law to yield one or more additional conservation laws appears in Bluman, Temuerchaolu & Anco (2006). A summary of results connecting symmetries and conservation laws appears in Bluman (2005).

In practice, the PDE systems of physical interest can be written in a solved form with respect to some leading derivatives of dependent variables. For any such PDE system (whether it admits a Cauchy–Kovalevskaya form or not), essentially all of its local conservation laws arise from multipliers, and the direct method [Anco & Bluman (1997a), (2002a,b)] then reduces the problem of finding conservation laws to a computation of solving an over-determined system of linear equations for multipliers. Physically important conservation law multipliers, such as those for mass, energy, momentum, and angular momentum, are distinguished by depending on derivatives of dependent variables at most of order $k-1$, in addition to the dependent and independent variables themselves, for a given kth-order PDE system. In general, one can solve the multiplier determining equations to obtain all conservation law multipliers of any order strictly less than the order of the PDE system by means of the same algorithmic method used to solve the determining equations for point or contact symmetries [Bluman & Anco (2002)].

Determination of higher-order multipliers requires that one impose some extra conditions on the form of multipliers. In particular, for any kth order PDE system, there exist (1) singular multipliers yielding divergence expressions that do not vanish on any solutions of the PDE system and hence fail to yield a conservation law; and (2) trivial multipliers yielding divergence expressions with fluxes that vanish for all solutions of the PDE system and hence yield a trivial conservation law. Such divergences obviously do not provide any useful information about the PDE system. These difficulties can be avoided if one seeks multipliers whose form does not involve leading derivatives and differential consequences. Such a restriction, in general, may entail a loss of completeness in finding all conservation law multipliers *except* when the PDE system is of Cauchy-Kovalevskya form. The same considerations arise for finding conservation laws from higher-order variational symmetries through Noether's theorem. A full discussion of these issues is presented in Anco (2009).

In comparing the direct method with Noether's theorem, it is important to reiterate that conservation laws arise from multipliers for both approaches. But unlike Noether's theorem, which is restricted to self-adjoint PDE systems, the direct method can be used on any PDE system. In particular, one can apply the direct method whether or not a PDE system has a solved form.

Integral (homotopy) formulas to obtain fluxes of conservation laws directly from multipliers, with the restrictions that $u=0$ is a solution of a given PDE system $\mathbf{R}\{x\,;u\}$ and that there is convergence of the integrals in these formulas, appear in Olver (1986), Anco & Bluman (1997a) and Hereman

(2005). Integral formulas that do not have such restrictions are presented in Anco & Bluman (2002b), Anco (2009) and Cheviakov (2009b).

In the case of ODEs, multipliers of local conservation laws are called *integrating factors*. For a detailed discussion on how to use and find integrating factors of ODEs, see Anco & Bluman [(1998)]. [See also Bluman & Anco (2002) where in addition there is a detailed discussion on how to find and use higher-order symmetries of ODEs.]

All local conservation laws, arising from local multipliers of the form $\Lambda(x, t, U)$, have been found for variable-coefficient nonlinear telegraph equations of the form [Huang & Ivanova (2007)]

$$f(x)u_{tt} - [f(u)u_x]_x - G(u)u_x = 0$$

and for variable coefficient diffusion-convection equations of the form [Ivanova, Popovych & Sophocleous (2008a)]

$$f(x)u_t - [g(x)f(u)u_x]_x - H(x)G(u)u_x = 0.$$

Ivanova & Sophocleous (2008) classify local conservation laws for a class of diffusion equations of the form

$$\begin{aligned} u_t &= [f(u, v)u_x]_x, \\ v_t &= [g(u, v)u_x]_x. \end{aligned}$$

All local conservation laws arising from multipliers $\Lambda(x, t, u, u_x, u_{xx})$ have been classified [Anco & Bluman (2002a)] for the family of generalized Korteweg–de Vries equations

$$u_t + u^n u_x + u_{xxx} = 0$$

with parameter $n \neq 0$. Anco & Bluman (2002a) give a complete classification of nonlinear wave equations

$$u_{tx} = g(u)$$

that admit higher-order local conservation laws.

Anco & Ivanova (2007) classify all local conservation laws with densities and fluxes depending on up to first-order spatial derivatives for parametric families of several types of semilinear radial wave equations in $n > 1$ spatial dimensions including a radial nonlinear hyperbolic equation, a radial nonlinear Schrödinger equation and its derivative variant, and radial generalizations of modified Korteweg–de Vries equations, as well as Hamiltonian variants.

For Maxwell's equations and other fundamental relativistic wave equations in Minkowski space, Anco & Pohjanpelto [(2001), (2003), (2004), (2008)] have classified all local conservation laws with densities and fluxes of arbitrary order, as well as all point symmetries, first- and higher-order local symmetries.

Cheviakov & Anco (2008) find all local conservation laws arising from multipliers that are linear in first derivatives for static isotropic and anisotropic MHD plasma models in three-dimensional space.

For Euler's equation of incompressible fluid flow with a barotropic equation of state (for pressure as a constitutive function of density) in $n > 1$ spatial dimensions, all local conservation laws have been classified in two cases of physical interest [Anco & Dar (2009)]: (i) kinematic conserved densities depending on the fluid density and velocity, in addition to the time and space coordinates; (ii) vorticity conserved densities depending essentially on the curl of the fluid velocity.

In Chapter 2, it is shown how to use the local symmetries of a given PDE and a target PDE to determine whether there exists a mapping (invertible or non-invertible, i.e, an isomorphism or homomorphism) that relates the PDEs. Two important situations are considered in detail: to determine whether a given nonlinear PDE system can be mapped invertibly into some linear PDE system and to determine whether a given linear PDE with variable coefficients can be mapped invertibly into a linear PDE with constant coefficients. In particular, for a given nonlinear PDE system, it is shown that knowledge of its point (contact) symmetries or its local conservation law multipliers determines whether it can be mapped invertibly to a linear PDE system. Moreover, when such a linearization mapping exists, it is shown how to find it systematically in terms of either the symmetries or conservation law multipliers of the given PDE system.

In Chapter 3, it is shown how to use local conservation laws of a given PDE system, with two independent variables, to find equivalent nonlocally related PDE systems. Applications of such nonlocally related systems are considered in Chapter 4. Such applications include extending the results presented in Chapters 1 and 2 to include the computation of nonlocal symmetries and nonlocal conservation laws of a given PDE system. It is also shown how to use the nonlocal symmetries and the multipliers yielding the nonlocal conservation laws to obtain non-invertible mappings to target equations of interest.

In Chapter 5, it is shown how to use local symmetries of a given PDE system (and nonlocal symmetries realized as local symmetries of a nonlocally related PDE system) to find invariant and other specific solutions of the given PDE system. The use of local symmetries to find invariant solutions through local symmetry reductions (classical method due to Lie) is extended to the nonclassical method and further refinements for obtaining solutions of PDEs. The work presented in Chapter 3 is extended to the situation of a given PDE system with three or more independent variables. Finally, there is a discussion of symbolic computation software for symmetry and conservation law calculations.

Chapter 2
Construction of Mappings Relating Differential Equations

2.1 Introduction

A symmetry of a PDE is a transformation (mapping) of its solution manifold into itself, i.e., it is a transformation that maps any solution of the PDE into another solution of the same PDE. *Invariant solutions* (*similarity solutions*) are solutions that map into themselves. If a symmetry of a given PDE is a *point symmetry*, then invariant solutions arise constructively from a reduced differential equation with fewer independent variables [Ovsiannikov [(1962), (1982)]; Bluman & Cole (1974); Olver (1986); Bluman & Kumei (1989); Stephani (1989); Bluman & Anco (2002); Cantwell (2002)].

In this chapter, we consider the problem of determining whether there exists a mapping of a given PDE into a target PDE of interest and to construct such a mapping when it exists. A target PDE is either a specific PDE or a member of a class of PDEs. The target PDE is *locally equivalent* to the given PDE if the mapping is invertible. The invertible mapping is not necessarily unique if a target PDE is a member of a class of PDEs. It is shown that the situation for showing existence and then finding such a mapping is especially fruitful when the target PDE (or target class of PDEs) is completely characterized by a class of contact symmetries (which only exist as point symmetries in the case of a system of PDEs).

If the mapping is invertible, then any infinitesimal generator of a symmetry of the given PDE maps into an infinitesimal generator of a symmetry of the target PDE. Moreover, the invertible mapping must establish an isomorphism between every Lie subalgebra of infinitesimal symmetry generators of the given PDE and corresponding Lie subalgebras of infinitesimal symmetry generators of the target PDE.

If the mapping is not invertible, then the mapping is a homomorphism of the given PDE to the target PDE. Here the mapping must take any infinitesimal generator of a symmetry of the given PDE into an infinitesimal

G.W. Bluman et al., *Applications of Symmetry Methods to Partial Differential Equations*, Applied Mathematical Sciences 168, DOI 10.1007/978-0-387-68028-6_2,

generator (which could be the null generator) of a symmetry of the target PDE. In particular, the mapping must establish a homomorphism between any Lie algebra of infinitesimal generators of the given PDE and a Lie algebra of infinitesimal generators of the target PDE.

Suppose a Lie algebra of infinitesimal symmetry generators uniquely determines a target PDE (or target class of PDEs). In this case, if one constructs an invertible mapping that transforms a subalgebra of the Lie algebra of infinitesimal symmetry generators of the given PDE into the Lie algebra of infinitesimal symmetry generators uniquely determining the target PDE (or a member of the target class of PDEs), then it naturally follows that such a mapping transforms any solution of the given PDE to a solution of the target PDE (or a member of the target class of PDEs).

When a target class of PDEs is uniquely characterized by a Lie algebra of infinitesimal symmetry generators, there should exist an algorithm to determine whether there exists an invertible mapping of a given PDE to some PDE in the target class of PDEs. Moreover, such an algorithm should construct the mapping when it exists. From this point of view, algorithms are presented to determine whether there exists an invertible mapping (as well as to construct such a mapping when one exists)

(i) of a given nonlinear scalar PDE to a linear scalar PDE;
(ii) of a given nonlinear system of PDEs to a linear system of PDEs;
(iii) of a given linear scalar PDE with variable coefficients to a linear scalar PDE with constant coefficients.

For a linear PDE that can be mapped invertibly into one with constant coefficients, the problem of finding the most general invertible mapping is also considered.

An alternative algorithm is also presented to determine whether a given nonlinear scalar PDE (or nonlinear system of PDEs) can be mapped invertibly to a linear PDE. This algorithm builds on work presented in the first chapter on multipliers for conservation laws. In particular, from the form of the multipliers (playing the analogue of the form of infinitesimal symmetry generators) yielding conservation laws of a given nonlinear PDE, one can determine whether an invertible mapping to a nonlinear PDE exists and construct such a mapping when it exists. Here it turns out that the determining equations for the multipliers yield the adjoint of a target linear PDE (whereas the determining equations for symmetries yield a target linear PDE) when such a mapping exists.

In Chapter 3, we consider important extensions of the results in this chapter to include non-invertible mappings through consideration of nonlocally related but equivalent systems for a given PDE.

2.2 Notations; Mappings of Infinitesimal Generators

For a given system of PDEs $\mathbf{R}\{x\,;u\}$ with n independent variables $x = (x^1, \ldots, x^n)$ and m dependent variables $u = (u^1, \ldots, u^m)$, the following notations are used:

G : the local symmetries of $\mathbf{R}\{x\,;u\}$;
L : the Lie algebra of G;
$\mathcal{G}$: a subgroup of G;
$\mathcal{L}$: the Lie algebra of $\mathcal{G}$;
g_ε : a one-parameter local transformation from $\mathcal{G}$, given by

$$(x^*)^i = x^i + \varepsilon\xi^i(x, u, \partial u, \ldots, \partial^k u) + O(\varepsilon^2), \tag{2.1a}$$
$$(u^*)^\nu = u^\nu + \varepsilon\eta^\nu(x, u, \partial u, \ldots, \partial^k u) + O(\varepsilon^2); \tag{2.1b}$$

X: the infinitesimal generator of g_ε, $\mathrm{X} \in \mathcal{L}$, given by

$$\mathrm{X} = \xi^i[u]\frac{\partial}{\partial x^i} + \eta^\nu[u]\frac{\partial}{\partial u^\nu}. \tag{2.2}$$

For a *target system* of PDEs $\mathbf{S}\{z\,;w\}$ with n independent variables $z = (z^1, \ldots, z^n)$ and m dependent variables $w = (w^1, \ldots, w^m)$, the following notations are used:

H : the local symmetries of $\mathbf{S}\{z\,;w\}$;
M : the Lie algebra of H;
$\mathcal{H}$: a subgroup of H;
$\mathcal{M}$: the Lie algebra of $\mathcal{H}$;
h_ε : a one-parameter local transformation from $\mathcal{H}$, given by

$$(z^*)^i = z^i + \varepsilon\zeta^i(z, w, \partial w, \ldots, \partial^K w) + O(\varepsilon^2), \tag{2.3a}$$
$$(w^*)^\nu = w^\nu + \varepsilon\omega^\nu(z, w, \partial w, \ldots, \partial^K w) + O(\varepsilon^2); \tag{2.3b}$$

Z: the infinitesimal generator of h_ε, $\mathrm{Z} \in \mathcal{M}$, given by

$$\mathrm{Z} = \zeta^i[w]\frac{\partial}{\partial z^i} + \omega^\nu[w]\frac{\partial}{\partial w^\nu}. \tag{2.4}$$

Note that one-parameter local transformations of the form (2.1) and (2.3) (*local symmetries*) include one-parameter Lie groups of point transformations (*point symmetries*), one-parameter Lie groups of contact transformations (*contact symmetries*), and one-parameter higher-order local transformations (*higher-order symmetries*).

Let μ denote a mapping (assuming that one exists), that transforms any solution $u = U(x)$ of $\mathbf{R}\{x\,;u\}$ to a solution $w = W(z)$ of $\mathbf{S}\{z\,;w\}$. In seeking

μ (which may not be invertible), a priori μ is restricted to be a local mapping of the form

$$z = \phi(x, u, \partial u, \dots, \partial^l u), \tag{2.5a}$$

$$w = \psi(x, u, \partial u, \dots, \partial^l u). \tag{2.5b}$$

Let T_l denote the set of mappings of the form (2.5) that depend at most on the lth partial derivatives of u. For any $u = U(x)$ for which the transformation (2.5a) is invertible, equations (2.5a) yield x as a function of z, and consequently, equations (2.5b) yield $w = W(z)$ and hence the components of $\partial w, \partial^2 w, \dots$.

Through the mapping $\mu \in T_l$ of the form (2.5), a one-parameter local transformation g_ε of $\mathcal{G}$ of the form (2.1) induces either a one-parameter local transformation h_ε of $\mathcal{H}$ of the form (2.3) or the identity transformation. The relationship among g_ε, μ, and h_ε is illustrated in Figure 2.1; the relationship among G, $\mathcal{G}$, H, $\mathcal{H}$, and μ is illustrated in Figure 2.2.

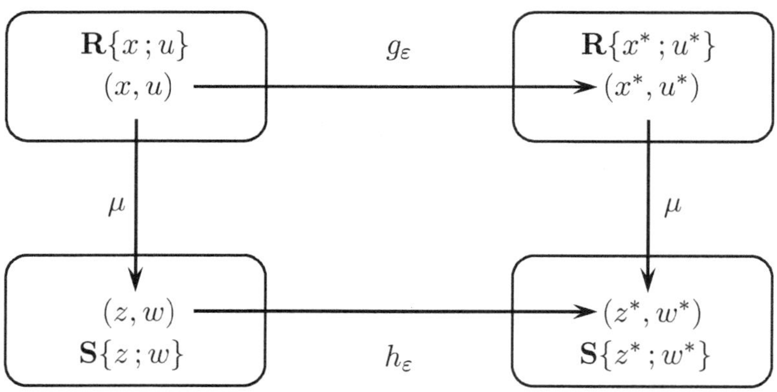

Fig. 2.1 The relationship amongg_ε, μ, and h_ε.

In particular, for the existence of a mapping μ, the relationship between g_ε and h_ε must be such that the composition transformations $\mu \circ g_\varepsilon$ and $h_\varepsilon \circ \mu$ yield the same action on (x, u)-space. Specifically,

$$\begin{aligned} g_\varepsilon(x, u) &= (x^*, u^*) \\ &= (x + \varepsilon\xi(x, u, \partial u, \dots, \partial^k u) + O(\varepsilon^2), \\ &\quad\ u + \varepsilon\eta(x, u, \partial u, \dots, \partial^k u) + O(\varepsilon^2)); \end{aligned} \tag{2.6}$$

$$\mu(x, u) = (z, w) = (\phi(x, u, \partial u, \dots, \partial^l u), \psi(x, u, \partial u, \dots, \partial^l u)), \tag{2.7}$$

and hence,

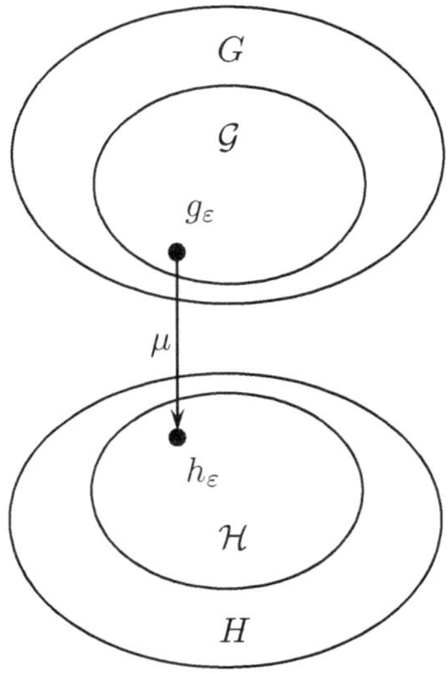

Fig. 2.2 The relationship among G, $\mathcal{G}$, H, $\mathcal{H}$, and μ.

$$\begin{aligned}\mu \circ g_\varepsilon(x,u) &= \mu(x^*,u^*)\\ &= (\phi(x^*,u^*,\partial u^*,\ldots,\partial^l u^*),\\ &\quad\ \psi(x^*,u^*,\partial u^*,\ldots,\partial^l u^*)).\end{aligned} \tag{2.8}$$

Through satisfaction of the contact conditions, a component $(u^*)^\nu_{i_1\cdots i_j}$ of $\partial^j u^*$ is given by

$$(u^*)^\nu_{i_1\cdots i_j} = u^\nu_{i_1\cdots i_j} + \varepsilon\eta^{(j)\nu}_{i_1\cdots i_j}(x,u,\partial u,\ldots,\partial^{k+j}u) + O(\varepsilon^2),$$

with extended infinitesimals

$$\begin{aligned}\eta^{(j)\nu}_{i_1\cdots i_j}[u] &= \mathrm{D}_{i_j}\eta^{(j-1)\nu}_{i_1\cdots i_{j-1}}[u] - (\mathrm{D}_{i_j}\xi^q[u])u^\nu_{i_1\cdots i_{j-1}q},\\ \eta^{(1)\nu}_i[u] &= \mathrm{D}_i\eta^\nu[u] - (\mathrm{D}_i\xi^q[u])u^\nu_q,\end{aligned}$$

in terms of total derivative operators

$$\mathrm{D}_i = \frac{\partial}{\partial x^i} + u^\sigma_i\frac{\partial}{\partial u^\sigma} + u^\sigma_{ip}\frac{\partial}{\partial u^\sigma_p} + \cdots \quad .$$

Note that

$$\begin{aligned}g_\varepsilon(x,u,\partial u,\ldots,\partial^l u) &= (x^*,u^*,\partial u^*,\ldots,\partial^l u^*)\\ &= (e^{\varepsilon \mathrm{X}^\infty}x, e^{\varepsilon \mathrm{X}^\infty}u, e^{\varepsilon \mathrm{X}^\infty}\partial u,\ldots e^{\varepsilon \mathrm{X}^\infty}\partial^l u)\end{aligned}$$

in terms of the extension of the infinitesimal generator (2.2) given by

$$\mathrm{X}^{\infty}=\xi^{i}[u]\frac{\partial}{\partial x^{i}}+\eta^{\sigma}[u]\frac{\partial}{\partial u^{\sigma}}+\eta_{i}^{(1)\sigma}[u]\frac{\partial}{\partial u_{i}^{\sigma}}+\cdots+\eta_{i_{1}\cdots i_{l}}^{(l)\sigma}[u]\frac{\partial}{\partial u_{i_{1}\cdots i_{l}}^{\sigma}}+\cdots. \quad (2.9)$$

Moreover, from the group property, it follows that

$$\mu(x^{*},u^{*})=e^{\varepsilon\mathrm{X}^{\infty}}(\phi(x,u,\partial u,\ldots,\partial^{l}u),\psi(x,u,\partial u,\ldots,\partial^{l}u)). \quad (2.10)$$

On the other hand,

$$\begin{aligned}h_{\varepsilon}\circ\mu(x,u) &= h_{\varepsilon}(z,w)\big|_{(z,w)=(\phi,\psi)}\\ &= (z^{*},w^{*})=e^{\varepsilon\mathrm{Z}^{\infty}}(z,w)\big|_{(z,w)=(\phi,\psi)},\end{aligned} \quad (2.11)$$

in terms of the corresponding extension of the infinitesimal generator (2.4). Equating the $O(\varepsilon)$ terms of (2.10) and (2.11), respectively, one obtains the following necessary conditions that the mapping μ, given by (2.5), must satisfy:

$$\begin{aligned}&\xi^{i}[u]\frac{\partial\phi}{\partial x^{i}}+\eta^{\sigma}[u]\frac{\partial\phi}{\partial u^{\sigma}}+\eta_{i}^{(1)\sigma}[u]\frac{\partial\phi}{\partial u_{i}^{\sigma}}+\cdots+\eta_{i_{1}\cdots i_{l}}^{(l)\sigma}[u]\frac{\partial\phi}{\partial u_{i_{1}\cdots i_{l}}^{\sigma}}\\ &=\zeta(\phi,\psi,\partial\psi,\ldots,\partial^{K}\psi),\end{aligned} \quad (2.12a)$$

$$\begin{aligned}&\xi^{i}[u]\frac{\partial\psi}{\partial x^{i}}+\eta^{\sigma}[u]\frac{\partial\psi}{\partial u^{\sigma}}+\eta_{i}^{(1)\sigma}[u]\frac{\partial\psi}{\partial u_{i}^{\sigma}}+\cdots+\eta_{i_{1}\cdots i_{l}}^{(l)\sigma}[u]\frac{\partial\psi}{\partial u_{i_{1}\cdots i_{l}}^{\sigma}}\\ &=\omega(\phi,\psi,\partial\psi,\ldots,\partial^{K}\psi).\end{aligned} \quad (2.12b)$$

The mapping equations (2.12) can be expressed concisely in the form

$$\mathrm{X}^{(l)}\phi=\mathrm{Z}z\big|_{(z,w,\partial w,\ldots,\partial^{K}w)=(\phi,\psi,\partial\psi,\ldots,\partial^{K}\psi)}, \quad (2.13a)$$

$$\mathrm{X}^{(l)}\psi=\mathrm{Z}w\big|_{(z,w,\partial w,\ldots,\partial^{K}w)=(\phi,\psi,\partial\psi,\ldots,\partial^{K}\psi)}, \quad (2.13b)$$

where $\mathrm{X}^{(l)}$ is the lth extension of X. Note that if $\mathrm{X}^{(l)}\phi\equiv 0$, $\mathrm{X}^{(l)}\psi\equiv 0$, then $\mathrm{Z}\equiv 0$. In this case, the mapping (2.5) is a differential invariant of g_{ε} and thus induces the identity transformation $h_{\varepsilon}\equiv\mathrm{I}$ on (z,w)-space.

The *mapping equations* (2.13) play the essential role in the mapping algorithms that are presented in the rest of this chapter. For the existence of a mapping μ of the form (2.5) of a given system of PDEs $\mathbf{R}\{x\,;u\}$ to a target system of PDEs $\mathbf{S}\{z\,;w\}$, it is necessary that for each infinitesimal generator X of a local symmetry of $\mathbf{R}\{x\,;u\}$ there must correspond some infinitesimal generator Z (which could be the null generator) of a local symmetry of $\mathbf{S}\{z\,;w\}$ and, moreover, through such a correspondence the mapping components (ϕ,ψ) of μ must satisfy (2.13). The mapping equations (2.13) are *necessary conditions* that μ must satisfy in terms of local symmetries of $\mathbf{R}\{x\,;u\}$ and $\mathbf{S}\{z\,;w\}$. In turn, if a mapping μ exists, these conditions significantly restrict how the components of (ϕ, ψ) can depend on $(x, u, \partial u, \ldots, \partial^l u)$. In the absence of such restrictions, it is tedious and usually impossible to determine systematically whether there exists a mapping $\mu \in T_l$ of a given system of PDEs $\mathbf{R}\{x\,;u\}$ to a target system of PDEs $\mathbf{S}\{z\,;w\}$. Later in this chapter, it is shown that if the mapping μ is restricted to being an invertible mapping and a target system $\mathbf{S}\{z\,;w\}$ can be completely characterized in terms of its point or contact symmetries, then, in terms of such characterizing symmetries, the mapping equations (2.13) yield a mapping μ when one exists.

2.2.1 Theorems on invertible mappings

If one seeks a one-to-one (invertible) mapping of a given PDE to a target PDE, then the following two theorems on invertible mappings show that the mapping (2.5) is restricted to a contact transformation in the case of a scalar PDE ($l = 1$) or to a point transformation in the case of a system of PDEs ($l = 0$).

Theorem 2.2.1 (Case of one dependent variable u ($m = 1$), i.e., a scalar PDE). *If $m = 1$, then a mapping μ defines an invertible mapping of $(x, u, \partial u, \ldots, \partial^p u)$-space to $(z, w, \partial w, \ldots, \partial^p w)$- space for any fixed p if and only if μ is a* one-to-one contact transformation *of the form*

$$z = \phi(x, u, \partial u), \tag{2.14a}$$

$$w = \psi(x, u, \partial u), \tag{2.14b}$$

$$\partial w = \partial\psi(x, u, \partial u). \tag{2.14c}$$

[Cf. Theorem 1.2.1, i.e., equations (1.15), for the conditions on the components of (ϕ, ψ) so that $\partial\psi$ depends at most on ∂u.]

Proof. The proof depends on showing that if (ϕ, ψ) has an essential dependence on the components of $\partial^q u$ for some $q \geq 2$, then $\partial^r \psi$ has an essential dependence on the components of $\partial^{q+r} u$ for any $r \geq 1$. The completion of the proof was left to Exercise 1.2.3. □

Theorem 2.2.1 is due to Bäcklund (1876). Note that if (ϕ, ψ) does not depend on the components of ∂u, then (2.14a), (2.14b) define a point transformation.

Theorem 2.2.2 (Case of more than one dependent variable ($m \geq 2$), i.e., u has at least two components as would normally be the case for a system of PDEs). *If $m \geq 2$, then a mapping μ defines an invertible mapping of $(x, u, \partial u, \ldots, \partial^p u)$-space to $(z, w, \partial w, \ldots, \partial^p w)$-space for any fixed p if and only if μ is a* one-to-one point transformation *of the form*

$$z = \phi(x, u), \tag{2.15a}$$

$$w = \psi(x, u). \tag{2.15b}$$

Proof. Here the proof depends on showing that if (ϕ, ψ) has an essential dependence on the components of $\partial^q u$ for some $q \geq 1$, then $\partial^r \psi$ has an essential dependence on the components of $\partial^{q+r} u$ for any $r \geq 1$. The completion of the proof was left to Exercise 1.2.3. □

Theorem 2.2.2 is due to Müller & Matschat (1962).

The proof of the following theorem is left to Exercise 2.2.2.

Theorem 2.2.3. *Suppose a given system of PDEs* $\mathbf{R}\{x\,;u\}$ *has* r *local symmetries whose infinitesimal generators form an* r*-dimensional Lie algebra* $\mathcal{L}$ *spanned by* $\mathrm{X}_1, \ldots, \mathrm{X}_r$*. Let* μ *be an invertible mapping of* $\mathbf{R}\{x\,;u\}$ *to* $\mathbf{S}\{z\,;w\}$*. Then* $\mathbf{S}\{z\,;w\}$ *has* r *local symmetries whose infinitesimal generators form an* r*-dimensional Lie algebra* $\mathcal{M}$ *spanned by* $\mathrm{Z}_1, \ldots, \mathrm{Z}_r$*. Moreover, the Lie algebra* $\mathcal{M}$ *is isomorphic to the Lie algebra* $\mathcal{L}$*. In particular, the mapping* μ *preserves the commutation relations of* $\mathcal{L}$*, i.e., if* $[\mathrm{X}_\alpha, \mathrm{X}_\beta] = C^\gamma_{\alpha\beta}\mathrm{X}_\gamma$*, then* $[\mathrm{Z}_\alpha, \mathrm{Z}_\beta] = C^\gamma_{\alpha\beta}\mathrm{Z}_\gamma$*, with the same structure constants* $C^\gamma_{\alpha\beta}$*,* $\alpha, \beta, \gamma = 1, \ldots, r$*.*

This isomorphism of commutation relations yields a *second set of necessary conditions* for an invertible mapping μ from $\mathbf{R}\{x\,;u\}$ *to* $\mathbf{S}\{z\,;w\}$.

Exercises 2.2

2.2.1. Discuss the situation for Theorem 2.2.1 in the case of one independent variable.

(a) Show that when the mapping is not necessarily an invertible mapping of an ODE, then the mapping is restricted to a contact transformation.
(b) How can the situation change when one is considering the invertible mapping of an ODE?

2.2.2. Prove Theorem 2.2.3.

2.3 Mapping of a Given PDE to a Specific Target PDE

Consider the problem of seeking to map a given PDE to a *specific* target PDE by a restricted local mapping of the form (2.5) by means of local symmetries (2.2) and (2.4) of the given and target PDEs, respectively. Through examples, two types of such mapping problems are considered:

(i) the mapping μ is not invertible;

(ii) the mapping μ is an invertible point transformation.

2.3.1 Construction of non-invertible mappings

(1) Mapping of the heat equation to Burgers' equation (a derivation of the Hopf–Cole transformation)
Hopf (1950) and Cole (1951) independently showed that the non-invertible mapping μ given by

$$z^1 = \phi^1 = x^1 = x, \tag{2.16a}$$
$$z^2 = \phi^2 = x^2 = t, \tag{2.16b}$$
$$w = \psi = -2u^{-1}u_1 = -2u^{-1}u_x, \tag{2.16c}$$

transforms any solution $u = \Theta(x,t)$ of the heat equation

$$u_t = u_{xx} \tag{2.17}$$

to the solution $w = -2\Theta^{-1}\Theta_x$ of Burgers' equation

$$w_t + ww_x = w_{xx}. \tag{2.18}$$

It is now shown how to derive the Hopf–Cole transformation (2.16) with the aid of the point symmetries of the given heat equation (2.17) and the target Burgers' equation (2.18) [Bluman (1974)].

The point symmetries of the given PDE (2.17) have the infinitesimal generators [Bluman (1967); Bluman & Cole (1969)]

$$\begin{aligned} &\mathrm{X}_1 = \frac{\partial}{\partial x}, \quad \mathrm{X}_2 = \frac{\partial}{\partial t}, \quad \mathrm{X}_3 = x\frac{\partial}{\partial x} + 2t\frac{\partial}{\partial t}, \\ &\mathrm{X}_4 = xt\frac{\partial}{\partial x} + t^2\frac{\partial}{\partial t} - [\tfrac{1}{4}x^2 + \tfrac{1}{2}t]u\frac{\partial}{\partial u}, \quad \mathrm{X}_5 = t\frac{\partial}{\partial x} - \tfrac{1}{2}xu\frac{\partial}{\partial u}, \\ &\mathrm{X}_6 = u\frac{\partial}{\partial u}, \quad \mathrm{X}_\infty = f(x,t)\frac{\partial}{\partial u}; \end{aligned} \tag{2.19}$$

$f(x,t)$ is any function satisfying $f_t - f_{xx} = 0$.

The point symmetries of the target PDE (2.18) have the infinitesimal generators [Bluman (1974)]

$$\begin{aligned} &\mathrm{Z}_1 = \frac{\partial}{\partial x}, \quad \mathrm{Z}_2 = \frac{\partial}{\partial t}, \quad \mathrm{Z}_3 = x\frac{\partial}{\partial x} + 2t\frac{\partial}{\partial t} - w\frac{\partial}{\partial w}, \\ &\mathrm{Z}_4 = xt\frac{\partial}{\partial x} + t^2\frac{\partial}{\partial t} - [x - tw]\frac{\partial}{\partial w}, \quad \mathrm{Z}_5 = t\frac{\partial}{\partial x} + \frac{\partial}{\partial w}. \end{aligned} \tag{2.20}$$

Since the given PDE (2.17) has more point symmetries than the target PDE (2.18) and a point transformation maps a point symmetry to a point symmetry, it immediately follows that no point transformation μ maps the heat equation (2.17) into Burgers' equation (2.18). After comparing the coefficients of the infinitesimal generators (2.19) and (2.20), one sees that the infinitesimal generators (2.19) of the given PDE (2.17) have the same action on (x,t)-space as the infinitesimal generators (2.20) of the target PDE (2.18). One also sees that the infinitesimal generator X_6 must map into the null generator $\mathrm{Z}_6 \equiv 0$. Moreover, one can show that the commutator relations for the Lie algebra $\mathcal{L}$ with basis set $\{\mathrm{X}_1, \ldots, \mathrm{X}_5\}$ are isomorphic to the commutator relations for the Lie algebra $\mathcal{M}$ with basis set $\{\mathrm{Z}_1, \ldots, \mathrm{Z}_5\}$. In particular, for these five infinitesimal generators, the commutators

$$[\mathrm{Z}_\alpha, \mathrm{Z}_\beta] = C^\gamma_{\alpha\beta} \mathrm{Z}_\gamma$$

and

$$[\mathrm{X}_\alpha, \mathrm{X}_\beta] = C^\gamma_{\alpha\beta} \mathrm{X}_\gamma$$

have the *same* structure constants $\{C^\gamma_{\alpha\beta}\}$. These observations lead one to seek a transformation μ that could map (2.17) to (2.18). Since the point symmetries of these two PDEs have the same group action on their respective spaces of independent variables and since such a mapping must be non-invertible, the simplest form for a transformation μ (2.5) that could map the given PDE (2.17) to the target PDE (2.18) is given by

$$z^1 = x, \tag{2.21a}$$

$$z^2 = t, \tag{2.21b}$$

$$w = \psi(x, t, u, u_t, u_x). \tag{2.21c}$$

In terms of each of the point symmetries of the given PDE (2.17) and the target PDE (2.18), respectively, one now imposes the necessary condition (2.13b) on the mapping function ψ; the necessary conditions (2.13a) are satisfied by a mapping μ of the form (2.21) from the above remarks about the common group action on the respective spaces of independent variables. In order to impose the necessary condition (2.13b) on the mapping function ψ, one must first determine the once-extended infinitesimal generators $\{\mathrm{X}_i^{(1)}\}$, $i = 1, \ldots, 6$. In particular,

$$\begin{aligned}
\mathrm{X}_1^{(1)} &= \mathrm{X}_1, \quad \mathrm{X}_2^{(1)} = \mathrm{X}_2, \quad \mathrm{X}_3^{(1)} = \mathrm{X}_3 - u_x\frac{\partial}{\partial u_x} - 2u_t\frac{\partial}{\partial u_t},\\
\mathrm{X}_4^{(1)} &= \mathrm{X}_4 - [\tfrac{1}{2}xu + (\tfrac{3}{2}t + \tfrac{1}{4}x^2)u_x]\frac{\partial}{\partial u_x} - [\tfrac{1}{2}u + xu_x + (\tfrac{5}{2}t + \tfrac{1}{4}x^2)u_t]\frac{\partial}{\partial u_t},\\
\mathrm{X}_5^{(1)} &= \mathrm{X}_5 - \tfrac{1}{2}[u + xu_x]\frac{\partial}{\partial u_x} - [u_x + \tfrac{1}{2}xu_t]\frac{\partial}{\partial u_t},\\
\mathrm{X}_6^{(1)} &= \mathrm{X}_6 + u_x\frac{\partial}{\partial u_x} + u_t\frac{\partial}{\partial u_t}.
\end{aligned}$$

Then the necessary condition (2.13b) yields

$$\mathrm{X}_\alpha^{(1)}\psi = \mathrm{Z}_\alpha w|_{w=\phi}, \quad \alpha = 1, \ldots, 6. \tag{2.22}$$

with $\mathrm{Z}_6 \equiv 0$. Consequently, if a mapping μ of the form (2.21) exists, then from (2.22) it follows that ψ must satisfy the over-determined system of six PDEs given by

$$\frac{\partial\psi}{\partial x} = 0, \tag{2.23a}$$

$$\frac{\partial\psi}{\partial t} = 0, \tag{2.23b}$$

$$x\frac{\partial\psi}{\partial x} + 2t\frac{\partial\psi}{\partial t} - u_x\frac{\partial\psi}{\partial u_x} - 2u_t\frac{\partial\psi}{\partial u_t} = -\psi, \tag{2.23c}$$

$$\begin{aligned}
&xt\frac{\partial\psi}{\partial x} + t^2\frac{\partial\psi}{\partial t} - [\tfrac{1}{2}t + \tfrac{1}{4}x^2]u\frac{\partial\psi}{\partial u} - [\tfrac{1}{2}xu + (\tfrac{3}{2}t + \tfrac{1}{4}x^2)]u_x\frac{\partial\psi}{\partial u_x}\\
&-[\tfrac{1}{2}u + xu_x + (\tfrac{1}{4}x^2 + \tfrac{5}{2}t)u_t]\frac{\partial\psi}{\partial u_t} = x - t\psi,
\end{aligned} \tag{2.23d}$$

$$t\frac{\partial\psi}{\partial x} - \tfrac{1}{2}tu\frac{\partial\psi}{\partial u} - \tfrac{1}{2}[u + xu_x]\frac{\partial\psi}{\partial u_x} - [u_x + \tfrac{1}{2}xu_t]\frac{\partial\psi}{\partial u_t} = 1, \tag{2.23e}$$

$$u\frac{\partial\psi}{\partial u} + u_x\frac{\partial\psi}{\partial u_x} + u_t\frac{\partial\psi}{\partial u_t} = 0. \tag{2.23f}$$

From equations (2.23a), (2.23b), it follows that $\psi = \psi(u, u_x, u_t)$. Consequently, equations (2.23c)–(2.23f) reduce to

$$\begin{aligned}
&\psi - u_x\frac{\partial\psi}{\partial u_x} - 2u_t\frac{\partial\psi}{\partial u_t} = 0,\\
&u\frac{\partial\psi}{\partial u} + u_x\frac{\partial\psi}{\partial u_x} + u_t\frac{\partial\psi}{\partial u_t} = 0,\\
&1 + \tfrac{1}{2}u\frac{\partial\psi}{\partial u_x} + u_x\frac{\partial\psi}{\partial u_t} = 0,\\
&u\frac{\partial\psi}{\partial u_t} = 0.
\end{aligned} \tag{2.24}$$

The last equation in (2.24) leads to $\psi = \psi(u, u_x)$. Then it is easy to show that the unique solution of the over-determined system (2.24) is given by the Hopf–Cole transformation

$$\psi = -2\frac{u_x}{u}. \tag{2.25}$$

[In conversation with George Bluman, Julian Cole indicated that by inspection he noticed that both the heat equation and Burgers' equation were invariant under translations in x and t as well as scalings and a Galilean transformation. In turn, this led him to consider the mapping (2.25) that satisfies the property of relating these four symmetries common to the two PDEs.]

(2) Mapping of the modified KdV equation to the KdV equation (a derivation of the Miura transformation)
Miura (1968) showed that the mapping μ given by

$$z^1 = \phi^1 = x^1 = x, \tag{2.26a}$$
$$z^2 = \phi^2 = x^2 = t, \tag{2.26b}$$
$$w = \psi = u^2 \pm i\sqrt{6}u_1 = u^2 \pm i\sqrt{6}u_x, \tag{2.26c}$$

transforms any solution $u = \Theta(x,t)$ of the modified KdV equation

$$u_t + u^2 u_x + u_{xxx} = 0 \tag{2.27}$$

to the solutions $w = \Theta^2 \pm i\sqrt{6}\Theta_x$ of the KdV equation

$$w_t + w w_x + w_{xxx} = 0. \tag{2.28}$$

The point symmetries of the given PDE (2.27) are given by the infinitesimal generators

$$\mathrm{X}_1 = \frac{\partial}{\partial x}, \quad \mathrm{X}_2 = \frac{\partial}{\partial t}, \quad \mathrm{X}_3 = x\frac{\partial}{\partial x} + 3t\frac{\partial}{\partial t} - u\frac{\partial}{\partial u}. \tag{2.29}$$

The point symmetries of the target PDE (2.28) are given by the infinitesimal generators

$$\mathrm{Z}_1 = \frac{\partial}{\partial x}, \quad \mathrm{Z}_2 = \frac{\partial}{\partial t}, \quad \mathrm{Z}_3 = x\frac{\partial}{\partial x} + 3t\frac{\partial}{\partial t} - 2w\frac{\partial}{\partial w}, \quad \mathrm{Z}_4 = t\frac{\partial}{\partial x} + \frac{\partial}{\partial w}. \tag{2.30}$$

Since the given PDE (2.27) has fewer point symmetries than the target PDE (2.28), it immediately follows that there exists no point transformation mapping the modified KdV equation (2.27) into the KdV equation (2.28). Moreover, since the group actions of X_1, X_2, X_3 and Z_1, Z_2, Z_3 are the same on (x,t)-space, and since such a mapping must be non-invertible, the simplest form for a transformation μ (2.5) that could map the given PDE (2.27) to the target PDE (2.28) is given again by (2.21). In terms of each of the three point symmetries (2.29) of the given PDE (2.27) and the corresponding ones (2.30) for the target PDE (2.28), the necessary condition (2.13b) is now imposed on the mapping function ψ [Note that $\mathrm{X}_1^{(1)} = \mathrm{X}_1$, $\mathrm{X}_2^{(1)} = \mathrm{X}_2$, $\mathrm{X}_3^{(1)} = \mathrm{X}_3 - 2u_x\,\partial/\partial u_x - 4u_t\,\partial/\partial u_t$.]; the necessary conditions (2.13a) are satisfied by a mapping μ of the form (2.21) from the above remarks about the common group action of the symmetry generators on their respective spaces of independent variables. Consequently, if a mapping μ of the form (2.21) exists, then the mapping function ψ must satisfy the over-determined system of three PDEs given by

$$\frac{\partial\psi}{\partial x} = 0, \tag{2.31a}$$

$$\frac{\partial\psi}{\partial t} = 0, \tag{2.31b}$$

$$u\frac{\partial\psi}{\partial u} + 2u_x\frac{\partial\psi}{\partial u_x} + 4u_t\frac{\partial\psi}{\partial u_t} = 2\psi. \tag{2.31c}$$

The solution of (2.31) yields

$$\psi = u^2 F\left(\frac{u_x}{u^2}, \frac{u_t}{u^4}\right), \tag{2.32}$$

where F is an arbitrary function of its two arguments. The reader should check that the substitution of (2.32) into the KdV equation (2.28) yields the Miura transformation (2.26).

2.3.2 Construction of an invertible mapping by a point transformation

Now consider the construction of an invertible mapping by a point transformation through establishing an isomorphism of the point symmetries of a given PDE and a target PDE. [Of course, this only provides necessary conditions for the existence of such a mapping.] As an example, a point transformation is found that invertibly relates the cylindrical KdV equation and the KdV equation.

In 1979, Zal'mez [cf. Korobeinikov (1982)] showed that the point transformation ($t > 0$)

$$z^1 = \phi^1 = xt^{-1/2}, \tag{2.33a}$$
$$z^2 = \phi^2 = -2t^{-1/2}, \tag{2.33b}$$
$$w = \psi = tu - \tfrac{1}{2}x, \tag{2.33c}$$

invertibly transforms a solution $u = \Theta(x,t)$ of the cylindrical KdV equation

$$u_t + uu_x + u_{xxx} + \tfrac{1}{2}t^{-1}u = 0 \tag{2.34}$$

to the solution

$$w = \Psi(z^1, z^2) = 4(z^2)^{-2}\Theta(-2(z^2)^{-1}z^1, 4(z^2)^{-2}) + (z^2)^{-1}z^1 \tag{2.35}$$

of the KdV equation

$$w_{z^2} + ww_{z^1} + w_{z^1z^1z^1} = 0. \tag{2.36}$$

The point symmetries of the given PDE (2.34) have infinitesimal generators

$$\begin{aligned} \mathrm{X}_1 &= 2t^{1/2}\frac{\partial}{\partial x} + t^{-1/2}\frac{\partial}{\partial u}, \\ \mathrm{X}_2 &= 2t^{1/2}x\frac{\partial}{\partial x} + 4t^{3/2}\frac{\partial}{\partial t} + [t^{-1/2}x - 4t^{1/2}u]\frac{\partial}{\partial u}, \\ \mathrm{X}_3 &= x\frac{\partial}{\partial x} + 3t\frac{\partial}{\partial t} - 2u\frac{\partial}{\partial u}, \quad \mathrm{X}_4 = \frac{\partial}{\partial x}. \end{aligned} \tag{2.37}$$

The infinitesimal generators for the point symmetries of the target PDE (2.36) are given by

$$\begin{aligned} &\mathrm{Z}_1 = \frac{\partial}{\partial z^1}, \quad \mathrm{Z}_2 = \frac{\partial}{\partial z^2}, \quad \mathrm{Z}_3 = z^1\frac{\partial}{\partial z^1} + 3z^2\frac{\partial}{\partial z^2} - 2w\frac{\partial}{\partial w}, \\ &\mathrm{Z}_4 = z^2\frac{\partial}{\partial z^1} + \frac{\partial}{\partial w}. \end{aligned} \tag{2.38}$$

The nonzero commutators for the Lie algebra arising from (2.37) are given by

$$\begin{aligned} &[\mathrm{X}_1, \mathrm{X}_3] = -\tfrac{1}{2}\mathrm{X}_1, \quad [\mathrm{X}_2, \mathrm{X}_3] = -\tfrac{3}{2}\mathrm{X}_2, \\ &[\mathrm{X}_2, \mathrm{X}_4] = -\mathrm{X}_1, \quad [\mathrm{X}_3, \mathrm{X}_4] = -\mathrm{X}_4, \end{aligned} \tag{2.39}$$

whereas the nonzero commutators for the Lie algebra arising from (2.38) are given by

$$\begin{aligned} &[\mathrm{Z}_1, \mathrm{Z}_3] = \mathrm{Z}_1, \quad [\mathrm{Z}_2, \mathrm{Z}_3] = 3\mathrm{Z}_2, \\ &[\mathrm{Z}_2, \mathrm{Z}_4] = \mathrm{Z}_1, \quad [\mathrm{Z}_3, \mathrm{Z}_4] = 2\mathrm{X}_4. \end{aligned} \tag{2.40}$$

Now consider the scalings $\mathrm{Z}_i = \alpha_i\tilde{\mathrm{Z}}_i$, $i = 1, \ldots, 4$. Then the commutators for the Lie algebra arising from $\{\tilde{\mathrm{Z}}_i\}$ are the same as (2.39) if and only if

$$\alpha_3 = -2, \quad \alpha_2\alpha_4 = -\alpha_1. \tag{2.41}$$

As a special case, let $\alpha_1 = -1$, $\alpha_2 = 1$, and hence $\alpha_4 = 1$. This *suggests* seeking a one-to-one point transformation mapping μ of the given PDE (2.34) to a target PDE such that it maps each X_i to the corresponding $\tilde{\mathrm{Z}}_i$. In particular, μ is of the form

$$z^1 = \phi^1(x,t,u), \quad z^2 = \phi^2(x,t,u), \quad w = \psi(x,t,u), \tag{2.42}$$

and satisfies the necessary conditions (2.13) which here become the 12 equations

$$\begin{gathered}\mathrm{X}_j\phi_i = \tilde{\mathrm{Z}}_j z_i\Big|_{(z,w)=(\phi,\psi)}, \ \mathrm{X}_j\psi = \tilde{\mathrm{Z}}_j w\Big|_{(z,w)=(\phi,\psi)}, \\ i = 1,2, \ j = 1,\ldots,4.\end{gathered} \tag{2.43}$$

From the commutation relations, one can see that the three equations for $j = 1$ are redundant but for convenience they are included. The resulting explicit equations are given by

$$2t^{1/2}\frac{\partial\phi^1}{\partial x} + t^{-1/2}\frac{\partial\phi^1}{\partial u} = -1, \tag{2.44a}$$

$$2t^{1/2}x\frac{\partial\phi^1}{\partial x} + 4t^{3/2}\frac{\partial\phi^1}{\partial t} + [t^{-1/2}x - 4t^{1/2}u]\frac{\partial\phi^1}{\partial u} = 0, \tag{2.44b}$$

$$x\frac{\partial\phi^1}{\partial x} + 3t\frac{\partial\phi^1}{\partial t} - 2u\frac{\partial\phi^1}{\partial u} = -\tfrac{1}{2}\phi^1, \tag{2.44c}$$

$$\frac{\partial\phi^1}{\partial x} = \phi^2, \tag{2.44d}$$

$$2t^{1/2}\frac{\partial\phi^2}{\partial x} + t^{-1/2}\frac{\partial\phi^2}{\partial u} = 0, \tag{2.44e}$$

$$2t^{1/2}x\frac{\partial\phi^2}{\partial x} + 4t^{3/2}\frac{\partial\phi^2}{\partial t} + [t^{-1/2}x - 4t^{1/2}u]\frac{\partial\phi^2}{\partial u} = 1, \tag{2.44f}$$

$$x\frac{\partial\phi^2}{\partial x} + 3t\frac{\partial\phi^2}{\partial t} - 2u\frac{\partial\phi^2}{\partial u} = -\tfrac{3}{2}\phi^2, \tag{2.44g}$$

$$\frac{\partial\phi^2}{\partial x} = 0, \tag{2.44h}$$

$$2t^{1/2}\frac{\partial\psi}{\partial x} + t^{-1/2}\frac{\partial\psi}{\partial u} = 0, \tag{2.44i}$$

$$2t^{1/2}x\frac{\partial\psi}{\partial x} + 4t^{3/2}\frac{\partial\psi}{\partial t} + [t^{-1/2}x - 4t^{1/2}u]\frac{\partial\psi}{\partial u} = 0, \tag{2.44j}$$

$$x\frac{\partial\psi}{\partial x} + 3t\frac{\partial\psi}{\partial t} - 2u\frac{\partial\psi}{\partial u} = \psi, \tag{2.44k}$$

$$\frac{\partial\psi}{\partial x} = 1. \tag{2.44l}$$

From equations (2.44e) and (2.44h), one sees that $\phi^2 = \phi^2(t)$. Then (2.44f), (2.44g) reduce to

$$4t^{3/2}[\phi^2(t)]' = 1, \quad 2t[\phi^2(t)]' = -\phi^2(t).$$

Hence

$$\phi^2 = -\tfrac{1}{2}t^{-1/2}. \tag{2.45}$$

Substituting (2.45) into (2.44d) and comparing the resulting equation with equation (2.44a), one sees that

$$\frac{\partial \phi^1}{\partial u} = 0, \quad \frac{\partial \phi^1}{\partial x} = -\tfrac{1}{2}t^{-1/2}.$$

Consequently, one obtains

$$\phi^1 = -\tfrac{1}{2}t^{-1/2}x + A(t), \tag{2.46}$$

for arbitrary $A(t)$. Substitution of equation (2.46) into equations (2.44b), (2.44c) yields $A(t) \equiv 0$. Hence

$$\phi^1 = -\tfrac{1}{2}t^{-1/2}x. \tag{2.47}$$

Equation (2.44l) leads to

$$\psi = x + B(t, u), \tag{2.48}$$

for arbitrary $B(t, u)$. Substitution of equation (2.48) into equation (2.44i) yields

$$\psi = x - 2tu + C(t), \tag{2.49}$$

for arbitrary $C(t)$. Substitution of equation (2.49) into equations (2.44j), (2.44k) yields $C(t) \equiv 0$. Thus, one obtains

$$\psi = x - 2tu. \tag{2.50}$$

From equations (2.45), (2.47) and (2.50), one sees that the point transformation

$$z^1 = -\tfrac{1}{2}t^{-1/2}x, \quad z^2 = -\tfrac{1}{2}t^{-1/2}, \quad w = x - 2tu, \tag{2.51}$$

defines an invertible mapping ($t > 0$) of the cylindrical KdV equation (2.34) to a target PDE whose point symmetries are the same as the point symmetries of the KdV equation (2.36). It is easy to show that the mapping (2.51) transforms any solution $u = \Theta(x,t)$ of the given PDE (2.34) to the solution

$$w = \Psi(z^1, z^2) = (z^2)^{-1}z^1 - \tfrac{1}{2}(z^2)^{-2}\Theta((z^2)^{-1}z^1, \tfrac{1}{4}(z^2)^{-2})$$

of the target PDE

$$w_{z^2} + ww_{z^1} - \tfrac{1}{2}w_{z^1z^1z^1} = 0. \tag{2.52}$$

Note that the PDE (2.52) is not the KdV equation (2.36). Why? This follows from showing that the KdV equation (2.36) is not the unique third-order PDE that has the point symmetries (2.38). In particular, this can be seen from the fact that there is a two-parameter family of scalings (2.41) that establishes an isomorphism between the commutator relations for the point symmetries of the cylindrical KdV equation (2.34) and those of the KdV equation (2.36). It is easy to see that the scaling

$$\tilde{z}^1 = \lambda z^1, \quad \tilde{z}^2 = -\frac{1}{2}\lambda^3 z^2, \quad \tilde{w} = -2\lambda^{-2}w \tag{2.53}$$

maps the PDE (2.52) into the KdV equation

$$\tilde{w}_{\tilde{z}^2} + \tilde{w}\tilde{w}_{\tilde{z}^1} + \tilde{w}_{\tilde{z}^1\tilde{z}^1\tilde{z}^1} = 0, \tag{2.54}$$

for *any* $\lambda \neq 0$. One can show that a scaling (2.53) corresponds to $\alpha_1 = -\lambda$, $\alpha_2 = -\frac{1}{2}\lambda^3$, $\alpha_4 = -2\lambda^{-2}$ in (2.41). As a consequence, one obtains the mapping

$$\tilde{z}^1 = -\tfrac{1}{2}\lambda t^{-1/2}x, \quad \tilde{z}^2 = \tfrac{1}{4}\lambda^3 t^{-1/2}, \quad \tilde{w} = -2\lambda^{-2}[x - 2tu], \tag{2.55}$$

that transforms any solution $u = \Theta(x,t)$ of the cylindrical KdV equation (2.34) to the solution

$$w = (\tilde{z}^2)^{-1}\tilde{z}^1 + \tfrac{1}{4}\lambda^4(\tilde{z}^2)^{-2}\Theta\left(-\tfrac{1}{2}\lambda^2(\tilde{z}^2)^{-1}\tilde{z}^1, \tfrac{1}{16}\lambda^6(\tilde{z}^2)^{-2}\right)$$

of the KdV equation (2.54). Zal'mez's transformation (2.33) corresponds to the special case $\lambda = 2$.

Exercises 2.3

2.3.1. Show that the Hopf–Cole transformation that maps the heat equation to Burgers' equation can be determined from the invariances of these two PDEs under the point symmetries X_2, X_3, X_5 and Z_2, Z_3, Z_5, respectively.

2.3.2. Find the most general second-order PDE of the form

$$u_{xx} = K(x, t, u, u_t, u_x, u_{tt}, u_{xt})$$

that has the point symmetries of Burgers' equation [Bluman (1974)].

2.3.3. From the commutation relations of the Lie algebra generated by (2.19):

(a) Show that the necessary conditions to obtain the Hopf–Cole transformation reduce to $\mathrm{X}_\alpha^{(1)}\psi = \mathrm{Z}_\alpha w|_{w=\phi}$, $\alpha = 1,\ 2,\ 4$.
(b) Accordingly, reduce the sets of equations (2.23), (2.24).
(c) Derive the Hopf–Cole transformation from these reduced sets of equations.
(d) Discuss the relationship between this problem and the first problem [Exercise 2.3.1].

2.3.4. Show that there exists no mapping of the form

$$z^1 = x, \quad z^2 = t, \quad w = \psi(x, t, u_x, u_t),$$

that maps Burgers' equation $u_t + uu_x = u_{xx}$ to the heat equation $w_t = w_{xx}$.

2.3.5. Show that if a given PDE is invariant under translations in x^1, then a mapping μ of the form $z = \phi = x$, $w = \psi(x, u, \partial u, \ldots, \partial^k u)$, from the given PDE to a target PDE is such that $\partial\psi/\partial x^1 \equiv 0$. Find a corresponding result if a given PDE is invariant under scalings in u and a target PDE is invariant under scalings in w.

2.3.6. Find the most general third-order PDE of the form

$$u_{xxx} = K(x, t, u, u_x, u_t, u_{xx}, u_{tt}, u_{xt}, u_{xxt}, u_{xtt}, u_{ttt})$$

that has the point symmetries of the KdV equation.

2.3.7. Show that there exists no mapping of the form

$$z^1 = x, \quad z^2 = t, \quad w = \psi(x, t, u, u_x),$$

that maps the KdV equation $u_t + uu_x + u_{xxx} = 0$ to the modified KdV equation $w_t + w^2 w_x + w_{xxx} = 0$.

2.3.8. Show explicitly that the necessary conditions (2.43), restricted to $j = 2, 3, 4$, yield the mapping equations (2.44).

2.3.9. In the mapping of the cylindrical KdV equation to a target PDE, use the two-parameter family defined by the relations (2.41) to find the most general target PDE obtained by a mapping (2.42) that satisfies the corresponding 12 equations arising from the necessary conditions (2.43).

2.4 Invertible Mappings of Nonlinear PDEs to Linear PDEs Through Symmetries

Clearly, an important mapping question is: when can a nonlinear system of PDEs be mapped to some linear system of PDEs? For example, the well-known hodograph transformation maps any first-order quasilinear system of two PDEs with two independent and two dependent variables to a linear system of PDEs.

For any such mapping, one is interested in the situation when any solution of the linear system yields a solution of the nonlinear system and, conversely, any solution of the nonlinear system yields a solution of the linear system. Note that such a situation does not require the mapping to be an invertible mapping. However, it does require the nonlinear and linear systems to be equivalent in the sense of having such a relationship between their solutions.

In this section it is shown how to determine whether there exists an invertible mapping of a given nonlinear PDE to some linear PDE from the point symmetries (systems case) or contact symmetries (scalar case) of the nonlinear PDE. Moreover, it is shown how to construct explicitly such an invertible mapping when one exists from the symmetries of the nonlinear PDE.

In Section 2.6, it is shown how to consider this question from knowledge of the local conservation law multipliers of a given nonlinear PDE. In the next chapter, the work presented in this chapter is extended to include non-invertible mappings of nonlinear PDEs to linear PDEs where the respective PDEs are nonetheless equivalent.

In Kumei & Bluman (1982) and Bluman & Kumei (1990a,b), necessary and sufficient conditions are given for when a given nonlinear PDE system with $n \geq 2$ independent variables and $m \geq 1$ dependent variable(s) can be transformed to some linear PDE system by an invertible mapping μ. These papers present a symmetry-based algorithm to determine whether these conditions hold for a given nonlinear PDE system and, when these conditions do hold, another symmetry-based algorithm to construct such a mapping μ. To apply these algorithms to a given nonlinear PDE system, it is unnecessary to know a specific target linear PDE system. A specific target linear system (when one exists) arises naturally from the symmetries of the given nonlinear PDE system (first algorithm). Moreover, the symmetries yield the equations to construct a specific mapping μ (second algorithm).

From Theorem 2.2.1, it immediately follows that in the case of a nonlinear scalar PDE ($m = 1$), an invertible mapping μ to a linear PDE must be an *invertible contact transformation* [cf. Section 1.1]:

$$z = \phi(x, u, \partial u), \tag{2.56a}$$
$$w = \psi(x, u, \partial u), \tag{2.56b}$$
$$\partial w = \partial\psi(x, u, \partial u). \tag{2.56c}$$

From Theorem 2.2.2, it immediately follows that in the case of a nonlinear PDE system ($m \geq 2$), an invertible mapping μ to a linear PDE system must be an *invertible point transformation*:

$$z = \phi(x, u), \tag{2.57a}$$
$$w = \psi(x, u). \tag{2.57b}$$

The starting point, leading to the symmetry-based invertible mapping algorithms relating nonlinear and linear PDEs, is the observation that a linear PDE system $\mathbf{S}\{z\,;w\}$, defined in terms of a linear operator $\mathrm{L}[z]$, i.e.,

$$\mathrm{L}[\mathrm{z}]w = g(z), \tag{2.58}$$

for some inhomogeneous term $g(z)$, which could equal zero, is completely characterized by its infinite-parameter set of point symmetries of the form

$$\mathrm{Z} = \omega\frac{\partial}{\partial w}, \tag{2.59}$$

where $\omega = f(z)$ is any function satisfying

$$\mathrm{L}[\mathrm{z}]f = 0. \tag{2.60}$$

Consequently, here the mapping equations (2.13) become necessary and sufficient conditions for the existence and construction of a mapping from a nonlinear PDE to a linear PDE in terms of a class of symmetries of the nonlinear PDE ($l = 1$ in the case of a scalar PDE; $l = 0$ in the case of a nonlinear PDE system).

A contact transformation is mapped to another contact transformation (which could be a point transformation) under any specific contact transformation whereas a point transformation is mapped to another point transformation under any specific point transformation. Hence, from the mapping equations (2.13), in the case of a scalar PDE, the infinite-parameter set of point symmetries (2.59) (which are point transformations) of a target linear PDE (2.58) must correspond to *an infinite-parameter set of contact symmetries of a given nonlinear scalar PDE* $\mathbf{R}\{x\,;u\}$ for an invertible mapping μ to exist; in the case of a system of PDEs the infinite-parameter set of point symmetries (2.59) of a target system of linear PDEs (2.58) must correspond to *an infinite-parameter set of point symmetries of a given nonlinear PDE system* $\mathbf{R}\{x\,;u\}$ for an invertible mapping μ to exist.

Consequently, in the case of a given nonlinear scalar PDE, *if the Lie algebra of the infinitesimal generators of its contact symmetries is at most finite-dimensional, then there exists no invertible mapping to a linear PDE*; in the case of a given nonlinear PDE system, *if the Lie algebra of the infinitesimal generators of its point symmetries is at most finite-dimensional, then there exists no invertible mapping to a linear PDE system.*

2.4.1 Invertible mappings of nonlinear PDE systems (with at least two dependent variables) to linear PDE systems

Theorem 2.4.1 (Necessary conditions for the existence of an invertible linearization mapping of a nonlinear PDE system). *If there exists an invertible mapping μ of a given nonlinear PDE system $\mathbf{R}\{x\,;u\}$ $(m \geq 2)$ to some linear PDE system $\mathbf{S}\{z\,;w\}$, then*

(i) μ is a point transformation of the form

$$z^j = \phi^j(x,u), \quad j = 1,\ldots,n, \tag{2.61a}$$

$$w^\gamma = \psi^\gamma(x,u), \quad \gamma = 1,\ldots,m; \tag{2.61b}$$

(ii) $\mathbf{R}\{x\,;u\}$ has an infinite set of point symmetries given by an infinitesimal generator

$$\mathrm{X} = \xi^i(x,u)\frac{\partial}{\partial x^i} + \eta^\nu(x,u)\frac{\partial}{\partial u^\nu} \tag{2.62}$$

with infinitesimals $\xi_i(x,u)$, $\eta^\nu(x,u)$ of the form

$$\xi^i(x,u) = \alpha^i_\sigma(x,u)F^\sigma(x,u), \tag{2.63a}$$

$$\eta^\nu(x,u) = \beta^\nu_\sigma(x,u)F^\sigma(x,u), \tag{2.63b}$$

where $\alpha^i_\sigma(x,u)$, $\beta^\nu_\sigma(x,u)$, $i = 1,\ldots,n$; $\nu;\sigma = 1,\ldots,m$, are specific functions of x and u, and where $F = (F^1,\ldots,F^m)$ is an arbitrary solution of some linear PDE system

$$\mathrm{L}[X]F = 0 \tag{2.64}$$

in terms of some linear differential operator $\mathrm{L}[X]$ and specific independent variables $X = (X^1(x,u),\ldots,X^n(x,u)) = (\phi^1,\ldots,\phi^n)$.

Proof. Necessary condition (i) follows directly from Theorem 2.2.2. Suppose there exists an invertible mapping μ from a given nonlinear PDE system $\mathbf{R}\{x\,;u\}$ to some linear system of PDEs $\mathbf{S}\{z\,;w\}$, represented by

$$\mathrm{L}[z]w = g(z), \tag{2.65}$$

with linear differential operator $\mathrm{L}[z]$ in terms of independent variables $z = (z^1, \ldots, z^n)$ with an allowed inhomogeneous term $g(z)$. Then $\mathbf{S}\{z\,; w\}$ has an infinite set of point symmetries given by the infinitesimal generator

$$\mathrm{Z} = f^{\gamma}(z)\frac{\partial}{\partial w^{\gamma}}, \tag{2.66}$$

where $f(z) = (f^1(z), \ldots, f^m(z))$ is an arbitrary solution of the linear homogeneous system $\mathrm{L}[z]f(z) = 0$. Corresponding to Z, there must exist an infinite set of point symmetries, in terms of an infinitesimal generator

$$\mathrm{X} = \xi^i(x,u)\frac{\partial}{\partial x^i} + \eta^{\nu}(x,u)\frac{\partial}{\partial u^{\nu}}, \tag{2.67}$$

of $\mathbf{R}\{x\,; u\}$ such that the components $\phi^j(x,u)$, $\psi^{\gamma}(x,u)$ of an invertible mapping μ and the infinitesimal coefficients $\xi^i(x,u)$, $\eta^{\nu}(x,u)$ of X satisfy relations (2.13) which here become

$$\xi^i(x,u)\frac{\partial \phi^j}{\partial x^i} + \eta^{\nu}(x,u)\frac{\partial \phi^j}{\partial u^{\nu}} = 0, \quad j = 1, \ldots, n, \tag{2.68a}$$

$$\xi^i(x,u)\frac{\partial \psi^{\gamma}}{\partial x^i} + \eta^{\nu}(x,u)\frac{\partial \psi^{\gamma}}{\partial u^{\nu}} = f^{\gamma}(\phi), \quad \gamma = 1, \ldots, m. \tag{2.68b}$$

From the invertibility of μ, it follows that the Jacobian $\partial(\phi, \varphi)/\partial(x,u) \neq 0$. Thus one can solve (2.68) for $\xi^i(x,u)$, $\eta^{\nu}(x,u)$, yielding expressions that are linear homogeneous in the components of f. In particular, the solution of the relations (2.68) is of the form

$$\xi^i(x,u) = \alpha^i_{\sigma}(x,u)f^{\sigma}(\phi(x,u)), \tag{2.69a}$$

$$\eta^{\nu}(x,u) = \beta^{\nu}_{\sigma}(x,u)f^{\sigma}(\phi(x,u)), \tag{2.69b}$$

where $\alpha^i_{\sigma}(x,u)$, $\beta^{\nu}_{\sigma}(x,u)$ are specific functions of x and u. If one sets $X(x,u) = \phi(x,u)$, then

$$F = (F^1(x,u), \ldots, F^m(x,u)) = (f^1(X(x,u)), \ldots, f^m(X(x,u)))$$

satisfies $\mathrm{L}[X]F = 0$. □

Theorem 2.4.2 (Sufficient conditions for the existence of an invertible linearization mapping of a nonlinear PDE system). *Suppose a given nonlinear PDE system* $\mathbf{R}\{x\,; u\}$ $(m \geq 2)$ *has an infinitesimal generator of point symmetries* (2.62) *with infinitesimal coefficients of the form* (2.63) *where* $F(X)$ *is an arbitrary solution of a linear system* (2.64) *with specific independent* *variables* $X = (X^1(x,u), \ldots, X^n(x,u))$. *If the linear homogeneous system of* m *first order PDEs for a scalar* Φ *given by*

$$\alpha^i_\sigma(x,u)\frac{\partial \Phi}{\partial x^i} + \beta^\nu_\sigma \frac{\partial \Phi}{\partial u^\nu} = 0, \quad \sigma = 1, \ldots, m, \tag{2.70}$$

has $X^1(x,u), \ldots, X^n(x,u)$ *as* n *functionally independent solutions, and the linear inhomogeneous system of* m^2 *first-order PDEs*

$$\alpha^i_\sigma(x,u)\frac{\partial \Psi^\gamma}{\partial x^i} + \beta^\nu_\sigma \frac{\partial \Psi^\gamma}{\partial u^\nu} = \delta^\gamma_\sigma, \tag{2.71}$$

where δ^γ_σ *is the Kronecker symbol,* $\gamma, \sigma = 1, \ldots, m$, *has a particular solution* $\Psi = \psi = (\psi^1(x,u), \psi^2(x,u), \ldots, \psi^m(x,u))$, *then the mapping* μ *given by*

$$z^j = \phi^j(x,u) = X^j(x,u), \quad j = 1, \ldots, n, \tag{2.72a}$$

$$w^\gamma = \psi^\gamma(x,u), \quad \gamma = 1, \ldots, m \tag{2.72b}$$

is invertible and transforms $\mathbf{R}\{x\,; u\}$ *to the linear PDE system* $\mathbf{S}\{z\,; w\}$ *given by*

$$\mathrm{L}[z]w = g(z), \tag{2.73}$$

which may involve an inhomogeneous term $g(z)$.

Proof. By construction, the mapping μ defined by (2.72) through a solution of the linear systems (2.70), (2.71) is invertible. Let $f^\sigma(z) = F^\sigma(x,u)$, $\sigma = 1, \ldots, m$. Then μ transforms the point symmetry infinitesimal generator X of the nonlinear PDE system $\mathbf{R}\{x\,; u\}$ to the point symmetry infinitesimal generator Z of a target system $\mathbf{S}\{z\,; w\}$ where Z is of the form

$$\mathrm{Z} = f^\gamma(z)\frac{\partial}{\partial w^\gamma} \tag{2.74}$$

for any solution $f(z) = (f^1(z), \ldots, f^m(z))$ satisfying the linear PDE system $\mathrm{L}[z]f(z) = 0$. But since only a linear system of the form (2.73) can be characterized as having a point symmetry of the form (2.74), and the mapping μ defined by the equations (2.72) is invertible, it follows that a target system of the mapping μ must be of the form (2.73). □

Remark 2.4.1. Note that the general solution of (2.71) yields an arbitrary inhomogeneous term $g(z)$ in (2.73), including $g(z) = 0$.

Now consider two examples.

(1) Linearization by a hodograph transformation
The first-order quasilinear PDE system ($m = n = 2$) $\mathbf{R}\{x\,; u\}$, with independent variables $(x^1, x^2) = (x, t)$ and dependent variables $(u^1, u^2) = (u, v)$, given by

$$a(u,v)u_x + b(u,v)u_t + c(u,v)v_x + d(u,v)v_t = 0, \tag{2.75a}$$

$$p(u,v)u_x + q(u,v)u_t + r(u,v)v_x + s(u,v)v_t = 0, \tag{2.75b}$$

with Jacobian $\partial(u,v)/\partial(x,t) \neq 0$, has an infinite set of point symmetries given by the infinitesimal generator

$$\mathrm{X} = \xi \frac{\partial}{\partial x} + \tau \frac{\partial}{\partial t}, \tag{2.76}$$

where

$$\xi = F^1(u,v), \quad \tau = F^2(u,v), \tag{2.77}$$

is an arbitrary solution of the linear PDE system

$$d(u,v)\frac{\partial F^1}{\partial u} - b(u,v)\frac{\partial F^1}{\partial v} - c(u,v)\frac{\partial F^2}{\partial u} + a(u,v)\frac{\partial F^2}{\partial v} = 0, \tag{2.78a}$$

$$s(u,v)\frac{\partial F^1}{\partial u} - q(u,v)\frac{\partial F^1}{\partial v} - r(u,v)\frac{\partial F^2}{\partial u} + p(u,v)\frac{\partial F^2}{\partial v} = 0. \tag{2.78b}$$

Clearly, the point symmetry given by (2.76), (2.77) and (2.78) satisfies the necessary conditions of Theorem 2.4.1 for the existence of an invertible mapping μ that transforms the quasilinear system (2.75) to some linear system. In particular, from (2.77), one sees that $\alpha_1^1 = \alpha_2^2 = 1$, $\alpha_1^2 = \alpha_2^1 = \beta_1^1 = \beta_1^2 = \beta_2^1 = \beta_2^2 = 0$. Moreover, from equations (2.78), it follows that $X^1 = u$, $X^2 = v$.

The existence and construction of μ follow from applying Theorem 2.4.2. Here equations (2.70) given by

$$\frac{\partial \Phi}{\partial x} = 0, \quad \frac{\partial \Phi}{\partial t} = 0, \tag{2.79}$$

have functionally independent solutions $\Phi = X^1 = u$, $\Phi = X^2 = v$, and equations (2.71) become

$$\frac{\partial \Psi^1}{\partial x} = 1, \quad \frac{\partial \Psi^1}{\partial t} = 0, \quad \frac{\partial \Psi^2}{\partial x} = 0, \quad \frac{\partial \Psi^2}{\partial t} = 1. \tag{2.80}$$

Clearly, a particular solution of the PDE system (2.80) is given by

$$\Psi^1 = \psi^1 = x, \quad \Psi^2 = \psi^2 = t. \tag{2.81}$$

Consequently, the invertible mapping μ given by

$$z^1 = u, \quad z^2 = v, \quad w^1 = x, \quad w^2 = t, \tag{2.82}$$

transforms the quasilinear PDE system (2.75) to the linear PDE system $\mathbf{S}\{z\,; w\}$ given by

$$d(u,v)\frac{\partial x}{\partial u} - b(u,v)\frac{\partial x}{\partial v} - c(u,v)\frac{\partial t}{\partial u} + a(u,v)\frac{\partial t}{\partial v} = 0, \tag{2.83a}$$

$$s(u,v)\frac{\partial x}{\partial u} - q(u,v)\frac{\partial x}{\partial v} - r(u,v)\frac{\partial t}{\partial u} + p(u,v)\frac{\partial t}{\partial v} = 0. \tag{2.83b}$$

The mapping (2.82) is the well-known hodograph transformation (involving an interchange of the roles of dependent and independent variables) that linearizes the quasilinear PDE system (2.75).

(2) Linearization of a nonlinear telegraph system
The nonlinear telegraph system $\mathbf{R}\{x\,;u\}$ given by

$$\begin{aligned} u_x - v_t &= 0, \\ u_t - u^2 v_x - u(1-u) &= 0, \end{aligned} \tag{2.84}$$

with Jacobian $\partial(u,v)/\partial(x,t) \neq 0$, has an infinite set of point symmetries given by the infinitesimal generator

$$\mathrm{X} = \xi\frac{\partial}{\partial x} + \tau\frac{\partial}{\partial t} + \eta\frac{\partial}{\partial u} + \zeta\frac{\partial}{\partial v}, \tag{2.85}$$

where

$$\xi = \zeta = F^1(X^1, X^2), \quad \tau = e^{-t}F^2(X^1, X^2), \quad \eta = e^{-t}uF^2(X^1, X^2) \tag{2.86}$$

with

$$X^1 = x - v, \quad X^2 = t - \log u, \tag{2.87}$$

and $F = (F^1, F^2)$ is any solution of the linear PDE system

$$\begin{aligned} \frac{\partial F^2}{\partial X^2} - e^{X^2}\frac{\partial F^1}{\partial X^1} &= 0, \\ \frac{\partial F^2}{\partial X^1} - e^{X^2}\frac{\partial F^1}{\partial X^2} &= 0. \end{aligned} \tag{2.88}$$

Clearly, the point symmetry (2.85)–(2.87) satisfies the necessary conditions of Theorem 2.4.1 for the existence of an invertible mapping μ that transforms the nonlinear telegraph system (2.84) to some linear system. In particular, from (2.86), one sees that $\alpha_1^1 = \beta_1^2 = 1$, $\alpha_2^2 = e^{-t}$, $\beta_2^1 = e^{-t}u$, $\alpha_1^2 = \alpha_2^1 = \beta_1^1 = \beta_2^2 = 0$.

For the existence and construction of μ through the application of Theorem 2.4.2, one sees that here the equations (2.70) become

$$\frac{\partial \Phi}{\partial x} + \frac{\partial \Phi}{\partial v} = 0, \quad \frac{\partial \Phi}{\partial t} + u\frac{\partial \Phi}{\partial u} = 0. \tag{2.89}$$

It is easy to see that $\Phi = X^1 = x - v$, $\Phi = X^2 = t - \log u$ are functionally independent solutions of the linear PDE system (2.89). Then equations (2.71) here become

$$\begin{aligned} &\frac{\partial \Psi^1}{\partial x} + \frac{\partial \Psi^1}{\partial v} = 1, \quad \frac{\partial \Psi^1}{\partial t} + u\frac{\partial \Psi^1}{\partial u} = 0, \\ &\frac{\partial \Psi^2}{\partial x} + \frac{\partial \Psi^2}{\partial v} = 0, \quad \frac{\partial \Psi^2}{\partial t} + u\frac{\partial \Psi^2}{\partial u} = e^t. \end{aligned} \tag{2.90}$$

A particular solution of (2.90) is given by

$$\Psi^1 = \psi^1 = x, \quad \Psi^2 = \psi^2 = e^t. \tag{2.91}$$

Consequently, the invertible mapping μ given by

$$z^1 = x - v, \quad z^2 = t - \log u, \quad w^1 = x, \quad w^2 = e^t, \tag{2.92}$$

transforms the nonlinear telegraph system (2.84) to the linear PDE system $\mathbf{S}\{z\,; w\}$ given by

$$\begin{aligned} &\frac{\partial w^2}{\partial z^2} - e^{z^2}\frac{\partial w^1}{\partial z^1} = 0, \\ &\frac{\partial w^2}{\partial z^1} - e^{z^2}\frac{\partial w^1}{\partial z^2} = 0. \end{aligned} \tag{2.93}$$

Varley & Seymour (1985) found a hodograph-type transformation equivalent to (2.92) that linearizes the nonlinear telegraph system (2.84) through a different procedure than the one presented in this section. Their clever procedure for linearization is not symmetry-based. It only applies to specific types of nonlinear PDEs and, in general, cannot determine whether a given nonlinear PDE system can be linearized by an invertible mapping.

2.4.2 Invertible mappings of nonlinear PDE systems (with one dependent variable) to linear PDE systems

The work presented in Section 2.4.1 can be further extended in the case when a given nonlinear PDE system $\mathbf{R}\{x\,; u\}$ has only one dependent variable. Usually, in this situation $\mathbf{R}\{x\,; u\}$ is a nonlinear scalar PDE. This is the case for the presented examples. For the rest of this section, $\mathbf{R}\{x\,; u\}$ is considered to be a scalar PDE but the presented theorems apply to the situation when $\mathbf{R}\{x\,; u\}$ is a nonlinear PDE system with one dependent variable.

First of all, it is easy to see that Theorems 2.4.1 and 2.4.2 hold for the existence and construction of invertible mappings of a nonlinear scalar PDE to a linear PDE when the nonlinear PDE has an infinite number of point symmetries that satisfy the criteria of these theorems for the situation where $m = 1$ and the mapping μ is restricted to an invertible point transformation.

As an illustrative example, consider the nonlinear diffusion equation

$$(u_x)^2 u_t - u_{xx}. \tag{2.94}$$

The scalar PDE (2.94) has an infinite set of point symmetries given by the infinitesimal generator

$$\mathrm{X} = \xi \frac{\partial}{\partial x}, \tag{2.95}$$

where

$$\xi = F(u,t) \tag{2.96}$$

is any solution of the linear heat equation

$$F_t - F_{uu} = 0. \tag{2.97}$$

Clearly, the point symmetries (2.95)–(2.97) satisfy the necessary conditions of Theorem 2.4.1 for the existence of an invertible point transformation μ that transforms the nonlinear diffusion equation (2.94) to some linear scalar PDE. In particular, from (2.96), one sees that $\alpha_1^1 = 1$, $\alpha_1^2 = \beta_1^1 = 0$. From (2.96) and (2.97), it follows that $X^1 = u$, $X^2 = t$.

For the existence and construction of a mapping μ through the use of Theorem 2.4.2, here equation (2.70) becomes

$$\frac{\partial \Phi}{\partial x} = 0. \tag{2.98}$$

Clearly, $\Phi = X_1 = u, \Phi = X_2 = t$ are functionally independent solutions of (2.98). Then equation (2.71) becomes

$$\frac{\partial \Psi}{\partial x} = 1, \tag{2.99}$$

with an obvious particular solution

$$\Psi = \psi = x. \tag{2.100}$$

Consequently, the invertible mapping μ given by $z^1 = u$, $z^2 = t$, $w = x$, transforms the nonlinear diffusion equation (2.94) to the linear heat equation

$$x_t - x_{uu} = 0.$$

This result was previously found in Bluman & Cole [1974, Section 2.15] through the use of symmetry methods in a more ad-hoc manner.

As shown in following two theorems, in the case of a scalar PDE, one can extend the results presented in Theorems 2.4.1 and 2.4.2 to include linearization by a contact transformation.

Theorem 2.4.3 (Necessary conditions for the existence of an invertible linearization mapping of a nonlinear scalar PDE). *If there exists an invertible mapping* μ *of a given nonlinear scalar PDE* $\mathbf{R}\{x\,;u\}$ $(m=1)$ *to some linear scalar PDE* $\mathbf{S}\{z\,;w\}$*, then*

(i) μ *is a contact transformation of the form*

$$z^j = \phi^j(x,u,\partial u), \tag{2.101a}$$

$$w = \psi(x,u,\partial u), \tag{2.101b}$$

$$\frac{\partial w}{\partial z^j} = \partial\psi_j(x,u,\partial u), \quad j=1,\ldots,n; \tag{2.101c}$$

(ii) $\mathbf{R}\{x\,;u\}$ *has an infinite set of contact symmetries given by an infinitesimal generator*

$$\mathrm{X} = \xi^i(x,u,\partial u)\frac{\partial}{\partial x^i} + \eta(x,u,\partial u)\frac{\partial}{\partial u} + \eta_i^{(1)}(x,u,\partial u)\frac{\partial}{\partial u_i} \tag{2.102}$$

with infinitesimals $\xi^i(x,u,\partial u)$, $\eta(x,u,\partial u)$, *and first extended infinitesimals* $\eta_i^{(1)}(x,u,\partial u)$ *of the form*

$$\xi^i(x,u,\partial u) = \alpha^i(x,u,\partial u)F(x,u,\partial u) \tag{2.103a}$$
$$+\alpha^{ij}(x,u,\partial u)H_j(x,u,\partial u),$$

$$\eta(x,u,\partial u) = \beta(x,u,\partial u)F(x,u,\partial u) \tag{2.103b}$$

$$+\beta^j(x,u,\partial u)H_j(x,u,\partial u), \tag{2.103c}$$

$$\eta_i^{(1)}(x,u,\partial u) = \lambda_i(x,u,\partial u)F(x,u,\partial u) \tag{2.103d}$$

$$+\lambda_i^j(x,u,\partial u)H_j(x,u,\partial u), \tag{2.103e}$$

where α^i, α^{ij}, β, β^j, λ_i, λ_i^j, $i,j=1,\ldots,n$, *are specific functions of* x, u, *and the components of* ∂u, *and* $F(x,u,\partial u)$ *is an* arbitrary *solution of some linear scalar PDE*

$$\mathrm{L}[X]F = 0 \tag{2.104}$$

in terms of some linear differential operator $\mathrm{L}[X]$, *and specific independent variables*

$$X = (X^1(x,u,\partial u),\ldots,X^n(x,u,\partial u)) = (\phi^1,\ldots\phi^n);$$

and $H_j(x, u, \partial u)$ *satisfies*

$$H_j = \frac{\partial F}{\partial X^j}, \quad j = 1, \ldots, n. \tag{2.105}$$

Proof. Necessary condition (i) follows directly from Theorem 2.2.1. Suppose there exists an invertible mapping μ from a given nonlinear scalar PDE $\mathbf{R}\{x\,;u\}$ to some linear scalar PDE $\mathbf{S}\{z\,;w\}$, represented by

$$\mathrm{L}[z]w = g(z),$$

with linear differential operator $\mathrm{L}[z]$ in terms of independent variables $z = (z^1, \ldots, z^n)$ and an allowed inhomogeneous term $g(z)$. Then $\mathbf{S}\{z\,;w\}$ has an infinite set of point symmetries given by the infinitesimal generator

$$\mathrm{Z} = f(z)\frac{\partial}{\partial w}, \tag{2.106}$$

where $f(z)$ is an arbitrary solution of the linear homogeneous scalar PDE $\mathrm{L}[z]f(z) = 0$. Corresponding to Z, there must exist an infinite set of contact symmetries, given by an infinitesimal generator of the form (2.102), of $\mathbf{R}\{x\,;u\}$ such that the components ϕ^j, ψ, ψ_j of an invertible mapping μ and the infinitesimal coefficients ξ^i, $\eta, \eta_i^{(1)}$ of X satisfy relations (2.13) which here become

$$\xi^i\frac{\partial \phi^j}{\partial x^i} + \eta\frac{\partial \phi^j}{\partial u} + \eta_i^{(1)}\frac{\partial \phi^j}{\partial u_i} = 0, \tag{2.107a}$$

$$\xi^i\frac{\partial \psi}{\partial x^i} + \eta\frac{\partial \psi}{\partial u} + \eta_i^{(1)}\frac{\partial \psi}{\partial u_i} = f(\phi), \tag{2.107b}$$

$$\xi^i\frac{\partial \psi_j}{\partial x^i} + \eta\frac{\partial \psi_j}{\partial u} + \eta_i^{(1)}\frac{\partial \psi_j}{\partial u_i} = \left.\frac{\partial f(z)}{\partial z^j}\right|_{z=\phi(x,u,\partial u)}, \tag{2.107c}$$

$j = 1, \ldots, n$. From the invertibility of μ, it follows that the Jacobian $\partial(\phi, \varphi, \partial\psi)/\partial(x, u, \partial u) \neq 0$. Thus one can solve equations (2.107) for ξ^i, η, $\eta_i^{(1)}$, yielding expressions that are linear homogeneous in f. In particular, the solution of the equations (2.107) is of the form

$$\xi^i = \alpha^i(x, u, \partial u)f(\phi) + \alpha^{ij}(x, u, \partial u)\left.\frac{\partial f(z)}{\partial z^j}\right|_{z=\phi}, \tag{2.108a}$$

$$\eta = \beta(x, u, \partial u)f(\phi) + \beta^j(x, u, \partial u)\left.\frac{\partial f(z)}{\partial z^j}\right|_{z=\phi}, \tag{2.108b}$$

$$\eta_i^{(1)} = \lambda_i(x, u, \partial u)f(\phi) + \lambda_i^j(x, u, \partial u)\left.\frac{\partial f(z)}{\partial z^j}\right|_{z=\phi}, \tag{2.108c}$$

where α^i, α^{ij}, β, β^j, λ_i, λ_i^j are specific functions of x, u, and the components of ∂u. If one sets $X(x,u,\partial u) = \phi(x,u,\partial u)$, then $F(x,u,\partial u) = f(X(x,u,\partial u))$ satisfies $\mathrm{L}[X]F = 0$. □

Theorem 2.4.4 (Sufficient conditions for the existence of an invertible linearization mapping of a nonlinear scalar PDE). *Suppose a given nonlinear scalar PDE* $\mathbf{R}\{x\,;u\}$ $(m=1)$ *has an infinitesimal generator of contact symmetries* (2.102) *with infinitesimal coefficients of the form* (2.103) *where* $F(X)$ *is an arbitrary solution of a linear scalar PDE* (2.104) *and* $H_j = \partial F/\partial X^j$, *with specific independent variables* $X = (X^1(x,u,\partial u),\ldots,X^n(x,u,\partial u))$. *Suppose the following four conditions hold.*

(i) The linear homogeneous system of $n+1$ *first-order PDEs for a scalar* $\Phi(x,u,\partial u)$ *given by*

$$\begin{aligned} \alpha^i\frac{\partial\Phi}{\partial x^i} + \beta\frac{\partial\Phi}{\partial u} + \lambda_i\frac{\partial\Phi}{\partial u_i} &= 0, \\ \alpha^{ij}\frac{\partial\Phi}{\partial x^i} + \beta^j\frac{\partial\Phi}{\partial u} + \lambda_i^j\frac{\partial\Phi}{\partial u_i} &= 0, \quad j=1,\ldots,n, \end{aligned} \tag{2.109}$$

has $X^1(x,u,\partial u),\ldots,X^n(x,u,\partial u)$ *as* n *functionally independent solutions.*

(ii) The linear inhomogeneous system of $n+1$ *first-order PDEs*

$$\begin{aligned} \alpha^i\frac{\partial\Psi}{\partial x^i} + \beta\frac{\partial\Psi}{\partial u} + \lambda_i\frac{\partial\Psi}{\partial u_i} &= 1, \\ \alpha^{ij}\frac{\partial\Psi}{\partial x^i} + \beta^j\frac{\partial\Psi}{\partial u} + \lambda_i^j\frac{\partial\Psi}{\partial u_i} &= 0, \quad j=1,\ldots,n, \end{aligned} \tag{2.110}$$

has a particular solution $\Psi = \psi(x,u,\partial u)$.

(iii) The linear inhomogeneous system of $n(n+1)$ *first-order PDEs*

$$\begin{aligned} \alpha^i\frac{\partial\Psi_j}{\partial x^i} + \beta\frac{\partial\Psi_j}{\partial u} + \lambda_i\frac{\partial\Psi_j}{\partial u_i} &= 1, \quad j=1,\ldots,n, \\ \alpha^{ik}\frac{\partial\Psi_j}{\partial x^i} + \beta^k\frac{\partial\Psi_j}{\partial u} + \lambda_i^k\frac{\partial\Psi_j}{\partial u_i} &= \delta_j^k, \quad j,k=1,\ldots,n, \end{aligned} \tag{2.111}$$

where δ_j^k *is the Kronecker symbol, has a particular solution*

$$(\Psi_1,\Psi_2,\ldots,\Psi_n) = \partial\psi = (\psi_1(x,u,\partial u),\psi_2(x,u,\partial u),\ldots,\psi_n(x,u,\partial u)).$$

(iv) There exists a particular solution $\partial\psi$ *of* (2.111) *such that* $(z,w,\partial w) = (X(x,u,\partial u),\psi(x,u,\partial u),\partial\psi(x,u,\partial u))$ *defines a contact transformation.*

Then the mapping μ *given by*

$$\begin{aligned} z^j &= \phi^j(x, u, \partial u) = X^j(x, u, \partial u), \\ w &= \psi(x, u, \partial u), \\ w_j &= \psi_j(x, u, \partial u), \end{aligned} \tag{2.112}$$

$j = 1, \ldots, n$, is invertible and transforms $\mathbf{R}\{x\,; u\}$ to the linear scalar PDE $\mathbf{S}\{z\,; w\}$ given by

$$\mathrm{L}[z]w = g(z), \tag{2.113}$$

which may involve an inhomogeneous term $g(z)$.

Proof. By construction, the mapping μ defined by (2.112) through a solution of (2.109)–(2.111) is invertible. Let $f(z) = F(x, u, \partial u)$. Then μ transforms the contact symmetry infinitesimal generator X of the scalar PDE $\mathbf{R}\{x\,; u\}$ to the point symmetry infinitesimal generator Z, of a target scalar PDE $\mathbf{S}\{z\,; w\}$ where Z is of the form

$$\mathrm{Z} = f(z)\frac{\partial}{\partial w} \tag{2.114}$$

for any solution $f(z)$ satisfying the linear scalar PDE $\mathrm{L}[z]f(z) = 0$. But since only a linear PDE of the form (2.113) can be characterized as having a point symmetry of the form (2.114) and since the mapping μ defined by the transformation (2.112) is invertible, it follows that a target PDE of the mapping μ must be of the form (2.113). □

Remark 2.4.2. Using the properties of a contact transformation (1.15), one can more readily determine the components of $\partial w = \partial\psi$ of the contact transformation (2.112) through replacing the sufficiency conditions (iii) and (iv) in Theorem 2.4.4 by the much simpler equations given by

$$\begin{aligned} \frac{\partial\psi}{\partial u_i} - \psi_j\frac{\partial\phi^j}{\partial u_i} &= 0, \\ \frac{\partial\psi}{\partial x^i} + u_i\frac{\partial\psi}{\partial u} &= \psi_j\left(\frac{\partial\phi^j}{\partial x^i} + u_i\frac{\partial\phi^j}{\partial u}\right), \end{aligned} \tag{2.115}$$

$i = 1, \ldots, n$. In equations (2.115), one has $\phi^j = X^j$, $j = 1, \ldots, n$, and ψ is determined from the sufficiency conditions (i) and (ii), respectively.

Now consider two examples of mapping nonlinear scalar PDEs to linear PDEs through contact transformations that are not point transformations.

(1) Linearization by a Legendre transformation

The second-order quasilinear PDE $\mathbf{R}\{x\,; u\}$, with independent variables $(x^1, x^2) = (x, t)$, given by

$$a(u_x, u_t)u_{xx} + 2b(u_x, u_t)u_{xt} + c(u_x, u_t)u_{tt} = 0, \tag{2.116}$$

has an infinite set of contact symmetries in terms of the evolutionary infinitesimal generator

$$F(x,u,\partial u)\frac{\partial}{\partial u}, \tag{2.117}$$

where $F(x,u,\partial u)=F(u_x,u_t)$ is any solution of the second-order linear PDE

$$a(u_x,u_t)\frac{\partial^2 F}{\partial u_x^2}-2b(u_x,u_t)\frac{\partial^2 F}{\partial u_x\partial u_t}+c(u_x,u_t)\frac{\partial^2 F}{\partial u_t^2}=0. \tag{2.118}$$

From Theorem 1.2.5, it follows that the symmetry given by (2.117) and (2.118) is uniquely equivalent to an infinitesimal generator of a contact symmetry $\mathrm{X}=\xi\,\partial/\partial x+\tau\,\partial/\partial t+\eta\,\partial/\partial u+\eta^x\,\partial/\partial u_x+\eta^t\,\partial/\partial u_t$ with

$$\xi=-\frac{\partial F}{\partial u_x},\ \tau=-\frac{\partial F}{\partial u_t},\ \eta=F-u_x\frac{\partial F}{\partial u_x}-u_t\frac{\partial F}{\partial u_t},\ \eta^x=\eta^t=0. \tag{2.119}$$

Clearly, the contact symmetry (2.117)–(2.119) satisfies the necessary conditions of Theorem 2.4.3 for the existence of an invertible mapping μ that transforms the quasilinear PDE (2.116) to some linear PDE. In particular, from (2.119), one sees that $\alpha^1=\alpha^2=\alpha^{12}=\alpha^{21}=0$, $\alpha^{11}=\alpha^{22}=-1, \beta=1$, $\beta^1=-u_x, \beta^2=-u_t$, $\lambda_i=\lambda_i^j=0$, $i,j=1,2$. Moreover, from the form of PDE (2.118), it follows that $X^1=u_x$, $X^2=u_t$.

The existence and construction of μ follow from applying Theorem 2.4.4. Here equations (2.109) become

$$\frac{\partial\Phi}{\partial u}=0,\quad \frac{\partial\Phi}{\partial x}+u_x\frac{\partial\Phi}{\partial u}=0,\quad \frac{\partial\Phi}{\partial t}+u_t\frac{\partial\Phi}{\partial u}=0, \tag{2.120}$$

with obvious functionally independent solutions

$$\Phi=\phi^1=X^1=u_x, \Phi=\phi^2=X^2=u_t, \tag{2.121}$$

and equations (2.110) become

$$\frac{\partial\Psi}{\partial u}=1,\quad \frac{\partial\Psi}{\partial x}+u_x\frac{\partial\Psi}{\partial u}=0,\quad \frac{\partial\Psi}{\partial t}+u_t\frac{\partial\Psi}{\partial u}=0. \tag{2.122}$$

It is easy to determine that a particular solution of the linear PDE system (2.122) is given by

$$\Psi=\psi=u-xu_x-tu_t. \tag{2.123}$$

Then, after substituting the solutions (2.121) and (2.123) into the corresponding four equations (2.115), one directly obtains

$$\psi_1=-x,\quad \psi_2=-t. \tag{2.124}$$

Consequently, the invertible mapping μ given by the contact transformation

$$z^1 = u_x, \quad z^2 = u_t, \quad w = u - xu_x - tu_t, \quad w_1 = -x, \quad w_2 = -t, \tag{2.125}$$

transforms the quasilinear PDE (2.116) to the linear PDE $\mathbf{S}\{z\,;w\}$ given by

$$a(u_x, u_t)\frac{\partial^2 w}{\partial u_x^2} - 2b(u_x, u_t)\frac{\partial^2 w}{\partial u_x \partial u_t} + c(u_x, u_t)\frac{\partial^2 w}{\partial u_t^2} = 0. \tag{2.126}$$

The mapping (2.125) is the well-known Legendre transformation that linearizes the quasilinear PDE (2.116).

(2) Linearization of an equation arising in a fluid flow problem
Sukharev (1967) showed that for any constant p, the first-order nonlinear PDE system

$$\frac{\partial v^2}{\partial t} + \frac{\partial v^1}{\partial x} = 0, \tag{2.127a}$$

$$v^1\frac{\partial v^2}{\partial t} - (v^2)^p = 0, \tag{2.127b}$$

which describes a fluid flow through a long pipeline, has the infinite set of point symmetries given by the infinitesimal generator

$$\mathrm{X} = g(t, v^2)\frac{\partial}{\partial x} + \left[(v^2)^p\frac{\partial g(t, v^2)}{\partial v^2}\right]\frac{\partial}{\partial v^1}, \tag{2.128}$$

where $g(t, v^2)$ is any solution of the linear diffusion equation

$$\frac{\partial}{\partial v^2}\left((v^2)^p\frac{\partial g}{\partial v^2}\right) = \frac{\partial g}{\partial t}. \tag{2.129}$$

The use of (2.128) and (2.129) to linearize the nonlinear PDE system (2.127) by an invertible point transformation is left to Exercise 2.4.2.

Here, the conservation law (2.127a) is used to introduce a potential variable u through setting

$$v^1 = -u_t, \quad v^2 = u_x. \tag{2.130}$$

Then the system of PDEs (2.127) is equivalent to the nonlocally related scalar PDE $\mathbf{R}\{x\,;u\}$ given by

$$u_t u_{xx} + u_x^p = 0. \tag{2.131}$$

[The problem of finding equivalent nonlocally related systems of PDEs through conservation laws is fully considered in Chapter 3.]

The nonlinear scalar PDE (2.131) has an infinite set of contact symmetries in terms of an evolutionary infinitesimal generator of the form (2.117) where

$F(x, u, \partial u) = F(t, u_x)$ is any solution of the second-order linear PDE

$$\frac{\partial F}{\partial t} - u_x^p \frac{\partial^2 F}{\partial u_x^2} = 0. \tag{2.132}$$

Thus the symmetry given by (2.117) and (2.132) is uniquely equivalent to an infinite set of contact symmetries with an infinitesimal generator

$$\mathrm{X} = \xi \frac{\partial}{\partial \mathrm{x}} + \tau \frac{\partial}{\partial t} + \eta \frac{\partial}{\partial u} + \eta^x \frac{\partial}{\partial u_x} + \eta^t \frac{\partial}{\partial u_t},$$

where

$$\xi = -\frac{\partial F}{\partial u_x}, \ \tau = 0, \ \eta = F - u_x \frac{\partial F}{\partial u_x} - u_t \frac{\partial F}{\partial u_t}, \ \eta^x = 0, \quad \eta^t = \frac{\partial F}{\partial t}. \tag{2.133}$$

Clearly, the contact symmetries given by (2.117), (2.132), and (2.133) satisfy the necessary conditions of Theorem 2.4.3 for the existence of an invertible mapping μ that transforms the nonlinear PDE (2.129) to some linear PDE. In particular, from equations (2.133), one sees that $\alpha^1 = \alpha^2 = \alpha^{12} = \alpha^{21} = \alpha^{22} = 0$, $\alpha^{11} = -1$, $\beta = 1$, $\beta^1 = -u_x, \beta^2 = 0$, $\lambda_1 = \lambda_2 = \lambda_1^1 = \lambda_1^2 = \lambda_2^1 = 0$, $\lambda_2^2 = 1$. Moreover, from (2.132), it follows that $X^1 = t$, $X^2 = u_x$.

The existence and construction of μ follow from applying Theorem 2.4.4. Here equations (2.109) become

$$\frac{\partial \Phi}{\partial u} = 0, \quad \frac{\partial \Phi}{\partial x} + u_x \frac{\partial \Phi}{\partial u} = 0, \quad \frac{\partial \Phi}{\partial u_t} = 0, \tag{2.134}$$

with functionally independent solutions

$$\Phi = \phi^1 = X^1 = t, \Phi = \phi^2 = X^2 = u_x, \tag{2.135}$$

and equations (2.110) become

$$\frac{\partial \Psi}{\partial u} = 1, \quad \frac{\partial \Psi}{\partial x} + u_x \frac{\partial \Psi}{\partial u} = 0, \quad \frac{\partial \Psi}{\partial u_t} = 0. \tag{2.136}$$

A particular solution of (2.136) is given by

$$\Psi = \psi = u - x u_x. \tag{2.137}$$

Then, after substituting the solutions (2.135) and (2.137) into the corresponding equations (2.115), one directly obtains

$$\psi_1 = u_t, \quad \psi_2 = -x. \tag{2.138}$$

Consequently, the invertible mapping μ given by the contact transformation

$$z^1 = t, \quad z^2 = u_x, \quad w = u - xu_x, \quad w_1 = u_t, \quad w_2 = -x, \tag{2.139}$$

transforms the nonlinear PDE (2.129) to the linear PDE $\mathbf{S}\{z\,; w\}$ given by

$$u_x^p \frac{\partial^2 w}{\partial u_x^2} - \frac{\partial w}{\partial t} = 0. \tag{2.140}$$

The *non-invertible* transformation given by the composition of μ^{-1} with the non-invertible transformation (2.130) maps any solution of the second-order linear PDE (2.140) to a solution of the nonlinear system of PDEs (2.127). This linearization first appeared in Kumei (1981).

It should be remarked that the authors are unaware of any example that satisfies the necessary conditions but does not satisfy the sufficient conditions of the four theorems presented in this section. The work in this section first appeared in a less general and less structured form in Kumei & Bluman (1982). In particular, the work presented in this section follows and updates the presentations in Bluman & Kumei [(1989), (1990a,b)].

Exercises 2.4

2.4.1. Show that the following nonlinear scalar PDEs have point symmetries that lead to their linearizations and find the corresponding point transformations that yield their linearizations. [References to their linearizations are listed.]

(a) $u_t + \frac{1}{2}u_x^2 - u_{xx} = 0$ [Kumei & Bluman (1982)].

(b) $u_t + u_x - u_{xt} + u_x u_t = 0$ [Thomas (1944)].

2.4.2. Show that the point symmetries given by (2.128) and (2.129) satisfy the criteria of Theorems 2.4.1 and 2.4.2, and hence find an invertible mapping that linearizes the nonlinear PDE system given by (2.127).

2.4.3. Consider the nonlinear heat conduction equation

$$u_t - (u^{-2}u_x)_x = 0. \tag{2.141}$$

(a) Show that the nonlinear scalar PDE (2.141) cannot be linearized by an invertible contact transformation.

(b) Consider the nonlocally related PDE system [see Chapter 3]

$$\begin{aligned} v_t &= u^{-2}u_x, \\ v_x &= u. \end{aligned} \tag{2.142}$$

Show that the nonlinear system (2.142) has the point symmetries given by the infinitesimal generator

$$X = g(t,v)\frac{\partial}{\partial x} - \left[u^2 g_v(t,v)\right]\frac{\partial}{\partial u}, \tag{2.143}$$

where $g(t, v)$ is any solution of the linear heat equation

$$g_t - g_{vv} = 0. \tag{2.144}$$

Use Theorems 2.4.1 and 2.4.2 to linearize the PDE system (2.142) by an invertible point transformation.

(c) Find a relationship between the PDE systems (2.142) and (2.127).

(d) Show that $X = g(t, v)\,\partial/\partial x$ is an infinitesimal generator of an infinite set of point symmetries of $v_x^2 v_t - v_{xx}$ when $g(t, v)$ satisfies the linear heat equation (2.144).

(e) Hence find a mapping that transforms any solution of the linear heat equation to a solution of the nonlinear heat conduction equation (2.141). Is the mapping invertible? [Storm (1950); Rosen (1979); Bluman & Kumei (1980); Bluman, Kumei, & Reid (1988); Bluman & Kumei (1990b)].

2.4.4. Consider Burgers' equation

$$u_t + uu_x - u_{xx} = 0. \tag{2.145}$$

(a) Show that the nonlocally related nonlinear PDE system [Chapter 3]

$$\begin{aligned} v_x &= 2u, \\ v_t &= 2u_x - u^2, \end{aligned}$$

has an infinite set of point symmetries given by the infinitesimal generator $X = e^{u^2/4}\{[2g_x + ug]\,\partial/\partial u + 4g\,\partial/\partial v\}$ where $g(x, t)$ is any solution of the linear heat equation $g_t - g_{xx} = 0$ [Vinogradov & Krasil'shchik (1984); Kersten (1987)].

(b) Hence, appropriately apply Theorems 2.4.1 and 2.4.2 to derive the Hopf–Cole transformation that linearizes Burgers' equation (2.145).

2.4.5. For the second-order quasilinear PDE (2.116), let $v = u_x$, $w = u_t$. Now find a first-order quasilinear PDE system that is locally related to the quasilinear PDE (2.116). Consequently, use an appropriate hodograph transformation to find the Legendre transformation that linearizes (2.116).

2.4.6. Consider the PDE

$$u_t - u_{xx} - u_{yy} - 2(1-u)^{-1}(u_x^2 + u_y^2) - u + u^2 = 0. \tag{2.146}$$

(a) Show that the nonlinear PDE (2.146) has the infinite set of point symmetries given by the infinitesimal generator $X = (1-u)^2 F(x, y, t)\,\partial/\partial u$ where $F(x, y, t)$ is any solution of the linear PDE $F_t - F_{xx} - F_{yy} - F = 0$.

(b) Use the point symmetries exhibited in (a) to linearize the PDE (2.146).

2.4.7. The nonlinear PDE

$$x(u_x u_{xy} - u_y u_{xx}) + y(u_x u_{yy} - u_y u_{xy}) \tag{2.147}$$

arises in economics [Wagener (2004)].

(a) Find the point symmetries of the PDE (2.147) that satisfy the necessary conditions for its linearization.
(b) Use these point symmetries to linearize (2.147).
(c) Find the general solution of the PDE (2.147).

2.4.8. Consider the nonlinear reaction-diffusion equation [Bluman (1993)]

$$u_t - u^2 u_{xx} - 2bu^2 = 0, \quad b = \text{const} \neq 0. \tag{2.148}$$

(a) Show that the nonlinear PDE (2.148) cannot be linearized by a contact transformation.
(b) Show that the equivalent nonlocally related system (obtained from a conservation law of (2.148); see Chapter 3) given by

$$\begin{aligned} v_x &= u^{-1}, \\ v_t &= -u_x - 2bx, \end{aligned} \tag{2.149}$$

cannot be linearized by a point transformation.
(c) Show that the equivalent nonlocally related system given by

$$\begin{aligned} v_x &= u^{-1}, \\ w_x &= v, \\ w_t &= -u - bx^2, \end{aligned} \tag{2.150}$$

has the infinite set of point symmetries given by the infinitesimal generator

$$\begin{aligned} \mathrm{X} = e^{b(w-xv)}\{&(F(t,v) - bxH(t,v))\frac{\partial}{\partial x} \\ &+(G(t,v) - 2bxF(t,v) + (b^2x^2 - bu)H(t,v))\frac{\partial}{\partial u} \\ &+(vF(t,v) - (1+bxv)H(t,v))\frac{\partial}{\partial w}\}, \end{aligned}$$

where $(F(t,v), G(t,v), H(t,v))$ is an arbitrary solution of the linear PDE system $\partial H(t,v)/\partial v = F(t,v)$, $\partial H(t,v)/\partial t = G(t,v)$, $\partial F(t,v)/\partial v = G(t,v)$.
(d) Consequently, use Theorems 2.4.1 and 2.4.2 to linearize the PDE system (2.150) by a point transformation.

(e) Use Theorems 2.4.1 and 2.4.2 to find a point transformation that linearizes the equivalent nonlocally related subsystem given by

$$\begin{aligned} w_x &= v, \\ w_t &= -\left(\frac{1}{v_x} + bx^2\right). \end{aligned}$$

(f) Use Theorems 2.4.1 and 2.4.2 to find a contact transformation that linearizes the equivalent nonlocally related scalar PDE given by

$$w_t = -\left(\frac{1}{w_{xx}} + bx^2\right).$$

2.4.9. Consider the fourth-order nonlinear PDE [Broadbridge & Tritscher (1994)]

$$u_t = [u^{-1}(u^{-3}u_x)_x]_{xx}. \tag{2.151}$$

(a) Show that the PDE (2.151) cannot be linearized by a contact transformation.
(b) Apply Theorems 2.4.1 and 2.4.2 to linearize the equivalent nonlocally related system given by

$$\begin{aligned} v_x &= u, \\ v_t &= [u^{-1}(u^{-3}u_x)_x]_x. \end{aligned}$$

2.5 Invertible Mappings of Linear PDEs to Linear PDEs with Constant Coefficients

If a linear PDE has constant coefficients, then there is an arsenal of techniques (including use of Fourier series, Fourier and Laplace transforms) to find appropriate Green's functions and solve various posed boundary value problems. This leads to two obvious questions.

(i) Can one map a given linear PDE with variable coefficients to some linear PDE with constant coefficients by an invertible point transformation?
(ii) What is the most general point transformation that yields such a mapping?

The second question is connected with the problem of finding all domains that yield the possibility of Fourier or Laplace transform analysis for a given linear PDE. As in the case for invertible mappings of nonlinear PDEs to linear PDEs, both of these mapping questions can be formulated in terms of a class of point symmetries of the given PDE since a target class of constant coefficient linear PDEs is completely characterized by the point symmetries connected with its linearity and invariance under the abelian group of translations of its independent variables. Consequently, one is able to establish necessary and sufficient conditions for the existence of a mapping of a given variable coefficient linear PDE to some constant coefficient linear PDE. In particular, an algorithm is presented that determines whether such conditions hold for a given linear PDE, and finds an explicit mapping when one exists. The presented formulation and examples are restricted to linear scalar PDEs, but one should be able to extend the presented results to linear systems. This work first appeared in Bluman (1983) and in a more comprehensive form in Bluman & Kumei (1990a). In the next chapter, extensions of these results are made to non-invertible mappings of variable coefficient linear PDEs to equivalent, but nonlocally related, constant coefficient linear PDEs.

Consider a given pth-order linear PDE $\mathbf{R}\{x\,;u\}$ with n independent variables $x=(x^1,\ldots,x^n)$ and dependent variable u given by

$$a(x)u + a^i(x)u_i + \cdots + a^{i_1\cdots i_p}(x)u_{i_1\cdots i_p} = 0, \tag{2.152}$$

defined on a domain $D \subset \mathbb{R}^n$. The aim is to determine whether the variable coefficient linear PDE (2.152) can be mapped invertibly by some point transformation μ to a constant coefficient linear PDE $\mathbf{S}\{z\,;w\}$ of the form

$$bw + b^i w_i + \cdots + b^{i_1\cdots i_p}w_{i_1\cdots i_p} = 0, \tag{2.153}$$

with independent variables $z=(z^1,\ldots,z^n)$ and dependent variable w, and find such a mapping μ when one exists.

In order to preserve linearity, such a mapping μ must be a point transformation of the form

$$\begin{aligned} z^i &= \phi^i(x), \quad i=1,\ldots,n, \\ w &= \psi(x,u) = G(x)u; \end{aligned} \tag{2.154}$$

$G(x)$ is the *multiplier* of the mapping. The mapping (2.154) is invertible if and only if

$$\det\left|\frac{\partial \phi^i}{\partial x^j}\right| \neq 0 \text{ in } D. \tag{2.155}$$

A constant coefficient linear PDE $\mathbf{S}\{z\,;w\}$ with n independent variables z is invariant under the n-parameter Lie group $\mathcal{H}$ consisting of translations of its independent variables. Hence, since $\mathcal{H}$ is an abelian group, it is necessary that the given PDE $\mathbf{R}\{x\,;u\}$ admit an n-parameter abelian Lie group of point transformations $\mathcal{G}$ for the existence of an invertible mapping of $\mathbf{R}\{x\,;u\}$ to a constant coefficient linear PDE. Moreover, $\mathcal{G}$ must also be an n-parameter abelian Lie group of point transformations when its action is projected onto the space of n independent variables x since the mapping must preserve the commutation relations of the abelian Lie algebra $\mathcal{M}$ of the Lie group $\mathcal{H}$. An algorithm is now presented that yields such a mapping μ through establishing the existence of an n-dimensional abelian Lie algebra $\mathcal{L}$ for a set of n point symmetries of the given PDE $\mathbf{R}\{x\,;u\}$.

In particular, a constant coefficient linear PDE $\mathbf{S}\{z\,;w\}$ with n independent variables has n infinitesimal generators of translation symmetries given by

$$\mathrm{Z}_\alpha = \frac{\partial}{\partial z^\alpha}, \quad \alpha = 1, \ldots, n. \tag{2.156}$$

Consequently, in order for a mapping μ to exist, the given linear PDE $\mathbf{R}\{x\,;u\}$ must have n point symmetries in terms of n infinitesimal generators of the form [Bluman (1990)]

$$\mathrm{X}_\alpha = \xi^j_\alpha(x)\frac{\partial}{\partial x^j} + f_\alpha(x)u\frac{\partial}{\partial u}, \quad \alpha = 1, \ldots, n, \tag{2.157}$$

that satisfy the commutation relations [Theorem 2.2.3]

$$[\mathrm{X}_\alpha, \mathrm{X}_\beta] = 0, \quad \alpha, \beta = 1, \ldots, n. \tag{2.158}$$

From the mapping equations (2.13), it follows that *the first set of necessary conditions* for a mapping μ is given by

$$\begin{aligned} \mathrm{X}_\alpha\phi^i &= \mathrm{Z}_\alpha z^i = \delta^i_\alpha, \\ \mathrm{X}_\alpha(G(x)u) &= \mathrm{Z}_\alpha w = 0, \quad \alpha, i = 1, \ldots, n; \end{aligned} \tag{2.159}$$

δ^i_α is the Kronecker symbol. More explicitly, the mapping equations (2.159) are given by

$$\xi^j_\alpha \frac{\partial \phi^i}{\partial x^j} = \delta^i_\alpha, \quad \alpha, i = 1, \ldots, n; \tag{2.160}$$

$$\xi^j_\alpha \frac{\partial G}{\partial x^j} + f_\alpha G = 0, \quad \alpha = 1, \ldots, n. \tag{2.161}$$

From the mapping equations (2.160), it immediately follows that

$$\frac{\partial \phi^j}{\partial x^\alpha}\xi^i_j = \delta^i_\alpha, \quad \alpha, i = 1, , \ldots, n. \tag{2.162}$$

Hence from the invertibility condition (2.155) and the mapping equations (2.160), one sees that μ is invertible if and only if

$$\det|\xi^j_i(x)| \neq 0 \text{ in } D. \tag{2.163}$$

From the commutation relations (2.158), it follows that *the second set of necessary conditions* for a mapping μ is given by

$$\begin{aligned}\xi^k_\beta \frac{\partial \xi^j_\alpha}{\partial x^k} &= \xi^k_\alpha \frac{\partial \xi^j_\beta}{\partial x^k},\\ \xi^k_\beta \frac{\partial f_\alpha}{\partial x^k} &= \xi^k_\alpha \frac{\partial f_\beta}{\partial x^k}, \quad \alpha, \beta, j = 1, \ldots, n.\end{aligned} \tag{2.164}$$

The following theorem shows that the first and second sets of necessary conditions (2.159) and (2.164) are also sufficient conditions to determine the mapping μ.

Theorem 2.5.1. *If a given linear PDE* $\mathbf{R}\{x\,;u\}$ *has* n *point symmetries with infinitesimal generators of the form* (2.157) *whose components* $\{\xi^j_i(x), f_i(x)\}$ *satisfy the equations* (2.164) *and the invertibility condition* (2.163), *then there exists a solution* $\{\phi^i(x), G(x)\}$ *of the equations* (2.160), (2.161) *that defines an invertible mapping of* $\mathbf{R}\{x\,;u\}$ *to some constant coefficient linear PDE* $\mathbf{S}\{z\,;w\}$.

Proof. The proof is accomplished by showing that any set of functions $\{\phi^i(x), G(x)\}$ which solves the equations (2.160), (2.161), whose coefficients are defined by (2.163) and (2.164), satisfies the integrability conditions given by

$$\frac{\partial^2 \phi^i}{\partial x^j \partial x^k} = \frac{\partial^2 \phi^i}{\partial x^k \partial x^j}, \quad i, j, k = 1, \ldots, n; \tag{2.165a}$$

$$\frac{\partial^2 G}{\partial x^j \partial x^k} = \frac{\partial^2 G}{\partial x^k \partial x^j}, \quad j, k = 1, \ldots, n. \tag{2.165b}$$

It is now shown that equations (2.165a) are satisfied; the verification that equations (2.165b) are satisfied is analogous and left to Exercise 2.5.1.

Taking $\frac{\partial}{\partial x^k}$ of any equation of (2.160), one obtains $\xi^j_\alpha \frac{\partial^2 \phi^i}{\partial x^k \partial x^j} = -\frac{\partial \xi^j_\alpha}{\partial x^k}\frac{\partial \phi^i}{\partial x^j}$. Consequently, $\frac{\partial \phi^\alpha}{\partial x^l}\xi^j_\alpha \frac{\partial^2 \phi^i}{\partial x^k \partial x^j} = -\frac{\partial \phi^\alpha}{\partial x^l}\frac{\partial \xi^j_\alpha}{\partial x^k}\frac{\partial \phi^i}{\partial x^j}$, and hence from equations (2.162), one has

$$\frac{\partial^2 \phi^i}{\partial x^k \partial x^l} = -\frac{\partial \phi^\alpha}{\partial x^l}\frac{\partial \xi^j_\alpha}{\partial x^k}\frac{\partial \phi^i}{\partial x^j}. \tag{2.166}$$

Furthermore, from equations (2.162), one obtains

$$\frac{\partial^2 \phi^i}{\partial x^k \partial x^l} = \frac{\partial \phi^\alpha}{\partial x^k}\xi^m_\alpha \frac{\partial^2 \phi^i}{\partial x^m \partial x^l}. \tag{2.167}$$

After substituting the right-hand side of equations (2.166) into the right-hand side of equations (2.167) and rearranging the order of terms, one obtains

$$\frac{\partial^2 \phi^i}{\partial x^k \partial x^l} = -\frac{\partial \phi^\alpha}{\partial x^k}\frac{\partial \phi^\beta}{\partial x^l}\xi^m_\alpha \frac{\partial \xi^j_\beta}{\partial x^m}\frac{\partial \phi^i}{\partial x^j}. \tag{2.168}$$

After appropriately substituting the first set of equations in (2.164) into the right-hand side of (2.168) and then using (2.162), one finds that

$$\frac{\partial^2 \phi^i}{\partial x^k \partial x^l} = -\frac{\partial \phi^\alpha}{\partial x^k} \frac{\partial \xi_\alpha^j}{\partial x^l} \frac{\partial \phi^i}{\partial x^j}. \tag{2.169}$$

Comparing equations (2.166) and (2.169), one sees that the integrability conditions (2.165a) are satisfied. □

Now we summarize the mapping algorithm that firstly determines whether a given linear PDE $\mathbf{R}\{x\,;u\}$ can be mapped invertibly to a constant coefficient linear PDE $\mathbf{S}\{z\,;w\}$ and secondly yields such a mapping μ when one exists. The procedure is as follows.

(i) Set up the determining equations for the infinitesimals of the point symmetries of $\mathbf{R}\{x\,;u\}$. [*Note that it is unnecessary to solve explicitly the determining equations!*]

(ii) Use the determining equations to check if the coefficients of the given linear equation $\mathbf{R}\{x\,;u\}$ are such that the second set of necessary conditions (2.164) has a nontrivial solution for which $\det|\xi_i^j(x)| \neq 0$ in some domain D. If the system of equations (2.164) only has trivial solutions for which $\det|\xi_{ij}(x)| \equiv 0$, then no such invertible mapping μ exists.

(iii) Solve the system of equations (2.160) to find $\phi(x)$.

(iv) Find the multiplier $G(x)$ by solving either the system of equations (2.161) or, equivalently, the system of equations given by

$$G^{-1}\frac{\partial G}{\partial x^k} = -f_\alpha \frac{\partial \phi^\alpha}{\partial x^k}, \quad k = 1, \ldots, n. \tag{2.170}$$

In the case of two independent variables, we use the notations $x = x^1$, $y = x^2$, $\xi_1 = \xi_1^1$, $\xi_2 = \xi_2^1$, $\eta_1 = \xi_1^2$, $\eta_2 = \xi_2^2$. Then equations (2.164) become

$$\begin{aligned}
\xi_2 \frac{\partial \xi_1}{\partial x} + \eta_2 \frac{\partial \xi_1}{\partial y} &= \xi_1 \frac{\partial \xi_2}{\partial x} + \eta_1 \frac{\partial \xi_2}{\partial y}, \\
\eta_1 \frac{\partial \eta_2}{\partial y} + \xi_1 \frac{\partial \eta_2}{\partial x} &= \eta_2 \frac{\partial \eta_1}{\partial y} + \xi_2 \frac{\partial \eta_1}{\partial x}, \\
\xi_2 \frac{\partial f_1}{\partial x} + \eta_2 \frac{\partial f_1}{\partial y} &= \xi_1 \frac{\partial f_2}{\partial x} + \eta_1 \frac{\partial f_2}{\partial y};
\end{aligned} \tag{2.171}$$

equations (2.160) become the mapping equations

$$\begin{aligned}\xi_1\frac{\partial \phi^1}{\partial x}+\eta_1\frac{\partial \phi^1}{\partial y}&=1,\\ \xi_2\frac{\partial \phi^1}{\partial x}+\eta_2\frac{\partial \phi^1}{\partial y}&=0,\\ \xi_2\frac{\partial \phi^2}{\partial x}+\eta_2\frac{\partial \phi^2}{\partial y}&=1,\\ \xi_1\frac{\partial \phi^2}{\partial x}+\eta_1\frac{\partial \phi^2}{\partial y}&=0;\end{aligned} \tag{2.172}$$

equations (2.170) become

$$\begin{aligned}G^{-1}\frac{\partial G}{\partial x}&=-\left[f_1\frac{\partial \phi^1}{\partial x}+f_2\frac{\partial \phi^2}{\partial x}\right],\\ G^{-1}\frac{\partial G}{\partial y}&=-\left[f_1\frac{\partial \phi^1}{\partial y}+f_2\frac{\partial \phi^2}{\partial y}\right];\end{aligned} \tag{2.173}$$

and the determinant condition (2.163) becomes

$$\xi_1\eta_2 \neq \xi_2\eta_1 \quad \text{in some domain } D. \tag{2.174}$$

2.5.1 Examples of mapping variable coefficient linear PDEs to constant coefficient linear PDEs through invertible point transformations

Now consider two examples of mapping variable coefficient linear PDEs to constant coefficient linear PDEs.

(1) Parabolic equation
For any linear parabolic equation

$$\frac{\partial^2 u'}{\partial x'^2}+\alpha(x',y')\frac{\partial u'}{\partial x'}+\beta(x',y')\frac{\partial u'}{\partial y'}+\gamma(x',y')u'=0, \tag{2.175}$$

the transformation $x=X(x',y')=\int^{x'}[\beta(t,y')]^{1/2}dt$, $y=y'$, $u=e^{-C(x,y)}u'$ where $C_x=-\frac{1}{2}[X_{y'}+\alpha(x',y')[\beta(x',y')^{-1/2}+\frac{1}{2}\beta^{-3/2}\beta_{x'}]$, maps (2.175) to the standard form

$$u_{xx}+u_y+V(x,y)u=0 \tag{2.176}$$

with $V(x,y)=C_y+C_{xx}+C_x^2+\gamma(x',y')/\beta(x',y')$.

In Bluman (1980), it was shown explicitly that a PDE of the form (2.176) can be mapped by an invertible point transformation to the "backward" heat equation $w_{z^1z^1}+w_{z^2}=0$ if and only if $V(x,y)$ is of the form

$$V(x,y) = q_0(y) + q_1(y)x + q_2(y)x^2 \tag{2.177}$$

for arbitrary coefficients $q_i(y)$. This was accomplished by mapping the nontrivial point symmetries of (2.176) into the six nontrivial point symmetries of the backward heat equation. It is now shown that the mapping algorithm presented in Section 2.5 achieves this result in a simpler way and in a more general context.

If a linear PDE (2.176) has two nontrivial point symmetries with infinitesimal generators $\mathrm{X}_i = \xi_i(x,y)\,\partial/\partial x + \eta_i(x,y)\,\partial/\partial y + f_i(x,y)u\,\partial/\partial u$, then one can show that the determining equations for their infinitesimal components (ξ_i, η_i, f_i) reduce to

$$\begin{aligned}
\eta_i &= \eta_i(y),\\
\xi_i &= \tfrac{1}{2}\eta_i'(y)x + A_i(y),\\
f_i &= \tfrac{1}{8}\eta_i''(y)x^2 + \tfrac{1}{2}A_i'(y)x + B_i(y),
\end{aligned} \tag{2.178}$$

where $\eta_i(y), A_i(y), B_i(y), V(x,y)$ satisfy the identity

$$\begin{aligned}
&\tfrac{1}{8}\eta_i'''(y)x^2 + \tfrac{1}{2}A_i''(y)x + \tfrac{1}{4}\eta_i''(y) + B_i'(y)\\
&\quad+[\tfrac{1}{2}\eta_i'(y)x + A_i(y)]V_x + \eta_i(y)V_y + \eta_i'(y)V \equiv 0, \quad i=1,2.
\end{aligned} \tag{2.179}$$

In order to satisfy the determinant (invertibility) condition (2.174), one must have

$$\eta_2(y)[\eta_1'(y)x + 2A_1(y)] \neq \eta_1(y)[\eta_2'(y)x + 2A_2(y)] \tag{2.180}$$

in some domain D.

The first two equations of (2.171) here become $\eta_1\eta_2' = \eta_2\eta_1'$, $A_2\eta_1' + 2A_1'\eta_2 = A_1\eta_2' + 2A_2'\eta_1$, with their general solution given by

$$\eta_1 = k\eta_2, \quad A_1 = kA_2 + l\eta_2^{1/2}, \tag{2.181}$$

for arbitrary constants k, l. In terms of (2.181), the invertibility condition (2.180) becomes $l\eta_2^{3/2} \neq 0$. Hence it is necessary that $l \neq 0$, $\eta_2 \neq 0$. Since the linear transformation $\bar{z}^1 = lz^1$, $\bar{z}^2 = z^2 + kz^1$, maps any constant coefficient linear PDE to a constant coefficient linear PDE, without loss of generality, one can correspondingly use $\bar{\mathrm{X}}_2 = \mathrm{X}_2$, $\bar{\mathrm{X}}_1 = l^{-1}[\mathrm{X}_1 - k\mathrm{X}_2]$ as the infinitesimal generators of point symmetries of PDE (2.176). Then $\bar{\xi}_2 = \xi_2$, $\bar{\eta}_2 = \eta_2$, $\bar{\xi}_1 = l^{-1}[\xi_1 - k\xi_2]$, $\bar{\eta}_1 = l^{-1}[\eta_1 - k\eta_2]$, $\bar{\phi}^1 = l\phi^1$, $\bar{\phi}^2 = \phi^2 + k\phi^1$. Moreover, $(\bar{\xi}_1, \bar{\eta}_1, \bar{\xi}_2, \bar{\eta}_2, \bar{\phi}^1, \bar{\phi}^2)$ satisfy the mapping equations (2.172) if and only if $(\xi_1, \eta_1, \xi_2, \eta_2, \phi^1, \phi^2)$ do. Now relabel the barred quantities by unbarring them. Hence, without loss of generality, in equations (2.181) one can set $k = 0$, $l = 1$, i.e.,

$$\eta_1 = 0, \quad A_1 = \eta_2^{1/2}. \tag{2.182}$$

Thus for $i = 1$, the reduced determining equation (2.179) becomes

$$A_1''(y)x + 2[B_1'(y) + A_1(y)V_x] \equiv 0. \tag{2.183}$$

Hence it is necessary that $V_{xxx} = 0$. This yields the quadratic condition (2.177).

Now assume that $V(x,y)$ is of the form (2.177). Then from the identity (2.183), one obtains

$$A_1'' + 4q_2(y)A_1 = 0, \tag{2.184a}$$

$$B_1' = -q_1(y)A_1. \tag{2.184b}$$

The third equation of (2.171), the third equation of (2.178), and equations (2.182) lead to

$$A_1A_2' - A_2A_1' = 2B_1'\eta_2 = 2B_1'A_1^2. \tag{2.185}$$

Consequently,

$$A_2 = 2B_1A_1 \tag{2.186}$$

is a solution of (2.185). Finally, for $i = 2$, equation (2.179) leads to $B_2(y)$ satisfying

$$B_2' = -\tfrac{1}{4}\eta_2'' - [q_0(y)\eta_2]' - q_1(y)A_2. \tag{2.187}$$

If $\eta_2(y)$ is known, then $A_1(y), A_2(y), B_1(y), B_2(y)$ can be determined from equations (2.182), (2.184), (2.186), and (2.187). Hence, *from Theorem* 2.5.1, *it follows that a linear parabolic PDE* (2.176) *can be mapped invertibly by a point transformation to a constant coefficient linear PDE if and only if* $V(x,y)$ *is of the quadratic form given by* (2.177).

Now the mapping is constructed by solving the mapping equations (2.172) and (2.173). From equations (2.182), the fourth equation of (2.172) yields $\phi^2 = \phi^2(y)$. Consequently, the third equation of (2.172) leads to

$$(\phi^2)' = \eta_2(y)^{-1}. \tag{2.188}$$

From equations (2.182), (2.184a), and (2.188), one sees that $\phi^2(y) = T(y)$ is any solution of the ODE

$$2T'''T' - 3T''^2 - 16q_2(y)T'^2 = 0. \tag{2.189}$$

Let $M(y) = T''(y)/T'(y)$. Then the solution of the nonlinear third-order ODE (2.189) reduces to solving the Riccati equation given by

$$2M' = M^2 + 16q_2(y), \tag{2.190}$$

followed by two obvious quadratures [Bluman (1980)]. From the Riccati equation (2.190), one sees that on a domain where $q_2(y) \geq 0$, it follows that $T'(y)$

is a monotone function. Hence on such a domain, a solution of the ODE (2.189) yields an invertible point transformation. Note that the Riccati transformation $M = 2Y'/Y$ maps any solution of the second-order linear ODE $Y'' - 4q_2(y)Y = 0$ to a solution of the Riccati equation (2.190).

Having found $T'(y)$, one now proceeds as follows. The first two equations of (2.172) lead to $\phi^1(x,y) = [T'(y)]^{1/2}x + D(y)$ where $D'(y) = -A_2(y)[T'(y)]^{3/2}$. Finally, equations (2.173) yield the multiplier $G(x,y)$ through the equation

$$\log G(x,y) = -\tfrac{1}{4}(A_1'(y)/A_1(y))x^2 + (B_1(y)/A_1(y))x + W(y),$$

where $W'(y) = B_2(y)T'(y) - A_2(y)B_1(y)[T'(y)]^{3/2}$.

Thus $\phi^1(x,y), \phi^2(x,y) = \phi^2(y)$ and $G(x,y)$ have been determined. One can then show that the resulting constant coefficient linear PDE $\mathbf{S}\{z\,; w\}$ is given by

$$\frac{\partial^2 w}{\partial (z^1)^2} + \frac{\partial w}{\partial z^2} + rw = 0,$$

where $r = \frac{1}{2}A_1'(y)A_1(y) - [B_1(y)]^2 + B_2(y) + q_0(y)[A_1(y)]^2 = \text{const}.$

(2) Hyperbolic equation

One can show that the linear hyperbolic PDE $\mathbf{R}\{x\,; u\}$ given by

$$u_{xy} + \alpha(x,y)u_x + \beta(x,y)u_y + \gamma(x,y)u = 0 \tag{2.191}$$

has two nontrivial point symmetries given by the infinitesimal generators $\mathrm{X}_j = \xi_j(x,y)\,\partial/\partial x + \eta_j(x,y)\,\partial/\partial y + f_j(x,y)u\,\partial/\partial u$ if and only if

$$\xi_j = \xi_j(x), \tag{2.192a}$$
$$\eta_j = \eta_j(y), \tag{2.192b}$$
$$\frac{\partial f_j}{\partial x} = -[\beta\xi_j'(x) + \beta_x\xi_j(x) + \beta_y\eta_j(y)], \tag{2.192c}$$
$$\frac{\partial f_j}{\partial y} = -[\alpha\eta_j'(y) + \alpha_x\xi_j(x) + \alpha_y\eta_j(y)], \tag{2.192d}$$

$$\frac{\partial^2 f_j}{\partial x\partial y} + \alpha\frac{\partial f_j}{\partial x} + \beta\frac{\partial f_j}{\partial y} + \gamma[\xi_j'(x) + \eta_j'(y)] + \gamma_x\xi_j(x) + \gamma_y\eta_j(y) = 0, \tag{2.192e}$$

$j = 1, 2$. In order to satisfy the determinant condition (2.174), it is necessary that $\xi_1(x)\eta_2(y) \neq \xi_2(x)\eta_1(y)$ in some domain D. Here the first two equations of (2.171) become $\eta_2(y) = k\eta_1(y)$, $\xi_1(x) = l\xi_2(x)$, for arbitrary constants k, l. Let $\bar{z}^1 = z^1 + kz^2$, $\bar{z}^2 = z^2 + lz^1$. Then, without loss of generality, one can set $k = l = 0$. Hence one obtains

$$\xi_1(x) = 0, \quad \eta_2(y) = 0. \tag{2.193}$$

From equations (2.192c), (2.192d) and (2.193), one now has

$$\frac{\partial f_1}{\partial x} = -\beta_y \eta_1(y), \tag{2.194a}$$

$$\frac{\partial f_1}{\partial y} = -[\alpha\eta_1'(y) + \alpha_y \eta_1(y)], \tag{2.194b}$$

$$\frac{\partial f_2}{\partial x} = -[\beta\xi_2'(x) + \beta_x \xi_2(x)], \tag{2.194c}$$

$$\frac{\partial f_2}{\partial y} = -\alpha_x \xi_2(x). \tag{2.194d}$$

Next, the third equation of (2.171) and equations (2.194a), (2.194d) lead to the *first necessary condition*

$$\beta_y = \alpha_x. \tag{2.195}$$

Let

$$\delta(x,y) = \alpha_x + \alpha\beta - \gamma. \tag{2.196}$$

Then after substituting equations (2.194) into equation (2.192e) for $j = 1, 2$, and using equation (2.196), one obtains

$$\begin{aligned} \delta_y \eta_1(y) + \delta\eta_1'(y) = 0, \\ \delta_x \xi_2(x) + \delta\xi_2'(x) = 0. \end{aligned} \tag{2.197}$$

This yields the *second necessary condition* $[\log\delta]_{xy} = 0$, i.e., $\delta(x,y)$ must be of the separable form

$$\delta(x,y) = mA(x)B(y), \tag{2.198}$$

for some $A(x)$, $B(y)$ with $m = 0$ if $\delta \equiv 0$, and $m = 1$ if $\delta \not\equiv 0$. If $m = 1$, then from equations (2.197), one obtains $\eta_1(y) = [B(y)]^{-1}$, $\xi_2(x) = [A(x)]^{-1}$. If $m = 0$, then $\eta_1(y)$ and $\xi_2(x)$ can be arbitrary functions of their respective arguments. A solution of equations (2.194) is given by $f_1(x,y) = -\alpha(x,y)\eta_1(y)$, $f_2(x,y) = -\beta(x,y)\xi_2(x)$. Consequently, from Theorem 2.5.1, it follows that the variable coefficient linear PDE (2.191) can be mapped invertibly by a point transformation to a constant coefficient linear PDE if and only if its coefficients $\alpha(x,y)$, $\beta(x,y)$, and $\gamma(x,y)$ satisfy the equations

$$\beta_y = \alpha_x, \quad \alpha_x + \alpha\beta - \gamma = mA(x)B(y), \tag{2.199}$$

for some functions $A(x)$, $B(y)$, and constant m.

For any coefficients $\alpha(x,y)$, $\beta(x,y)$, and $\gamma(x,y)$ satisfying equations (2.199), we now construct a mapping μ of the linear hyperbolic PDE (2.191) to some constant coefficient linear PDE $\mathbf{S}\{z\,;w\}$. Here the mapping equations (2.172) lead to

$$\phi^1(x,y) = \phi^1(y) = \int \frac{1}{\eta_1(y)}dy, \quad \phi^2(x,y) = \phi^2(x) = \int \frac{1}{\xi_2(x)}dx,$$

and the mapping equations (2.173) yield a multiplier $G(x,y)$ satisfying

$$\log G(x,y) = \int \alpha(x,y)\,dy.$$

Thus $z^1 = \phi^1(x,y) = \phi^1(y)$, $z^2 = \phi^2(x,y) = \phi^2(x)$, and the multiplier $G(x,y)$ have been determined. The corresponding constant coefficient linear PDE $\mathbf{S}\{z\,;w\}$ is given by

$$w_{z^1z^2} - mw = 0. \tag{2.200}$$

In summary, the following theorem has been proved.

Theorem 2.5.2. *The linear hyperbolic PDE* (2.191) *can be mapped invertibly by a point transformation to a constant coefficient linear PDE if and only if its coefficients $\alpha(x,y)$, $\beta(x,y)$, and $\gamma(x,y)$ satisfy equations* (2.199) *for some functions $A(x)$, $B(y)$, and constant m.*

(i) If $m = 0$, the mapping μ given by the invertible point transformation

$$z^1 = x, \quad z^2 = y, \quad w = u\exp\left[\int \alpha(x,y)dy\right],$$

transforms the PDE (2.191) *to the wave equation $w_{z^1z^2} = 0$.*

(ii) If $m = 1$, the mapping μ given by the invertible point transformation

$$z^1 = \int B(y)\,dy, \quad z^2 = \int A(x)\,dx, \quad w = u\exp\left[\int \alpha(x,y)dy\right],$$

transforms the PDE (2.191) *to the Klein–Gordon equation $w_{z^1z^2} - w = 0$.*

2.5.2 Example of finding the most general mapping of a given constant coefficient linear PDE to some constant coefficient linear PDE

Now consider the problem of finding the most general invertible point transformation that can map a given constant coefficient linear PDE $\mathbf{R}\{x\,;u\}$ to some constant coefficient linear PDE $\mathbf{S}\{z\,;w\}$. For a given linear PDE, this allows one to find all domains that yield Fourier or Laplace transform analyses. To accomplish this, one modifies the mapping algorithm presented in Section 2.5 to find the general solution of the second set of necessary conditions (2.164) and then the general solution of the mapping equations (2.160), (2.161). As an example, the most general invertible point transformation is found that maps the biharmonic equation

$$\Delta^2 u \equiv \nabla^2(\nabla^2)u = \left(\frac{\partial^2}{\partial x^2} + \frac{\partial^2}{\partial y^2}\right)^2 u = 0 \tag{2.201}$$

to a constant coefficient linear PDE [Bluman & Gregory (1985)]. One can show that the biharmonic equation (2.201) has the nontrivial point symmetries given by the infinitesimal generators

$$\begin{aligned} &\mathrm{X}_1 = (x^2 - y^2)\frac{\partial}{\partial x} + 2xy\frac{\partial}{\partial y} + 2xu\frac{\partial}{\partial u}, \\ &\mathrm{X}_2 = 2xy\frac{\partial}{\partial x} + (y^2 - x^2)\frac{\partial}{\partial y} + 2yu\frac{\partial}{\partial u}, \quad \mathrm{X}_3 = x\frac{\partial}{\partial x} + y\frac{\partial}{\partial y}, \\ &\mathrm{X}_4 = y\frac{\partial}{\partial x} - x\frac{\partial}{\partial y}, \quad \mathrm{X}_5 = \frac{\partial}{\partial x}, \quad \mathrm{X}_6 = \frac{\partial}{\partial y}, \quad \mathrm{X}_7 = u\frac{\partial}{\partial u}. \end{aligned} \tag{2.202}$$

Let $z = x + iy$. Then from the set of infinitesimal generators (2.202), one can determine that the set of nontrivial point symmetries of the biharmonic equation (2.201) is given by the group of point transformations

$$z^* = \frac{az + b}{cz + d}, \tag{2.203a}$$

$$u^* = \left|\frac{dz^*}{dz}\right|^{-1} u, \tag{2.203b}$$

where a, b, c, d are arbitrary complex constants with $ad - bc \neq 0$. Note that the group of point transformations (2.203a) is the group of bilinear transformations (Möbius group).

To find the most general mapping of the biharmonic equation (2.201) to a constant coefficient linear PDE, one first finds the most general two-parameter abelian group of point transformations that can result from the set of infinitesimal generators (2.202). Let

$$\mathrm{Y}_1 = \sum_{i=1}^{6} a_i \mathrm{X}_i, \quad \mathrm{Y}_2 = \sum_{i=1}^{6} b_i \mathrm{X}_i,$$

for arbitrary constants a_i, b_i. Then one can show that the commutator $[\mathrm{Y}_1, \mathrm{Y}_2] = 0$ if and only if the constants a_i, b_i satisfy the bilinear relations

$$\mathsf{Ba} = 0, \tag{2.204}$$

where

$$\mathbf{a} = \begin{bmatrix} a_1 \\ a_2 \\ \vdots \\ a_6 \end{bmatrix} \tag{2.205}$$

and

$$\mathsf{B} = \begin{bmatrix} -b_3 & b_4 & b_1 & -b_2 & 0 & 0 \\ -b_4 & -b_3 & b_2 & b_1 & 0 & 0 \\ b_5 & -b_6 & 0 & 0 & -b_1 & b_2 \\ b_6 & b_5 & 0 & 0 & -b_2 & -b_1 \\ 0 & 0 & -b_5 & b_6 & b_3 & -b_4 \\ 0 & 0 & -b_6 & -b_5 & b_4 & b_3 \end{bmatrix}. \tag{2.206}$$

In order to satisfy the determinant condition (2.174), one must have $\mathsf{B} \not\equiv 0$. One can show that rank $\mathsf{B} = 4$, and that two linearly independent solutions of equation (2.204) are given by

$$\mathbf{a} = \mathbf{a}^{(1)} = \begin{bmatrix} b_1 \\ b_2 \\ b_3 \\ b_4 \\ b_5 \\ b_6 \end{bmatrix} \tag{2.207}$$

and

$$\mathbf{a} = \mathbf{a}^{(2)} = \begin{bmatrix} b_2 \\ -b_1 \\ b_4 \\ -b_3 \\ b_6 \\ -b_5 \end{bmatrix}. \tag{2.208}$$

Now the mapping equations (2.172) are solved. One can show that in the case when $\mathbf{a} = \mathbf{a}^{(2)}$, the mapping functions $z^1 = \phi^1(x, y)$, $z^2 = \phi^2(x, y)$ satisfy the Cauchy-Riemann equations $\partial\phi^1/\partial x = \partial\phi^2/\partial y$, $\partial\phi^1/\partial y = -\partial\phi^2/\partial x$. Hence, in this case the mapping μ is conformal. Let

$$z = x + iy, \quad \zeta = z^1 + iz^2,$$
$$\alpha = a_1 + ia_2, \quad \beta = \tfrac{1}{2}(a_3 + ia_4), \quad \gamma = a_5 + ia_6.$$

Then the mapping equations (2.172) and (2.173) reduce to

$$\frac{d\zeta}{dz} = \frac{1}{\alpha z^2 + 2\beta z + \gamma}, \tag{2.209a}$$

$$G(x, y) = \left|\frac{d\zeta}{dz}\right|^{-1}. \tag{2.209b}$$

One can show that the general solution of equation (2.209a) can be represented in the form

$$\zeta = A^{-1}\log(1 + AZ) \text{ with } Z = \frac{az+b}{cz+d}, \tag{2.210}$$

where $\{A, a, b, c, d\}$ is a set of arbitrary complex constants with $ad - bc \neq 0$. The corresponding real constant coefficient linear PDE is given by

$$\begin{aligned}&\left(\frac{\partial^2}{\partial(z^1)^2} + \frac{\partial^2}{\partial(z^2)^2}\right)^2 w + [A^2 + \bar{A}^2]\left(\frac{\partial^2}{\partial(z^2)^2} - \frac{\partial^2}{\partial(z^1)^2}\right) w \\ &\quad +2i[\bar{A}^2 - A^2]\frac{\partial^2 w}{\partial z^1 \partial z^2} + |A|^4 w = 0,\end{aligned} \tag{2.211}$$

where $\bar{A}$ is the complex conjugate of A.

Note that in the limiting case $A = 0$, the mapping (2.210), (2.209b) yields the conformal mapping (2.203) that leaves invariant the biharmonic equation (2.201).

Exercises 2.5

2.5.1. Verify the integrability conditions (2.165b) of Theorem 2.5.1.

2.5.2.

(a) Show that if $\gamma \equiv 0$, then the linear hyperbolic PDE (2.191) can be mapped invertibly by a point transformation to the wave equation $w_{z^1z^2} = 0$ if and only if its coefficients $\alpha(x,y)$ and $\beta(x,y)$ are of the form

$$\alpha(x,y) = \frac{D'(y)}{C(x) + D(y)}, \quad \beta(x,y) = \frac{C'(x)}{C(x) + D(y)},$$

where $C(x)$ and $D(y)$ are arbitrary differentiable functions of their respective arguments.

(b) Find the mapping μ [Bluman (1983)].

2.5.3. Consider the class of linear hyperbolic PDEs given by

$$u_{tt} - c^2(x,t)u_{xx} = 0. \tag{2.212}$$

(a) Show that a PDE (2.212) can be mapped invertibly by a point transformation to the wave equation $w_{z^1z^1} - w_{z^2z^2} = 0$ if and only if $c(x,t)$ is of the form

$$c(x,t) = \frac{a_0 + 2a_1x + a_2x^2}{b_0 + 2b_1t + b_2t^2},$$

where the constants a_i, b_i are related by $a_1^2 - a_0a_2 = b_1^2 - b_0b_2 = \Delta$.

(b) Find the mapping μ [Distinguish between the cases $\Delta < 0$, $\Delta > 0$, $\Delta = 0$.] [Bluman (1983)].

2.5.4. Consider the class of linear hyperbolic PDEs given by

$$u_{tt} - c^2(x)u_{xx} = 0. \tag{2.213}$$

(a) Show that a PDE (2.213) can be mapped invertibly by a point transformation to a constant coefficient linear PDE if and only if $c(x)$ satisfies the fourth-order ODE

$$\left[\frac{c^2 c'''}{2cc'' - c'^2}\right]' = 0. \tag{2.214}$$

(b) By inspection, find three point symmetries of the ODE (2.214). Show that the Lie algebra, formed from their infinitesimal generators, is solvable [Bluman & Anco (2002)]. Hence, use these point symmetries to reduce the ODE (2.214) to a first-order ODE and three quadratures.
(c) Find the mapping μ [Bluman (1983)].

2.5.5. Consider the class of linear elliptic PDEs given by

$$u_{xx} + u_{yy} + \alpha(x,y)u_x + \beta(x,y)u_y + \gamma(x,y)u = 0. \tag{2.215}$$

(a) Show that a PDE (2.215) can be mapped invertibly by a point transformation μ to a constant coefficient linear PDE if and only if its coefficients $\alpha(x,y)$, $\beta(x,y)$, and $\gamma(x,y)$ satisfy the equations $\beta_x = \alpha_y$, $2(\alpha_x + \beta_y) + \alpha^2 + \beta^2 - 4\gamma = |K(z)|^2$, for some analytic function $K(z)$ of the complex variable $z = x + iy$.
(b) Find a mapping μ and show that the resulting constant coefficient linear PDE is equivalent to

  (i) Laplace's equation $w_{z^1z^1} + w_{z^2z^2} = 0$ if $K(z) \equiv 0$;
  (ii) the Helmholtz equation $w_{z^1z^1} + w_{z^2z^2} - w = 0$ if $K(z) \not\equiv 0$.

2.5.6. Show that the most general invertible point transformation mapping the biharmonic equation (2.201) into a linear PDE with constant coefficients is given by

$$x = X(pz^1 + qz^2, rz^1 + sz^2), \quad y = Y(pz^1 + qz^2, rz^1 + sz^2),$$

in terms of any conformal transformation, $x = X(z^1, z^2), y = Y(z^1, z^2)$, which is the inverse of the conformal transformation given by (2.210); p, q, r, s are arbitrary real constants with $ps - qr \neq 0$. Find the corresponding multiplier $G(x,y)$ [Bluman & Gregory (1985)].

2.5.7. Find all forms of the coefficient $h(x)$ so that the linear PDE

$$h(x)\varphi_{xx} + \varphi_x - \varphi_{yy} = 0$$

can be mapped into the linear wave equation by some invertible point transformation [Carbonaro (1991)].

2.6 Invertible Mappings of Nonlinear PDEs to Linear PDEs Through Conservation Law Multipliers

In this section, an alternative algorithm is presented for invertibly mapping a given nonlinear PDE system to some linear PDE system by means of a class of conservation law multipliers of the nonlinear PDE system. The premise for this algorithm is the observation that a linear PDE system possesses an infinite set of conservation law multipliers satisfying its adjoint system. This feature of a linear PDE system isexploited to detect whether a given nonlinear system can be linearized by an invertible transformation and, when such a linearization mapping exists, to obtain the explicit form of the linearizing transformation from an associated conservation law identity.

In particular, using the determining equations for the conservation law multipliers, one can see whether an invertible linearization mapping exist, and also find the adjoint of a target linear system as well as the independent variables of a target linear system whenever a linearization mapping is possible [Bluman & Doran-Wu (1995)]. Furthermore, it is shown that through a conservation law identity coming from the multiplier equations of an augmented system consisting of the given nonlinear PDE system and the adjoint of the target linear system, one can determine the dependent variables of the target linear system as well as find an explicit invertible linearization mapping (as is the case, discussed in Section 2.4, for linearization through an appropriate class of symmetries of a given nonlinear PDE system).

The work presented in this section appears in Anco, Bluman & Wolf (2008). In particular, this work allows one to detect and construct linearization mappings completely by the use of algorithmic methods for obtaining multipliers of conservation laws that were presented in Chapter 1 [Anco & Bluman [(1997a), (2002a,b)]; Wolf (2002a,b)].

The starting point, leading to an alternative invertible mapping algorithm relating nonlinear and linear PDEs through conservation law multipliers, is to notice that for any linear operator L and its adjoint operator L*, the formal relation $VLW - WL^*V$ is a divergence expression. Consider a kth-order linear PDE system $\mathbf{S}\{z\,;w\}$ with n independent variables $z = (z^1, \ldots, z^n)$ and m dependent variables $w = (w^1, \ldots, w^m)$, given by

$$\mathrm{L}^{\sigma}_{\alpha}[z]w^{\alpha} = 0, \tag{2.216}$$

in terms of linear operators

$$\begin{aligned}\mathrm{L}_\alpha^\sigma[z] &= b_\alpha^\sigma(z) + b_{\alpha i}^\sigma(z)\frac{\partial}{\partial z^i} + \cdots + b_{\alpha i_1\cdots i_k}^\sigma(z)\frac{\partial^k}{\partial z^{i_1}\cdots\partial z^k}, \\ \sigma &= 1,\ldots,M, \quad \alpha = 1,\ldots,m.\end{aligned} \tag{2.217}$$

The corresponding adjoint linear system is given by

$$\mathrm{L}_\alpha^{*\sigma}[z]v_\sigma = 0, \tag{2.218}$$

where, for arbitrary functions $V(z) = (V_1(z),\ldots,V_M(z))$, one has

$$\begin{aligned}\mathrm{L}_\alpha^{*\sigma}[z]V_\sigma = b_\alpha^\sigma(z)V_\sigma &- \frac{\partial}{\partial z^i}(b_{\alpha i}^\sigma(z)V_\sigma) + \cdots \\ &+(-1)^k\frac{\partial^k}{\partial z^{i_1}\cdots\partial z^{i_k}}(b_{\alpha i_1\cdots i_k}^\sigma(z)V_\sigma), \quad \alpha = 1,\ldots,m.\end{aligned} \tag{2.219}$$

Then for arbitrary functions $V(z)$ and $W(z) = (W^1(z),\ldots,W^m(z))$, which can be viewed as sets of multipliers $\{V_\sigma(z)\}$ and $\{W^\alpha(z)\}$ for the augmented linear system consisting of the linear system (2.216) and the adjoint system (2.218), there is a conservation law identity

$$\mathrm{D}_{z^i}\Lambda^i \equiv V_\sigma\mathrm{L}_\alpha^\sigma[z]W^\alpha - W^\alpha\mathrm{L}_\alpha^{*\sigma}[z]V_\sigma, \tag{2.220}$$

holding for some specific functions $\Lambda^i(z)$ that have a bilinear dependence on the components of the two sets of multipliers and their derivatives.

Now consider a nonlinear PDE system $\mathbf{R}\{x\,;u\}$ with n independent variables $x = (x^1,\ldots,x^n)$ and m dependent variables $u = (u^1,\ldots,u^m)$ given by

$$R^\sigma[u] = R^\sigma(x,u,\partial u,\ldots,\partial^k u) = 0, \quad \sigma = 1,\ldots,M. \tag{2.221}$$

Suppose the nonlinear PDE system $\mathbf{R}\{x\,;u\}$ (2.221) can be invertibly mapped to some linear PDE system $\mathbf{S}\{z\,;w\}$ by a point transformation (more generally, a contact transformation if $m=1$),

$$z = \phi(x,u), \quad w = \psi(x,u).$$

Then for some set of nontrivial factors $\{Q_\nu^\sigma[U]\}$, one must have

$$Q_\nu^\sigma[U]R^\nu[U] \equiv \mathrm{L}_\alpha^\sigma[z]W^\alpha, \quad \sigma = 1,\ldots,M, \tag{2.222}$$

for arbitrary functions $U(x) = (U^1(x),\ldots,U^m(x))$, where the functions $W(z) = (W^1(z),\ldots,W^m(z))$, are given by the linearization mapping

$$z = \phi(x,U(x)), \quad W(z) = \psi(x,U(x)). \tag{2.223}$$

Through the point transformation (2.223), the conservation law identity (2.220) becomes

$$\mathrm{D}_{x^i}\lambda^i \equiv \mathrm{J}[U](V_\sigma Q^\sigma_\nu[U]R^\nu[U] - W^\alpha \mathrm{L}^{*\,\sigma}_\alpha[z]V_\sigma), \tag{2.224}$$

where $\mathrm{D}_{x^i}\lambda^i = |\mathrm{D}z/\mathrm{D}x|\,\mathrm{D}_{z^i}\Lambda^i$ in terms of the non-vanishing Jacobian factor

$$\mathrm{J}[U] = \left|\frac{\mathrm{D}z}{\mathrm{D}x}\right| = \det\left(\frac{\mathrm{D}z^i}{\mathrm{D}x^j}\right). \tag{2.225}$$

[See Bluman (2005) or Section 1.5.1 for the explicit expressions for λ^i.]

This leads to the following two theorems for linearization from conservation law multipliers which are the counterparts to Theorems 2.4.1 and 2.4.2 for linearization through point symmetries.

Theorem 2.6.1 (Necessary conditions for the existence of an invertible linearization mapping). *If there exists an invertible point transformation* (2.61) *under which a given kth-order nonlinear PDE system $\mathbf{R}\{x\,;u\}$* (2.221) *is mapped to some linear PDE system $\mathbf{S}\{z\,;w\}$, then $\mathbf{R}\{x\,;u\}$ must have a set of conservation law multipliers of the form*

$$\mu_\nu[U] = \mathrm{J}[U]v_\sigma(X)Q^\sigma_\nu[U], \tag{2.226}$$

where $Q^\sigma_\nu[U]$, $\sigma,\nu = 1,\ldots,M$, are specific functions of the components of x, U, and ∂U; $v(X) = (v_1(X),\ldots,v_M(X))$ is any solution of some kth order linear PDE system

$$\begin{aligned}\tilde{\mathrm{L}}^\sigma_\alpha[X]v_\sigma = \tilde{b}^\sigma_\alpha(X)v_\sigma + \tilde{b}^\sigma_{\alpha i}(X)\frac{\partial v_\sigma}{\partial X^i}&\\ + \cdots + \tilde{b}^\sigma_{\alpha i_1\cdots i_k}(X)\frac{\partial^k v_\sigma}{\partial X^{i_1}\cdots\partial X^{i_k}} &= 0,\end{aligned} \tag{2.227}$$

$\alpha = 1,\ldots,m$, in terms of specific independent variables $X = (X^1(x,U),\ldots, X^n(x,U))$.

Proof. The existence of the set of conservation law multipliers (2.226) follows from equation (2.224) with X playing the role of z, $\tilde{\mathrm{L}}^\sigma_\alpha[X]$ playing the role of $\mathrm{L}^{*\,\sigma}_\alpha[z]$ and with $V_\sigma = v_\sigma$ satisfying the linear PDE system (2.227). □

Theorem 2.6.2 (Sufficient conditions for the existence of an invertible linearization mapping). *Suppose a given nonlinear PDE system $\mathbf{R}\{x\,;u\}$* (2.221) *has conservation law multipliers of the form* (2.226) *where the components of v are dependent variables of some linear PDE system* (2.227) *with specific independent variables $X = (X^1(x,U),\ldots,X^n(x,U))$. Let $\tilde{\mathrm{L}}^*[X]$ be the adjoint of the linear operator $\tilde{\mathrm{L}}[X]$ in* (2.227). *Consider the augmented system of PDEs consisting of the given nonlinear PDE system* (2.221) *and the linear*

PDE system (2.227). *Then, from the identity* (2.224), *there exists an infinite set of conservation law multipliers*

$$\{\mu_\nu[U], \tilde{\Lambda}^\alpha[U]\} = \{\mathrm{J}[U]V_\sigma(X(x,U))Q^\sigma_\nu(x,U,\partial U),\ -\mathrm{J}[U]W^\alpha(x,U)\},$$

$\nu = 1,\ldots,M$, $\alpha = 1,\ldots,m$, yielding a conservation law identity

$$\mu_\nu[U]R^\nu[U] - \tilde{\Lambda}^\alpha[U]\tilde{\mathrm{L}}^\sigma_\alpha[X(x,U)]V_\sigma \equiv \mathrm{D}_{x^i}\Theta^i \tag{2.228}$$

for some specific fluxes $\Theta^i(x)$, *in terms of the Jacobian determinant*

$$\mathrm{J}[U] = \left|\frac{\mathrm{D}X}{\mathrm{D}x}\right| = \det\left(\frac{\mathrm{D}X^i(x,U)}{\mathrm{D}x^j}\right). \tag{2.229}$$

The conservation law identity (2.228) *is equivalent to the identity*

$$V_\sigma Q^\sigma_\nu(x,U,\partial U)R^\nu[U] - W^\alpha(x,U)\tilde{\mathrm{L}}^\sigma_\alpha[X(x,U)]V_\sigma \equiv \mathrm{D}_{X^i}\Gamma^i, \tag{2.230}$$

holding for some functions Γ^i. *If the variables* X^i, $i = 1,\ldots,n$, *are functionally independent and if the factors* Q^σ_ν, ν, $\sigma = 1,\ldots,M$, *are non-degenerate, then the point transformation given by*

$$z = X(x,u), \quad w = W(x,u) \tag{2.231}$$

maps the nonlinear PDE system $\mathbf{R}\{x\,;u\}$ (2.221) *invertibly into the linear system* $\mathbf{S}\{z\,;w\}$ *given by*

$$\begin{aligned}\tilde{\mathrm{L}}^{*\sigma}_\alpha[z]w^\alpha &= \tilde{b}^\sigma_\alpha(z)w^\alpha - \frac{\partial}{\partial z^i}(\tilde{b}^\sigma_{\alpha i}(z)w^\alpha) + \cdots \\ &+(-1)^k\frac{\partial^k}{\partial z^{i_1}\cdots\partial z^{i_k}}(b^\sigma_{\alpha i_1\cdots i_k}(z)w^\alpha) = 0, \\ \sigma &= 1,\ldots,M.\end{aligned} \tag{2.232}$$

Proof. Since $\tilde{\mathrm{L}}[X]$ is a linear operator, the identity (2.220) yields

$$W^\alpha(x,U)\tilde{\mathrm{L}}^\sigma_\alpha[X(x,U)]V_\sigma \equiv V_\sigma\tilde{\mathrm{L}}^{*\sigma}_\alpha[X(x,U)]W^\alpha(x,U) + \mathrm{D}_{X^i}\theta^i, \tag{2.233}$$

for some specific functions $\theta^i[U,V,W]$. Consequently, the identity (2.230) becomes

$$V_\sigma(Q^\sigma_\nu[U]R^\nu[U] - \tilde{\mathrm{L}}^{*\sigma}_\alpha[X(x,U)]W^\alpha(x,U)) \equiv \mathrm{D}_{X^i}(\Gamma^i + \theta^i). \tag{2.234}$$

Now apply the Euler operators with respect to V_σ, i.e.,

$$\mathrm{E}_{V_\sigma} = \frac{\partial}{\partial V_\sigma} - \mathrm{D}_{X^i}\frac{\partial}{\partial(\frac{\partial V_\sigma}{\partial X^i})} + \cdots, \quad \sigma = 1,\ldots,M,$$

to each side of equation (2.234). Each of these Euler operators annihilates the right-hand side of the identity (2.234) and hence one obtains the identity

$$Q_\nu^\sigma[U]R^\nu[U] \equiv \tilde{\mathrm{L}}_\alpha^{*\,\sigma}[X(x,U)]W^\alpha(x,U), \qquad (2.235)$$

holding for arbitrary functions $U(x)$. Now suppose $U(x) = u(x)$ solves the given nonlinear PDE system $\mathbf{R}\{x\,;u\}$ (2.221). It then follows that $w = W(x, u(x))$ solves the linear system given by (2.232). Consequently, one obtains the invertible point transformation (2.231) that maps the nonlinear PDE system (2.221) into the linear PDE system (2.232). □

2.6.1 Computational steps

We now outline the computational steps involved in applying Theorems 2.6.1 and 2.6.2 to linearize a given nonlinear PDE system through a set of conservation law multipliers.

Step 1: For a given nonlinear system of M PDEs $\mathbf{R}\{x\,;u\}$(2.221) of order k, solve the determining equations

$$\mathrm{E}_{U^\sigma}(\mu_\nu[U]R^\nu[U]) = 0, \quad \sigma = 1, \ldots, m,$$

to obtain sets of conservation law multipliers $\{\mu_\nu[U]\}$ depending on the independent variables $x = (x^1, \ldots, x^n)$, the dependent variables replaced by arbitrary functions $U(x) = (U^1(x), \ldots, U^m(x))$ and their first derivatives $\partial U(x)$, i.e., $\mu_\nu[U] = \mu_\nu(x, U, \partial U)$, $\nu = 1, \ldots, M$. Two cases arise, depending on whether $\mathbf{R}\{x\,;u\}$ has a set of conservation law multipliers of the form (2.226) with functions $v(X) = (v_1(X), \ldots, v_M(X))$ satisfying some linear PDE system of the form (2.227) in terms of specific independent variables $X = (X^1(x,U), \ldots, X^n(x,U))$.

Case I. There exists no set of conservation law multipliers of the required form. Here, from Theorem 2.6.1, one concludes that $\mathbf{R}\{x\,;u\}$ cannot be mapped invertibly by any point transformation to a linear PDE system.

Case II. There is an infinite set of conservation law multipliers of the required form. Typically in this case, the independent variables $X(x,U)$ are found directly from the specific form of the multiplier determining equations, either by inspection (see the first two examples of Section 2.6.2) or, more generally, through integration of the characteristic first-order linear PDEs contained in the system of multiplier determining equations (see the third example of Section 2.6.2). Then from the set of conservation law multipliers (2.226), one can read off (through Theorem 2.6.1):

(a) the independent variable part of a linearizing point transformation (2.231);
(b) a specific target linear PDE system which is the adjoint of the linear system (2.227).

Step 2: Assume that the necessary conditions of Theorem 2.6.1 hold *(Case II)*. Consider the augmented PDE system consisting of the given nonlinear PDE system $\mathbf{R}\{x\,;u\}$ with dependent variables $u(x)$ and the linear system (2.227) with dependent variables $v(X(x,u))$ and independent variables $X=(X^1(x,u),\ldots,X^n(x,u))$. Seek conservation law multipliers for the augmented PDE system to obtain the identity (2.228) in the hypothesis of Theorem 2.6.2 where a solution $u(x)$ of $\mathbf{R}\{x\,;u\}$ is replaced by an arbitrary function $U(x)$ and a solution $v(X(x,u))$ of the linear system (2.227) is replaced by a corresponding arbitrary function $V(X(x,U))$. Use any method [Anco & Bluman (2002a,b); Wolf (2002a,b), or see Section 1.3.7] to seek the set of multipliers that yield the conservation law identity (2.228). From the components of the set of multipliers for the linear system (2.227) in this identity, one directly obtains:

(c) the dependent variable part $w=W(x,u)$ of the linearization mapping to a target linear system.

In summary, when there exists a linearization of a given nonlinear PDE system $\mathbf{R}\{x\,;u\}$ by an invertible point transformation, then the necessary conditions stated in Theorem 2.6.1 yield a target linear PDE system along with the independent variables of this system, whereas the sufficiency conditions stated in Theorem 2.6.2 yield the dependent variables of this target linear PDE system, which completes the linearizing transformation. Furthermore, if the necessary conditions do not hold, then no linearizing point transformation exists for $\mathbf{R}\{x\,;u\}$.

It is straightforward to extend these results to include the linearization of scalar PDEs ($m=1$) by invertible contact transformations, where X and W can now depend on first-order derivatives of $U(x)$. Since in this case it follows that the Jacobian determinant $\mathrm{J}[U]$ and the factors $Q^{\sigma}_{\nu}[U]$ may have a dependence on second-order derivatives of $U(x)$, one must consequently seek sets of conservation law multipliers $\{\mu_\nu[U]\}$ depending on the components of $x, U, \partial U$ and $\partial^2 U$. Then in Step 1, conservation law multipliers of the form (2.226) will yield the independent variable part of a linearizing contact transformation, i.e., $X(x,u,\partial u)$; similarly, in Step 2, the resulting conservation law identity (2.230) will yield the dependent variable part of this mapping, i.e., $W(x,u,\partial u)$. If a given scalar PDE $\mathbf{R}\{x\,;u\}$ has no such multipliers, then no linearizing contact transformation exists.

2.6.2 *Examples of linearizations of nonlinear PDEs through conservation law multipliers*

To illustrate the conservation law multiplier approach for obtaining linearizations, three examples are considered.

(1) Linearization of Burgers' equation
As a first example, consider the nonlinear system $\mathbf{R}\{x\,;u\}$, with independent variables $(x^1, x^2) = (x, t)$ and dependent variables (u^1, u^2), given by

$$\begin{aligned} R^1[u] &= \frac{\partial u^2}{\partial x} - 2u^1 = 0, \\ R^2[u] &= \frac{\partial u^2}{\partial t} - 2\frac{\partial u^1}{\partial x} + (u^1)^2 = 0. \end{aligned} \tag{2.236}$$

Then $u^1 = u$ satisfies Burgers' equation

$$u_t + uu_x - u_{xx} = 0. \tag{2.237}$$

By a straightforward computation, one can show that the nonlinear system (2.236) has an infinite set of conservation law multipliers of the form $\mu_i[U] = \mu_i(x, t, U)$ given by

$$\mu_1[U] = v^1\left(\tfrac{1}{2}U^1 e^{-U^2/4}\right) + v^2 e^{-U^2/4}, \quad \mu_2[U] = v^1 e^{-U^2/4}, \tag{2.238}$$

where $v(x,t) = (v^1(x,t),\ v^2(x,t))$ is any solution of the linear system

$$\begin{aligned} &\frac{\partial v^1}{\partial x} - v^2 = 0, \\ &\frac{\partial v^2}{\partial x} + \frac{\partial v^1}{\partial t} = 0. \end{aligned} \tag{2.239}$$

Hence, the necessary conditions for the existence of an invertible mapping of the nonlinear system (2.236) to a linear PDE system are satisfied, where a target linear system has the same independent variables as the given system (2.236).

In the conservation law arising from the set of multipliers (2.238), one now replaces (u, v) by arbitrary functions (U, V). This leads to the following conservation law identity for the augmented system, consisting of the given nonlinear system (2.236) and the linear system (2.239):

$$\begin{aligned}&\left[V^1\left(\tfrac{1}{2}U^1e^{-U^2/4}\right)+V^2e^{-U^2/4}\right]R^1[U]+V^1e^{-U^2/4}R^2[U]\\&-2U^1e^{-U^2/4}\left[\mathrm{D}_xV^1-V^2\right]-4e^{-U^2/4}\left[\mathrm{D}_xV^2+\mathrm{D}_tV^1\right]\\&\equiv\mathrm{D}_x\left[e^{-U^2/4}(-4V^2-2U^1V^1)\right]+\mathrm{D}_t\left[-4V^1e^{-U^2/4}\right].\end{aligned}\tag{2.240}$$

Consequently, after directly comparing the identity (2.240) with the identity (2.228) in Theorem 2.6.2, the sufficiency conditions of Theorem 2.6.2 yield an invertible mapping of the nonlinear system (2.236) to a linear system that is the adjoint of the linear system (2.239). In particular, it follows that the invertible point transformation

$$w^1=2u^1e^{-u^2/4},\quad w^2=4e^{-u^2/4},\tag{2.241}$$

with no change of independent variables, maps the nonlinear Burgers system (2.236) to the linear system

$$\begin{aligned}&\frac{\partial w^2}{\partial x}+w^1=0,\\&\frac{\partial w^2}{\partial t}+\frac{\partial w^1}{\partial x}=0,\end{aligned}\tag{2.242}$$

which is the adjoint of the linear system (2.239).

Note that $w^2(x,t)$ satisfies the linear heat equation $\partial w^2/\partial t-\partial^2w^2/\partial x^2=0$. Consequently, for any solution $w^2(x,t)$ of the linear heat equation, one obtains the Hopf–Cole transformation $u=u^1=\frac{1}{2}w^1e^{u^2/4}=2w^1/w^2=-2\partial w^2/\partial x/w^2$ that yields a solution of Burgers' equation (2.237).

(2) Linearization of a pipeline flow equation

As a second example, consider the pipeline flow equation

$$R[u]=u_tu_{xx}+u_x^p=0.\tag{2.243}$$

One can show that the nonlinear scalar PDE (2.243) has an infinite set of conservation law multipliers of the form $\mu[U]=\mu(x,t,U,U_x,U_t)$ given by

$$\mu[U]=v(X^1,X^2)=v(U_x,t),\tag{2.244}$$

where $v(X^1,X^2)$ is any solution of the linear scalar PDE

$$\frac{\partial v}{\partial X^2}+\frac{\partial^2(X_1^pv)}{\partial(X^1)^2}=0.\tag{2.245}$$

By inspection, since v depends on two variables, one sees that the necessary conditions for the existence of an invertible contact transformation that linearizes the nonlinear PDE (2.243) are satisfied with the target linear system being the adjoint of (2.245) and having $X^1 = u_x$, $X^2 = t$ as independent variables.

In the conservation law arising from the multipliers (2.244), one replaces (u, v) by arbitrary functions (U, V). This leads to the following conservation law identity for the augmented system consisting of the given nonlinear PDE (2.243) and the linear PDE (2.245):

$$\begin{aligned} VG[U] - \mathrm{J}[U](xU_x - U)\left[\frac{\partial V}{\partial X^2} + \frac{\partial^2((X^1)^p V)}{\partial (X^1)^2}\right] \\ \equiv \mathrm{D}_x\left[(xU_x - U)(U_{tx}V + U_x^p V_{X^1}) + ((1-p)xU_x + pU)U_x^{p-1}V\right] \\ +\mathrm{D}_t\left[U_{xx}(U - xU_x)V\right], \end{aligned} \tag{2.246}$$

where the Jacobian determinant

$$\mathrm{J}[U] = \left|\frac{\mathrm{D}X}{\mathrm{D}x}\right| = \det\begin{bmatrix} U_{xx} & U_{xt} \\ 0 & 1 \end{bmatrix} = U_{xx}. \tag{2.247}$$

For verifying the identity (2.246), note that $V_x = V_{X^1}U_{xx}$, $V_t = V_{X^1}U_{xt} + V_{X^2}$. Consequently, the sufficiency conditions of Theorem 2.6.2 hold for the existence of an invertible mapping by a contact transformation of the nonlinear PDE (2.243) to a linear PDE which is the adjoint of the linear PDE (2.245). In particular, from a comparison of the identity (2.246) with the identity (2.230), it follows that the invertible contact transformation given by

$$\begin{aligned} z^1 = X^1 = u_x, \quad z^2 = X^2 = t, \quad w = xu_x - u, \\ w_{z^1} = w_{X^1} = x, \quad w_{z^2} = w_{X^2} = -u_t, \end{aligned} \tag{2.248}$$

maps the nonlinear pipeline flow equation (2.243) invertibly into the linear PDE

$$\frac{\partial w}{\partial X^2} - (X^1)^p \frac{\partial^2 w}{\partial (X^1)^2} = 0. \tag{2.249}$$

which is the adjoint of the linear PDE given by (2.245).

(3) Linearization of a nonlinear telegraph system
As a final example, consider the nonlinear telegraph (NLT) system given by

$$\begin{aligned} R^1[u] &= \frac{\partial u^2}{\partial t} - \frac{\partial u^1}{\partial x} = 0, \\ R^2[u] &= \frac{\partial u^1}{\partial t} + u^1(u^1 - 1) - (u^1)^2 \frac{\partial u^2}{\partial x} = 0, \end{aligned} \tag{2.250}$$

with independent variables $(x^1, x^2) = (x, t)$ and dependent variables (u^1, u^2). Here $u^1 = u$ satisfies the nonlinear telegraph equation

$$(u^{-2}u_t + 1 - u^{-1})_t - u_{xx} = 0.$$

One can show that the NLT system (2.250) has an infinite set of conservation law multipliers of the form $\mu_i[U] = \mu_i(x, t, U^1, U^2)$ that yields its linearization by an invertible point transformation as follows. After some integrability analysis of the multiplier determining equations (e.g., using a computer algebra package such as `CRACK` [Wolf (2004)]), one first obtains

$$\mu_1[U] = f_{U^2}, \quad \mu_2[U] = f_{U^1}, \tag{2.251}$$

in terms of any function $f(x, t, U^1, U^2)$ satisfying the linear PDE system

$$f_x + f_{U^2} = 0, \quad f_t + U^1 f_{U^1} = 0, \tag{2.252a}$$

$$(U^1)^2 f_{U^1U^1} + 2U^1 f_{U^1} - f_{U^2U^2} = 0. \tag{2.252b}$$

To proceed, one integrates the pair of first-order PDEs (2.252a), which yields a reduction of the number of independent variables in $f(x, t, U^1, U^2)$. In particular, an arbitrary function $f(x, t, U^1, U^2) = f(X, T)$ yields the general solution of (2.252a), where

$$X = x - U^2, \quad T = t - \log U^1. \tag{2.253}$$

Then the second order linear PDE (2.252b) combined with the equations (2.251) yields the infinite set of conservation multipliers

$$\mu_1[U] = -f_X(X, T), \quad \mu_2[U] = -f_T(X, T)/U^1, \tag{2.254}$$

where $f(X, T)$ is any solution of the linear PDE

$$f_{XX} - f_{TT} + f_T = 0. \tag{2.255}$$

Let

$$v = (v_1, v_2) = (-f_X, -f_T). \tag{2.256}$$

After comparing (2.254) with (2.226) in Theorem 2.6.1, one sees that the necessary conditions for the existence of an invertible mapping of the nonlinear PDE system (2.250) to a target linear PDE system are satisfied, with the adjoint system of the target PDE system being given by

$$\frac{\partial v_1}{\partial X} - \frac{\partial v_2}{\partial T} + v_2 = 0, \quad \frac{\partial v_2}{\partial X} - \frac{\partial v_1}{\partial T} = 0. \tag{2.257}$$

In the conservation law arising from the set of multipliers (2.254)–(2.256), one now replaces (u, v) by arbitrary functions $(U(x, t), V(X, T))$. This leads to the following conservation law identity for the augmented system consisting of the given NLT system (2.250) and the linear system (2.257):

$$\begin{aligned}
&V_1R^1[U]+V_2(U^1)^{-1}R^2[U]\\
&\quad -U^1\mathrm{J}[U]\left(\frac{\partial V_1}{\partial X}-\frac{\partial V_2}{\partial T}+V_2\right)-x\mathrm{J}[U]\left(\frac{\partial V_2}{\partial X}-\frac{\partial V_1}{\partial T}\right)\\
&\equiv \mathrm{D}_x\left[-V_1\left(x\frac{\partial U^2}{\partial t}+\frac{\partial U^1}{\partial t}-U^1\right)\right.\\
&\quad\left.+V_2\left(x-x(U^1)^{-1}\frac{\partial U^1}{\partial t}-U^1\frac{\partial U^2}{\partial t}\right)\right)\\
&\quad +\mathrm{D}_t\left[-V_1\left(x-x\frac{\partial U^2}{\partial x}+\frac{\partial U^1}{\partial t}\right)\right.\\
&\quad\left.+V_2\left(x(U^1)^{-1}\frac{\partial U^1}{\partial x}+U^1\frac{\partial U^2}{\partial x}-U^1\right)\right),
\end{aligned} \tag{2.258}$$

where, from (2.253), one has the Jacobian determinant

$$\mathrm{J}[U]=\left|\frac{\mathrm{D}(X,T)}{\mathrm{D}(x,t)}\right|=(U^1)^{-1}\left(\left(1-\frac{\partial U^2}{\partial x}\right)\left(U^1-\frac{\partial U^1}{\partial t}\right)-\frac{\partial U^2}{\partial t}\frac{\partial U^1}{\partial x}\right).$$

Consequently, the sufficiency conditions of Theorem 2.6.2 hold for the existence of an invertible mapping by a point transformation of the nonlinear PDE system (2.250) to a target linear PDE system which is given by the adjoint of the linear PDE system (2.257). In particular, from a comparison of the identity (2.258) with the identity (2.230), it follows that the point transformation

$$z^1=x-u^2,\quad z^2=t-\log(u^1),\quad w^1=x,\quad w^2=u^1, \tag{2.259}$$

maps the nonlinear telegraph system (2.250) invertibly to the linear PDE system

$$\begin{aligned}
\frac{\partial w^1}{\partial z^1}-\frac{\partial w^2}{\partial z^2}-w^2=0,\\
\frac{\partial w^2}{\partial z^1}-\frac{\partial w^1}{\partial z^2}=0.
\end{aligned} \tag{2.260}$$

Note that the point transformation $\tilde{w}^1=w^1$, $\tilde{w}^2=e^{z^2}w^2$ maps the linear PDE system (2.260) to the equivalent linear system (2.93) obtained from the linearization of the NLT system (2.250) through its infinite set of point symmetries (2.85)–(2.87).

Exercises 2.6

2.6.1. Consider the nonlinear diffusion equation

$$R[u] = u_x^2 u_t - u_{xx} = 0. \tag{2.261}$$

(a) Show that $\mu[U] = v(X^1, X^2) = v(U, t)$ is a conservation law multiplier of the PDE (2.261) if $v(U, t)$ satisfies the linear PDE

$$v_{X^1X^1} + v_{X^2} = 0.$$

(b) Show that for arbitrary functions U and V, one has the identity

$$VR[U] - xU_x\left[\frac{\partial V}{\partial X^2} + \frac{\partial^2 V}{\partial (X^1)^2}\right]$$
$$\equiv \mathrm{D}_t(xU_xV) + \mathrm{D}_x(-xU_tV - U_x^{-1}V + xV_{X^1}).$$

(c) Hence derive the point transformation $X^1 = u$, $X^2 = t$, $w = x$ that maps the nonlinear PDE (2.261) to the linear heat equation

$$\frac{\partial^2 w}{\partial (X^1)^2} - \frac{\partial w}{\partial X^2} = 0.$$

2.6.2. Use the method of conservation law multipliers to derive the hodograph transformation that linearizes the quasilinear system (2.75).

2.6.3. Use the method of conservation law multipliers to linearize the nonlinear heat conduction system (2.142).

2.7 Discussion

An interesting discussion on mappings of a given PDE to a specific target PDE by comparing the symmetry groups of a given PDE and the target PDE appears in Ibragimov (1980).

Nonlinear PDEs that admit recursion operators invariably seem to be related to linear PDEs. For many nonlinear evolution equations that exhibit such behaviour and are not equivalent to linear PDEs through invertible (or non-invertible [Section 4.3.1]) transformations, one can transform initial value problems to inverse scattering problems involving linear PDEs [Gardner et al. (1967); Lax (1968); Zakharov & Shabat (1971)]. In particular, for such nonlinear evolution equations, there exist recursion operators that are related to linear operators arising in eigenvalue problems of the associated scattering problems [Ablowitz et al. (1974)]. Konopelchenko (1987) comprehensively reviews work on recursion operators and such integrability of nonlinear evolution equations.

Mikhailov, Shabat & Yamilov (1987) [see also Mikhailov, Shabat & Sokolov (1991)] give an extensive overview of the work of the Russian school on how to use invariance under local symmetries combined with differential substitutions to obtain wide classes of linearizable equations. Their papers include extensive lists of integrable PDEs.

Kingston & Sophocleous (1998) consider the problem of finding point transformations (including discrete ones) that preserve the general form of a wide class of (1+1)-dimensional PDEs. Their results are generalized in Tsaousi (2008).

Tsaousi & Sophocleous (2008) [see also Tsaousi (2008)] use equivalence transformations to derive differential invariants for the general class of hyperbolic equations $u_{xt} = F(x, t, u, u_x, u_t)$ as well as two subclasses. These invariants are used to construct equations that can be linearized through local mappings.

Momoniat (2007) uses the methods presented in this chapter to show that the generalized thin film equation on a moving substrate given by

$$\frac{\partial u}{\partial t} + \frac{\partial}{\partial x}\left(\alpha u^n \frac{\partial^m u}{\partial x^m} - v(t)u\right) = 0 \tag{2.262}$$

can be mapped by an invertible point transformation to the generalized thin film equation

$$\frac{\partial u}{\partial t} + \frac{\partial}{\partial x}\left(\alpha u^n \frac{\partial^m u}{\partial x^m}\right) = 0, \tag{2.263}$$

where m, n, and α are constants. This is accomplished by first showing that each of the PDEs (2.262) and (2.263) has four point symmetries with the same Lie algebra commutation structure and then solving the corresponding mapping equations (2.13) [(2.22)].

An important question to consider is the following. Which of the two methods presented in Sections 2.4 and 2.6, respectively, is better for linearization? It would appear that the method [Section 2.6] based on using local conservation law multipliers is better computationally, since the solution space of the local multiplier determining equations, in general, is expected to be smaller than that for point (contact) symmetries. A way of seeing this is to consider the situation when a given PDE system is variational, i.e., its linearization operator (Fréchet derivative) is self-adjoint. For such a system, as shown in Chapter 1, the determining equations for local multipliers that have a linear dependence on first derivatives are more over-determined than those for point symmetries since they include the point symmetry determining equations as a subset; and, in the case of a given scalar PDE, the determining equations for first-order local multipliers are more over-determined than those for contact symmetries since they include the contact symmetry determining equations as a subset.

In Chapter 4, it is shown how to extend the work presented in this chapter to include non-invertible mappings of a given nonlinear PDE system to an equivalent linear PDE system as well as non-invertible mappings of a given linear PDE with variable coefficients to an equivalent linear PDE with constant coefficients.

Chapter 3
Nonlocally Related PDE Systems

3.1 Introduction

Up to now, for a given PDE system, we have considered the calculation and application of its local symmetries (point, contact or higher-order) as well as the calculation of its local conservation laws. In particular, it has been shown how to use local symmetries to map solutions to other solutions; how to use local symmetries of given and target PDEs as an aid in relating them; how to use point or contact symmetries to determine whether a given PDE system can be mapped invertibly to some PDE system belonging to a target class of PDE systems that is completely characterized by its point symmetries as well as determine an explicit mapping when one exists; how to use multipliers yielding local conservation laws to determine whether a given nonlinear PDE system can be mapped invertibly to some linear PDE system as well as determine a specific mapping when one exists. Moreover, as it is well known, local symmetries can be used to find specific solutions (*invariant solutions*) of PDEs; this application is considered and extended in Chapter 5.

In this chapter, a framework is introduced to find nonlocally related systems for a given PDE system. This framework allows one to extend the calculations and applications of local symmetries and local conservation laws, that were presented in Chapters 1 and 2, to include nonlocal symmetries and nonlocal conservation laws which is considered in Chapter 4. In general, a symmetry of a PDE system is any transformation of its solution manifold into itself, i.e., a symmetry transforms (maps) any solution of a PDE system to another solution of the same PDE system. Hence continuous symmetry transformations (which are essentially deformations of solutions) are defined topologically and are thus not restricted to local transformations acting on the space of independent and dependent variables and their derivatives (even if this space is infinite-dimensional as is the case for the global action

G.W. Bluman et al., *Applications of Symmetry Methods to Partial Differential Equations*, Applied Mathematical Sciences 168, DOI 10.1007/978-0-387-68028-6_3,

of higher-order symmetries). So, in principle, from this point of view any nontrivial PDE system has symmetries. The problem is to find systematic procedures to find and use continuous symmetries beyond the local ones obtained through a direct application of Lie's algorithm. In particular, a direct application of Lie's algorithm only allows one to calculate local symmetries whose infinitesimals depend at most on a finite number of derivatives of the dependent variables.

For a given PDE system, the introduced framework leads to nonlocally related systems with the property that any solution of a nonlocally related PDE system yields a solution of the given PDE system and, conversely, any solution of the given PDE system yields a solution of a nonlocally related PDE system.

A way to find such nonlocally related PDE systems, in order to extend the calculation and use of continuous symmetries to include the calculation and use of nonlocal symmetries of a given PDE system, is to embed the given PDE system in an augmented PDE system. In such an embedding, it is important that each solution of an augmented PDE system projects onto a solution of the given PDE system and, conversely, that each solution of the given PDE system yields a solution of the augmented PDE system. Consequently, the solution of any boundary value problem posed for the given PDE system is embedded in the solution of a boundary value problem posed for the augmented PDE system and the converse also holds. Moreover, in order to be able to calculate further conservation laws and/or symmetries of the given PDE system, it is necessary that the relationship between the given PDE system and such an augmented PDE system is nonlocal, i.e., there is not a one-to-one local transformation connecting the solutions of the given PDE system and the augmented PDE system. Within such a relationship, it follows that a symmetry (conservation law) of the augmented PDE system yields a symmetry (conservation law) of the given PDE system, and the converse statement also holds. Moreover, since their solutions are not related by a local mapping, it also follows that a local symmetry or local conservation law of the augmented PDE system could yield, respectively, a nonlocal symmetry or nonlocal conservation law of the given PDE system; conversely, a local symmetry or local conservation law of the given PDE system could yield, respectively, a nonlocal symmetry or nonlocal conservation law of the augmented PDE system. More importantly, it turns out that nonlocal symmetries of a given PDE system can be calculated through applying Lie's algorithm to the calculation of local symmetries of such augmented PDE systems and, similarly, nonlocal conservation laws of a given PDE system can be calculated through any local procedure such as the direct method applied to such augmented PDE systems. Perhaps, most importantly, since such nonlocally related PDE systems have the same solution sets, it follows that any general method of analysis (qualitative, perturbation, numerical, etc.) that fails to work for a given PDE system, especially a method that is not coordinate-dependent, could turn out to be successful when applied to such a nonlocally related PDE system.

In this chapter, the discussion is restricted to finding nonlocally related PDE systems in the case of two independent variables. [The situation for three or more independent variables is more complex and is treated in Chapter 5.]

It turns out that a natural way to find such nonlocally related augmented PDE systems is through the use of local conservation laws of a given PDE system. In the case of two independent variables, say x and t, a local conservation law of a given PDE system directly yields an augmented PDE system consisting of the given PDE system and a pair of PDEs with a potential variable v, arising from the conservation law. Satisfaction of the integrability condition $v_{tx} = v_{xt}$ leads to the same solution sets for the given and augmented PDE systems.

After seeing that each local conservation law of a given PDE system yields a potential variable and, as a consequence, could yield an equivalent nonlocally related augmented PDE system called a *potential system*, it turns out that one can consider the systematic construction of other nonlocally related, but distinct, PDE systems. In particular, it is shown that if a given PDE system has n local conservation laws which, in turn, respectively yield n potential variables, then one could obtain a *tree* of up to $2^n - 1$ nonlocally related PDE systems by considering the obtained potential systems one-by-one (n singlets, each with one potential variable), in pairs ($\frac{1}{2}n(n-1)$ couplets, each with two potential variables), ..., and all together (one n-plet containing all n potential variables).

Moreover, for any PDE system contained in such a tree of nonlocally related PDE systems, one can calculate and use its local conservation laws (which now have multipliers that can depend on the obtained potential variables) to obtain further potential systems and their combinations. In addition, in the situation when the given PDE system has *precisely* n local conservation laws, one can show that this extending procedure could yield a further distinct nonlocally related PDE system only if a set of multipliers for such local conservation laws has an *essential* dependence on potential (nonlocal) variables.

Furthermore, it is shown that another way to obtain a nonlocally related PDE system for a given PDE system is through consideration of *subsystems of the given PDE system*. In particular, one can obtain a *subsystem* from a given PDE system if one is able to exclude one of the dependent variables from the given PDE system through some means that connects each solution of the given PDE system and subsystem. Moreover, such subsystems could also arise following an interchange of one or more independent and dependent variables.

A subsystem is nonlocally related to the given PDE system if the excluded dependent variable cannot be locally expressed in terms of the independent variables, the remaining dependent variables and their derivatives in the given PDE system.

As a consequence of using the various systematic procedures outlined above, one can obtain an *extended tree* of nonlocally related PDE systems for a given PDE system. Moreover, the given PDE system could be any PDE system within such an extended tree!

The nonlocally related PDE systems within such an extended tree are related to each other through nonlocal mappings. In particular, such nonlocally related PDE systems are equivalent in the sense that the solution set of each PDE system in an extended tree yields the solution set of any other PDE system within the extended tree.

The above outlined procedures for obtaining nonlocally related PDE systems are illustrated through many examples. A particularly important example is the system of equations for planar gas dynamics. Here it is shown that for such a PDE system given in its Eulerian formulation, a corresponding potential system arises from the continuity equation (conservation of mass). After an interchange of the potential variable (which turns out to be the Lagrangian mass coordinate) and the spatial variable in the potential system that results from the continuity equation, one obtains a PDE system with velocity, pressure, density, and the spatial variable as dependent variables and with the time and the potential variable as independent variables. The nonlocally related subsystem obtained by excluding the spatial variable turns out to be the system of planar gas dynamics equations in its Lagrangian formulation. Here a resulting extended tree of nonlocally related PDE systems includes the Euler and Lagrange systems as well as other nonlocally related PDE systems. Moreover, for specific constitutive functions (depending on the pressure and density variables) related to the entropy function, one obtains further extended trees than in the case of an arbitrary constitutive function.

Through several illustrative examples in Chapter 4, it is shown that through the study of particular PDE systems in an extended tree of nonlocally related PDE systems that are constructed through each of the above outlined procedures, one is able to calculate further symmetries as well as non-invertible linearizations of a given PDE system. In the case of the system of equations for planar gas dynamics, it is shown that for specific constitutive functions, one is able to obtain nonlocal symmetries of the Euler (Lagrange) system from point symmetries of the Lagrange (Euler) system and/or other nonlocally related PDE systems in an extended tree. In Chapter 4, it is also shown that through such nonlocally related PDE systems, one can extend the classes of variable coefficient linear PDEs that can be mapped into constant coefficient linear PDEs beyond those obtained through invertible mappings.

Much of the work presented in this chapter has appeared in Bluman & Kumei [(1987), (1989), (1990a,b)], Bluman, Kumei & Reid (1988), Bluman (1993), Bluman & Doran-Wu (1995), Bluman & Temuerchaolu (2005a), Bluman & Cheviakov [(2005), (2007)], and Bluman, Cheviakov & Ivanova (2006). Closely related work has also appeared in Akhatov, Gazizov & Ibragimov (1991), Pukhnachev (1987), Meirmanov, Pukhnachov & Shmarev (1997), Sjöberg & Mahomed (2004), Popovych & Ivanova (2005b), and Ma (1991).

3.2 Nonlocally Related Potential Systems and Subsystems in Two Dimensions

Now we initiate the development of a framework for obtaining nonlocally related systems of partial differential equations for the case of two independent variables $(x^1, x^2) = (x, t)$. In Chapter 5, generalizations are considered for the case of more than two independent variables.

Consider a scalar PDE $\mathbf{R}\{x, t\,; u\}$ of order k with one dependent variable u and two independent variables (x, t), which is given in the conservation law form

$$\mathrm{D}_t\Psi(x, t, u, \partial u, \ldots, \partial^{k-1}u) + \mathrm{D}_x\Phi(x, t, u, \partial u, \ldots, \partial^{k-1}u) = 0. \tag{3.1}$$

In equation (3.1), the total derivative operators are given by

$$\mathrm{D}_i = \frac{\partial}{\partial x^i} + u_i\frac{\partial}{\partial u} + u_{ij_1}\frac{\partial}{\partial u_{j_1}} + \cdots + u_{ij_1j_2\cdots j_{k-1}}\frac{\partial}{\partial u_{j_1j_2\cdots j_{k-1}}}, \quad i, j_l = 1, 2,$$

where $\mathrm{D}_1 = \mathrm{D}_x$, $\mathrm{D}_2 = \mathrm{D}_t$.

The conservation law (3.1) yields a pair of *potential equations* $\mathbf{S}\{x, t\,; u, v\}$ given by

$$\mathcal{P}: \quad \begin{cases} v_x = \Psi(x, t, u, \partial u, \ldots, \partial^{k-1}u), \\ v_t = -\Phi(x, t, u, \partial u, \ldots, \partial^{k-1}u) \end{cases} \tag{3.2}$$

for some auxiliary *potential variable* $v = v(x, t)$.

In (3.2), the potential variable v is *a nonlocal variable*, i.e. it cannot be expressed as a local function of the given variables (x, t, u) and partial derivatives of u. In particular, v is determined to within an integration constant by performing a path integral in the xt-plane.

The *potential system* $\mathbf{S}\{x, t\,; u, v\}$ (3.2) has essentially the same solution set as that of the scalar PDE $\mathbf{R}\{x, t\,; u\}$ (3.1). In particular, if $u = \Theta(x, t)$ is a solution of the PDE $\mathbf{R}\{x, t\,; u\}$ (3.1), then due to the satisfaction of the integrability condition $v_{xt} = v_{tx}$, it follows that there is a corresponding solution $v = \Xi(x, t)$ of the potential system $\mathbf{S}\{x, t\,; u, v\}$ (3.2), unique to within an arbitrary constant, i.e., if $(u,v) = (\Theta(x,t), \Xi(x,t))$ is a solution of the potential system $\mathbf{S}\{x,t\,;u,v\}$ (3.2), then so is $(u,v) = (\Theta(x,t), \Xi(x,t)+C)$ for any constant C. Conversely, if $(u,v) = (\Theta(x,t), \Xi(x,t))$ solves the potential system $\mathbf{S}\{x,t\,;u,v\}$ (3.2), then by projection, $u = \Theta(x,t)$ solves the scalar PDE $\mathbf{R}\{x,t\,;u\}$ (3.1). Consequently, through this relationship between their solution sets, the potential system $\mathbf{S}\{x,t\,;u,v\}$ (3.2) is nonlocally equivalent to the scalar PDE $\mathbf{R}\{x,t\,;u\}$ (3.1) and the mapping that relates the PDE systems (3.2) and (3.1) is non-invertible.

3.2.1 Potential systems

Consider a given PDE system $\mathbf{R}\{x,t\,;u\}$ of order k, with m dependent variables $u = (u^1,\dots,u^m)$ and two independent variables (x,t), which consists of N equations

$$R^{\sigma}(x,t,u,\partial u,\dots,\partial^k u) = 0, \quad \sigma = 1,\dots,N.$$

Suppose a conservation law (3.1) is known for $\mathbf{R}\{x,t\,;u\}$. As discussed previously, the conservation law (3.1) yields a set of potential equations (3.2).

Definition 3.2.1. A system of PDEs consisting of a given PDE system $\mathbf{R}\{x,t\,;u\}$ and the pair of potential equations $\mathcal{P}$ given by (3.2) that follows from a conservation law (3.1) of $\mathbf{R}\{x,t\,;u\}$, is a *potential system* denoted by $\mathbf{S}\{x,t\,;u,v\} = \mathbf{R}\{x,t\,;u\} \cup \mathcal{P}$.

Remark 3.2.1. Note that in principle, it is possible for the potential variable v to depend only on x,t,u and/or partial derivatives of u, and thus be a local variable. See Exercise 3.2.4 for details.

If the conservation law (3.1) used to yield a pair of potential equations is itself one of the equations of the given PDE system $\mathbf{R}\{x,t\,;u\}$, then in the potential system it is clearly not necessary to include this redundant equation. In contrast, if the conservation law (3.1) arises from a multiplier that has an essential dependence on u, it may be necessary to retain all equations of $\mathbf{R}\{x,t\,;u\}$ in $\mathbf{S}\{x,t\,;u,v\}$ in order to be assured that every solution of $\mathbf{S}\{x,t\,;u,v\}$ projects onto a solution of $\mathbf{R}\{x,t\,;u\}$.

The following important theorem holds, which concerns potential variables arising from equivalent conservation laws. [See Definition 1.3.3.]

Theorem 3.2.1. *Suppose two equivalent conservation laws are known for a PDE system $\mathbf{R}\{x,t\,;u\}$. Then the corresponding potential variables v^1 and v^2 are locally related to each other. In particular, $v^2 = v^1 + f(x,t,u,\partial u,\dots,\partial^s u)$ holds for some function $f(x,t,u,\partial u,\dots,\partial^s u)$, $s \geq 0$.*

Proof. See Exercise 3.2.1. □

As an example of potential systems, let $\mathbf{R}\{x,t\,;u\}$ be the nonlinear diffusion equation

$$u_t = (L(u))_{xx}, \tag{3.3}$$

where $L(u)$ is an arbitrary function.

Since the scalar PDE (3.3) is a conservation law as it stands, one can introduce a potential variable v and obtain the potential system $\mathbf{S}\{x,t\,;u,v\}$ given by

$$\begin{aligned} v_x &= u, \\ v_t &= (L(u))_x. \end{aligned} \tag{3.4}$$

Since the second equation of the potential system (3.4) is also a conservation law, one can introduce a second potential variable w and obtain another potential system $\mathbf{T}\{x,t\,;u,v,w\}$ given by

$$\begin{aligned} v_x &= u, \\ w_x &= v, \\ w_t &= L(u). \end{aligned} \tag{3.5}$$

By construction, the three PDE systems $\mathbf{R}\{x,t\,;u\}$, $\mathbf{S}\{x,t\,;u,v\}$, and $\mathbf{T}\{x,t;\,u,v,w\}$ are nonlocally related to each other.

3.2.2 Nonlocally related subsystems

Another important way of obtaining PDE systems that are nonlocally related to a given PDE system $\mathbf{R}\{x,t\,;u\}$ is through the construction of appropriate subsystems. Suppose $\mathbf{R}\{x,t\,;u\}$ has m dependent variables $u=(u^1,\ldots,u^m)$. A subsystem of $\mathbf{R}\{x,t\,;u\}$ is a PDE system that can be obtained from $\mathbf{R}\{x,t\,;u\}$ by excluding one or more of its dependent variables with the properties that (1) any solution of the subsystem yields a solution of $\mathbf{R}\{x,t\,;u\}$; and (2) that the solutions of the subsystem yield all solutions of $\mathbf{R}\{x,t\,;u\}$. Hence in this sense the subsystem is equivalent to $\mathbf{R}\{x,t\,;u\}$. Subsystems can arise directly through the elimination of one or more of the given dependent variables of $\mathbf{R}\{x,t\,;u\}$ as well as indirectly through the elimination of one or more of the resulting dependent variables following a point transformation that involves an interchange of one or more of the dependent and independent variables of $\mathbf{R}\{x,t\,;u\}$.

As a first example, let $\mathbf{R}\{x,t\,;u,v\}$ be the system of nonlinear telegraph (NLT) equations given by

$$\begin{aligned} u_t - v_x &= 0, \\ v_t - F(u)u_x - G(u) &= 0, \end{aligned} \tag{3.6}$$

where (u, v) are dependent variables, (x, t) are independent variables, and $F(u), G(u)$ are arbitrary constitutive functions. For arbitrary $F(u)$ and $G(u)$, one cannot exclude u from (3.6). However v may be excluded, using $v_{tx} = v_{xt}$, to obtain the subsystem $\underline{\mathbf{R}}\{x, t\,; u\}$ given by

$$u_{tt} - (F(u)u_x)_x - (G(u))_x = 0. \tag{3.7}$$

The subsystem $\underline{\mathbf{R}}\{x, t\,; u\}$ (3.7) is obviously nonlocally related to $\mathbf{R}\{x, t\,; u, v\}$ (3.6) since $\mathbf{R}\{x, t\,; u, v\}$ is a potential system of $\underline{\mathbf{R}}\{x, t\,; u\}$, with potential variable v.

As a second example, let $\mathbf{R}\{x, t\,; u, v, w\}$ be the system of PDEs given by

$$\begin{aligned} v_x - u &= 0, \\ w_x - v &= 0, \\ w_t + \left(\frac{1}{u} + bx^2\right) &= 0, \end{aligned} \tag{3.8}$$

related to a nonlinear reaction-diffusion equation. Here $b = \text{const}$ is an arbitrary parameter. Through the first two equations of (3.8), one can eliminate either one of the dependent variables u and v to obtain, respectively, subsystems $\underline{\mathbf{R}_1}\{x, t\,; v, w\}$ given by

$$\begin{aligned} w_x - v &= 0, \\ w_t + \left(\frac{1}{v_x} + bx^2\right) &= 0 \end{aligned} \tag{3.9}$$

and $\underline{\mathbf{R}_2}\{x, t\,; u, w\}$ given by

$$\begin{aligned} w_{xx} - u &= 0, \\ w_t + \left(\frac{1}{u} + bx^2\right) &= 0. \end{aligned} \tag{3.10}$$

The subsystem $\underline{\mathbf{R}_1}\{x, t\,; v, w\}$ (3.9) is obviously locally related to $\mathbf{R}\{x, t;$ $u, v, w\}$ (3.8). In particular, for any solution $(v, w) = (\Xi(x, t), \Gamma(x, t))$ of $\underline{\mathbf{R}_1}\{x, t\,; v, w\}$, the corresponding solution of $\mathbf{R}\{x, t\,; u, v, w\}$ is locally expressed through $(u, v, w) = (\Xi_x(x, t), \Xi(x, t), \Gamma(x, t))$, since $u = v_x$, and, conversely, any solution of $\mathbf{R}\{x, t\,; u, v, w\}$ projects onto a solution of $\underline{\mathbf{R}_1}\{x, t\,; v, w\}$. Similarly, $\underline{\mathbf{R}_2}\{x, t\,; u, w\}$ (3.10) is a locally related subsystem of $\mathbf{R}\{x, t\,; u, v, w\}$.

Moreover, one can eliminate w from the system $\mathbf{R}\{x, t\,; u, v, w\}$ (3.8) through equating the mixed partials $w_{xt} = w_{tx}$. The resulting subsystem

$\underline{\mathbf{R}}_3\{x,t\,;u,v\}$ given by

$$\begin{aligned} v_x - u &= 0, \\ v_t + \left(\frac{1}{u} + bx^2\right)_x &= 0 \end{aligned} \tag{3.11}$$

is nonlocally related to $\mathbf{R}\{x,t\,;u,v,w\}$ since the given system (3.8) is the potential system resulting from the conservation law expressed by the second equation of (3.11).

One can obtain additional subsystems of $\mathbf{R}\{x,t\,;u,v,w\}$ (3.8) through elimination of pairs of dependent variables (equivalent to excluding a single dependent variable from any one of the subsystems (3.9), (3.10), or (3.11)). This yields the subsystems $\underline{\underline{\mathbf{R}}}_1\{x,t\,;u\}$ given by

$$u_t + \left(\frac{1}{u} + bx^2\right)_{xx} = 0; \tag{3.12}$$

$\underline{\underline{\mathbf{R}}}_2\{x,t\,;v\}$ given by

$$v_t + \left(\frac{1}{v_x} + bx^2\right)_x = 0; \tag{3.13}$$

and $\underline{\underline{\mathbf{R}}}_3\{x,t\,;w\}$ given by

$$w_t + \left(\frac{1}{w_{xx}} + bx^2\right) = 0. \tag{3.14}$$

The subsystems (3.12) and (3.13) are nonlocally related to $\mathbf{R}\{x,t\,;u,v,w\}$ (3.8), whereas the subsystem (3.14) is locally related to $\mathbf{R}\{x,t\,;u,v,w\}$ [Exercise 3.2.2].

The above examples illustrate that such subsystems of a PDE system $\mathbf{R}\{x,t\,;u\}$ may be locally or nonlocally related to it. In applications, one is interested in nonlocally related subsystems since for a given PDE system only nonlocally related subsystems could yield new results for a particular method of analysis. The following theorem is concerned with obtaining nonlocally related subsystems of a given PDE system.

Theorem 3.2.2. *A subsystem $\underline{\mathbf{R}}\{x,t\,;u^1,\ldots,u^{m-1}\}$, obtained from a system of PDEs $\mathbf{R}\{x,t\,;u\}$ with m dependent variables by excluding a dependent variable, say u^m, is* nonlocally related *to $\mathbf{R}\{x,t\,;u\}$ if u^m cannot be directly expressed from the equations of $\mathbf{R}\{x,t\,;u\}$ in terms of its independent variables and its remaining dependent variables $u^1,\ldots,u^{m-1}$, and their derivatives. Otherwise the subsystem $\underline{\mathbf{R}}\{x,t\,;u^1,\ldots,u^{m-1}\}$ is locally related to $\mathbf{R}\{x,t\,;u\}$.*

Proof. See Exercise 3.2.3. □

So far the exhibited examples of subsystems have been obtained by the elimination of one or more of the given dependent variables of a given PDE system $\mathbf{R}\{x,t\,;u\}$. Now examples are considered of nonlocally related subsystems that are obtained directly from potential systems of $\mathbf{R}\{x,t\,;u\}$ through exclusion of a dependent variable of $\mathbf{R}\{x,t\,;u\}$ from the potential system; obtained from $\mathbf{R}\{x,t\,;u\}$ or potential systems of $\mathbf{R}\{x,t\,;u\}$ after an interchange of one or more independent and dependent variables.

(1) Example of a nonlocally related subsystem obtained from a potential system
Let $\mathbf{R}\{x,t\,;u\}$ be the linear wave equation with a variable wave speed $c(x)$ given by

$$u_{tt} = c^2(x)u_{xx}. \tag{3.15}$$

The scalar PDE (3.15) can be rewritten as a conservation law $\mathrm{D}_t(c^{-2}(x) \times u_t) - \mathrm{D}_x(u_x) = 0$, therefore one can introduce a potential variable v and obtain the potential system $\mathbf{S}\{x,t\,;u,v\}$ given by

$$\begin{aligned} v_x &= c^{-2}(x)u_t, \\ v_t &= u_x. \end{aligned} \tag{3.16}$$

Excluding u from (3.16) after cross-differentiation, one obtains the nonlocally related subsystem $\underline{\mathbf{S}}\{x,t\,;v\}$ given by

$$v_{tt} = (c^2(x)v_x)_x. \tag{3.17}$$

The subsystem $\underline{\mathbf{S}}\{x,t\,;v\}$ (3.17) is clearly nonlocally related to the given PDE $\mathbf{R}\{x,t\,;u\}$ (3.15).

(2) Example of a nonlocally related subsystem obtained after an interchange of variables
Let $\mathbf{U}\{x,t\,;u\}$ be the nonlinear diffusion equation

$$u_t - (K(u)u_x)_x = 0. \tag{3.18}$$

Since the PDE (3.18) is a conservation law as written, one can directly obtain the corresponding potential system $\mathbf{UV}\{x,t\,;u,v\}$ given by

$$\begin{aligned} v_x &= u, \\ v_t &= K(u)u_x. \end{aligned} \tag{3.19}$$

Two obvious subsystems of $\mathbf{UV}\{x,t\,;u,v\}$ include the given PDE $\mathbf{U}\{x,t\,;u\}$ itself and its locally related subsystem $\underline{\mathbf{UV}}\{x,t\,;v\}$ given by

$$v_t = K(v_x)v_{xx}. \tag{3.20}$$

However, from the potential system $\mathbf{UV}\{x,t\,;u,v\}$ (3.19), one can obtain another subsystem that is nonlocally related to $\mathbf{U}\{x,t\,;u\}$. Consider a point transformation that involves an interchange of a dependent and independent variable of $\mathbf{UV}\{x,t\,;u,v\}$. In particular, choose (u,t) as independent variables, and $x = x(u,t)$, $v = v(u,t)$ as dependent variables. This non-degenerate point transformation invertibly maps the potential system $\mathbf{UV}\{x,t\,;u,v\}$ (3.19) to the nonlinear PDE system $\mathbf{XV}\{u,t\,;x,v\}$ given by

$$\begin{aligned} v_u &= u x_u, \\ v_t x_u - x_t v_u &= K(u). \end{aligned} \tag{3.21}$$

Hence, it is easy to see that the PDE system (3.21) is nonlocally related to the diffusion equation $\mathbf{U}\{x,t\,;u\}$ (3.18).

Observe that from the PDE system $\mathbf{XV}\{u,t\,;x,v\}$ (3.21), one can exclude the dependent variable x through cross-differentiation, and thus obtain the subsystem $\mathbf{V}\{u,t\,;v\}$ given by

$$v_{uu} = \frac{v_u}{u^2 K(u)}(u^2 K'(u) + v_t v_u). \tag{3.22}$$

The subsystem (3.22) is nonlocally related to both the PDE system $\mathbf{XV}\{u,t;\,x,v\}$ (3.21) and the given nonlinear diffusion equation $\mathbf{U}\{x,t\,;u\}$ (3.18).

Similarly, one can exclude the dependent variable v from (3.21) through cross-differentiation, and obtain a nonlocally related subsystem $\mathbf{X}\{u,t\,;x\}$ given by

$$x_{uu} = \frac{x_u}{K(u)}(K'(u) + x_t x_u). \tag{3.23}$$

However it is easy to see that the PDE (3.23) is locally related to the given nonlinear diffusion equation $\mathbf{U}\{x,t\,;u\}$ (3.18), since $\mathbf{X}\{u,t\,;x\}$ is obtained through elimination of the nonlocal variable v. Indeed, the equation $\mathbf{X}\{u,t\,;x\}$ (3.23), through its indicated notation, is related to $\mathbf{U}\{x,t\,;u\}$ (3.18) by a hodograph-type point transformation, with (u,t) as the independent variables, and $x = x(u,t)$ as the dependent variable.

Exercises 3.2

3.2.1. Prove Theorem 3.2.1.

3.2.2. Show that the subsystems (3.12) and (3.13) are nonlocally related to the PDE system (3.8), and that the subsystem (3.14) is locally related to (3.8).

3.2.3. Prove Theorem 3.2.2.

3.2.4.

(a) Consider the PDE system $\mathbf{R}\{x,t\,;u,v\}$ given by

$$\begin{aligned} u_t &= v_{xx}, \\ v_t &= u^{-1}(v-u)_x v_x + v_{xx}. \end{aligned} \tag{3.24}$$

Use the first equation of (3.24) to introduce a potential variable w. Show that w is a local variable by showing that w is a function of $u-v$.

(b) Show that if a PDE system $\mathbf{R}\{x,t\,;u\}$ with two independent variables x,t has a local conservation law (3.1), then the potential variable $v(x,t)$ in the potential equations (3.2) is a local variable (i.e., v is functionally dependent on the local variables x,t,u and the partial derivatives of u) if and only if there exists a function $g=g(x,t,u,\partial u,\ldots,\partial^l u)$, $l\geq 0$, such that the equation

$$\Psi \mathrm{D}_t g + \Phi \mathrm{D}_x g = 0 \tag{3.25}$$

holds, where Ψ and Φ are, respectively, the density and the flux of the conservation law (3.1).

(c) For the PDE system $\mathbf{R}\{x,t\,;u,v\}$ (3.24), find Ψ, Φ and g.

(d) Consider a linear advection equation (conservation law) $u_t+u_x=0$. Using (3.25), show that the resulting potential variable is a local variable.

3.2.5. Consider a PDE system in 3D space given by

$$\operatorname{curl}\mathbf{B}\times\mathbf{B} = \operatorname{grad} p, \quad \operatorname{div}\mathbf{B}=0. \tag{3.26}$$

The system (3.26) has two important physical applications. Firstly, it describes the static equilibrium of ideal plasmas; here $\mathbf{B}$ is the magnetic field, and p is the plasma pressure. Secondly, the PDE system (3.26) describes a time-independent (equilibrium) flow of an incompressible fluid; here $\mathbf{V}=\mathbf{B}$ is the fluid velocity vector field, and $P=-p-\mathbf{V}^2/2$ is the fluid pressure.

(a) In applications, the axially-symmetric reduction of the equations (3.26) is often considered. Assume that $\mathbf{B}$ and p are both independent of the polar angle ϕ, i.e., $\mathbf{B} = b^1(r,z)\mathbf{e}_r + b^2(r,z)\mathbf{e}_\phi + b^3(r,z)\mathbf{e}_z$, $p=p(r,z)$. Explicitly write down the axially symmetric PDE system $\mathbf{BP}\{r,z\,;b^1,b^2,b^3,p\}$ consisting of four scalar equations.

(b) Show that the system $\mathbf{BP}\{r,z\,;b^1,b^2,b^3,p\}$ has a family of conservation laws

$$\mathrm{D}_r\left[rb^1F(rb^2)\right] + \mathrm{D}_z\left[rb^3F(rb^2)\right] = 0, \tag{3.27}$$

where F is an arbitrary smooth function of rb^2. Find a set of multipliers that yields the conservation law (3.27).

(c) From (b), it follows that any choice for the arbitrary function $F(rb^2)$ yields a pair of potential equations

$$\begin{aligned} v_z^F &= rb^1 F(rb^2), \\ v_r^F &= -rb^3 F(rb^2). \end{aligned} \tag{3.28}$$

For any two choices for $F(rb^2)$ given by $F_1(rb^2)$, $F_2(rb^2)$, show that the corresponding potential variables $v^{F_1}(r,z)$ and $v^{F_2}(r,z)$ are *functionally dependent.*

(d) Write down the potential system $\mathbf{BPV^F}\{r,z\,;b^1,b^2,b^3,p,v^F\}$ for an arbitrary $F(rb^2)$.
For any $F(rb^2)$, show that $b^2(r,z) = r^{-1}I(v(r,z))$, $p(r,z) = P(v(r,z))$, where $v(r,z)$ is the corresponding potential variable in (3.28), and I, P are arbitrary sufficiently smooth functions of their arguments. Thus the potential variable $v(r,z)$ is a *local variable.* Hence the potential system $\mathbf{BPV}\{r,z\,;b^1,b^2,b^3,p,v\}$ is equivalent and locally related to the given PDE system $\mathbf{BP}\{r,z\,;b^1,b^2,b^3,p\}$.

(e) Let $F(rb^2) = 1$. From the corresponding potential system $\mathbf{BPV}\{r,z;$ $b^1,b^2,b^3,p,v\}$, exclude dependent variables b^1,b^2,b^3,p, and obtain the subsystem $\mathbf{V}\{r,z\,;v\}$. Show that it is given by the scalar PDE

$$v_{rr} - \frac{v_r}{r} + v_{zz} + I(v)I'(v) = -r^2 P'(v), \tag{3.29}$$

i.e., the famous *Bragg–Hawthorne* or *Grad–Shafranov* equation [Bragg & Hawthorne (1950); Grad & Rubin (1958); Shafranov (1958)]. This demonstrates that the Bragg–Hawthorne equation (3.29) is *locally related* to the given system $\mathbf{BP}\{r,z\,;b^1,b^2,b^3,p\}$ (3.26), in spite of the fact that it involves a potential variable.

3.3 Trees of Nonlocally Related PDE Systems

The framework presented in Section 3.2, for constructing nonlocally related PDE systems, is now extended. The extended framework leads one to consider trees of nonlocally related PDE systems. Within each tree, all PDE systems will be equivalent in the sense that the solution set of any PDE system in a tree can be obtained from the solution set of any other PDE system in the same tree.

Consider a given PDE system $\mathbf{R}\{x,t\,;u\}$, with n known local conservation laws and n corresponding potential systems denoted by $\mathbf{S}^i\{x,t\,;u,v^i\}$ ($i = 1,\ldots,n$). A natural way to extend this set of n nonlocally related systems for the given PDE system $\mathbf{R}\{x,t\,;u\}$ is to consider potential systems for each of the potential systems $\mathbf{S}^i\{x,t\,;u,v^i\}$, obtained from the conservation laws of the latter to introduce further potential variables. A systematic procedure that uses this idea is outlined in Section 3.3.1 and illustrated with examples.

In Section 3.5, the extended framework for obtaining nonlocally related PDE systems that is presented in this section is still further extended. In particular, it is shown that if a given PDE system $\mathbf{R}\{x,t\,;u\}$ has n known local conservation laws, then one can immediately obtain up to 2^n-1 distinct nonlocally related PDE systems from the n local conservation laws. Such nonlocally related systems are obtained by considering the potential systems $\mathbf{S}^i\{x,t\,;u,v^i\}$ one-by-one (n singlets, each with one potential variable), in pairs ($\frac{1}{2}n(n-1)$ couplets, each with two potential variables),..., and all together (one n-plet containing all n potential variables). Moreover, for any such nonlocally related PDE system, through its local conservation laws (which have fluxes that depend on the corresponding potential variables, and thus may be nonlocal conservation laws of the given PDE system $\mathbf{R}\{x,t\,;u\}$; see Section 3.4), one could obtain additional potential systems and accordingly still more nonlocally related PDE systems through combinations of potential systems.

3.3.1 Basic procedure of tree construction

A procedure is now outlined for constructing a hierarchy (tree) of nonlocally related potential systems and subsystems for a given PDE system $\mathbf{R}\{x,t\,;u\}$, based on the work presented in Section 3.2.

1. **Construction of potential systems.** Suppose $\mathbf{R}\{x,t\,;u\}$ includes explicit conservation laws. For each explicit conservation law, introduce a potential variable and construct the corresponding potential system. For any such potential system, use its explicit conservation laws to obtain further potential systems, etc., until the obtained set of potential systems includes no more explicit conservation laws. Let $\mathcal{T}_1$ denote the resulting tree of nonlocally related PDE systems. [If $\mathbf{R}\{x,t\,;u\}$ does not include explicit conservation laws as written, then $\mathcal{T}_1$ is simply $\mathbf{R}\{x,t\,;u\}$.]
2. **Additional conservation laws. Tree extension.** For each PDE system in $\mathcal{T}_1$, use any method to find further local conservation laws. Use these additional conservation laws to obtain more potential systems. Attempt to delete locally related systems. Continue until no further conservation laws are found for any nonlocally related potential system. Assuming that all locally related systems have been deleted, this yields a tree of nonlocally related PDE systems denoted by $\mathcal{T}_2$.

3. **Construction of subsystems.** For all PDE systems in the tree $\mathcal{T}_2$, exclude where possible, one by one, dependent variables, to generate all subsystems of the systems in the tree $\mathcal{T}_2$. Eliminate locally related subsystems. In addition, generate nonlocally related subsystems obtained after an interchange of one or more independent and dependent variables. [See the last example in Section 3.2.2.] This yields a possibly larger tree of nonlocally related systems denoted by $\mathcal{T}_3$.
4. **Continuation.** Continue until no further local conservation laws are found for any nonlocally related potential system or subsystem. This yields a further extended tree denoted by $\mathcal{T}_4$.

In the extended tree $\mathcal{T}_4$, all PDE systems are equivalent in the terms of their related solution sets and typically will be nonlocally related. Moreover, any PDE system in $\mathcal{T}_4$ can serve as a "given" system since the same tree $\mathcal{T}_4$ can hold for any PDE system in $\mathcal{T}_4$. Furthermore, it should be noted that if a given PDE system contains one or more constitutive functions with some degrees of arbitrariness, then corresponding extended trees of nonlocally related PDE systems can be further enlarged for different forms of the constitutive functions since the obtained conservation laws as well as the obtained subsystems for the given system could depend on the form of a constitutive function. In particular, some branches of a tree could be "general" (i.e., holding for all constitutive functions), whereas other branches could depend on the specific forms of constitutive functions determined from conservation law and/or subsystem construction. This construction of further branches is related to the problem of the classification of conservation laws of PDE systems with constitutive functions.

It should be noted that sometimes it may be difficult to determine whether a newly constructed system in an extended tree is nonlocally related to all other systems in the tree. However, from the point of view of computations for a particular method of analysis, although redundant (locally related) PDE systems in a tree will lead to additional computations, they do not lead to incorrect results.

Note that nonlocally related subsystems do not yield further conservation laws, since a local conservation law of a subsystem of a PDE system $\mathbf{R}\{x,t\,;u\}$ (obtained directly or after an interchange of one or more independent and independent variables) must be a local conservation law of $\mathbf{R}\{x,t\,;u\}$ [Exercise 3.4.2].

3.3.2 A tree for a nonlinear diffusion equation

The procedure presented in Section 3.3.1 is now used to construct an extended tree of potential systems and subsystems for the nonlinear diffusion equation $\mathrm{U}\{x,t\,;u\}$ (3.18) with an arbitrary constitutive function $K(u)$. As shown previously, the PDE (3.18) has the potential system $\mathbf{UV}\{x,t\,;u,v\}$ (3.19).

Let the conductivity $K(u) = L'(u)$ and observe that the second equation of the potential system $\mathbf{UV}\{x,t\,;u,v\}$ (3.19) is also a conservation law. Accordingly, one obtains another potential variable w and the potential system $\mathbf{UVW}\{x,t\,;u,v,w\}$ given by

$$\begin{aligned} v_x &= u, \\ w_x &= v, \\ w_t &= L(u). \end{aligned} \tag{3.30}$$

Since neither $\mathbf{UV}\{x,t\,;u,v\}$ (3.19) nor $\mathbf{UVW}\{x,t\,;u,v,w\}$ (3.30) have further obvious local conservation laws, one seeks conservation laws of systems $\mathbf{U}\{x,t\,;u\}$, $\mathbf{UV}\{x,t\,;u,v\}$ and $\mathbf{UVW}\{x,t\,;u,v,w\}$ that arise from sets of multipliers that are functions of their respective dependent and independent variables.

For the nonlinear diffusion equation $\mathbf{U}\{x,t\,;u\}$ (3.18), for arbitrary $K(u)$, the only such multipliers are $\Lambda^{(1)} = 1$, yielding the potential system $\mathbf{UV}\{x,t;u,v\}$ (3.19), and $\Lambda^{(2)} = x$ leading to the conservation law

$$\mathrm{D}_t(xu) - \mathrm{D}_x(x(L(u))_x - L(u)) = 0. \tag{3.31}$$

From (3.31), one can introduce a potential variable α and obtain the potential system $\mathbf{UA}\{x,t\,;u,\alpha\}$ given by

$$\begin{aligned} \alpha_x &= xu, \\ \alpha_t &= x(L(u))_x - L(u). \end{aligned} \tag{3.32}$$

For the potential system $\mathbf{UV}\{x,t\,;u,v\}$ (3.19), for arbitrary $K(u)$, the only set of multipliers is given by $(\Lambda_1, \Lambda_2) = (0,1)$ and this yields the known potential system $\mathbf{UVW}\{x,t\,;u,v,w\}$ (3.30).

For the potential system $\mathbf{UA}\{x,t\,;u,\alpha\}$ (3.32), for arbitrary $K(u)$, the only set of multipliers is given by $(\Lambda_1, \Lambda_2) = (0, x^{-2})$. This yields the conservation law

$$\mathrm{D}_t(x^{-2}\alpha) - \mathrm{D}_x(x^{-1}L(u)) = 0. \tag{3.33}$$

From the conservation law (3.33), one can introduce another potential variable β and obtain the potential system $\mathbf{UAB}\{x,t\,;u,\alpha,\beta\}$ given by

$$\begin{aligned} \alpha_x &= xu, \\ \beta_x &= x^{-2}\alpha, \\ \beta_t &= x^{-1}L(u). \end{aligned} \tag{3.34}$$

It turns out that the potential systems $\mathbf{UVW}\{x,t\,;u,v,w\}$ (3.30) and $\mathbf{UAB}\{x,t\,;u,\alpha,\beta\}$ (3.34) are locally related through the transformation

$$v = x^{-1}\alpha + \beta, \quad w = x\beta \tag{3.35}$$

[Exercise 3.3.1].

Seeking local conservation laws of the potential system $\mathbf{UVW}\{x,t\,;u,v,w\}$ (3.30), for arbitrary $K(u)$, with multipliers that depend on its dependent and independent variables, one only finds the trivial set of multipliers $(\Lambda_1,\Lambda_2,\Lambda_3) = (0,0,0)$. Hence, the construction of local conservation laws ends at this stage. [In general, there might exist additional local conservation laws with multipliers depending on derivatives of its dependent variables, but here such multipliers are not considered.] Note also that local conservation laws have only been sought that hold for arbitrary $K(u)$. For particular forms of $K(u)$, additional local conservation laws arise to yield further potential systems and hence a tree extension [Exercise 3.3.2].

Now consider direct subsystems. Excluding the dependent variable u from the potential system $\mathbf{UV}\{x,t\,;u,v\}$ (3.19), or excluding the dependent variables u and v from the potential system $\mathbf{UVW}\{x,t\,;u,v,w\}$ (3.30), one only obtains locally related subsystems [Exercise 3.3.3].

Now subsystems are sought that are obtained after an interchange of independent and dependent variables. As shown previously, the interchange of a dependent and independent variable in the potential system $\mathbf{UV}\{x,t\,;u,v\}$ (3.19) (namely, treating (u,t) as independent variables, and $x = x(u,t), v = v(u,t)$ as dependent variables) yields an invertibly equivalent PDE system $\mathbf{XV}\{u,t\,;x,v\}$ (3.21). After separately excluding the dependent variables x and v through cross-differentiation, the PDE system $\mathbf{XV}\{u,t\,;x,v\}$ yields the subsystems $\mathbf{V}\{u,t\,;v\}$ and $\mathbf{X}\{u,t\,;x\}$ given by (3.22) and (3.23), respectively. The PDE $\mathbf{V}\{u,t\,;v\}$ is nonlocally related to the given nonlinear diffusion equation $\mathbf{U}\{x,t\,;u\}$ (3.18), whereas the PDE $\mathbf{X}\{u,t\,;x\}$ is locally related to it.

Next, consider a second interchange of variables for the potential system $\mathbf{UV}\{x,t\,;u,v\}$ (3.19): let (u,v) be treated as the independent variables, and $x = x(u,v), t = t(u,v)$ as the dependent variables. This point transformation leads to the invertibly equivalent PDE system $\mathbf{XT}\{u,v\,;x,t\}$ given by

$$\begin{aligned} x_u &= K(u)t_v, \\ t_u(ux_v - 1) &= uK(u)t_v^2, \end{aligned} \tag{3.36}$$

and hence locally related to $\mathbf{UV}\{x,t\,;u,v\}$ (3.19). The PDE system (3.36) has two subsystems obtained by excluding x and t, respectively: $\mathbf{X}\{u,v\,;x\}$ (not to be confused with $\mathbf{X}\{t,u\,;x\}$ (3.23)) and $\mathbf{T}\{u,v\,;t\}$. These subsystems are nonlocally related to both $\mathbf{XT}\{u,v\,;x,t\}$ (3.36) and the given nonlinear diffusion equation $\mathbf{U}\{x,t\,;u\}$ (3.18) [Exercise 3.3.3]. However, since the scalar PDE $\mathbf{T}\{u,v\,;t\}$ is obviously locally related to the PDE $\mathbf{V}\{u,t\,;v\}$ (3.22) [Their sets of variables coincide.], it is not included in the tree of nonlocally related systems that includes the PDE $\mathbf{V}\{u,t\,;v\}$.

Further subsystems arise after interchanging and then excluding variables in the potential system $\mathbf{UA}\{x,t\,;u,\alpha\}$ (3.32). Consider an interchange of a dependent and independent variable: treat (x,u) as the independent variables, and $t=t(x,u), \alpha=\alpha(x,u)$ as the dependent variables. The resulting PDE system $\mathbf{TA}\{x,u\,;t,\alpha\}$ given by

$$\begin{aligned} a_x - \frac{t_x a_u}{t_u} &= xu, \\ \frac{a_u}{t_u} &= -L(u) - \frac{xL(u)t_x}{t_u} \end{aligned} \tag{3.37}$$

is locally related to $\mathbf{UA}\{x,t\,;u,\alpha\}$ (3.32). The PDE system (3.37) has two subsystems $\mathbf{T}\{x,u\,;t\}$ and $\mathbf{A}\{x,u\,;\alpha\}$, obtained by excluding α and t, respectively. It is obvious that the scalar PDE $\mathbf{T}\{x,u\,;t\}$ is invertibly equivalent to the given nonlinear diffusion equation $\mathbf{U}\{x,t\,;u\}$ (3.18). However, the PDE $\mathbf{A}\{x,u\,;\alpha\}$ given by

$$a_{xx} = u + \frac{L(u)}{L'(u)} + (a_x - xu)^2\left(\frac{L''(u)}{L'(u)a_u} - \frac{a_{uu}}{a_u^2}\right) + 2(a_x - xu)\frac{a_{xu}}{a_u} \tag{3.38}$$

is nonlocally related to both the potential system $\mathbf{TA}\{x,u\,;t,\alpha\}$ (3.37) and the given nonlinear diffusion equation $\mathbf{U}\{x,t\,;u\}$ (3.18).

Other nonlocally related subsystems that may arise after further interchanges of variables (in particular, from point transformations that involve compositions of dependent and independent variables) in potential systems $\mathbf{UV}\{x,t\,;u,v\}$, $\mathbf{UA}\{x,t\,;u,\alpha\}$, $\mathbf{UVW}\{x,t\,;u,v,w\}$ and $\mathbf{UAB}\{x,t\,;u,\alpha,\beta\}$ are not considered here.

The resulting tree of nonlocally related potential systems and subsystems is exhibited in Figure 3.1.

3.3.3 A tree for planar gas dynamics (PGD) equations

The two fundamental PDE systems that describe non-stationary planar (1+1)-dimensional gas motions are the Euler and Lagrange systems. In the

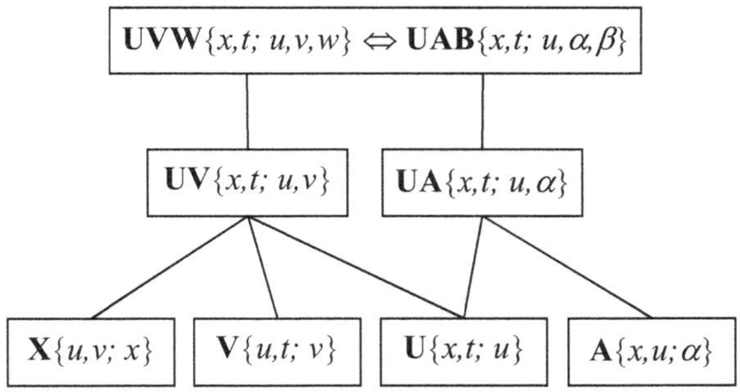

Fig. 3.1 A tree of nonlocally related PDE systems for the nonlinear diffusion equation (3.18) (for arbitrary $K(u)$).

Eulerian description, x is a cartesian coordinate in a fixed coordinate frame. The *Euler system* $\mathbf{E}\{x,t\,;v,p,\rho\}$ is given by

$$\begin{aligned}\rho_t + (\rho v)_x &= 0,\\ \rho(v_t + vv_x) + p_x &= 0,\\ \rho(p_t + vp_x) + B(p,1/\rho)v_x &= 0.\end{aligned} \tag{3.39}$$

In (3.39), v is the gas velocity, ρ is the gas density, and p is the gas pressure. In terms of the entropy density $S(p,\rho)$, the constitutive function $B(p,1/\rho)$ is given by

$$B(p,1/\rho) = -\rho^2 S_\rho / S_p.$$

Following the tree construction procedure, one obtains an extended tree of nonlocally related PDE systems, with the Euler system (3.39) as the given system. In particular, it is shown that the Lagrange system of gas dynamics naturally arises as a nonlocally related system within this tree. Since the first equation of (3.39) is a conservation law as written, one can introduce a potential variable (mass coordinate) α^1 and obtain the corresponding potential system $\mathbf{EA}^1\{x,t\,;v,p,\rho,\alpha^1\}$ given by

$$\begin{aligned}\alpha^1_x - \rho &= 0,\\ \alpha^1_t + \rho v &= 0,\\ \rho(v_t + vv_x) + p_x &= 0,\\ \rho(p_t + vp_x) + B(p,1/\rho)v_x &= 0.\end{aligned} \tag{3.40}$$

Two obvious subsystems arise by excluding the density ρ and the velocity v, respectively, from the potential system $\mathbf{EA}^1\{x,t\,;v,p,\rho,\alpha^1\}$ (3.40). However, these subsystems are not of interest since both of them are locally related to $\mathbf{EA}^1\{x,t\,;v,p,\rho,\alpha^1\}$. Consider a local (point) coordinate transformation of the potential system (3.40) with $\alpha^1 = y$, $t = s$ treated as the independent variables, and x, v, p, ρ as the dependent variables (an interchange of the variables α^1 and x variables). Without loss of generality, $\rho \neq 0$. Letting $q = 1/\rho$, one then obtains the invertibly equivalent PDE system $\mathbf{LX}\{y, s\,; v, p, q, x\}$ given by

$$\begin{aligned} q - x_y &= 0, \\ v - x_s &= 0, \\ v_s + p_y &= 0, \\ p_s + B(p,q)v_y &= 0. \end{aligned} \tag{3.41}$$

It turns out that the subsystem of $\mathbf{LX}\{y, s\,; v, p, q, x\}$, obtained by excluding x through the integrability condition $x_{sy} = x_{ys}$, is the Lagrange system of gas dynamics $\mathbf{L}\{y, s\,; v, p, q\}$ given by

$$\begin{aligned} q_s - v_y &= 0, \\ v_s + p_y &= 0, \\ p_s + B(p,q)v_y &= 0. \end{aligned} \tag{3.42}$$

The Lagrange system (3.42), with the time variable s and the Lagrange mass coordinate $y = \int_{x_0}^{x} \rho(\xi, t)d\xi$ as independent variables, is used in many applications. The mass coordinate y essentially enumerates the fluid particles; its domain does not change with time. The time derivative $\partial/\partial s = \partial/\partial t + v\partial/\partial x$ is the material derivative. The use of Lagrange coordinates often significantly facilitates the formulation of boundary conditions. The Euler and Lagrange systems of PGD equations arise as nonlocally related PDE systems within a tree of nonlocally related PDE systems obtained through the tree construction procedure. Consequently, one can obtain additional equivalent descriptions of the PGD equations.

The construction of a tree of potential systems and subsystems for the PGD equations for a general constitutive function $B(p, 1/\rho)$ is now continued. First one seeks further potential systems arising from local conservation laws of the potential system $\mathbf{EA^1}\{x, t\,; v, p, \rho, \alpha^1\}$ (3.40). Note that for this system, conservation of momentum holds and is given by the conservation law equation

$$\mathrm{D}_t(\rho v) + \mathrm{D}_x(p + \rho v^2) = 0. \tag{3.43}$$

Accordingly, one can introduce a potential variable α^2 to obtain the potential system $\mathbf{EA^1A^2}\{x, t\,; v, p, \rho, \alpha^1, \alpha^2\}$ given by

$$\begin{aligned}
\alpha^1_x - \rho &= 0,\\
\alpha^1_t + \rho v &= 0,\\
\alpha^2_x + \alpha^1_t &= 0,\\
\alpha^2_t + p + \rho v^2 &= 0,\\
\rho(v_t + v v_x) + p_x &= 0,\\
\rho(p_t + v p_x) + B(p, 1/\rho) v_x &= 0.
\end{aligned} \tag{3.44}$$

The third equation of the PDE system (3.44) is a conservation law as written, hence one can introduce another potential variable α^3 to obtain the additional potential system $\mathbf{EA^1A^2A^3}\{x, t\,; v, p, \rho, \alpha^1, \alpha^2, \alpha^3\}$ given by

$$\begin{aligned}
\alpha^1_x - \rho &= 0,\\
\alpha^1_t + \rho v &= 0,\\
\alpha^3_t - \alpha^2 &= 0,\\
\alpha^3_x + \alpha^1 &= 0,\\
\alpha^2_t + p + \rho v^2 &= 0,\\
\rho(v_t + v v_x) + p_x &= 0,\\
\rho(p_t + v p_x) + B(p, 1/\rho) v_x &= 0.
\end{aligned} \tag{3.45}$$

Next, one analyzes nonlocally related subsystems. By exclusion of dependent variables, the only direct nonlocally related subsystems of the PDE systems (3.44) and (3.45) arise from excluding the dependent variable α^1. On the other hand, in the Lagrange system (3.42), one can exclude the velocity v, to obtain a nonlocally related subsystem $\underline{\mathbf{L}}\{y, s\,; p, q\}$ given by

$$\begin{aligned}
q_{ss} + p_{yy} &= 0,\\
p_s + B(p, q) q_s &= 0.
\end{aligned} \tag{3.46}$$

The resulting tree of nonlocally related PDE systems is shown in Figure 3.2. [Later, this tree is extended through finding further local conservation laws for the Lagrange system $\mathbf{L}\{y, s\,; v, p, q\}$.]

Exercises 3.3

3.3.1. For the nonlinear diffusion equation (3.18), show that the transformation of dependent variables

$$v = x^{-1}\alpha + \beta, \quad w = x\beta,$$

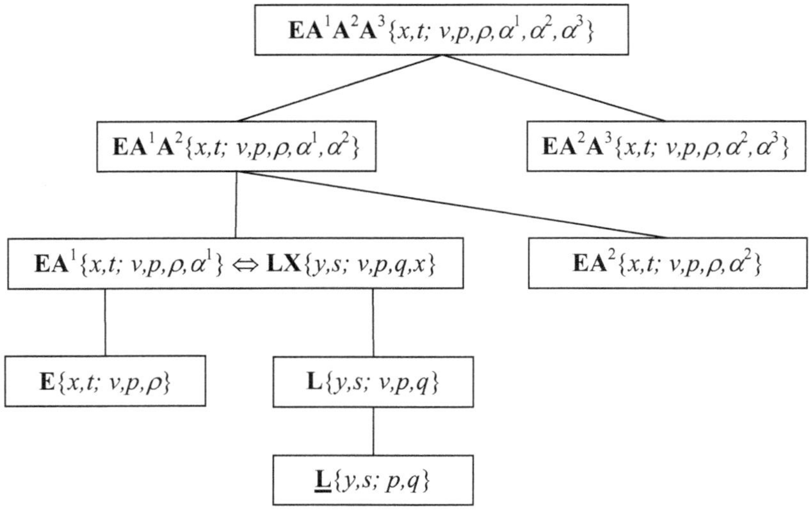

Fig. 3.2 A tree of nonlocally related PDE systems for the planar gas dynamics equations for an arbitrary constitutive function $B(p,q)$ with $q = 1/\rho$.

maps its potential system (3.30) into the potential system (3.34) [Bluman & Doran-Wu (1995)]. Thus the potential systems (3.30) and (3.34) are locally related, and only one of them should be included in a tree of nonlocally related PDE systems.

3.3.2. Assuming the dependence of multipliers on only the dependent and independent variables of a given PDE system, classify the local conservation laws of the potential systems (3.19) and (3.32) of the nonlinear diffusion equation (3.18) with respect to the constitutive function $K(u)$. For the cases where nonlocal conservation laws of the nonlinear diffusion equation (3.18) arise, obtain extensions of the tree of nonlocally related PDE systems constructed in Section 3.3.2 for the nonlinear diffusion equation (3.18). In particular, show that for $K(u) = u^{-2}$, there exists an infinite set of nonlocal conservation laws that lead to the linearization of the nonlinear diffusion equation (3.18) by a nonlocal transformation (as described in Chapter 2) [Bluman & Doran-Wu (1995)].

3.3.3.

(a) Consider the potential systems (3.19) and (3.30) of the nonlinear diffusion equation (3.18). Show that exclusion of the dependent variable u from the potential system (3.19), as well as exclusion of the dependent variables u and v from the potential system (3.30), yield only locally related subsystems.

(b) Exclude the dependent variables t and x, respectively, to obtain the subsystems $\mathbf{X}\{u, v\,; x\}$ and $\mathbf{T}\{u, v\,; t\}$ of the PDE system (3.36). Show that these subsystems are nonlocally related to both the PDE system (3.36) and the nonlinear diffusion equation $\mathbf{U}\{x, t\,; u\}$ (3.18).

3.4 Nonlocal Conservation Laws

Suppose a given PDE system $\mathbf{R}\{x, t\,; u\}$ has a potential system $\mathbf{S}\{x, t\,; u, v\}$ with potential variable v. An important problem to consider is the comparison of sets of local conservation laws of $\mathbf{R}\{x, t\,; u\}$ (with fluxes and densities depending on components of x, u, and partial derivatives of u) and local conservation laws of $\mathbf{S}\{x, t\,; u, v\}$ (with fluxes and densities depending on components of x, u, v, and partial derivatives of u and v).

Clearly every local conservation law of $\mathbf{R}\{x, t\,; u\}$ is also a local conservation law of the potential system $\mathbf{S}\{x, t\,; u, v\}$, since the equations of $\mathbf{R}\{x, t\,; u\}$ that yield this conservation law appear explicitly in $\mathbf{S}\{x, t\,; u, v\}$. Now consider a local conservation law

$$\mathrm{D}_t\Psi[u, v] + \mathrm{D}_x\Phi[u, v] = 0 \tag{3.47}$$

of the potential system $\mathbf{S}\{x, t\,; u, v\}$. If neither the density Ψ nor the flux Φ depends explicitly on the nonlocal variable v, then obviously (3.47) must be a local conservation law of $\mathbf{R}\{x, t\,; u\}$. On the other hand, if Ψ and/or Φ explicitly depends on the nonlocal variable v, one of the following two situations occurs.

1. The conservation law (3.47) of $\mathbf{S}\{x, t\,; u, v\}$ can be expressed as a linear combination of the local conservation laws of the PDE system $\mathbf{R}\{x, t\,; u\}$ and trivial conservation laws.
2. The conservation law (3.47) of $\mathbf{S}\{x, t\,; u, v\}$ is *not* expressible as a linear combination of the local conservation laws of $\mathbf{R}\{x, t\,; u\}$ and trivial conservation laws, i.e. the flux and/or density in (3.47) have an essential dependence on the components of the potential variable v.

In the first situation, the conservation law (3.47) can be rewritten so that neither Φ nor Ψ has an essential dependence on the nonlocal variable v, and hence this conservation law is not a new conservation law of $\mathbf{R}\{x, t\,; u\}$ (assuming that all of its local conservation laws are known). In the second situation, the conservation law (3.47) is not equivalent to any local conservation law and is hence a *nonlocal conservation law of the given PDE system* $\mathbf{R}\{x, t\,; u\}$.

In general, nonlocal conservation laws for a system $\mathbf{R}\{x,t\,;u\}$ can be found from related potential systems, as well as other nonlocally related systems. This leads to the following definition.

Definition 3.4.1. A conservation law of a PDE system nonlocally related to a given PDE system $\mathbf{R}\{x,t\,;u\}$ is a *nonlocal conservation law* of $\mathbf{R}\{x,t\,;u\}$ if it is not equivalent to a linear combination of local conservation laws (3.1) of $\mathbf{R}\{x,t\,;u\}$.

Similar to the situation for local conservation laws of a given PDE system $\mathbf{R}\{x,t\,;u\}$, nonlocal conservation laws are useful for:

- Direct applications in PDE problem analysis [see Chapters 1, 2].
- Obtaining further PDE systems nonlocally related to $\mathbf{R}\{x,t\,;u\}$.

The second application is considered in detail in Section 3.5.

3.4.1 Conservation laws arising from nonlocally related systems

Suppose a given PDE system $\mathbf{R}\{x,t\,;u\}$

$$R^\sigma[u] = R^\sigma(x,t,u,\partial u,\ldots,\partial^k u) = 0, \quad \sigma = 1,\ldots,N, \tag{3.48}$$

has a local conservation law given by

$$\mathrm{D}_t\Phi[u] + \mathrm{D}_x\Psi[u] = 0, \tag{3.49}$$

where $\Phi[u] = \Phi(x,t,u,\partial u,\ldots,\partial^{k-1}u), \Psi[u] = \Psi(x,t,u,\partial u,\ldots,\partial^{k-1}u)$.

Introducing the corresponding potential variable v and potential equations

$$\mathcal{P}: \quad \begin{cases} v_x = \Phi[u], \\ v_t = -\Psi[u], \end{cases} \tag{3.50}$$

one obtains the potential system $\mathbf{S}\{x,t\,;u,v\}$ given by

$$\begin{aligned} R^\sigma[u] &= 0, \quad \sigma = 1,\ldots,N', \\ v_x &= \Phi[u], \\ v_t &= -\Psi[u]. \end{aligned} \tag{3.51}$$

[Note that if all equations of $\mathbf{R}\{x,t\,;u\}$ (3.48) are kept in $\mathbf{S}\{x,t\,;u,v\}$ (3.51), one has $N = N'$. If one of the equations of $\mathbf{R}\{x,t\,;u\}$ (without loss of generality, $R^N[u] = 0$) is a differential consequence of the potential equations (3.50) and is not included in $\mathbf{S}\{x,t\,;u,v\}$, then $N' = N - 1$.]

Further (nonlocal) conservation laws are now sought for the given system $\mathbf{R}\{x,t\,;u\}$ through applying the direct method or other method for finding conservation laws of the potential system $\mathbf{S}\{x,t\,;u,v\}$. The following fundamental theorem [Kunzinger & Popovych (2008); Bluman, Cheviakov & Ivanova (2006)] prescribes the form of multipliers needed to obtain nonlocal conservation laws of $\mathbf{R}\{x,t\,;u\}$, i.e. conservation laws that are not expressible as a linear combination of local conservation laws of $\mathbf{R}\{x,t\,;u\}$.

Theorem 3.4.1. *Each conservation law of any potential system* $\mathbf{S}\{x,t\,;u,v\}$ (3.51), *arising from multipliers that do not essentially depend on the potential variable* v, *is equivalent to a local conservation law of the given system* $\mathbf{R}\{x,t\,;u\}$ (3.48).

Theorem 3.4.1 is a particular case of Theorem 5.3.1 that also holds for equations with three or more independent variables, and that is presented and proved in Section 5.3.3.

The converse statement to Theorem 3.4.1 is also true: if a conservation law (3.47) of the potential system $\mathbf{S}\{x,t\,;u,v\}$ is equivalent to a local conservation law of the given system $\mathbf{R}\{x,t\,;u\}$, then multipliers yielding (3.47) are independent of V [Exercise 3.4.1].

Note that Theorem 3.4.1 only holds for PDE systems for which *all* local conservation laws are found. However, for many PDE systems, this is not the situation. In particular, for a given PDE system, zeroth- and first-order conservation laws (i.e., conservation laws with density and flux(es) that depend at most on first derivatives of dependent variables) are usually found, and higher-order local conservation laws may be unknown or too difficult to determine. In such a case, through applying the direct method (or other method) to a potential system $\mathbf{S}\{x,t\,;u,v\}$, one may find previously unknown *local* conservation laws of the given PDE system $\mathbf{R}\{x,t\,;u\}$ even when a set of multipliers only has a dependence on local variables.

However, there exist classes of equations for which an upper bound is known for the order of local conservation laws. In particular, the following theorem holds [Ibragimov (1985)].

Theorem 3.4.2. *For any* $(1+1)$*-dimensional scalar evolution equation*

$$u_t = F(x,t,u,\partial_x u,\dots,\partial_x^{2l}u) \tag{3.52}$$

of even order $2l$ *with a single dependent variable* u *and two independent variables* t *and* x, *the flux* Φ *and the density* Ψ *of a local conservation law* (3.49) *(up to an equivalence transformation) depend only on* x,t,u *and derivatives of* u *with respect to* x, *and the maximal order of a derivative in a density* Ψ *is* l.

It follows that one can find all local conservation laws for evolution equations of the form (3.52).

In Section 3.4.2, nonlocal conservation laws are considered for diffusion-convection equations that belong to the class (3.52). For such equations, all local conservation laws are listed, and several nonlocal ones are presented. In Section 3.4.3, potential systems are used to find further conservation laws of nonlinear telegraph (NLT) equations for which several local conservation laws are known.

A new conservation law of the Lagrange system of planar gas dynamics equations is obtained in Exercise 3.5.6, from a potential system involving three potential variables. Nonlocal conservation laws of Maxwell's equations in $2+1$ dimensions are presented in Section 5.3.5.

3.4.2 Nonlocal conservation laws for diffusion-convection equations

Consider a class of diffusion-convection equations $\mathbf{R}\{x,t\,;u\}$ of the form

$$u_t = (A(u)u_x)_x + B(u)u_x, \tag{3.53}$$

where $A(u)$ and $B(u)$ are arbitrary smooth constitutive functions, and $A(u) \neq 0$. The linear case $A=1, B=0$ is excluded.

(1) Local conservation laws

Using Theorem 3.4.2, one can show that for an equation $\mathbf{R}\{x,t\,;u\}$ (3.53), one must have $\Phi = \Phi(x,t,u)$, $\Psi = \Psi(x,t,u,u_x)$ [Popovych & Ivanova (2005b)]. Hence one can compute the complete set of local conservation laws for diffusion-convection equations of the form $\mathbf{R}\{x,t\,;u\}$ (3.53), modulo the seven-parameter group of equivalence transformations given by

$$\widetilde{t} = a_4 t + a_1, \quad \widetilde{x} = a_5 x + a_7 t + a_2, \quad \widetilde{u} = a_6 u + a_3, \quad \widetilde{A} = \frac{a_5^2}{a_4}A, \quad \widetilde{B} = \frac{a_5^2}{a_4}B - a_7,$$

where $a_1, \dots, a_7$ are arbitrary constants, $a_4 a_5 a_6 \neq 0$.

The classification of linearly independent local conservation laws for $\mathbf{R}\{x,t\,;u\}$ (3.53) yields the following results [Popovych & Ivanova (2005b)]:

1. For arbitrary $A(u), B(u)$, the only local conservation law of (3.53) is given by
$$\mathrm{D}_t(u) - \mathrm{D}_x\Big(A(u)u_x + \int B(u)du\Big) = 0. \tag{3.54}$$
2. For arbitrary $A(u)$, and $B(u) = 0$, there are two local conservation laws of (3.53) given by (3.54) and

$$D_t(xu) + D_x\Big(\int A(u)du - xA(u)u_x\Big) = 0. \tag{3.55}$$

3. For arbitrary $A(u)$, and $B(u) = A(u)$, there are four local conservation laws of (3.53) given by (3.54) and

$$D_t\Big((e^x + \varepsilon)u\Big) - D_x\Big(\varepsilon \int A(u)du + (e^x + \varepsilon)uA(u)u_x\Big) = 0, \tag{3.56}$$
$$\varepsilon = 0, \pm 1.$$

The following potential systems arise from the conservation laws (3.54)–(3.56).

Case 1: Arbitrary $A(u), B(u)$. The conservation law (3.54) yields a potential variable v^1 and corresponding potential system $\mathbf{RV^1}\{x, t\,; u, v^1\}$ given by

$$\begin{aligned} v^1_x &= u, \\ v^1_t &= A(u)u_x + \textstyle\int B(u)du. \end{aligned} \tag{3.57}$$

Case 2: Arbitrary $A(u)$, $B(u) = 0$. The conservation law (3.55) yields an additional potential variable v^2 and corresponding potential system $\mathbf{RV^2}\{x, t;$ $u, v^2\}$ given by

$$\begin{aligned} v^2_x &= xu, \\ v^2_t &= \textstyle\int A(u)du - xA(u)u_x. \end{aligned} \tag{3.58}$$

Case 3: Arbitrary $A(u)$, $B(u) = A(u)$. The conservation laws (3.56) yield three additional potential variables w^i, $i = 1, 2, 3$ and corresponding potential systems $\mathbf{RW^i}\{x, t\,; u, w^i\}$ given by

$$\begin{aligned} w^i_x &= (e^x + \varepsilon)u, \\ w^i_t &= \varepsilon \textstyle\int A(u)du + (e^x + \varepsilon)A(u)u_x, \quad \varepsilon = 0, \pm 1. \end{aligned} \tag{3.59}$$

(2) Nonlocal conservation laws
From Theorem 3.4.1, it follows that in order to obtain nonlocal conservation laws of a diffusion-convection equation $\mathbf{R}\{x, t\,; u\}$ of the form (3.53), one should seek conservation laws of the potential systems (3.57), (3.58), or (3.59) that result from multipliers having an essential dependence on their respective potential variables. It can be shown that all local conservation laws of any one of the potential systems (3.57)–(3.59) are of order zero (i.e., their fluxes do not depend on the derivatives of u and/or the potential variables) [Popovych & Ivanova (2005a)].

It turns out that nonlocal conservation laws of a diffusion-convection equation $\mathbf{R}\{x, t\,; u\}$ (3.53) only result from local conservation laws of the potential system $\mathbf{RV}^1\{x,t\,;u,v^1\}$ (3.57). The four cases that arise are listed in Table 3.1 [Popovych & Ivanova (2005a)].

Table 3.1 Nonlocal conservation laws for the diffusion-convection equations (3.53)

Case	$A(u)$	$B(u)$	Multipliers	Nonlocal Conservation Law
(1)	Arbitrary	$\int A du + uA$	$\Lambda_1 = -e^v \int A du$, $\Lambda_2 = e^v$	$\mathrm{D}_t(e^v) - \mathrm{D}_x\left(e^v \int A du\right) = 0$ $\mathrm{D}_t(e^v) - \mathrm{D}_x\left(e^v \int A du\right) = 0$
(2)	u^{-2}	0	$\Lambda_1 = u^{-1}\sigma_{vv}$, $\Lambda_2 = \sigma_v$	$\mathrm{D}_t(\sigma) + \mathrm{D}_x(\sigma_v u^{-1}) = 0$ $\mathrm{D}_t(\sigma) + \mathrm{D}_x(\sigma_v u^{-1}) = 0$
(3)	u^{-2}	u^{-2}	$\Lambda_1 = e^x u^{-1}\sigma_{vv}$, $\Lambda_2 = e^x \sigma_v$	$\mathrm{D}_t(e^x\sigma) + \mathrm{D}_x(e^x\sigma_v u^{-1}) = 0$ $\mathrm{D}_t(e^x\sigma) + \mathrm{D}_x(e^x\sigma_v u^{-1}) = 0$
(4)	1	$2u$	$\Lambda_1 = (\alpha_x - \alpha u)e^v$, $\Lambda_2 = \alpha e^v$	$\mathrm{D}_t(\alpha e^v) + \mathrm{D}_x\left((\alpha_x - \alpha u)e^v\right) = 0$ $\mathrm{D}_t(\alpha e^v) + \mathrm{D}_x\left((\alpha_x - \alpha u)e^v\right) = 0$

In Table 3.1, for Case (1) [$A(u)$ arbitrary, $B(u) = \int A du + uA$], the diffusion-convection equation $\mathbf{R}\{x,t\,;u\}$ (3.53) is nonlinear and has a single nonlocal conservation law. In Cases (2) and (3), infinite sets of nonlocal conservation laws arise: $\sigma = \sigma(t,v)$ is an arbitrary solution to the linear backward heat equation $\sigma_t + \sigma_{vv} = 0$. It follows that these infinite sets of conservation laws of system $\mathbf{RV}^1\{x,t\,;u,v^1\}$ (3.57) have sufficient cardinality to apply Theorem 2.6.2 (sufficient conditions for the existence of an invertible mapping). Thus the PDE systems $\mathbf{RV}^1\{x,t\,;u,v^1\}$ with $A(u) = u^{-2}$ and $B(u) = 0$ or $B(u) = u^{-2}$ are linearizable by point transformations, and hence the corresponding diffusion-convection equations $\mathbf{R}\{x,t\,;u\}$ (3.53) are linearizable by nonlocal transformations [Exercise 3.4.3].

For Case (4) in Table 3.1, an infinite set of nonlocal conservation laws arises for the diffusion-convection equation $\mathbf{R}\{x,t\,;u\}$ (3.53) for $A(u) = 1$, $B(u) = 2u$: the function $\alpha(x,t)$ is an arbitrary solution of the linear backward heat equation $\alpha_t + \alpha_{xx} = 0$. From Theorem 2.6.2, it follows that the corresponding potential system $\mathbf{RV}^1\{x,t\,;u,v^1\}$ (3.57) is linearizable by a point transformation, and correspondingly the diffusion-convection equation $\mathbf{R}\{x,t\,;u\}$ is also linearizable by a nonlocal transformation [Exercise 3.4.3].

3.4.3 Additional conservation laws of nonlinear telegraph equations

Now consider nonlinear telegraph (NLT) equations $\mathbf{U}\{x,t\,;u\}$ of the form

$$u_{tt} - (F(u)u_x)_x - (G(u))_x = 0, \tag{3.60}$$

where $F(u)$ and $G(u)$ are arbitrary constitutive functions.

(1) Local conservation laws of (3.60)

For NLT equations $\mathbf{U}\{x, t\,; u\}$ of the form (3.60), linearly independent local conservation laws are constructed through seeking conservation law multipliers of the form $\Lambda(x, t, U)$.

First note that the class of PDEs of the form (3.60) has the group of equivalence transformations

$$\begin{aligned} &\widetilde{x} = a_1 x + a_4, \quad \widetilde{t} = a_2 t + a_5, \quad \widetilde{u} = a_3 u + a_6, \\ &\widetilde{F}(\widetilde{u}) = a_1^2 a_2^{-2} F(u), \quad \widetilde{G}(\widetilde{u}) = a_1 a_2^{-2} a_3 G(u) + a_7, \end{aligned} \tag{3.61}$$

where $a_1, \ldots, a_7$ are arbitrary constants, and $a_1 a_2 a_3 \neq 0$. Hence conservation laws of PDEs of the form (3.60) are classified modulo the equivalence transformations (3.61).

Obviously, the case where $G = u$ and $F = \text{const}$ is linear. Here the NLT equation (3.60) has an infinite number of local conservation laws. Only nonlinear cases are considered.

From the corresponding determining equations for local multipliers, it follows that $\Lambda = \Lambda(x, t)$ [Exercise 3.4.4]. The three distinct cases that arise are summarized in Table 3.2 [Bluman, Cheviakov & Ivanova (2006)].

Table 3.2 Local conservation laws of NLT equations (3.60)

Case	$F(u)$	$G(u)$	Multiplier Λ	Local Conservation Law
(a)	Arbitrary	Arbitrary	1	$\mathrm{D}_t u_t - \mathrm{D}_x(F(u)u_x + G(u)) = 0$
			t	$\mathrm{D}_t(tu_t - u)$ $-\mathrm{D}_x\big(t(F(u)u_x + G(u))\big) = 0$
(b)	Arbitrary	$G'(u) = F(u)$	e^x	$\mathrm{D}_t\left(e^x u_t\right) - \mathrm{D}_x\left(e^x F(u)u_x\right) = 0$
			te^x	$\mathrm{D}_t\left(e^x(tu_t - u)\right) - \mathrm{D}_x\left(te^x F(u)u_x\right) = 0$
(c)	Arbitrary	u	$x - \frac{1}{2}t^2$	$\mathrm{D}_t\left(\left(x - \frac{1}{2}t^2\right)u_t + tu\right)$ $-\mathrm{D}_x\Big(\left(x - \frac{1}{2}t^2\right)(F(u)u_x + u)$ $+\int F(u)du\Big) = 0$
	$(F(u) \neq$ const)		$xt - \frac{1}{6}t^3$	$\mathrm{D}_t\left(\left(tx - \frac{1}{6}t^3\right)u_t - \left(x - \frac{1}{2}t^2\right)u\right)$ $-\mathrm{D}_x\Big(\left(tx - \frac{1}{6}t^3\right)(F(u)u_x + u)$ $+t\int F(u)du\Big) = 0$

The following potential systems result from the conservation laws listed in Table 3.2.

Case (a): Arbitrary $F(u), G(u)$. The two conservation laws yield, respectively, the potential systems $\mathbf{UV^1}\{x,t\,;u,v^1\}$ given by

$$\begin{aligned} v^1_x &= u_t, \\ v^1_t &= F(u)u_x + G(u); \end{aligned} \tag{3.62}$$

and $\mathbf{UV^2}\{x,t\,;u,v^2\}$ given by

$$\begin{aligned} v^2_x &= tu_t - u, \\ v^2_t &= t(F(u)u_x + G(u)). \end{aligned} \tag{3.63}$$

Case (b): $G'(u) = F(u)$, $F(u)$ arbitrary. In addition to the potential systems (3.62) and (3.63), here one also has the potential systems $\mathbf{UB^3}\{x,t\,;u,b^3\}$ given by

$$\begin{aligned} b^3_x &= e^x u_t, \\ b^3_t &= e^x F(u)u_x; \end{aligned} \tag{3.64}$$

and $\mathbf{UB^4}\{x,t\,;u,b^4\}$ given by

$$\begin{aligned} b^4_x &= e^x(tu_t - u), \\ b^4_t &= te^x F(u)u_x. \end{aligned} \tag{3.65}$$

Case (c): $G(u) = u$, $F(u)$ arbitrary. In addition to the potential systems (3.62) and (3.63), here one also has the potential systems $\mathbf{UC^3}\{x,t\,;u,c^3\}$ given by

$$\begin{aligned} c^3_x &= \left(x - \tfrac{1}{2}t^2\right)u_t + tu, \\ c^3_t &= \left(x - \tfrac{1}{2}t^2\right)(F(u)u_x + u) - \textstyle\int F(u)du; \end{aligned} \tag{3.66}$$

and $\mathbf{UC^4}\{x,t\,;u,c^4\}$ given by

$$\begin{aligned} c^4_x &= \left(xt - \tfrac{1}{6}t^3\right)u_t - \left(x - \tfrac{1}{2}t^2\right)u, \\ c^4_t &= \left(xt - \tfrac{1}{6}t^3\right)(F(u)u_x + u) - t\textstyle\int F(u)du. \end{aligned} \tag{3.67}$$

(2) Additional conservation laws for NLT equations with power law nonlinearities

Further conservation laws are now sought for all singlet potential systems (3.62) – (3.67) of the NLT equation (3.60). In order to obtain further conservation laws, one requires a set of multipliers to have an essential dependence on potential variables. For power law nonlinearities $F(u) = u^\alpha$, $G(u) = u^\beta$, the additional conservation laws found for NLT equations of the form (3.60) are given in Tables 3.3a–c [Bluman, Cheviakov & Ivanova (2006)]. The results are presented modulo the corresponding equivalence transformations [Exercise 3.4.5].

Table 3.3 (a) Additional conservation laws of the NLT equations, arising from potential systems (3.62)–(3.67), for power law nonlinearities $F(u) = u^\alpha$, $G(u) = u^\beta$. Case (a): $F(u) = u^\alpha$, $G(u) = u^\beta$

System	Subcase	Multipliers	Conservation Law
$\mathbf{UV^1}$	$\beta = -1$	$\Lambda_1 = x + \frac{1}{2}(v^1)^2 + \frac{u^{\alpha+2}}{\alpha+2}$, $\Lambda_2 = uv^1$.	$\mathrm{D}_t\left(\left(x + \frac{u^{\alpha+2}}{(\alpha+2)(\alpha+3)} + \frac{1}{2}(v^1)^2\right)u\right) - \mathrm{D}_x\left(\left(x + \frac{u^{\alpha+2}}{\alpha+2} + \frac{1}{6}(v^1)^2\right)v^1\right) = 0$
		$\Lambda_1 = v^1, \Lambda_2 = u.$	$\mathrm{D}_t(t - uv^1) + \mathrm{D}_x\left(\frac{u^{\alpha+2}}{\alpha+2} + \frac{1}{2}(v^1)^2\right) = 0$
	$\alpha = -1$, $\beta = -1$	$\Lambda_1 = t + \frac{1}{3}(v^1)^3, +2(x+u)v^1,$ $\Lambda_2 = (2x + u + (v^1)^2)u.$	$\mathrm{D}_t\Big(\left(u + \frac{1}{3}(v^1)^2\right)uv^1 + 2xuv^1 - t(2x-u)\Big) - \mathrm{D}_x\left(\frac{1}{12}(v^1)^4 + (x+u)(v^1)^2 + tv^1 + tv^1 + \frac{1}{2}u^2 + 2xu\right) = 0$
		$\Lambda_1 = x^2 + 2xu + \frac{1}{2}u^2 + \frac{1}{12}(v^1)^4 + (x+u)(v^1)^2 + tv^1,$ $\Lambda_2 = (t + (2x+u)v^1)u + \frac{1}{3}(v^1)^3.$	$\mathrm{D}_t\Big(2x - \frac{1}{2}t^2 + \frac{1}{2}\left(\frac{1}{3}u + (v^1)^2\right)u^2 + \frac{1}{12}u(v^1)^4 + (t + xv^1)uv^1 + x^2u\Big) - \mathrm{D}_x\Big(\frac{1}{60}(v^1)^5 + \frac{1}{3}(x+u)(v^1)^3 + \frac{1}{2}(tv^1 + u^2)v^1 + (x+2u)xv^1 + tu\Big) = 0$
$\mathbf{UV^2}$	$\beta = -1$	$\Lambda_1 = -\frac{v^2}{t^2}$, $\Lambda_2 = \frac{u}{t}$.	$\mathrm{D}_t\left(\frac{uv^2 - t^2}{t}\right) - \mathrm{D}_x\left(\frac{(v^2)^2}{2t^2} + \frac{u^{\alpha+2}}{\alpha+2}\right) = 0$

No additional conservation laws are found for the potential system $\mathbf{UC^4}\{x, t; u, c^4\}$ (3.67).

Case (b) with $\alpha = -2$ is not considered in Table 3.3c since here the system $\mathbf{UV^1}\{x, t; u, v^1\}$ (3.62) is linearizable by a point transformation [Exercise 3.4.6].

Table 3.3 (b) Additional conservation laws of the NLT equations, arising from potential systems (3.62)–(3.67), for power law nonlinearities $F(u) = u^\alpha$, $G(u) = u^\beta$. Case (b): $F(u) = G'(u)$, $G(u) = u^{\alpha+1}$

System	Subcase	Multipliers	Conservation Law
$\mathbf{UV^1}$	$\alpha \neq -1$ $\alpha \neq -2$	$\Lambda_1 = e^x u^{\alpha+1}$, $\Lambda_2 = e^x v^1$.	$\mathrm{D}_t \left(e^x \left(\frac{u^{\alpha+2}}{\alpha+2} + \frac{1}{2}(v^1)^2 \right) \right)$ $-\mathrm{D}_x \left(e^x u^{\alpha+1} v^1 \right) = 0$
$\mathbf{UV^2}$	$\alpha = -4$	$\Lambda_1 = -e^x \frac{t}{u^3}$, $\Lambda_2 = e^x v^2$.	$\mathrm{D}_t \left(e^x \left(\frac{t^2}{u^2} - (v^2)^2 \right) \right) + \mathrm{D}_x \left(e^x \frac{tv^2}{u^3} \right) = 0$
$\mathbf{UB^3}$	$\alpha \neq -1$	$\Lambda_1 = -u^{\alpha+1}$, $\Lambda_2 = e^{-x} b^3$.	$\mathrm{D}_t \left(e^x \frac{u^{\alpha+2}}{\alpha+2} + e^{-x} \frac{(b^3)^2}{2} \right)$ $-\mathrm{D}_x \left(u^{\alpha+1} b^3 \right) = 0$
$\mathbf{UB^4}$	$\alpha = -4$	$\Lambda_1 = -\frac{t}{u^3}$, $\Lambda_2 = e^{-x} b^4$.	$\mathrm{D}_t \left(\frac{1}{2} e^{-x} (b^4)^2 - e^x \frac{t^2}{2u^2} \right) - \mathrm{D}_x \left(\frac{tb^4}{u^3} \right) = 0$

Note that one still needs to check which of the conservation laws presented in Tables 3.3a–c are nonlocal, and which can be expressed as linear combinations of some (unknown) local conservation laws of the NLT equations.

Table 3.3 (c) Additional conservation laws of the NLT equations, arising from potential systems (3.62)–(3.67), for power law nonlinearities $F(u) = u^\alpha$, $G(u) = u^\beta$. Case (c): $F(u) = u^\alpha$, $G(u) = u$

System	Subcase	Multipliers	Conservation Law
$\mathbf{UV^1}$	$\alpha = 1$	$\Lambda_1 = \frac{1}{12}t^4 - t^2 x + x^2$ $+tv^1 - \frac{1}{2}u^2$, $\Lambda_2 = -\frac{1}{3}t^3 + t(2x+u)$ $-v^1$.	$\mathrm{D}_t \Big(-\frac{1}{6}u^3 + (\frac{1}{12}t^4 + x^2 - xt^2 + tv^1)u$ $+(-\frac{1}{3}t^3 + 2xt - \frac{1}{2}v)v^1 \Big)$ $+\mathrm{D}_x \Big((-tx + \frac{1}{2}v^1 + \frac{1}{6}(t^3 - 2tu))u^2$ $-(tv^1 + \frac{1}{12}t^4 - t^2 x + x^2)v^1 \Big) = 0$
		$\Lambda_1 = \frac{1}{6}t^3 - tx + v^1$, $\Lambda_2 = -\frac{1}{2}t^2 + x + u$.	$\mathrm{D}_t \left((\frac{1}{3}t^3 - 2tx)u + (-t^2 + 2x + u)v^1 \right)$ $+\mathrm{D}_x \Big((\frac{1}{2}t^2 - x - \frac{1}{3}u)u^2$ $-(\frac{1}{3}t^3 - 2tx + \frac{1}{2}v^1)v^1 \Big) = 0$
$\mathbf{UV^2}$	$\alpha = 1$	$\Lambda_1 = \frac{1}{4}t^2 - x + \frac{v^2 - x^2}{t^2}$, $\Lambda_2 = t - \frac{2x+u}{t}$.	$\mathrm{D}_t \left(\frac{uv^2}{t} + \frac{(t^4 - 4x(t^2+x))u}{4t} + \frac{(2x - t^2)v^2}{t} \right)$ $-\mathrm{D}_x \Big(\frac{1}{3}u^3 - \frac{1}{2}(t^2 - 2x)u^2 + \frac{(v^2)^2}{2t^2}$ $+\frac{(t^4 - 4x(t^2+x))v^2}{4t^2} \Big) = 0$
$\mathbf{UC^3}$	$\alpha = 1$	$\Lambda_1 = -\frac{1}{80}(t^2 - 2x)$ $+\frac{2xt^2 + 5u^2}{40(t^2 - 2x)}$ $+\frac{4x^3 + 5tc^3}{10(t^2 - 2x)^2}$, $\Lambda_2 = \frac{3t^5 - 20c^3}{40(t^2 - 2x)^2}$. $-\frac{t(2x+u)}{4(t^2 - 2x)}$.	$\mathrm{D}_t \Big(\frac{1}{64}(t^4 - 4x^2)u + \frac{1}{96}(u^3 - 3t^4 u - 6tc^3)$ $+\frac{t(t^5 + 10c^3)u}{80(t^2 - 2x)} + \frac{(t^5 + 5c^3)c^3}{40(t^2 - 2x)^2} \Big)$ $+\mathrm{D}_x \Big(-\frac{1}{64}(t^2 - 2x)(tu^2 + 2c^3)$ $+\frac{1}{48}t(u^3 + 3tc^3)$ $+\frac{t^4(tu^2 - 10c^3) + 20u^2 c^3}{160(t^2 - 2x)} + \frac{t(t^5 + 5c^3)c^3}{40(t^2 - 2x)^2} \Big) = 0$

(3) Additional conservation laws following from the potential system $\mathbf{UV^1}\{x,t;\, u,v^1\}$ *(general nonlinearities)*

The above classification of conservation laws of the NLT equation $\mathbf{U}\{x,t\,;u\}$ (3.60) can be significantly extended in the case of general nonlinearities. Here one considers a classification of the conservation laws of NLT equations $\mathbf{U}\{x,t\,;u\}$ of the form (3.60) that arise as local conservation laws of the potential system $\mathbf{UV^1}\{x,t\,;u,v^1\}$ (3.62) from multipliers depending on x,t,U, and V^1 [Bluman & Temuerchaolu (2005a)].

First note that the potential system $\mathbf{UV^1}\{x,t\,;u,v^1\}$ (3.62) has the group of equivalence transformations (3.61) with in addition

$$\widetilde{v} = a_3 v + a_2 a_7 t + a_8.$$

The following results are given modulo the above equivalence transformations.

One can show that the potential system $\mathbf{UV^1}\{x,t\,;u,v^1\}$ is linearizable by a point transformation if and only if

$$F(u) = u^{-2},\ G(u) = u^{-1}, \quad \text{or} \quad F(u) \text{ arbitrary},\ G(u) = \text{const}$$

[Exercise 3.4.6]. For both of these cases, as well as for the linear case ($F(u) = \text{const}$, $G(u) = u$), the potential system $\mathbf{UV^1}\{x,t\,;u,v^1\}$ (3.62) has an infinite number of local conservation laws [Bluman & Doran-Wu (1995)]. These linearization cases are not included in the following computations.

The obvious conservation law of the system $\mathbf{UV^1}\{x,t\,;u,v^1\}$ given by its first equation

$$\mathrm{D}_t(u) - \mathrm{D}_x(v^1) = 0,$$

is excluded from further consideration.

Let $\Lambda_1(x,t,U,V)$ and $\Lambda_2(x,t,U,V)$ be conservation law multipliers for the first and second equations of the system $\mathbf{UV^1}\{x,t\,;u,v^1\}$ (3.62), respectively. From the determining equations for Λ_1 and Λ_2, additional nontrivial conservation laws of the potential system $\mathbf{UV^1}\{x,t\,;u,v^1\}$ arise in the following cases.

1. $F(u)$ arbitrary and $G(u) = 0,\ u$, or $1/u$.
2. $F(u)$ and $G(u)$ taking on one of the forms listed in Table 3.4.
3. $F(u) - G'(u) = G^2(u)$.
4. $F(u) - G'(u) = \mu$ for some constant μ.

Note that due to scaling invariance of the potential system $\mathbf{UV^1}\{x,t\,;u,v^1\}$, for Case 4 it suffices to consider $\mu = 0, 1$.

For example, if $F(u) - G'(u) = G^2(u)$, the system $\mathbf{UV^1}\{x,t\,;u,v^1\}$ (3.62) has the two local conservation laws

Table 3.4 $F(u)$ and $G(u)$ which yield additional conservation laws for the system $\mathbf{UV^1}\{x,t\,;u,v^1\}$ (3.62) of NLT equations (Case 2)

$F(u)$	$G(u)$
$\beta_1 u^2 + \beta_2 u + \beta_3$	u
$\beta_1/u^2 + \beta_2/u + \beta_3$	$1/u$
$\beta_1 e^{2u} + \beta_2 e^{u} + \beta_3$	e^u
$\beta_1 \operatorname{sech}^2 u + \beta_2 \tanh u + \beta_3$	$\tanh u$
$\beta_1 \sec^2 u + \beta_2 \tan u + \beta_3$	$\tan u$

$$\mathrm{D}_t\left(A(x,u)e^{\pm v^1}\right) \mp \mathrm{D}_x\left(G(u)A(x,u)e^{\pm v^1}\right) = 0, \tag{3.68}$$

arising from the two sets of multipliers

$$\Lambda_1 = A(x,U)G(U)e^{\pm V^1}, \quad \Lambda_2 = \pm A(x,U)e^{\pm V^1}, \tag{3.69}$$

where $A(x,U) = \exp\left(x + \int G(U)dU\right)$. The conservation laws (3.68) do not arise as local conservation laws of the corresponding NLT equations $\mathbf{U}\{x,t\,;u\}$ (3.60) from a multiplier of the form $\Lambda(x,t,U)$ [Section 3.4.3 (1)]. Since the sets of multipliers (3.69) explicitly depend on the nonlocal variable V, from Theorem 3.4.1 it follows that the conservation laws (3.68) may be nonlocal for the NLT equation $\mathbf{U}\{x,t\,;u\}$ (3.60). For another example, see Exercise 3.4.7.

This example shows that the seeking of local conservation laws of a potential system can yield a much richer set of conservation laws for a given PDE system than through local conservation analysis of the given PDE system itself.

Exercises 3.4

3.4.1. Prove the converse statement to Theorem 3.4.1: If a conservation law (3.47) of the potential system $\mathbf{S}\{x,t\,;u,v\}$ (3.51) is equivalent to a local conservation law of the given system $\mathbf{R}\{x,t\,;u\}$ (3.48), then the multipliers yielding the conservation law (3.47) are independent of the nonlocal variable V. [Provide the proof for cases $N' = N$ and $N' = N-1$ in (3.51).]

3.4.2. Let $\mathbf{R}\{x,t\,;u\}$ be a given PDE system, and suppose $\underline{\mathbf{R}}\{x,t\,;u\}$ is a subsystem of $\mathbf{R}\{x,t\,;u\}$ obtained either directly by exclusion of a dependent variable or indirectly by exclusion of a dependent variable after a point transformation of its variables (including interchanges of dependent and independent variables). Prove that each local conservation law of $\underline{\mathbf{R}}\{x,t\,;u\}$ is a local conservation law of $\mathbf{R}\{x,t\,;u\}$.

3.4.3. In each of the cases (2), (3) and (4) of Table 3.1, the potential system $\mathbf{UV^1}\{x,t\,;u,v^1\}$ (3.62) of the NLT equation $\mathbf{U}\{x,t\,;u\}$ (3.60) has an infinite number of local symmetries. Show that these conservation laws meet the sufficiency conditions for the existence of an invertible point transformation that maps the potential system $\mathbf{UV^1}\{x,t\,;u,v^1\}$ (3.62) into a linear PDE system [Theorem 2.6.2]. For each of the cases (2), (3) and (4), find an explicit linearizing transformation.

3.4.4. For an NLT equation $\mathbf{U}\{x,t\,;u\}$ of the form (3.60), show that the ansatz $\Lambda=\Lambda(x,t,U)$ for a conservation law multiplier reduces to $\Lambda=\Lambda(x,t)$.

3.4.5. Find the equivalence transformations of the class of potential NLT systems $\mathbf{UV^1}\{x,t\,;u,v^1\}$ of the form (3.62). Compare these equivalence transformations with the equivalence transformations of the class of NLT equations $\mathbf{U}\{x,t\,;u\}$ of the form (3.60).

3.4.6.

(a) Show that if

$$F(u)=\frac{c}{(au+b)^2},\qquad G(u)=\frac{d}{au+b}+f,$$

or

$$F(u)\text{ arbitrary},\qquad G(u)=\text{const},$$

the potential system $\mathbf{UV^1}\{x,t\,;u,v^1\}$ (3.62) of the nonlinear telegraph equation has an infinite number of point symmetries.

(b) Show that in these two cases, $\mathbf{UV^1}\{x,t\,;u,v^1\}$ can be mapped into a linear system by a point transformation.

(c) For each of these two cases, find a corresponding linearizing transformation and the resulting linear PDE system [Bluman & Kumei (1989), Bluman & Doran-Wu (1995)].

3.4.7.

(a) Show that for the case $F(u)-G'(u)=1$, the potential NLT system $\mathbf{UV^1}\{x,t\,;u,v^1\}$ (3.62) has local conservation laws arising from the multipliers $\Lambda_1=\mp\Lambda_2=e^{x\pm t}$.

(b) Find the fluxes and densities of the two conservation laws arising from these multipliers.

(c) Use Theorem 3.4.1 to show that these two conservation laws are equivalent to local conservation laws of the NLT equation $\mathbf{U}\{x,t\,;u\}$ (3.60). Find these local conservation laws.

3.5 Extended Tree Construction Procedure

We now simplify and further extend the framework for constructing a tree of nonlocally related PDE systems that was presented in Section 3.3.

Suppose a given PDE system $\mathbf{R}\{x,t\,;u\}$ has n known linearly independent and inequivalent local conservation laws, with corresponding potential equations $\mathcal{P}^i$ in terms of nonlocal (potential) variables $v^i, i=1,\dots,n$. In the tree construction procedure outlined in Section 3.3.1, these n potential variables were used to generate n potential systems $\mathbf{S}^{(1)}\{x,t\,;u,v^i\} = \mathbf{R}\{x,t\,;u\} \cup \mathcal{P}^i$. Through these n potential systems $\mathbf{S}^{(1)}\{x,t\,;u,v^i\}$, using the n potential variables v^i, one can directly generate more PDE systems, nonlocally related to $\mathbf{R}\{x,t\,;u\}$, by taking combinations of two or more potential systems $\mathbf{S}^{(1)}\{x,t\,;u,v^i\}$ to obtain a set of 2^n-1 potential systems with 1 to n potential variables.

Definition 3.5.1. A potential system

$$\mathbf{S}^{(k)}\{x,t\,;u,v^{i_1},\dots,v^{i_k}\} = \mathbf{R}\{x,t\,;u\} \cup \mathcal{P}^{i_1} \cup \cdots \cup \mathcal{P}^{i_k}, \quad 1 \le k \le n,$$

with k potential variables, is called a *k-plet potential system*. In particular, for $k=1,2,3,4$, we refer to such k-plet potential systems as *singlets*, *couplets*, *triplets*, and *quadruplets*, respectively.

Thus for a PDE system $\mathbf{R}\{x,t\,;u\}$ with n known linearly independent local conservation laws, a corresponding set of 2^n-1 potential systems (k-plets) arises.

- n singlets: $\mathbf{S}^{(1)}\{x,t\,;u,v^i\}, \quad i=1,\dots,n.$
- $\frac{1}{2}n(n-1)$ couplets: $\mathbf{S}^{(2)}\{x,t\,;u,v^i,v^j\}, \quad i=1,\dots,j-1, \quad j=2,\dots,n.$
- $\vdots$.
- One n-plet: $\mathbf{S}^{(n)}\{x,t\,;u,v^1,\dots,v^n\}$.

Definition 3.5.2. Suppose n linearly independent local conservation laws are known for a given PDE system $\mathbf{R}\{x,t\,;u\}$. In terms of the resulting potential variables $v^1,\dots,v^n$, the set of all corresponding 2^n-1 potential systems is called a *combination potential system* $\mathbb{P}_{v^1\dots v^n}$.

Now assume that one is able to obtain n linearly independent local conservation laws and hence the corresponding combination potential system $\mathbb{P}_{v^1\dots v^n}$. In order to obtain further potential systems, one could seek nonlocal conservation laws of $\mathbf{R}\{x,t\,;u\}$ through the consideration of local conservation laws for each of the potential systems in $\mathbb{P}_{v^1\dots v^n}$. Clearly, in the consideration of all potential systems in $\mathbb{P}_{v^1\dots v^n}$, , in order to avoid redundancies, one should apply a conservation law construction algorithm to the n-plet $\mathbf{S}^{(n)}\{x,t\,;u,v^1,\dots,v^n\}$ since it contains $\mathbf{R}\{x,t\,;u\}$ and all known potential equations, and thus has all local conservation laws of *each* of the potential systems in $\mathbb{P}_{v^1 \dots v^n}$.

If the *complete* set of linearly independent local conservation laws of $\mathbf{R}\{x,t\,;u\}$ is found, one may use the following generalization of Theorem 3.4.1 in order to avoid obtaining conservation laws linearly dependent on the known local ones.

Theorem 3.5.1. *Each conservation law of any potential system in* $\mathbb{P}_{v^1 \dots v^n}$*, arising from multipliers that depend only on local variables (i.e., x, t, u, and derivatives of u) of the given PDE system* $\mathbf{R}\{x,t\,;u\}$*, is linearly dependent on the local conservation laws of the given system* $\mathbf{R}\{x,t\,;u\}$.

Proof. The proof parallels that of Theorem 3.4.1 and is left to Exercise 3.5.1. □

Consequently, the following holds.

Corollary 3.5.1. *Suppose one finds the complete set of n local conservation laws for a given PDE system* $\mathbf{R}\{x,t\,;u\}$ *and then constructs the combination potential system* $\mathbb{P}_{v^1 \dots v^n}$*. It follows that if one starts with any one of the* $2^n - 1$ *potential systems in* $\mathbb{P}_{v^1 \dots v^n}$ *and seeks conservation laws from multipliers depending only on x, t, u, and derivatives of u, each of the resulting potential systems is locally equivalent to one of the* $2^n - 1$ *potential systems in* $\mathbb{P}_{v^1 \dots v^n}$.

However, often the complete set of linearly independent local conservation laws of a given PDE system $\mathbf{R}\{x,t\,;u\}$ is not known. In particular, when seeking local conservation laws through the direct method, one normally limits oneself to considering to rth-order multipliers of the form $\Lambda = \Lambda(x, u, \partial u, \dots, \partial^r u)$ for some fixed r.

From Theorem 3.5.1, it follows that further conservation laws of $\mathbf{S}^{(n)}\{x,t;\, u, v^1, \dots, v^n\}$, are more likely to arise from multipliers which have an essential dependence on the potential variables $v^1, \dots, v^n$. This is confirmed by experience. Indeed, in all studied examples, conservation laws of the n-plet $\mathbf{S}^{(n)}\{x,t;\, u, v^1, \dots, v^n\}$, arising from multipliers depending only on independent and the dependent variables of the given PDE system $\mathbf{R}\{x,t\,;u\}$ and derivatives of u, are linear combinations of known local conservation laws of $\mathbf{R}\{x,t\,;u\}$, as one might expect.

3.5.1 An extended tree construction procedure

The above discussion leads to the following extension of the tree construction procedure presented in Section 3.3.1.

Consider a given PDE system $\mathbf{R}\{x,t\,;u\}$.

1. **Construction of local conservation laws.** Find a set $\{\mathcal{K}_i\}$ of linearly independent and inequivalent local conservation laws for the given PDE system $\mathbf{R}\{x,t\,;u\}$. Let n be the number of such conservation laws that are found.
2. **Construction of potential systems.** Use the set of known local conservation laws $\{\mathcal{K}_i\}$ to introduce n potential variables v^i. Construct the corresponding combination potential system $\mathbb{P}_{v^1\ldots v^n}$ which contains 2^n-1 potential systems. Together with the given PDE system $\mathbf{R}\{x,t\,;u\}$, this yields a tree $\mathcal{T}_1$ with up to 2^n nonlocally related systems.
3. **Additional conservation laws.** In the tree $\mathcal{T}_1$, consider the n-plet potential system $\mathbf{S}^{(n)}\{x,t\,;u,v^1,\ldots,v^n\}$. For this n-plet, seek linearly independent conservation laws. [If using the direct method, normally seek multipliers that have an essential dependence on the potential variables $v^1,\ldots,v^n$.] Eliminate conservation laws that are linearly dependent on the set of known local conservation laws $\{\mathcal{K}_i\}$ of $\mathbf{R}\{x,t\,;u\}$. Let the number of newly obtained linearly independent conservation laws of $\mathbf{S}^{(n)}\{x,t\,;u,v^1,\ldots,v^n\}$ be n'. Introduce corresponding potential variables v^j, $j=n+1,\ldots,n+n'$. [By construction, the full set of potentials $\{v^1,\ldots,v^{n+n'}\}$ is linearly independent.]
4. **Tree extension.** Use the $n+n'$ potentials $\{v^i\}$ to construct the corresponding combination potential system $\mathbb{P}_{v^1\ldots v^{n+n'}}$. Together with the given system $\mathbf{R}\{x,t\,;u\}$, this yields an extended tree $\mathcal{T}_2$.
5. **Continuation.** Repeat Steps 3 and 4 for the tree $\mathcal{T}_2$, until no further linearly independent conservation laws are found for any nonlocally related potential system. This yields a possibly larger extended tree $\mathcal{T}_3$.
6. **Construction of subsystems.** For all systems in the tree $\mathcal{T}_3$, exclude where possible, one by one, dependent variables, to generate subsystems of the systems in the tree $\mathcal{T}_3$. Eliminate locally related subsystems. In addition, in the same manner generate nonlocally related subsystems obtained after an interchange of one or more independent and dependent variables. This yields a possibly larger extended tree of nonlocally related systems denoted by $\mathcal{T}_4$.

If the given PDE system $\mathbf{R}\{x,t\,;u\}$ includes arbitrary constitutive function(s), one may be able to still further extend the above procedure: in Steps 1, 3 and 6, classify conservation laws with respect to the constitutive function(s) to isolate cases for which additional conservation laws and/or additional nonlocally related subsystems arise. Trees for particular forms of the constitutive functions could be significantly different, although sharing branches that hold for arbitrary constitutive function(s).

The extended procedure outlined above is practically efficient. It improves and extends the framework presented in Section 3.3 in the following ways.

- One obtains a larger set ($2^n - 1$ compared to n) of potential systems from n known local conservation laws for any PDE system in a tree, at no extra computational cost.
- The extended tree does not contain PDE systems with linearly dependent potential variables.
- In computing conservation laws of potential systems of $\mathbf{R}\{x,t\,;u\}$, the number of computations is reduced since multipliers depending only on independent variables and designated dependent variables and their derivatives [Theorem 3.5.1] may not need to be considered.
- Normally, one can generate nonlocally related subsystems with minimal computational cost.

The extended tree construction procedure presented in this section assumes that the n known local conservation laws of a given PDE system $\mathbf{R}\{x,t\,;u\}$ are inequivalent and linearly independent. Indeed, equivalent conservation laws should not be considered since they yield locally related potential variables [Theorem 3.2.1].

For a given PDE system, often it seems to turn out that its simplest conservation laws are of greatest use to obtain new results.

3.5.2 *An extended tree for a nonlinear diffusion equation*

Again consider the example of Section 3.3.2. In particular, the extended tree construction procedure of Section 3.5.1 is applied to the nonlinear diffusion equation $\mathbf{U}\{x,t\,;u\}$ given by

$$u_t - (K(u)u_x)_x = 0 \tag{3.70}$$

with an arbitrary constitutive function $K(u)$.

One begins by seeking local conservation laws. The nonlinear diffusion equation $\mathbf{U}\{x,t\,;u\}$ (3.70) is a second-order evolution equation, a subcase of the family of nonlinear diffusion-convection equations (3.53) with $A(u) = K(u)$, $B(u) = 0$. As shown in Section 3.4.2, for arbitrary $K(u) = L'(u)$, the nonlinear diffusion equation (3.70) has exactly $n = 2$ linearly independent local conservation laws given by

$$\mathrm{D}_t(u) - \mathrm{D}_x\Big((L(u))_x\Big) = 0, \tag{3.71}$$

$$\mathrm{D}_t(xu) - \mathrm{D}_x\Big(x(L(u))_x - L(u)\Big) = 0 \tag{3.72}$$

[Bluman & Doran-Wu (1989)].

Next, one can construct the corresponding most general combination potential system $\mathbb{P}_{v\alpha}$ for $\mathbf{U}\{x,t\,;u\}$ (3.70) since all of its local conservation laws are known. In particular, $\mathbb{P}_{v\alpha}$ consists of the singlet potential systems $\mathbf{UV}\{x,t\,;u,v\}$ (3.19) and $\mathbf{UA}\{x,t\,;u,\alpha\}$ (3.32) and the couplet potential system $\mathbf{UVA}\{x,t\,;u,v,\alpha\}$ given by

$$\begin{aligned} v_x &= u, \\ v_t &= (L(u))_x, \\ \alpha_x &= xu, \\ \alpha_t &= x(L(u))_x - L(u). \end{aligned} \tag{3.73}$$

Note that the given nonlinear diffusion equation $\mathbf{U}\{x,t\,;u\}$ (3.70) need not be appended to the potential systems $\mathbf{UV}\{x,t\,;u,v\}$, $\mathbf{UA}\{x,t\,;u,\alpha\}$ and $\mathbf{UVA}\{x,t\,;u,v,\alpha\}$ since both conservation laws (3.71) and (3.72) arise from multipliers that do not vanish off of the solution space of the nonlinear diffusion equation $\mathbf{U}\{x,t\,;u\}$. Consequently, one obtains the trees

$$\begin{aligned} \mathcal{T}_1 &= \mathcal{T}_2 = \mathcal{T}_3 \\ &= \{\mathbf{U}\{x,t\,;u\}, \mathbf{UV}\{x,t\,;u,v\}, \mathbf{UA}\{x,t\,;u,\alpha\}, \mathbf{UVA}\{x,t\,;u,v,\alpha\}\}. \end{aligned}$$

[The system $\mathbf{UVA}\{x,t\,;u,v,\alpha\}$ (3.73) could be analyzed for further conservation laws to obtain additional potential systems. Comparing with Section 3.3.2, one notes that the potential systems $\mathbf{UVW}\{x,t\,;u,v,w\}$ (3.30) and $\mathbf{UAB}\{x,t\,;u,\alpha,\beta\}$ (3.34) are locally related to the system $\mathbf{UVA}\{x,t\,;u,v,\alpha\}$ (3.73). In particular, from the transformation (3.35) it follows that $w = xv - \alpha$, $\beta = v - x^{-1}\alpha$.]

Now one constructs nonlocally related subsystems for the potential systems in $\mathcal{T}_3$. Nonlocally related subsystems arising from the singlet potential systems $\mathbf{UV}\{x,t\,;u,v\}$ (3.19) and $\mathbf{UA}\{x,t\,;u,\alpha\}$ (3.32) were obtained in Sections 3.2.2 and 3.3.2. These include the scalar PDEs $\mathbf{V}\{u,t\,;v\}$ (3.22), $\mathbf{X}\{u,v\,;x\}$ [Exercise 3.3.3] and $\mathbf{A}\{x,u\,;\alpha\}$ (3.38). Subsystems arise from the couplet potential system $\mathbf{UVA}\{x,t\,;u,v,\alpha\}$ (3.73) as follows. In general, exclusion of the dependent variable u is not possible. [In cases when this is possible, one obtains a locally related subsystem $\mathbf{VA}\{x,t\,;v,\alpha\}$.] The exclusions of α and v, respectively, yield the known systems $\mathbf{UV}\{x,t\,;u,v\}$ (3.19) and $\mathbf{UA}\{x,t\,;u,\alpha\}$ (3.32). Consider interchanges of dependent and independent variables. For example, let $x = x(u,v), t = v(u,v), \alpha = \alpha(u,v)$ be treated as dependent variables and (u,v) as independent variables. This yields the PDE system $\mathbf{XTA}\{u,v\,;x,t,\alpha\}$ given by

$$\begin{aligned} t_u &= \frac{ux_u^2}{L'(u)(ux_v-1)}, \\ t_v &= \frac{x_u}{L'(u)}, \\ \alpha_u &= -K(u)t_u, \\ \alpha_v &= x - L'(u)x - \frac{L(u)}{L'(u)}x_u \end{aligned} \tag{3.74}$$

that is invertibly related to $\mathbf{UVA}\{x,t\,;u,v,\alpha\}$ (3.73). Exclusion of the dependent variable x does not yield a nonlocally related subsystem, since x appears explicitly in the PDE system (3.74). Exclusion of α leads to the known subsystem $\mathbf{XT}\{u,v\,;x,t\}$ (3.36) that is obviously locally related to $\mathbf{UV}\{x,t\,;u,v\}$. However, exclusion of the dependent variable t through cross-differentiation does lead to a new PDE system $\mathbf{XA}\{u,v\,;x,\alpha\}$ that is nonlocally related to all previously considered PDE systems [Exercise 3.5.3]. The only nonlocally related subsystems of the PDE system $\mathbf{XA}\{u,v\,;x,\alpha\}$ are the known PDEs $\mathbf{X}\{u,v\,;x\}$ [Exercise 3.3.3] and $\mathbf{A}\{x,u\,;\alpha\}$ (3.38).

Other interchanges of variables could yield additional nonlocally related subsystems.

In summary, one has obtained an extended tree of nonlocally related PDE systems given by

$$\mathcal{T}_4 = \mathcal{T}_3 \cup \{\mathbf{XA}\{u,v\,;x,\alpha\}, \mathbf{V}\{u,t\,;v\}, \mathbf{X}\{u,v\,;x\}, \mathbf{A}\{x,u\,;\alpha\}\}, \tag{3.75}$$

illustrated in Figure 3.3.

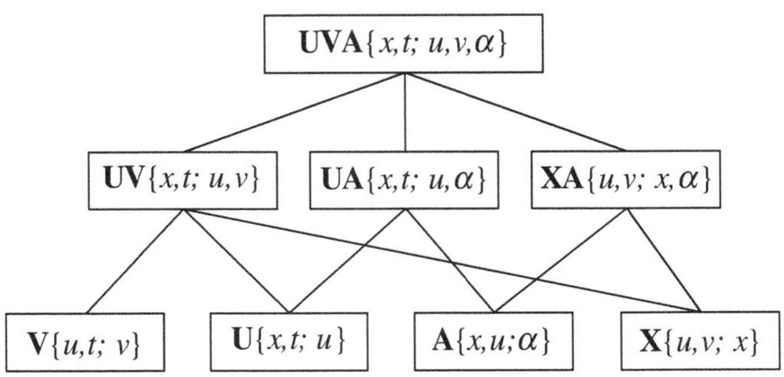

Fig. 3.3 An extended tree of nonlocally related PDE systems for the nonlinear diffusion equation (3.70) (for arbitrary $K(u)$).

Note that the tree $\mathcal{T}_4$ may be still further extended through considering conservation laws for particular forms of the constitutive function $K(u)$. In particular, for the case $K(u) = u^{-2}$, the system $\mathbf{UA}\{x,t\,;u,\alpha\}$ has an additional conservation law [Bluman & Doran-Wu (1995)].

3.5.3 An extended tree for a nonlinear wave equation

As another example, a tree of nonlocally related PDE systems is constructed for the nonlinear wave equation $\mathbf{U}\{x,t\,;u\}$ given by

$$u_{tt} = (c^2(u)u_x)_x, \tag{3.76}$$

holding for an arbitrary constitutive function $c(u)$.

First, local conservation laws are sought. Unlike the nonlinear diffusion equation (3.70), the nonlinear wave equation (3.76) is not a scalar evolution equation, and hence Theorem 3.4.2 does not hold. Conservation laws of the nonlinear wave equation (3.76) are sought that arise from multipliers of the form $\Lambda = \Lambda(x,t,U)$. For an arbitrary $c(u)$, there are exactly four such multipliers $\Lambda = 1, t, xt, x$, with a corresponding set of linearly independent local conservation laws given by

$$\mathrm{D}_t(u_t) - \mathrm{D}_x(c^2(u)u_x) = 0, \tag{3.77}$$

$$\mathrm{D}_t(tu_t - u) - \mathrm{D}_x(tc^2(u)u_x) = 0, \tag{3.78}$$

$$\mathrm{D}_t\Big(x[tu_t - u]\Big) - \mathrm{D}_x\Big(t\big[xc^2(u)u_x - \int c^2(u)du\big]\Big) = 0, \tag{3.79}$$

$$\mathrm{D}_t(xu_t) - \mathrm{D}_x\Big(xc^2(u)u_x - \int c^2(u)du\Big) = 0 \tag{3.80}$$

[Bluman & Cheviakov (2007)].

Since none of the multipliers vanishes off of the solution space of the nonlinear wave equation (3.76), it is redundant to retain the original nonlinear wave equation in corresponding potential systems. Introducing the potential variables v, w, α, β, one obtains four singlet potential systems:
$\mathbf{UV}\{x,t\,;u,v\}$ given by

$$\begin{aligned} v_x &= u_t, \\ v_t &= c^2(u)\,u_x; \end{aligned} \tag{3.81}$$

$\mathbf{UW}\{x,t\,;u,w\}$ given by

$$\begin{aligned} w_x &= tu_t - u, \\ w_t &= tc^2(u)u_x; \end{aligned} \tag{3.82}$$

$\mathbf{UA}\{x,t\,;u,\alpha\}$ given by

$$\begin{aligned}\alpha_x &= x[tu_t - u],\\ \alpha_t &= t\big[xc^2(u)u_x - \textstyle\int c^2(u)du\big]; \quad \text{and}\end{aligned} \tag{3.83}$$

$\mathbf{UB}\{x, t\,; u, \beta\}$ given by

$$\begin{aligned}\beta_x &= xu_t,\\ \beta_t &= xc^2(u)u_x - \textstyle\int c^2(u)du.\end{aligned} \tag{3.84}$$

The four conservation laws (3.77)–(3.80) yield up to $2^4 - 1 = 15$ nonlocally related PDE systems:

- Four singlet potential systems (3.81), (3.82), (3.83) and (3.84) involving single potential variables.
- Six couplets $\mathbf{UVW}\{x, t\,; u, v, w\}$, $\mathbf{UVA}\{x, t\,; u, v, \alpha\}$, $\mathbf{UVB}\{x, t; u, v, \beta\}$, $\mathbf{UWA}\{x, t\,; u, w, \alpha\}$, $\mathbf{UWB}\{x, t\,; u, w, \beta\}$ and $\mathbf{UAB}\{x, t\,; u, \alpha, \beta\}$, respectively, given by [(3.81), (3.82)], [(3.81), (3.83)], [(3.81), (3.84)], [(3.82), (3.83)], [(3.82), (3.84)] and [(3.83), (3.84)] that involve all pairs of potential variables.
- Four triplets $\mathbf{UVWA}\{x, t\,; u, v, w, \alpha\}$, $\mathbf{UVWB}\{x, t\,; u, v, w, \beta\}$, $\mathbf{UVAB}\{x, t;\ u, v, \alpha, \beta\}$, and $\mathbf{UWAB}\{x, t\,; u, w, \alpha, \beta\}$, respectively, given by (3.81)–(3.83), [(3.81), (3.82), (3.84)], [(3.81), (3.83), (3.84)] and (3.82)–(3.84) that involve all triplets of potential variables.
- One quadruplet $\mathbf{UVWAB}\{x, t\,; u, v, w, \alpha, \beta\}$ given by (3.81)–(3.84) that involves all four potential variables.

No nonlocally related subsystem other than the nonlinear wave equation (3.76) itself arises from the four singlet potential systems (3.81), (3.82), (3.83) or (3.84) through direct exclusion of dependent variables.

Now consider subsystems obtained from the potential systems after interchanges of dependent and independent variables. Here only one such interchange is considered. It leads to the linearization of the nonlinear wave equation (3.76) for an arbitrary nonlinearity $c(u)$. In particular, the point (hodograph) transformation $x = x(u, v)$, $t = t(u, v)$ transforms the PDE system (3.81) into an invertibly equivalent *linear* PDE system $\mathbf{XT}\{u, v\,; x, t\}$ given by

$$\begin{aligned}x_v &= t_u,\\ x_u &= c^2(u)\,t_v.\end{aligned} \tag{3.85}$$

The PDE system $\mathbf{XT}\{u, v\,; x, t\}$ (3.85) yields two linear subsystems: $\mathbf{X}\{u, v;$ $x\}$ given by

$$x_{vv} = (c^{-2}(u)x_u)_u \tag{3.86}$$

and $\mathbf{T}\{u, v\,; t\}$ given by

$$t_{uu} = c^2(u)t_{vv} \tag{3.87}$$

which are nonlocally related to both the potential system $\mathbf{XT}\{u, v\,; x, t\}$ (3.85) and the nonlinear wave equation (3.76) (as well as to each other). The equations $\mathbf{X}\{u, v\,; x\}$ (3.86) and $\mathbf{T}\{u, v\,; t\}$ (3.87) are two standard forms of the *linear wave equation* with a variable wave speed.

Let $\mathcal{T}_a$ denote the tree of nonlocally related systems obtained so far. To further extend the tree $\mathcal{T}_a$ one could follow Steps 3 and 4 of the tree construction procedure described in Section 3.5.1:

- Find additional linearly independent (most likely nonlocal) conservation laws of the given nonlinear wave equation (3.76) by seeking local conservation laws of the quadruplet potential system $\mathbf{UVWAB}\{x, t;\, u, v, w, \alpha, \beta\}$ given by (3.81)-(3.84).
- Introduce corresponding potential variables and singlet potential systems arising from each additional conservation law.
- Introduce additional k-plet potential systems involving previously known potentials v, w, α, β and the additional potentials, $k = 2, 3, \ldots$.
- Consider further nonlocally related subsystems.

In order to simplify computations, a slightly different procedure of tree extension is chosen as follows. Local conservation laws of the linear wave equation $\mathbf{T}\{u, v\,; t\}$ (3.87) are sought in terms of multipliers of the form $\Lambda = \Lambda(u, v, T)$. Since the PDE $\mathbf{T}\{u, v\,; t\}$ (3.87) is linear for any wave speed $c(u)$, there exists an infinite number of such conservation laws. Moreover, such multipliers $\Lambda(u, v, T)$ satisfy the same linear wave equation, i.e.,

$$\Lambda_{uu}(u, v) = c^2(u)\Lambda_{vv}(u, v),$$

since the PDE (3.87) is self-adjoint. However, it is easy to see that only four multipliers hold for *all* wave speeds (i.e, do not depend on the form of the wave speed $c(u)$), namely $\Lambda = 1, u, v, uv$. These local multipliers yield the four linearly independent local conservation laws respectively given by

$$\mathrm{D}_u(t_u) - \mathrm{D}_v\left(c^2(u)t_v\right) = 0, \tag{3.88}$$

$$\mathrm{D}_u(ut_u - t) - \mathrm{D}_v\left(uc^2(u)t_v\right) = 0, \tag{3.89}$$

$$\mathrm{D}_u(vt_u) - \mathrm{D}_v\left(c^2(u)(vt_v - t)\right) = 0, \tag{3.90}$$

$$\mathrm{D}_u\left(v[ut_u - t]\right) - \mathrm{D}_v\left(uc^2(u)[vt_v - t]\right) = 0. \tag{3.91}$$

The conservation law (3.88) leads to the known potential system $\mathbf{XT}\{u, v\,; x, t\}$ (3.85). The other three conservation laws (3.89), (3.90) and (3.91), respectively, yield singlet potential systems $\mathbf{TP}\{u, v\,; t, p\}$, $\mathbf{TQ}\{u, v\,; t, q\}$ and $\mathbf{TR}\{u, v\,; t, r\}$ given by

$$\begin{aligned} p_v &= ut_u - t, \\ p_u &= uc^2(u)t_v; \end{aligned} \tag{3.92}$$

$$\begin{aligned} q_v &= vt_u, \\ q_u &= c^2(u)(vt_v - t); \end{aligned} \tag{3.93}$$

$$\begin{aligned} r_v &= v(ut_u - t), \\ r_u &= uc^2(u)(vt_v - t). \end{aligned} \tag{3.94}$$

Thus the extended tree $\mathcal{T}_b$ that arises from the conservation laws (3.77)–(3.80) of the linear wave equation (3.87) includes the following PDE systems:

- Three additional singlet potential systems $\mathbf{TP}\{u,v\,;t,p\}$ (3.92), $\mathbf{TQ}\{u,v;\,t,q\}$ (3.93) and $\mathbf{TR}\{u,v\,;t,r\}$ (3.94).
- Six couplets $\mathbf{XTP}\{u,v\,;x,t,p\}$, $\mathbf{XTQ}\{u,v\,;x,t,q\}$, $\mathbf{XTR}\{u,v\,;x,t,r\}$, $\mathbf{TPQ}\{u,v\,;t,p,q\}$, $\mathbf{TPR}\{u,v\,;t,p,r\}$ and $\mathbf{TQR}\{u,v\,;t,q,r\}$, given by [(3.85), (3.92)], [(3.85), (3.93)], [(3.85), (3.94)], [(3.82), (3.83)], [(3.82), (3.84)] and [(3.83), (3.84)], respectively, involving pairs of potential variables.
- Four triplets $\mathbf{XTPQ}\{u,v\,;x,t,p,q\}$, $\mathbf{XTPR}\{u,v\,;x,t,p,r\}$, $\mathbf{XTQR}\{u,v;\,x,t,q,r\}$ and $\mathbf{TPQR}\{u,v\,;t,p,q,r\}$ given by [(3.85), (3.92), (3.93)], [(3.85), (3.92), (3.94)], [(3.85), (3.93), (3.94)] and [(3.92), (3.93), (3.94)], respectively.
- One quadruplet $\mathbf{XTPQR}\{u,v\,;x,t,p,q,r\}$ given by [(3.85), (3.92)–(3.94)].

Let $\mathcal{T}_c = \mathcal{T}_a \cup \mathcal{T}_b$ denote the resulting extended tree.

Now one looks for redundancies, i.e., for locally related systems in the tree $\mathcal{T}_c$. Treating x,t as independent variables and u,v as dependent variables in the potential systems $\mathbf{XTP}\{u,v\,;x,t,p\}$, $\mathbf{XTQ}\{u,v\,;x,t,q\}$ and $\mathbf{XTR}\{u,v\,;x,t,r\}$, one obtains the corresponding invertibly related systems $\mathbf{UVP}\{x,t\,;u,v,p\}$, $\mathbf{UVQ}\{x,t\,;u,v,q\}$ and $\mathbf{UVR}\{x,t\,;u,v,r\}$. A comparison of these systems with the couplets $\mathbf{UVW}\{x,t\,;u,v,w\}$, $\mathbf{UVA}\{x,t\,;u,v,\alpha\}$ and $\mathbf{UVB}\{x,t\,;u,v,\beta\}$ shows that the potential p is locally related to w, and the potential q is locally related to β [Exercise 3.5.4]. Hence the PDE systems $\mathbf{XTP}\{u,v\,;x,t,p\}$, $\mathbf{XTQ}\{u,v\,;x,t,q\}$ and $\mathbf{XTPQ}\{u,v\,;x,t,p,q\}$ in the tree $\mathcal{T}_c$ are redundant. The resulting extended tree

$$\mathcal{T}_d = \mathcal{T}_c \setminus \{\mathbf{XTP}\{u,v\,;x,t,p\}, \mathbf{XTQ}\{u,v\,;x,t,q\}, \mathbf{XTPQ}\{u,v\,;x,t,p,q\}\}$$

is exhibited in Figure 3.4.

None of the PDE systems in the tree $\mathcal{T}_d$ is obviously redundant (i.e., locally related). However, from the point of view of applications, even if some potential variables in a tree are linearly or functionally dependent, so that the corresponding systems are locally related, it may be worthwhile to analyze such redundant systems, since sometimes it may be too difficult computationally to determine whether two systems within a tree are nonlocally related (and the worst that could happen is repeated results).

In Section 4.2.2, nonlocal symmetries are sought for the nonlinear wave equation (3.76) through seeking local symmetries of PDE systems in the tree $\mathcal{T}_d$. It turns out that many nonlocal symmetries arise for the nonlinear wave equation (3.76) from point symmetries of distinct PDE systems within $\mathcal{T}_d$.

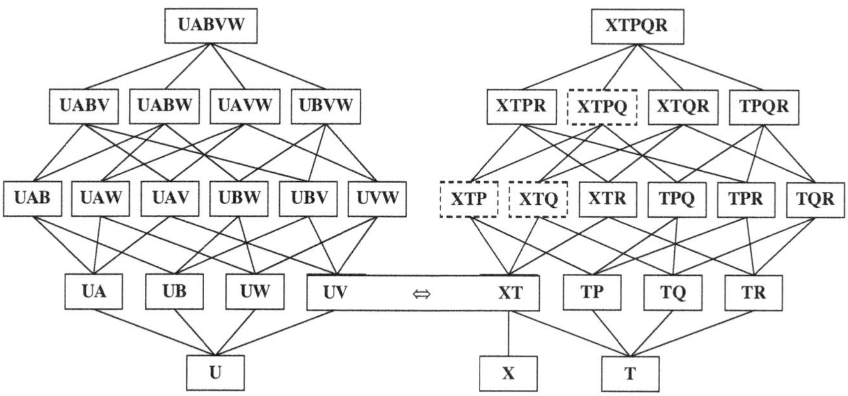

Fig. 3.4 An extended tree $\mathcal{T}_d$ of PDE systems for the nonlinear wave equation (3.76) (for arbitrary $c(u)$). The sets of dependent and independent variables are not shown. Redundant systems are outlined by dashed lines.

3.5.4 An extended tree for the planar gas dynamics equations

The tree construction procedure described in Section 3.5.1 is now used to extend the tree of nonlocally related systems for the planar gas dynamics (PGD) equations presented in Section 3.3.3 and exhibited in Figure 3.2. In particular, for both the Euler PGD system $\mathbf{E}\{x,t\,;v,p,\rho\}$ (3.39) and the Lagrange PGD system $\mathbf{L}\{y,s\,;v,p,q\}$ (3.42), the tree constructed in Section 3.3.3 is extended through finding local conservation laws to obtain resulting combination potential systems and nonlocally related subsystems.

In Section 3.3.3, it was shown that the potential systems (3.40) and (3.41) for the nonlocally related Euler and Lagrange systems, respectively, are related by a point transformation involving an interchange of dependent and independent variables. In this section, there is no consideration of nonlocally related subsystems arising from interchanges of dependent and independent variables.

(1) Nonlocally related systems arising from the Lagrange PGD system

Consider the Lagrange PDE system $\mathbf{L}\{y,s\,;v,p,q\}$ given by (3.42):

$$\begin{aligned} q_s - v_y &= 0,\\ v_s + p_y &= 0,\\ p_s + B(p,q)v_y &= 0, \end{aligned} \tag{3.95}$$

where $B(p,q) = S_q/S_p$ is the constitutive function.

Note that the PDE system (3.95) has the group of equivalence transformations

$$\begin{gathered} \tilde{s} = a_1 s + a_4, \quad \tilde{y} = a_2 y + a_5, \quad \tilde{v} = a_3 v + a_6,\\ \tilde{p} = \frac{a_2 a_3}{a_1} p + a_7, \quad \tilde{q} = \frac{a_1 a_3}{a_2} q + a_8, \quad \tilde{B}(\tilde{p},\tilde{q}) = \frac{a_2^2}{a_1^2} B(p,q) \end{gathered} \tag{3.96}$$

for arbitrary constants $a_1, \ldots, a_8$ with $a_1 a_2 a_3 \neq 0$. As usual, all further analysis is done modulo these equivalence transformations.

To construct PDE systems nonlocally related to $\mathbf{L}\{y,s\,;v,p,q\}$ (3.95) for an arbitrary constitutive function $B(p,q)$, one seeks conservation law multipliers of the form $\Lambda_i = \Lambda_i(y,s)$, $\ i = 1,2,3$. This leads to the three linearly independent conservation laws listed in Table 3.5.

Table 3.5 Local conservation laws and resulting potential equations for the Lagrange PGD system (3.42), arising from multipliers that are functions of independent variables

Multipliers $(\Lambda_1, \Lambda_2, \Lambda_3)$	Conservation Law	Potential Variable	Potential Equations
$(1,0,0)$	$\mathrm{D}_s(q) - \mathrm{D}_y(v) = 0$	w^1	$w^1_y = q,\ w^1_s = v$
$(0,1,0)$	$\mathrm{D}_s(v) + \mathrm{D}_y(p) = 0$	w^2	$w^2_y = v,\ w^2_s = -p$
$(y,s,0)$	$\mathrm{D}_s(sv + yq) + \mathrm{D}_y(sp - yv) = 0$	w^3	$w^3_y = sv + yq,$ $w^3_s = -sp + yv$

The three singlet potential systems that arise from these conservation laws are obtained as follows. The potential equations arising from the first conservation law replace the first equation of the Lagrange system (3.95); the potential equations arising from the second conservation law replace the second equation of (3.95); the potential equations arising from the third conservation law can equivalently replace either the first or second equation of (3.95).

The combination potential system $\mathbb{P}_{w^1w^2w^3}$ that follows from these three conservation laws contains the following seven nonlocally related systems.

- Three singlets: $\mathbf{LW^1}\{y, s\,; v, p, q, w^1\}$ given by

$$\begin{aligned} w^1_y &= q,\\ w^1_s &= v,\\ v_s + p_y &= 0,\\ p_s + B(p,q)v_y &= 0; \end{aligned} \tag{3.97}$$

$\mathbf{LW^2}\{y, s\,; v, p, q, w^2\}$ given by

$$\begin{aligned} q_s - v_y &= 0,\\ w^2_y &= v,\\ w^2_s &= -p,\\ p_s + B(p,q)v_y &= 0; \end{aligned} \tag{3.98}$$

and $\mathbf{LW^3}\{y, s\,; v, p, q, w^3\}$ given by

$$\begin{aligned} w^3_y &= vs + qy,\\ w^3_s &= -sp + vy,\\ v_s + p_y &= 0,\\ p_s + B(p,q)v_y &= 0. \end{aligned} \tag{3.99}$$

- Three couplets: $\mathbf{LW^1W^2}\{y, s\,; v, p, q, w^1, w^2\}$ given by

$$\begin{aligned} w^1_y &= q,\\ w^1_s &= v,\\ w^2_y &= v,\\ w^2_s &= -p,\\ p_s + B(p,q)v_y &= 0; \end{aligned} \tag{3.100}$$

$\mathbf{LW^1W^3}\{y, s\,; v, p, q, w^1, w^3\}$ given by

$$\begin{aligned} w^1_y &= q,\\ w^1_s &= v,\\ w^3_y &= vs + qy,\\ w^3_s &= -sp + vy,\\ p_s + B(p,q)v_y &= 0; \end{aligned} \tag{3.101}$$

and $\mathbf{LW^2W^3}\{y, s\,; v, p, q, w^2, w^3\}$ given by

$$\begin{aligned} w_y^2 &= v, \\ w_s^2 &= -p, \\ w_y^3 &= vs + qy, \\ w_s^3 &= -sp + vy, \\ p_s + B(p,q)v_y &= 0. \end{aligned} \tag{3.102}$$

- One triplet $\mathbf{LW^1W^2W^3}\{y, s\,; v, p, q, w^1, w^2, w^3\}$ given by

$$\begin{aligned} w_y^1 &= q, \\ w_s^1 &= v, \\ w_y^2 &= v, \\ w_s^2 &= -p, \\ w_y^3 &= vs + qy, \\ w_s^3 &= -sp + vy, \\ p_s + B(p,q)v_y &= 0. \end{aligned} \tag{3.103}$$

The tree $\mathcal{T}_1 = \mathbf{L}\{y, s\,; v, p, q\} \cup \mathbb{P}_{w^1w^2w^3}$ contains eight nonlocally related PDE systems. The only nonlocally related subsystem arising from the direct exclusion of a dependent variable (for an arbitrary constitutive function $B(p,q)$) is the subsystem $\underline{\mathbf{L}}\{y, s\,; p, q\}$ of the Lagrange system $\mathbf{L}\{y, s\,; v, p, q\}$ given by

$$\begin{aligned} q_{ss} + p_{yy} &= 0, \\ p_s + B(p,q)q_s &= 0. \end{aligned} \tag{3.104}$$

In considering subsystems, after relabelling $w^1 = x = \alpha^1$, note that the potential system $\mathbf{LW^1}\{y, s\,; v, p, q, w^1\}$ (3.97) is the PDE system $\mathbf{LX}\{y, s;$ $v, p, q, x\}$(3.41), and hence is locally equivalent (through an interchange of variables) to the system $\mathbf{EA}^1\{x, t\,; v, p, \rho, \alpha^1\}$ (3.40) ($w^1 = x = \alpha^1$). The Euler PGD system $\mathbf{E}\{x, t\,; v, p, \rho\}$ (3.39) is a nonlocally related subsystem of $\mathbf{EA}^1\{x, t\,; v, p, \rho, \alpha^1\}$.

Consequently, one obtains the extended tree

$$\mathcal{T}_a = \mathcal{T}_1 \cup \{\underline{\mathbf{L}}\{y, s\,; p, q\}, \mathbf{E}\{x, t\,; v, p, \rho\}\} \tag{3.105}$$

that contains ten nonlocally related PDE systems arising from the Lagrange system (3.95). This extended tree is exhibited in Figure 3.5 and extends the tree exhibited in Figure 3.2.

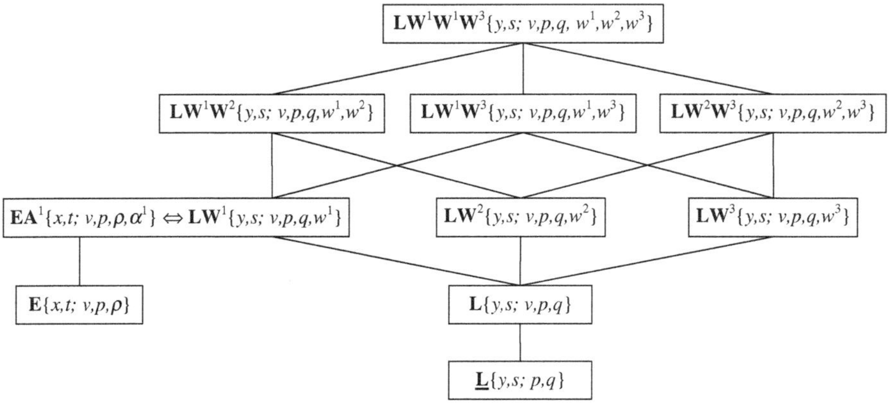

Fig. 3.5 The extended tree $\mathcal{T}_a$ of nonlocally related systems for the planar gas dynamics equations for an arbitrary constitutive function $B(p,q)$ with $q = 1/\rho$.

(2) Nonlocally related systems arising from the Euler PGD system
Now local conservation laws are sought for the Euler PGD system $\mathbf{E}\{x,t;\, v,p,\rho\}$ given by

$$\begin{aligned} \rho_t + (\rho v)_x &= 0, \\ \rho(v_t + v v_x) + p_x &= 0, \\ \rho(p_t + v p_x) + B(p, 1/\rho) v_x &= 0. \end{aligned} \tag{3.106}$$

For multipliers of the form $\Lambda_i = \Lambda_i(x,t,V)$, $i = 1,2,3$, one can show that the Euler system $\mathbf{E}\{x,t\,; v,p,\rho\}$ (3.106) has the three linearly independent local conservation laws listed in Table 3.6.

Table 3.6 Local conservation laws and resulting potential equations for the Euler PGD system (3.106) arising from multipliers of the form $\Lambda_i = \Lambda_i(x,t,V)$

Multipliers $(\Lambda_1, \Lambda_2, \Lambda_3)$	Conservation Law	Potential Variable	Potential Equations
$(1,0,0)$	$\mathrm{D}_t(\rho) + \mathrm{D}_x(\rho v) = 0$	α^1	$\alpha^1_x = \rho,\ \alpha^1_t = -\rho v$
$(V,1,0)$	$\mathrm{D}_t(\rho v) + \mathrm{D}_x(p + \rho v^2) = 0$	α^2	$\alpha^2_x = \rho v,\ \alpha^2_t = -(p + \rho v^2)$
$(tV - x, t, 0)$	$\mathrm{D}_t\big(\rho(tv - x)\big)$ $+\mathrm{D}_x\big(tp + \rho v(tv - x)\big) = 0$	α^3	$\alpha^3_x = \rho(tv - x)$, $\alpha^3_t = -\big(tp + \rho v(tv - x)\big)$

The first two conservation laws in Table 3.6 were used in the initial tree constructed in Section 3.3.3 to yield nonlocally related systems involving the potentials α^1 and α^2, respectively. At a first glance, the third conservation law appears to be new. However, it is equivalent (up to the addition of a trivial conservation law) to the known conservation law

$$\mathrm{D}_t(\alpha^1) + \mathrm{D}_x(\alpha^2) = 0 \tag{3.107}$$

in the potential system $\mathbf{EA^1A^2}\{x, t\,; v, p, \rho, \alpha^1, \alpha^2\}$ (3.44) [Exercise 3.5.7].

It is left to Exercise 3.5.8 to show that the potential variables α^1, α^2 and α^3 of the Euler PGD system coincide with the previously introduced potential variables w^1, w^2, w^3 of the Lagrange system and hence are redundant for a further extension of the tree exhibited in Figure 3.5.

(3) Additional conservation laws of the Lagrange PGD system leading to a further tree extension

To further extend the tree $\mathcal{T}_a$, one seeks additional linearly independent local conservation laws for the Lagrange PGD system $\mathbf{L}\{y, s\,; v, p, q\}$ (3.95). In order to accomplish this, one searches for conservation law multipliers of the more general form $\Lambda_i = \Lambda_i(y, s, V, P, Q)$, $i = 1, 2, 3$, with an essential dependence on V, P, and/or Q. One can show that the solution of the multiplier determining equations yields the additional multipliers [Exercise 3.5.9]:

$$\begin{aligned} \Lambda_1 &= -\beta P + B(P, Q)\Lambda_3, \\ \Lambda_2 &= \beta V, \\ \Lambda_3 &= \Lambda_3(y, P, Q), \end{aligned} \tag{3.108}$$

where β is an arbitrary real constant, and $\Lambda_3(y, p, q)$ is an arbitrary solution of the linear PDE

$$(\Lambda_3)_Q = (B(P, Q)\Lambda_3)_P - \beta. \tag{3.109}$$

One can show that these multipliers yield two new linearly independent local conservation laws. Indeed, since multipliers can always be scaled, in the equation (3.109) it suffices to consider three cases: $\beta = 0, \pm 1$.

In particular, the conservation law arising from $\beta = 0$ corresponds to the adiabatic process in Lagrangian coordinates:

$$\mathrm{D}_s(S(p, q)) = 0, \tag{3.110}$$

where the entropy $S(p, q)$ is a solution of the equation

$$S_q(p, q) = B(p, q)S_p(p, q). \tag{3.111}$$

The set of multipliers that yields the conservation law (3.110) is given by

$$\Lambda_1 = S_Q(P,Q), \quad \Lambda_2 = 0, \quad \Lambda_3 = S_P(P,Q).$$

A second conservation law ($\beta = 1$) represents the conservation of energy and is given by

$$\mathrm{D}_s\left(\tfrac{1}{2}v^2 + K(p,q)\right) + \mathrm{D}_y(pv) = 0, \tag{3.112}$$

where $K(p,q)$ is any solution of the equation

$$K_q(p,q) = B(p,q)K_p(p,q) - p. \tag{3.113}$$

The set of multipliers that yields the local conservation law (3.112) is given by

$$\Lambda_1 = K_Q(P,Q), \quad \Lambda_2 = V, \quad \Lambda_3 = K_P(P,Q).$$

The conservation law arising from $\beta = -1$ is given by

$$\mathrm{D}_s\left(-\tfrac{1}{2}v^2 + \tilde{K}(p,q)\right) - \mathrm{D}_y(pv) = 0, \tag{3.114}$$

where $\tilde{K}(p,q)$ satisfies the equation

$$\tilde{K}_q(p,q) = B(p,q)\tilde{K}_p(p,q) + p. \tag{3.115}$$

Each solution $\tilde{K}(p,q)$ of (3.115) is of the form $\tilde{K}(p,q) = S(p,q) - K(p,q)$, where $S(p,q)$ is a solution of (3.111) and $K(p,q)$ is a particular solution of (3.113). Consequently, the conservation law (3.114) is redundant since it is a linear combination (difference) of the conservation laws (3.110) and (3.112).

In principle, for any form of $B(p,q)$, the conservation laws (3.110) and (3.112), respectively, yield new linearly independent potential variables w^4 and w^5, defined by potential equations $\mathcal{P}^4$, given by

$$\begin{aligned} w^4_y &= S(p,q), \\ w^4_s &= 0, \\ S_q(p,q) &= B(p,q)S_p(p,q); \end{aligned} \tag{3.116}$$

and $\mathcal{P}^5$, given by

$$\begin{aligned} w^5_y &= \tfrac{1}{2}v^2 + K(p,q), \\ w^5_s &= -pv, \\ K_q(p,q) &= B(p,q)K_p(p,q) - p. \end{aligned} \tag{3.117}$$

In combination with the previous potential variables w^1, w^2 and w^3, this yields a correspondingly larger combination potential system $\mathbb{P}_{w^1 \ldots w^5} \supset \mathbb{P}_{w^1w^2w^3}$ containing up to 31 nonlocally related potential systems.

Note that the singlet potential system $\mathbf{LW^4}\{y, s\,; v, p, q, w^4\} = \mathcal{P}^4 \cup \mathbf{L}\{y, s\,; v, p, q\}$, obtained by excluding the dependent variable v, yields the nonlocally related subsystem $\underline{\mathbf{LW^4}}\{y, s\,; p, q, w^4\}$ given by

$$\begin{aligned} q_{ss} + p_{yy} &= 0, \\ w^4_y &= S(p, q), \\ w^4_s &= 0, \\ p_s + B(p, q) q_s &= 0, \\ S_q(p, q) &= B(p, q) S_p(p, q). \end{aligned} \tag{3.118}$$

Consequently, the tree $\mathcal{T}_a$ of 10 nonlocally related PGD equations given by (3.105), that is exhibited in Figure 3.5, can be significantly extended for an arbitrary constitutive function $B(p, q)$. In particular, one now has the extended tree

$$\begin{aligned} \mathcal{T}_b = \Big\{ &\mathbf{L}\{y, s\,; v, p, q\}, \underline{\mathbf{L}}\{y, s\,; p, q\}, \mathbf{E}\{x, t\,; v, p, \rho\}, \\ &\underline{\mathbf{LW^4}}\{y, s\,; p, q, w^4\} \Big\} \cup \mathbb{P}_{w^1 \ldots w^5} \end{aligned} \tag{3.119}$$

containing up to 35 nonlocally related PDE systems, including the Lagrange and Euler systems. This tree is exhibited in Figure 3.6.

It is important to note that the tree $\mathcal{T}_b$ can be extended even further, if additional independent conservation laws are found for the PDE systems within the tree $\mathcal{T}_b$. For example, one may show that for any form of the constitutive function $B(p, q)$, the Lagrange potential system $\mathbf{LW^1W^2W^3}\{y, s\,; v, p, q, w^1, w^2, w^3\}$ (3.103) has an additional conservation law that is independent of the five conservation laws that lead to the potential variables $w^1, \ldots, w^5$ [Exercise 3.5.6].

(4) Example: An extended tree for a polytropic gas

In the case of a polytropic gas, $B(p, q) = \gamma p / q$, where γ is an arbitrary constant, explicit computations can be performed. For the conservation law (3.110), one readily finds that

$$\Lambda_1 = \gamma P Q^{\gamma - 1}, \quad \Lambda_2 = 0, \quad \Lambda_3 = Q^\gamma, \text{ and } S(p, q) = S(p q^\gamma).$$

It suffices to take $S(p, q) = p q^\gamma$ [Exercise 3.5.10]. The corresponding singlet potential system $\mathbf{LW^4}\{y, s\,; v, p, q, w^4\}$ is given by

$$\begin{aligned} q_s - v_y &= 0, \\ v_s + p_y &= 0, \\ w^4_y &= p q^\gamma, \\ w^4_s &= 0. \end{aligned} \tag{3.120}$$

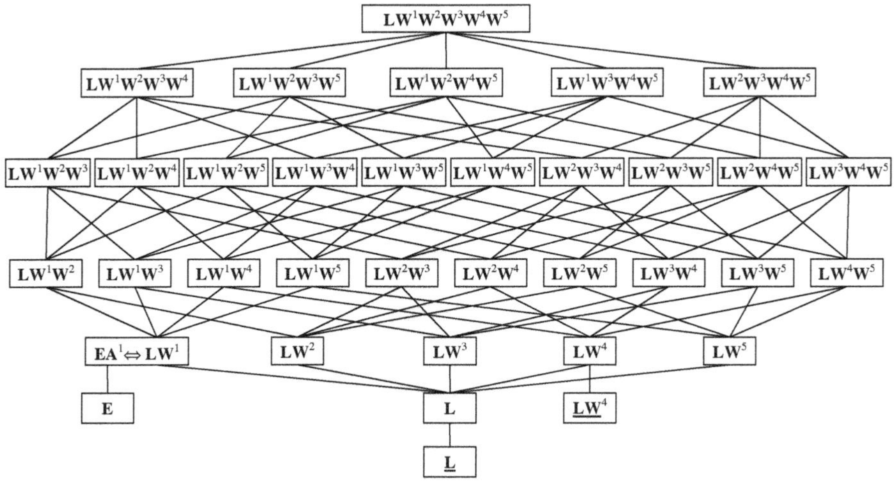

Fig. 3.6 The further extended tree $\mathcal{T}_b$ of up to 35 nonlocally related systems for the planar gas dynamics equations for an arbitrary constitutive function $B(p,q)$ with $q = 1/\rho$ [cf. the extended tree $\mathcal{T}_a$ exhibited in Figure 3.5].

[Here one can replace the third equation of the Lagrange system $\mathbf{L}\{y, s\,; v, p, q\}$ (3.95) by the potential equations since the multiplier $\Lambda_3 = Q^\gamma$ does not vanish for an arbitrary function $Q(y,s) \neq 0$.]

For the conservation law (3.112), one has

$$K(p,q) = \begin{cases} \dfrac{pq}{\gamma - 1} + T(pq^\gamma), & \gamma \neq 1; \\ pq \ln p + T(pq), & \gamma = 1. \end{cases} \tag{3.121}$$

Taking into account the potential equations for w^4, one can set $T(pq^\gamma) = 0$ [Exercise 3.5.10]. Hence, the singlet potential system $\mathbf{LW^5}\{y, s\,; v, p, q, w^5\}$ resulting from the conservation law (3.112) becomes

$$\begin{aligned} q_s - v_y &= 0, \\ v_s + p_y &= 0, \\ p_s + \gamma \frac{p}{q} v_y &= 0, \\ w^5_y &= \tfrac{1}{2} v^2 + K(p,q), \\ w^5_s &= -pv, \end{aligned} \tag{3.122}$$

with $K(p,q)$ given by (3.121).

The nonlocally related subsystem $\underline{\mathbf{LW}}^4\{y,s\,;p,q,w^4\}$ (3.118) here takes the form

$$\begin{aligned} q_{ss}+p_{yy}&=0,\\ w^4_y&=pq^{\gamma},\\ w^4_s&=0. \end{aligned} \tag{3.123}$$

Note that in the case $\gamma=-1$, i.e., when a polytropic gas is a Chaplygin gas, both the Euler and Lagrange PGD systems are nonlinear as they stand. But here the nonlocally related system (3.123) for $\gamma=-1$ becomes a linear PDE system in terms of the dependent variables p and q ($w^4=w^4(y)$ is an arbitrary function of y):

$$p=w^4(y)q,\quad q_{ss}+(w^4(y)q)_{yy}=0.$$

Exercises 3.5

3.5.1. Prove Theorem 3.5.1.

3.5.2. Suppose a PDE system $\mathbf{R}\{x,t\,;u\}$ has two linearly independent local conservation laws $\mathcal{K}_1$ and $\mathcal{K}_2$ that yield respective potential variables v^1 and v^2. Show that

(a) the linear combination of conservation laws $\mathcal{K}_1+c\mathcal{K}_2$, $c=\text{const}\neq 0$, yields a potential variable $w=v^1+cv^2$;
(b) the potential systems $\mathbf{RV}^1\{x,t\,;u,v^1\}$, $\mathbf{RV}^2\{x,t\,;u,v^2\}$, and $\mathbf{RW}\{x,t;u,w\}$ are nonlocally related to each other, to the couplet potential system $\mathbf{RV^1V^2}\{x,t\,;u,v^1,v^2\}$, and to the given PDE system $\mathbf{R}\{x,t\,;u\}$.

3.5.3. Consider the potential system $\mathbf{UVA}\{x,t\,;u,v,\alpha\}$ (3.73) of the nonlinear diffusion equation [Section 3.5.2].

(a) Find the PDE system $\mathbf{XTA}\{u,v\,;x,t,\alpha\}$ (3.74) and the subsystem $\mathbf{XA}\{u,v\,;x,\alpha\}$.
(b) Consider all other possible interchanges of dependent and independent variables for the potential system $\mathbf{UVA}\{x,t\,;u,v,\alpha\}$ (3.73). Find the corresponding PDE systems and additional nonlocally related subsystems.

3.5.4. Consider potential systems of the nonlinear wave equation (3.76) [Section 3.5.3]. Show that the potential system $\mathbf{XTP}\{u,v\,;x,t,p\}$ [(3.85), (3.92)] is locally related to the potential system $\mathbf{UVW}\{x,t\,;u,v,w\}$ [(3.81), (3.82)] [In particular, $p=-w$.], and the potential system $\mathbf{XTQ}\{u,v\,;x,t,p\}$ [(3.85),

(3.93)] is locally related to the potential system $\mathbf{UWB}\{x,t\,;u,w,\beta\}$ [(3.81), (3.84)].

3.5.5. Construct an extended tree of nonlocally related PDE systems for the nonlinear reaction-diffusion equation $u_t - u^2 u_{xx} - 2bu^2 = 0$ ($b = \text{const}$). Here $u = u(x,t)$. Show that for any value of b, this nonlinear reaction-diffusion equation can be linearized by a nonlocal transformation [Bluman (1993)].

3.5.6. Show that for any form of the constitutive function $B(p,q)$, the Lagrange potential system $\mathbf{LW^1W^2W^3}\{y,s\,;v,p,q,w^1,w^2,w^3\}$ (3.103) has the infinite set of local conservation laws

$$\mathrm{D}_s\left(w^1F(yw^1+sw^2-w^3)\right) - \mathrm{D}_y\left(w^2F(yw^1+sw^2-w^3)\right) = 0, \quad (3.124)$$

where F is an arbitrary function of its argument. Find the set of multipliers that yields the conservation laws (3.124). Prove that the conservation laws (3.124) are linearly independent of the set of five conservation laws that include the three conservation laws exhibited in Table 3.5 and the conservation laws given by equations (3.110) and (3.112). [Hint: Compute multipliers and compare with those for the five known conservation laws.]

3.5.7. Prove that the third conservation law listed in Table 3.6 for the Euler PGD system (3.106) is equivalent to the previously known conservation law $\mathrm{D}_t(\alpha^1) + \mathrm{D}_x(\alpha^2) = 0$ of the potential system $\mathbf{EA^1A^2}\{x,t\,;v,p,\rho,\alpha^1,\alpha^2\}$ (3.44). [Hint: Use the proof sequence of Theorem 3.5.1.]

3.5.8. Show that the potential variables α^1, α^2 and α^3 of the Euler PGD system (3.106) [Table 3.6] coincide with the potential variables w^1, w^2 and w^3 of the Lagrange system (3.95) [Table 3.5].

3.5.9. For the Lagrange PGD system (3.95), show that the sets of conservation law multipliers of the form $\Lambda_i = \Lambda_i(y,s,V,P,Q)$, $i = 1,2,3$, are given by (3.108) and (3.109).

3.5.10. Show that the potential $\tilde{w}^4$ arising from the conservation law (3.110) with $S(p,q) = S(pq^\gamma)$ (polytropic case) is functionally dependent on the potential w^4 arising from $S(p,q) = pq^\gamma$. Hence it is sufficient to consider $S(p,q) = pq^\gamma$.

3.6 Discussion

For diffusion-convection equations of the form

$$u_t - [f(u)u_x + k(u)]_x = 0,$$

Popovych & Ivanova (2005a,b) classify conservation laws and find resulting trees of potential systems.

In the procedure presented in Section 3.2.2, nonlocally related subsystems are obtained by excluding dependent variables of a given PDE system $\mathbf{R}\{x,t\,;u\}$ as *written*, or after the exclusion of one or more dependent variables following an interchange of its given independent and dependent variables. More generally, any invertible point transformation $U = U(x,t,u)$, $X = X(x,t,u)$, $T = T(x,t,u)$ could be used to exclude one or more of the m components of the resulting m dependent variables in U to obtain additional nonlocally related subsystems.

Note that the extended tree construction procedure [Section 3.5] can be generalized even further, as follows. Suppose one has a set of $n \geq 2$ linearly independent conservation laws of a PDE system $\mathbf{R}\{x,t\,;u\}$, yielding corresponding singlet potential systems $\mathbf{RV}^1\{x,t\,;u,v^1\}$, ..., $\mathbf{RV}^n\{x,t\,;u,v^n\}$. One can consider a linear combination of such conservation laws to yield a potential variable

$$w = \alpha_1 v^1 + \cdots + \alpha_n v^n. \tag{3.125}$$

As shown in Exercise 3.5.2, the singlet potential system $\mathbf{RW}\{x,t\,;u,w\}$ is nonlocally related to $\mathbf{R}\{x,t\,;u\}$ as well as $\mathbf{RV}^1\{x,t\,;u,v^1\}$, ..., $\mathbf{RV}^n\{x,t;\,u,v^n\}$, for any choice of the constants $\alpha_1,\ldots,\alpha_n$ (unless one of these constants is one and the others are zero). It follows that, in general, a set of n conservation laws directly yields *a spectrum of singlet potential systems*, rather than just n nonlocally related singlet potential systems. However, to date, there are few examples of useful applications of such potential systems. [One example is given in Remark 4.2.1 in Section 4.2.2.]

Applications of nonlocally related systems to obtain nonlocal symmetries and nonlocal conservation laws of a given PDE system are considered in the next chapter. In the following chapter it is also shown how to extend the work presented in Chapter 2, on the use of local symmetries and/or local conservation law multipliers to find an invertible mapping of a given PDE system to an equivalent target PDE system of interest, to include non-invertible mappings that result from consideration of local symmetries and/or local conservation law multipliers of nonlocally related systems.

The situation for PDE systems with three or more independent variables is much more complex and considered in Chapter 5. It is shown that in order to obtain an interesting potential system (i.e., one that yields nonlocal symmetries and/or nonlocal conservation laws through calculations of its local symmetries and/or local conservation laws, respectively) from a divergence-type conservation law of a given PDE system, one must introduce gauge constraints that relate the potential variables resulting from local conservation laws. To date it is not obvious which gauge constraint is of value for a particular application. However, it is shown that potential systems following from *lower-degree conservation laws* require fewer or no gauge constraints and can yield interesting results. Moreover, the use of subsystems of a given PDE system to obtain interesting nonlocally related PDE systems carries through for PDE systems with three or more independent variables.

Chapter 4
Applications of Nonlocally Related PDE Systems

4.1 Introduction

In Chapter 3, it was shown how one can systematically construct a set (*tree*) of PDE systems nonlocally related to a given PDE system. In particular, local conservation laws of a PDE system lead to augmented nonlocally related (potential) systems that explicitly include nonlocal (potential) variables. Moreover, further nonlocally related PDE systems (nonlocally related subsystems) arise when one or more dependent variables (including dependent variable(s) arising after a point transformation that involves an interchange of dependent and independent variable(s)) are excluded from a PDE system or its potential systems, through differential relations. In Section 3.5, an algorithm for the construction of an extended tree of nonlocally related systems was outlined. In particular, n local conservation laws of a given PDE system lead to a tree of up to 2^n-1 nonlocally related potential systems. A tree is further extended by considering subsystems of both the given PDE system and its nonlocally related potential systems as well as by considering potential systems arising from conservation laws (whose multipliers have an essential dependence on potential variables) of its nonlocally related potential systems.

Nonlocally related systems in such extended trees are important for applications since they are constructed systematically and each solution of any PDE system in such a tree yields a solution of any other PDE system in the tree, including the given PDE system. More importantly, there is not a one-to-one mapping between solutions of such nonlocally related systems. Consequently, the usefulness of standard methods of analysis, especially coordinate independent methods, can be enhanced when directly applied to different nonlocally related PDE systems. In particular, a method of analysis could be successful in achieving results when applied directly to a nonlocally related system in a tree even if it is unsuccessful in achieving results when directly applied to the given PDE system. Furthermore, from the simplicity

G.W. Bluman et al., *Applications of Symmetry Methods to Partial Differential Equations*, Applied Mathematical Sciences 168, DOI 10.1007/978-0-387-68028-6_4,

of the construction of the mappings that relate PDE systems in an extended tree, it is usually simple to transfer results achieved for a PDE system in such a tree to other PDE systems in the tree, including the given PDE system.

Applications that naturally can arise from the use of such nonlocally related systems include:

(1) The construction of nonlocal conservation laws of a given PDE system that arise as local conservation laws of nonlocally related PDE systems
This application was illustrated in Chapter 3 in the construction of nonlocally related PDE systems arising from local conservation laws of potential systems that in themselves arose from local conservation laws of the given PDE system. Such local conservation laws of potential systems can yield nonlocal conservation laws of the given PDE system, i.e., conservation laws whose fluxes and/or densities have an essential dependence on potential variables. Furthermore, such local conservation laws of potential systems may actually yield further local conservation laws of the given PDE system that had not been previously determined due to lack of completeness in the direct calculation of its local conservation laws.

(2) The construction of nonlocal symmetries of a given PDE system
In this chapter it is shown that point symmetries of a PDE system in a tree of nonlocally related systems can systematically yield nonlocal symmetries of a given PDE system.

A symmetry of a PDE system is defined topologically as a mapping (deformation) of its solution manifold into itself. From this point of view, essentially every PDE system has symmetries. The problem is to find such symmetries and to find those that have applications. In particular, to find explicit symmetries it is necessary to calculate them in some fixed coordinate system. Moreover, such calculations are simple to perform and the resulting symmetries are directly applicable if obtained through a direct application of Lie's algorithm, which yields only local symmetries of a PDE system. The infinitesimals of local symmetries depend at most on a finite number of derivatives of the dependent variables of the PDE system. However, such local symmetries constitute at most a small subset of the total set of symmetries of a PDE system.

In this chapter, it is shown that additional (nonlocal) symmetries of a given PDE system can be found by a direct application of Lie's algorithm to PDE systems in a tree of nonlocally related systems. For the computation of such nonlocal symmetries, it turns out that both nonlocally related potential systems and subsystems can separately yield new symmetries (contrary to the situation in the computation of nonlocal conservation laws, where all local conservation laws of a subsystem are included in the local conservation laws of a PDE system yielding the subsystem).

A point symmetry of a potential system yields a nonlocal symmetry of a given PDE system if at least one of its infinitesimal generator components for the dependent and independent variables of the given PDE system has an essential dependence on a potential (nonlocal) variable. On the other hand, in the case of a nonlocally related subsystem, in order to isolate a nonlocal symmetry of the given PDE system that arises from a point symmetry of the subsystem, one has to compare the local symmetries of both the given PDE system and the nonlocally related subsystem to determine whether a point symmetry of the subsystem yields a nonlocal symmetry of the given PDE system.

For a given PDE system that includes arbitrary constitutive functions and/or parameters, one is interested in the classification of its local and nonlocal symmetries with respect to such functions and/or parameters. In order to do this, one can classify the local symmetries (with respect to such functions and/or parameters) of PDE systems in a tree of nonlocally related systems constructed for a given PDE system. In this chapter, nonlinear diffusion equations, nonlinear wave equations and the equations of planar gas dynamics are considered as illustrative examples for such classifications. Comparisons are made of the point symmetries of various nonlocally related PDE systems in their respective trees to determine the point symmetries yielding nonlocal symmetries of particular systems in trees.

(3) The construction of solutions of a given PDE system that arise from symmetry reductions due to nonlocal symmetries but do not arise as invariant solutions from symmetry reductions due to point symmetries

For a given PDE system, an important application of nonlocal symmetries that arise from point symmetries of a nonlocally related system in a tree results from the construction of the corresponding invariant solutions of the nonlocally related system. In particular, such solutions are especially interesting when the corresponding solutions of the given PDE system are not invariant solutions that can be constructed from the point symmetries of the given PDE system. This application is considered in the next chapter [Section 5.2.3].

In Section 5.2.3, such solutions are constructed for the linear wave equation with a variable wave speed $c(x)$. It is shown that a potential system of such a linear wave equation has point symmetries that are nonlocal symmetries of the linear wave equation for an interesting special form of the constitutive function $c(x)$ corresponding to wave propagation in two-layered media with smooth transitions. These symmetries yield a countable infinite set of invariant solutions for initial value problems. Moreover, this set of solutions is complete and can be used to obtain Fourier series solutions for initial value problems with arbitrary piecewise smooth data in the infinite space domain.

A second example yields physical solutions for the Lagrange system of the planar gas dynamics equations that arise as invariant solutions obtained from nonlocal symmetries that are point symmetries of nonlocally related systems.

(4) The construction of non-invertible mappings relating PDEs
In Chapter 2, two important mapping problems were considered systematically: the invertible mapping of a given nonlinear PDE system to some linear PDE system (in terms of the point/contact symmetries or local conservation law multipliers of the nonlinear PDE system) and the invertible mapping of a given linear PDE with variable coefficients to a linear PDE with constant coefficients (in terms of the point symmetries of abelian type of the linear PDE with variable coefficients). Here these results are extended systematically to include non-invertible mappings.

Firstly, if a nonlocally related PDE system in a tree can be linearized by a point transformation whereas the given PDE system cannot be linearized by a point (contact) transformation, then one obtains a non-invertible mapping of the given PDE system to some linear system. Such non-invertible mappings arise from computing the point symmetries or local conservation law multipliers of a nonlocally related PDE system.

Secondly, suppose a given linear PDE system with variable coefficients cannot be mapped invertibly to a linear PDE system with constant coefficients. It turns out that for any given linear PDE system, it is straightforward to construct an infinite number of potential systems since any solution of the adjoint system of a given linear PDE system yields a set of conservation law multipliers. If one of the corresponding potential systems can be invertibly mapped into a constant coefficient linear PDE system, then as a consequence the given linear PDE is mapped non-invertibly to a constant coefficient linear PDE system. Such non-invertible mappings are constructed for linear parabolic equations with variable coefficients and lead to a significant extension of the classes of linear parabolic equations that can be mapped into the heat equation beyond those found in Section 2.5.1.

The results presented in this chapter have appeared in Bluman & Kumei [(1987), (1988), (1989)], Akhatov, Gazizov & Ibragimov (1991), Ames, Lohner & Adams (1981), Bluman & Cheviakov (2007), Kingston & Sophocleous (2001), Bluman, Temuerchaolu & Sahadevan (2005), Bluman, Cheviakov & Ivanova (2006), and Bluman & Shtelen [(1996a), (2004)].

4.2 Nonlocal Symmetries

Local symmetries of a nonlocally related system can yield explicit symmetries (*nonlocal symmetries*) of a given system of PDEs that do not arise as local symmetries by a direct application of Lie's algorithm to the given system. In particular, such nonlocal symmetries arise as local symmetries of nonlocally related systems with infinitesimal generators having an essential dependence on nonlocal potential variables in the case of nonlocally related systems that are not subsystems. It is shown that this significantly enhances the applicability of symmetry methods.

A symmetry of a system of differential equations is defined topologically as any transformation of its solution manifold into itself. Hence, symmetry transformations are not restricted to local transformations arising from infinitesimal generators whose coefficients are functions of the given system's independent and dependent variables and their derivatives to some finite order. Through many examples, it is demonstrated that local symmetries do not include all calculable (as well as useful) symmetries of a given PDE system.

Suppose a system of PDEs $\mathbf{R}\{x,t\,;u\}$ has a potential system (k-plet) $\mathbf{S}\{x,t\,;u,v\}$ that is invariant under the one-parameter (ϵ) Lie group of point transformations

$$\begin{aligned} x^* &= x+\epsilon\xi_S(x,t,u,v)+O(\epsilon^2),\\ t^* &= t+\epsilon\tau_S(x,t,u,v)+O(\epsilon^2),\\ u^* &= u+\epsilon\eta_S(x,t,u,v)+O(\epsilon^2),\\ v^* &= v+\epsilon\zeta_S(x,t,u,v)+O(\epsilon^2), \end{aligned} \tag{4.1}$$

with corresponding infinitesimal generator

$$\mathrm{X}=\xi_S^i(x,t,u,v)\frac{\partial}{\partial x^i}+\eta_S^\mu(x,t,u,v)\frac{\partial}{\partial u^\mu}+\zeta_S^p(x,t,u,v)\frac{\partial}{\partial v^p}; \tag{4.2}$$

ξ_S^i, $i=1,2$, are the infinitesimals corresponding to the independent variables $(x^1,x^2)=(x,t)$, η_S^μ are the infinitesimals corresponding to the dependent variables u^μ of $\mathbf{R}\{x,t\,;u\}$, $\mu=1,\ldots,m$, and ζ_S^p are the infinitesimals corresponding to the potential variables v^p, $p=1,\ldots,k$ of the k-plet potential system $\mathbf{S}\{x,t\,;u,v\}$.

The point symmetry (4.1) maps any solution of $\mathbf{S}\{x,t\,;u,v\}$ to a solution of $\mathbf{S}\{x,t\,;u,v\}$, and hence through projection, induces a mapping of any solution of $\mathbf{R}\{x,t\,;u\}$ to a solution of $\mathbf{R}\{x,t\,;u\}$. Thus (4.1) yields a symmetry of $\mathbf{R}\{x,t\,;u\}$. However, if the infinitesimals $(\xi_S(x,t,u,v),\eta_S(x,t,u,v))$ do not depend explicitly on the nonlocal potential variables v, i.e.,

$$\frac{\partial\xi_S^i}{\partial v}\equiv 0,\quad \frac{\partial\eta_S^\mu}{\partial v}\equiv 0,\quad i=1,2;\quad \mu=1,\ldots,m, \tag{4.3}$$

then (4.1) only yields a point symmetry of $\mathbf{R}\{x,t\,;u\}$, in terms of the infinitesimal generator given by

$$\mathrm{X}=\xi_S^i(x,t,u)\frac{\partial}{\partial x^i}+\eta_S^\mu(x,t,u)\frac{\partial}{\partial u^\mu}. \tag{4.4}$$

On the other hand, if the infinitesimals $(\xi_S(x,t,u,v), \eta_S(x,t,u,v))$ have an essential dependence on v, then the point symmetry (4.1) defines a nonlocal symmetry of $\mathbf{R}\{x,t\,;u\}$, since the potential variables v are nonlocal variables. This leads to the following definition and the proof of the subsequent theorem.

Definition 4.2.1. The point symmetry (4.1) of the potential system $\mathbf{S}\{x,t;\, u,v\}$ defines a *potential symmetry* of a PDE system $\mathbf{R}\{x,t\,;u\}$ if and only if the infinitesimals $(\xi_S(x,t,u,v), \tau_S(x,t,u,v), \eta_S(x,t,u,v))$ depend explicitly on one or more components of v.

Theorem 4.2.1. *A potential symmetry of* $\mathbf{R}\{x,t\,;u\}$ *is a nonlocal symmetry of* $\mathbf{R}\{x,t\,;u\}$.

Nonlocal symmetries of PDE systems can arise as potential symmetries (i.e., point symmetries of singlet or k-plet potential systems), as well as symmetries of nonlocally related subsystems, as discussed below. Related to this, it is important to note that a local symmetry of $\mathbf{R}\{x,t\,;u\}$ could yield a nonlocal symmetry of $\mathbf{S}\{x,t\,;u,v\}$. [By construction, $\mathbf{R}\{x,t\,;u\}$ is an obvious nonlocally related subsystem of $\mathbf{S}\{x,t\,;u,v\}$.]

Suppose $\mathbf{R}\{x,t\,;u\}$ is a given PDE system with m dependent variables, and $\underline{\mathbf{R}}\{x,t\,;u^{\mu_1},\ldots,u^{\mu_{m-p}}\}$ is a subsystem with $m-p$ dependent variables that is obtained by excluding p dependent variables u^α in $\mathbf{R}\{x,t\,;u\}$. Consider the problem of comparing the local symmetries of $\mathbf{R}\{x,t\,;u\}$ with those of its subsystem $\underline{\mathbf{R}}\{x,t\,;u^{\mu_1},\ldots,u^{\mu_{m-p}}\}$.

If the subsystem $\underline{\mathbf{R}}\{x,t\,;u^{\mu_1},\ldots,u^{\mu_{m-p}}\}$ is locally related to $\mathbf{R}\{x,t\,;u\}$ (in the sense of Theorem 3.2.2), then there is a one-to-one correspondence between solutions of the two systems. Consequently, the following theorem holds.

Theorem 4.2.2. *A local symmetry of a locally related subsystem* $\underline{\mathbf{R}}\{x,t\,;u^{\mu_1},\ldots,u^{\mu_{m-p}}\}$ *of a PDE system* $\mathbf{R}\{x,t\,;u\}$ *is a projection of some corresponding local symmetry of* $\mathbf{R}\{x,t\,;u\}$ *onto the space of variables of* $\underline{\mathbf{R}}\{x,t\,;u^{\mu_1},\ldots,u^{\mu_{m-p}}\}$.

Note that a *point* symmetry of a PDE system $\mathbf{R}\{x,t\,;u\}$ could project onto a *point* or *contact* (or, more generally, higher-order (*local*)) symmetry of a locally related subsystem $\underline{\mathbf{R}}\{x,t\,;u^{\mu_1},\ldots,u^{\mu_{m-p}}\}$.

The situation is different for a nonlocally related subsystem. Here, there is not a one-to-one correspondence between the solutions of a given PDE system and a nonlocally related subsystem. In particular, numerous examples exist where a local symmetry X of a nonlocally related subsystem $\underline{\mathbf{R}}\{x,t\,;u^{\mu_1},\ldots,u^{\mu_{m-p}}\}$ does not correspond to any local symmetry of $\mathbf{R}\{x,t\,;u\}$, and conversely, a local symmetry Y of $\mathbf{R}\{x,t\,;u\}$ does not correspond to a local symmetry of $\underline{\mathbf{R}}\{x,t\,;u^{\mu_1},\ldots,u^{\mu_{m-p}}\}$. For the rest of this chapter we only consider point symmetries of PDE systems. Correspondingly, one can modify the statements in this paragraph through replacing "local symmetry" by "point symmetry".

Summarizing the above discussion, one can isolate three different types of nonlocal symmetries that can be sought for a given PDE system $\mathbf{R}\{x, t\,; u\}$.

1. Nonlocal symmetries arising from point symmetry analysis of nonlocally related subsystems of $\mathbf{R}\{x, t\,; u\}$ obtained by excluding one or more of its dependent variables. [Recall that such nonlocally related subsystems could also arise through exclusion of a dependent variable that arises after an interchange of one or more independent and dependent variables of $\mathbf{R}\{x, t\,; u\}$.]
2. Nonlocal symmetries (potential symmetries) that arise as point symmetries of potential systems (including k-plet potential systems) of $\mathbf{R}\{x, t\,; u\}$.
3. Nonlocal symmetries that arise as point symmetries of nonlocally related subsystems of potential systems of $\mathbf{R}\{x, t\,; u\}$.

More generally, such nonlocal symmetries of $\mathbf{R}\{x, t\,; u\}$ can arise from seeking local symmetries of any PDE system in an extended tree of nonlocally related systems that includes $\mathbf{R}\{x, t\,; u\}$.

Among all such nonlocal symmetries of a PDE system $\mathbf{R}\{x, t\,; u\}$, the ones that explicitly involve nonlocal variables (Type 2 and, in part, Type 3) are easier to distinguish. In the case of finding Type 1 (and the remaining ones of Type 3) nonlocal symmetries of a PDE system $\mathbf{R}\{x, t\,; u\}$, in order to isolate nonlocal symmetries arising from a subsystem whose infinitesimal components for (x, t, u) do not involve nonlocal variables, one must find all point symmetries of $\mathbf{R}\{x, t\,; u\}$, and then see if a point symmetry of a considered nonlocally related system is included in the complete point symmetry analysis of $\mathbf{R}\{x, t\,; u\}$.

It often turns out, as is illustrated by several examples, that a given system $\mathbf{R}\{x, t\,; u\}$ with an arbitrary constitutive function(s) can have nonlocal symmetries for special forms of the constitutive function(s), arising as point symmetries of one or more systems in an extended tree of nonlocally related systems.

4.2.1 Nonlocal symmetries of a nonlinear diffusion equation

As a first example, consider a symmetry classification problem for the nonlinear diffusion equation $\mathbf{U}\{x, t\,; u\}$ given by

$$u_t - (K(u)u_x)_x = 0, \tag{4.5}$$

with an arbitrary constitutive function $K(u) = L'(u)$ [Bluman & Kumei (1989); Akhatov, Gazizov & Ibragimov (1991)]. All computations below are presented modulo the group of equivalence transformations of the class of PDEs (4.5), given by

$$\begin{aligned} &\widetilde{t} = a_4 t + a_1, \quad \widetilde{x} = a_5 x + a_2, \quad \widetilde{u} = a_6 u + a_3, \\ &\widetilde{K}(\widetilde{u}) = \frac{a_5^2}{a_4} K(u), \quad \widetilde{L}(\widetilde{u}) = \frac{a_5^2 a_6}{a_4} K(u) + a_7, \end{aligned} \tag{4.6}$$

where $a_1, \ldots, a_7$ are arbitrary constants with $a_4 a_5 a_6 \neq 0$.

An extended tree $\mathcal{T}_4$ of nonlocally related PDE systems for the nonlinear diffusion equation $\mathbf{U}\{x, t\,; u\}$ (4.5), holding for an arbitrary $K(u)$, was constructed in Section 3.5.2, and shown in Figure 3.3. This tree contains nine nonlocally related PDE systems that have equivalence transformations similar to those in (4.6) [Exercise 4.2.1]. It is also important to note that some of the systems within the tree $\mathcal{T}_4$, namely, $\mathbf{UVA}\{x, t\,; u, v, \alpha\}$ (3.73), $\mathbf{UV}\{x, t\,; u, v\}$ (3.19), $\mathbf{V}\{u, t\,; v\}$ (3.22), $\mathbf{X}\{u, v\,; x\}$ [Exercise 3.3.3] have an additional (projective) equivalence transformation

$$\begin{aligned} &\widetilde{t} = t, \quad \widetilde{x} = x - bv, \quad \widetilde{u} = \frac{u}{1 - bu}, \quad \widetilde{K}(\widetilde{u}) = (1 + b\widetilde{u})^{-2} K\left(\frac{\widetilde{u}}{1 + b\widetilde{u}}\right), \\ &\widetilde{L}(\widetilde{u}) = L\left(\frac{\widetilde{u}}{1 + b\widetilde{u}}\right); \end{aligned} \tag{4.7}$$

whereas the remaining nonlocally related systems, $\mathbf{UA}\{x, t\,; u, \alpha\}$ (3.73), $\mathbf{U}\{x, t\,; u\}$ (4.5), and $\mathbf{A}\{x, u\,; \alpha\}$ (3.38) do not have the equivalence transformation (4.7). [It is a nonlocal transformation of these systems!]

Before seeking nonlocal symmetries of $\mathbf{U}\{x, t\,; u\}$ (4.5), we present its point symmetry classification [Table 4.1] [Ovsiannikov (1959)]. One can show that no contact symmetries arise for any form of $K(u)$.

Table 4.1 Local (point) symmetries of the nonlinear diffusion equation $\mathbf{U}\{x, t\,; u\}$ (4.5)

$K(u)$	#	Point Symmetries
Arbitrary	3	$X_1 = \frac{\partial}{\partial x}$, $X_2 = \frac{\partial}{\partial t}$, $X_3 = x\frac{\partial}{\partial x} + 2t\frac{\partial}{\partial t}$.
u^{ν}	4	X_1, X_2, X_3, $X_4 = x\frac{\partial}{\partial x} + \frac{2}{\nu}u\frac{\partial}{\partial u}$.
e^{u}	4	X_1, X_2, X_3, $X_5 = x\frac{\partial}{\partial x} + 2\frac{\partial}{\partial u}$.
$u^{-4/3}$	5	X_1, X_2, X_3, $X_4\left(\nu = -\frac{4}{3}\right)$, $X_6 = x^2\frac{\partial}{\partial x} - 3xu\frac{\partial}{\partial u}$.

In principle, nonlocal symmetries of the nonlinear diffusion equation $\mathbf{U}\{x,t\,;u\}$ (4.5) can arise from any nonlocally related system within the tree $\mathcal{T}_4$ given by (3.75). In Table 4.2, we present the point symmetry classification of the two singlet potential systems $\mathbf{UV}\{x,t\,;u,v\}$ (3.19) and $\mathbf{UA}\{x,t\,;u,\alpha\}$ (3.32).

In comparison with Table 4.1, it is obvious that the point symmetry classification of the singlet potential system $\mathbf{UA}\{x,t\,;u,\alpha\}$ (3.32) yields no nonlocal symmetries of the nonlinear diffusion equation $\mathbf{U}\{x,t\,;u\}$ (4.5). On the other hand, the point symmetry classification of the singlet potential system $\mathbf{UV}\{x,t\,;u,v\}$ (3.19) yields potential symmetries of $\mathbf{U}\{x,t\,;u\}$.

In particular, when $K(u) = u^{-2}$, the system $\mathbf{UV}\{x,t\,;u,v\}$ has an infinite number of point symmetries that lead to the linearization of the system $\mathbf{UV}\{x,t\,;u,v\}$ by a point transformation [Section 2.4]; when $K(u) = e^{\lambda \tan^{-1} u}/(u^2+1)$ (corresponding to $L(u) = \lambda^{-1}e^{\lambda \tan^{-1} u}$), the system $\mathbf{UV}\{x, t\,;u,v\}$ has the point symmetry Y_9 that is obviously a nonlocal symmetry of the nonlinear diffusion equation $\mathbf{U}\{x,t\,;u\}$.

For all other distinguished cases, the point symmetry classification of $\mathbf{UV}\{x,t\,;u,v\}$ (3.19) is greatly simplified through use of the equivalence transformation (4.7). This readily leads to an additional point symmetry of $\mathbf{UV}\{x,t\,;u,v\}$ for

$$K(u) = u^{\nu}(1+bu)^{-(\nu+2)}, \quad K(u) = \frac{1}{(1+bu)^2}e^{u/(1+bu)},$$

and for

$$K(u) = \frac{1}{u^2+(1+bu)^2}\exp\left(\lambda \tan^{-1}\frac{u}{1+bu}\right).$$

These additional point symmetries of $\mathbf{UV}\{x,t\,;u,v\}$ are obviously nonlocal symmetries of $\mathbf{U}\{x,t\,;u\}$. Note that since $\widetilde{K}(\widetilde{u}) = \widetilde{u}^{-2}$ when $K(u) = u^{-2}$, no additional symmetries arise in the linearization case. The symmetry classification of system $\mathbf{UV}\{x,t\,;u,v\}$ first appeared in a different form in Bluman, Kumei & Reid (1988). This paper did not make use of the important simplifying equivalence transformation (4.7).

The symmetry classification of the couplet potential system $\mathbf{UVA}\{x,t;\, u,v,\alpha\}$ (3.73) is presented in Table 4.3.

Compared to the situation for the singlet potential system $\mathbf{UV}\{x,t\,;u,v\}$ (3.19), the couplet $\mathbf{UVA}\{x,t\,;u,v,\alpha\}$ (3.73) contains three additional distinguished cases: $K(u) = u^{-2/3}$, $K(u) = u^{-4/3}(1+bu)^{-2/3}$, and $K(u) = u^{-2/3}(1+bu)^{-4/3}$, with the respective point symmetries Z_9, Z_{15} and Z_{16}, which are nonlocal symmetries of all other PDE systems in the tree.

One can show that the point symmetry classification of each of the three remaining nonlocally related subsystems $\mathbf{A}\{x,u\,;\alpha\}$, $\mathbf{V}\{u,t\,;v\}$ and

Table 4.2 Point symmetries of singlet potential systems of the nonlinear diffusion equation (4.5)

$K(u)$	$\mathbf{UV}\{x,t\,;u,v\}$		$\mathbf{UA}\{x,t\,;u,\alpha\}$	
	#	Point Symmetries	#	Point Symmetries
Arbitrary	4	$\mathrm{Y}_1=\mathrm{X}_1,\quad \mathrm{Y}_2=\mathrm{X}_2,$ $\mathrm{Y}_3=\mathrm{X}_3+v\frac{\partial}{\partial v},\quad \mathrm{Y}_4=\frac{\partial}{\partial v}.$	3	$\hat{\mathrm{Y}}_1=\mathrm{X}_2,$ $\hat{\mathrm{Y}}_2=\mathrm{X}_3+2\alpha\frac{\partial}{\partial\alpha},$ $\hat{\mathrm{Y}}_3=\frac{\partial}{\partial\alpha}.$
u^{ν}	5	$\mathrm{Y}_1,\mathrm{Y}_2,\mathrm{Y}_3,\mathrm{Y}_4,$ $\mathrm{Y}_5=\mathrm{X}_4+\left(1+\frac{2}{\nu}\right)v\frac{\partial}{\partial v}.$	4	$\hat{\mathrm{Y}}_1,\hat{\mathrm{Y}}_2,\hat{\mathrm{Y}}_3,\hat{\mathrm{Y}}_4=\mathrm{X}_4$ $+2\left(1+\frac{1}{\nu}\right)\alpha\frac{\partial}{\partial\alpha}.$
e^{u}	5	$\mathrm{Y}_1,\mathrm{Y}_2,\mathrm{Y}_3,\mathrm{Y}_4,$ $\mathrm{Y}_6=\mathrm{X}_5+(2x+v)\frac{\partial}{\partial v}.$	4	$\hat{\mathrm{Y}}_1,\hat{\mathrm{Y}}_2,\hat{\mathrm{Y}}_3,\hat{\mathrm{Y}}_5=\mathrm{X}_5$ $+\left(x^2+2\alpha\right)\frac{\partial}{\partial\alpha}.$
$u^{-4/3}$	5	$\mathrm{Y}_1,\mathrm{Y}_2,\mathrm{Y}_3,\mathrm{Y}_4,\mathrm{Y}_5\ (\nu=-4/3).$	5	$\hat{\mathrm{Y}}_1,\hat{\mathrm{Y}}_2,\hat{\mathrm{Y}}_3,\hat{\mathrm{Y}}_4,$ $\hat{\mathrm{Y}}_6=\mathrm{X}_6.$
u^{-2}	∞	$\mathrm{Y}_1,\mathrm{Y}_2,\mathrm{Y}_3,\mathrm{Y}_4,\mathrm{Y}_5\ (\nu=-2),$ $\mathrm{Y}_7=-xv\frac{\partial}{\partial x}+(xu+v)u\frac{\partial}{\partial u}+2t\frac{\partial}{\partial v},$ $\mathrm{Y}_8=-x(2t+v^2)\frac{\partial}{\partial x}+4t^2\frac{\partial}{\partial t}$ $+u(6t+2xuv+v^2)\frac{\partial}{\partial u}+4tv\frac{\partial}{\partial v},$ $\mathrm{Y}_\infty=F^1(v,t)\frac{\partial}{\partial x}-u^2F^2(v,t)\frac{\partial}{\partial u},$ $(F^1(v,t),F^2(v,t))$ is an arbitrary solution of the linear system $F^1_t=F^2_v,\ F^1_v=F^2.$	4	$\hat{\mathrm{Y}}_1,\hat{\mathrm{Y}}_2,\hat{\mathrm{Y}}_3,\hat{\mathrm{Y}}_4.$
$(u^2+1)^{-1}$ $\times e^{\lambda\tan^{-1}u}$	5	$\mathrm{Y}_1,\mathrm{Y}_2,\mathrm{Y}_3,\mathrm{Y}_4,$ $\mathrm{Y}_9=v\frac{\partial}{\partial x}+\lambda t\frac{\partial}{\partial t}$ $-(u^2+1)\frac{\partial}{\partial u}-x\frac{\partial}{\partial v}.$	3	$\hat{\mathrm{Y}}_1,\hat{\mathrm{Y}}_2,\hat{\mathrm{Y}}_3.$
$u^{\nu}(1+bu)^{-(\nu+2)}$	5	$\mathrm{Y}_1,\mathrm{Y}_2,\mathrm{Y}_3,\mathrm{Y}_4,$ $\mathrm{Y}_{10}=bv\frac{\partial}{\partial x}+\nu t\frac{\partial}{\partial t}$ $-(1+bu)u\frac{\partial}{\partial u}-v\frac{\partial}{\partial v}.$	3	$\hat{\mathrm{Y}}_1,\hat{\mathrm{Y}}_2,\hat{\mathrm{Y}}_3.$
$(1+bu)^{-2}$ $\times e^{u/(1+bu)}$	5	$\mathrm{Y}_1,\mathrm{Y}_2,\mathrm{Y}_3,\mathrm{Y}_4,$ $\mathrm{Y}_{11}=b(2x+bv)\frac{\partial}{\partial x}$ $+(1+2b)t\frac{\partial}{\partial t}$ $-(1+bu)^2\frac{\partial}{\partial u}-x\frac{\partial}{\partial v}.$	3	$\hat{\mathrm{Y}}_1,\hat{\mathrm{Y}}_2,\hat{\mathrm{Y}}_3.$
$(u^2+(1+bu)^2)^{-1}$ $\times\exp\Big(\lambda$ $\times\tan^{-1}\frac{u}{1+bu}\Big)$	5	$\mathrm{Y}_1,\mathrm{Y}_2,\mathrm{Y}_3,\mathrm{Y}_4,$ $\mathrm{Y}_{12}=(2bx+(b^2+1)v)\frac{\partial}{\partial x}$ $+(\lambda+2b)t\frac{\partial}{\partial t}$ $-((1+bu)^2+u^2)\frac{\partial}{\partial u}-x\frac{\partial}{\partial v}.$	3	$\hat{\mathrm{Y}}_1,\hat{\mathrm{Y}}_2,\hat{\mathrm{Y}}_3.$

Table 4.3 Symmetries of the couplet potential system **UVA**$\{x,t\,;u,v,\alpha\}$ (3.73) of the nonlinear diffusion equation (4.5)

$K(u)$	#	Point Symmetries
Arbitrary	5	$\mathrm{Z}_1 = \mathrm{X}_1 + v\frac{\partial}{\partial\alpha},\quad \mathrm{Z}_2 = \mathrm{X}_2,\quad \mathrm{Z}_3 = \hat{\mathrm{Y}}_3,\quad \mathrm{Z}_4 = \mathrm{Y}_4,$ $\mathrm{Z}_5 = \mathrm{X}_3 + v\frac{\partial}{\partial v} + 2\alpha\frac{\partial}{\partial\alpha}.$
u^ν	6	$\mathrm{Z}_1, \mathrm{Z}_2, \mathrm{Z}_3, \mathrm{Z}_4, \mathrm{Z}_5,\ \mathrm{Z}_6 = \mathrm{X}_4 + \left(1+\frac{2}{\nu}\right)v\frac{\partial}{\partial v} + 2\left(1+\frac{1}{\nu}\right)\alpha\frac{\partial}{\partial\alpha}.$
e^u	6	$\mathrm{Z}_1, \mathrm{Z}_2, \mathrm{Z}_3, \mathrm{Z}_4, \mathrm{Z}_5,\ \mathrm{Z}_7 = \mathrm{X}_5 + (2x+v)\frac{\partial}{\partial v} + \left(x^2+2\alpha\right)\frac{\partial}{\partial\alpha}.$
$u^{-4/3}$	7	$\mathrm{Z}_1, \mathrm{Z}_2, \mathrm{Z}_3, \mathrm{Z}_4, \mathrm{Z}_5, \mathrm{Z}_6\ (\nu=-4/3),\ \mathrm{Z}_8 = \mathrm{X}_6 - \alpha\frac{\partial}{\partial v}.$
$u^{-2/3}$	7	$\mathrm{Z}_1, \mathrm{Z}_2, \mathrm{Z}_3, \mathrm{Z}_4, \mathrm{Z}_5, \mathrm{Z}_6\ (\nu=-2/3),$ $\mathrm{Z}_9 = (xv-\alpha)\frac{\partial}{\partial x} - 3uv\frac{\partial}{\partial u} - v^2\frac{\partial}{\partial v} - v\alpha\frac{\partial}{\partial\alpha}.$
u^{-2}	∞	$\mathrm{Z}_1, \mathrm{Z}_2, \mathrm{Z}_3, \mathrm{Z}_4, \mathrm{Z}_5, \mathrm{Z}_6\ (\nu=-2),$ $\mathrm{Z}_{10} = -(xv+\alpha)\frac{\partial}{\partial x} + (2xu+v)u\frac{\partial}{\partial u} + 2t\frac{\partial}{\partial v} - v\alpha\frac{\partial}{\partial\alpha},$ $\mathrm{Z}_{11} = -(6xt+xv^2+2va)\frac{\partial}{\partial x} + 4t^2\frac{\partial}{\partial t}$ $+u(10t+2u(2xv+a)+v^2)\frac{\partial}{\partial u} + 4tv\frac{\partial}{\partial v} - (2t+v^2)\alpha\frac{\partial}{\partial\alpha},$ $\mathrm{Z}_\infty = F^1(v,t)\frac{\partial}{\partial x} - u^2F^2(v,t)\frac{\partial}{\partial u} + F^3(v,t)\frac{\partial}{\partial\alpha},$ $(F^1(v,t), F^2(v,t), F^3(v,t))$ is an arbitrary solution of the linear system $F^3_v = F^1,\ F^3_t = F_2,\ F^1_v = F_2.$
$(u^2+1)^{-1}e^{\lambda\tan^{-1}u}$	6	$\mathrm{Z}_1, \mathrm{Z}_2, \mathrm{Z}_3, \mathrm{Z}_4, \mathrm{Z}_5,\ \mathrm{Z}_{12} = \mathrm{Y}_9 + \frac{v^2-x^2}{2}\frac{\partial}{\partial\alpha}.$
$u^\nu(1+bu)^{-(\nu+2)}$	6	$\mathrm{Z}_1, \mathrm{Z}_2, \mathrm{Z}_3, \mathrm{Z}_4, \mathrm{Z}_5,$ $\mathrm{Z}_{13} = bv\frac{\partial}{\partial x} + \nu t\frac{\partial}{\partial t} - (1+bu)u\frac{\partial}{\partial u} - v\frac{\partial}{\partial v} + (\frac{bv^2}{2}-\alpha)\frac{\partial}{\partial\alpha}.$
$(1+bu)^{-2}$ $\times e^{u/(1+bu)}$	6	$\mathrm{Z}_1, \mathrm{Z}_2, \mathrm{Z}_3, \mathrm{Z}_4, \mathrm{Z}_5,$ $\mathrm{Z}_{14} = b(2x+bv)\frac{\partial}{\partial x} + (1+2b)t\frac{\partial}{\partial t} - (1+bu)^2\frac{\partial}{\partial u}$ $-x\frac{\partial}{\partial v} + (\frac{b^2v^2-x^2}{2} + 2b\alpha)\frac{\partial}{\partial\alpha}.$
$u^{-4/3}(1+bu)^{-2/3}$	7	$\mathrm{Z}_1, \mathrm{Z}_2, \mathrm{Z}_3, \mathrm{Z}_4, \mathrm{Z}_5, \mathrm{Z}_{13}\ (\nu=-4/3),$ $\mathrm{Z}_{15} = (3b^2v^2 + 2b(2xv+\alpha) + 2x^2)\frac{\partial}{\partial x}$ $-6(x+bv)(1+bu)u\frac{\partial}{\partial u} - (bv^2+2\alpha)\frac{\partial}{\partial v} + (b^2v^+2b\alpha)v\frac{\partial}{\partial\alpha}.$
$u^{-2/3}(1+bu)^{-4/3}$	7	$\mathrm{Z}_1, \mathrm{Z}_2, \mathrm{Z}_3, \mathrm{Z}_4, \mathrm{Z}_5, \mathrm{Z}_{13}\ (\nu=-2/3),$ $\mathrm{Z}_{16} = (3bv^2 + 2(xv-\alpha))\frac{\partial}{\partial x} - 6(1+bu)uv\frac{\partial}{\partial u}$ $-2v^2\frac{\partial}{\partial v} + (bv^2-2\alpha)v\frac{\partial}{\partial\alpha}.$
$\frac{1}{u^2+(1+bu)^2}\times$ $\exp\left(\lambda\tan^{-1}\frac{u}{1+bu}\right)$	6	$\mathrm{Z}_1, \mathrm{Z}_2, \mathrm{Z}_3, \mathrm{Z}_4, \mathrm{Z}_5,\ \mathrm{Z}_{17} = (2bx+(b^2+1)v)\frac{\partial}{\partial x} + (\lambda+2b)t\frac{\partial}{\partial t}$ $-((1+bu)^2+u^2)\frac{\partial}{\partial u} - x\frac{\partial}{\partial v} + (\frac{(b^2+1)v^2-x^2}{2} + 2b\alpha)\frac{\partial}{\partial\alpha}.$

$\mathbf{X}\{u, v\,; x\}$ yields no new nonlocal symmetries of the nonlinear diffusion equation $\mathbf{U}\{x, t\,; u\}$ [Exercise 4.2.1].

Thus in this particular example, the point symmetry classification of the "grand" couplet potential system $\mathbf{UVA}\{x, t\,; u, v, \alpha\}$ yields all point symmetries of each of the other PDE systems in the tree $\mathcal{T}_4$.

4.2.2 Nonlocal symmetries of a nonlinear wave equation

As a second example, consider a symmetry classification problem for the nonlinear wave equation $\mathbf{U}\{x, t\,; u\}$ given by

$$u_{tt} = (c^2(u)u_x)_x, \tag{4.8}$$

with an arbitrary constitutive function $c(u)$ [Ames, Lohner & Adams (1981); Bluman & Kumei [(1987), (1988)]; Bluman & Cheviakov (2007)].

The group of equivalence transformations of $\mathbf{U}\{x, t\,; u\}$ (4.8) is given by

$$\widetilde{x} = a_1x + a_4, \quad \widetilde{t} = a_2t + a_5, \quad \widetilde{u} = a_3u + a_6, \quad \widetilde{c}(\widetilde{u}) = a_1a_2^{-1}c(u), \tag{4.9}$$

where $a_1, \ldots, a_6$ are arbitrary constants with $a_1a_2a_3 \neq 0$. The point symmetry classification of the nonlinear wave equation $\mathbf{U}\{x, t\,; u\}$ (4.8) [Ames, Lohner & Adams (1981)] is presented in Table 4.4 (modulo the equivalence transformations (4.9)).

Table 4.4 Point symmetries of the nonlinear wave equation $\mathbf{U}\{x, t\,; u\}$ (4.8)

$c(u)$	#	Point Symmetries
Arbitrary	3	$\mathrm{X}_1 = t\frac{\partial}{\partial t} + x\frac{\partial}{\partial x}$, $\mathrm{X}_2 = \frac{\partial}{\partial t}$, $\mathrm{X}_3 = \frac{\partial}{\partial x}$.
u^{ν}	4	$\mathrm{X}_1, \mathrm{X}_2, \mathrm{X}_3, \mathrm{X}_4 = \nu x\frac{\partial}{\partial x} + u\frac{\partial}{\partial u}$.
e^u	4	$\mathrm{X}_1, \mathrm{X}_2, \mathrm{X}_3, \mathrm{X}_5 = x\frac{\partial}{\partial x} + \frac{\partial}{\partial u}$.
u^{-2}	5	$\mathrm{X}_1, \mathrm{X}_2, \mathrm{X}_3, \mathrm{X}_4$ $(\nu = -2)$, $\mathrm{X}_6 = t^2\frac{\partial}{\partial t} + tu\frac{\partial}{\partial u}$.
$u^{-2/3}$	5	$\mathrm{X}_1, \mathrm{X}_2, \mathrm{X}_3, \mathrm{X}_4$ $(\nu = -2/3)$, $\mathrm{X}_7 = x^2\frac{\partial}{\partial x} - 3xu\frac{\partial}{\partial u}$.

An extended tree $\mathcal{T}_d$ of nonlocally related PDE systems for the nonlinear wave equation $\mathbf{U}\{x, t\,; u\}$ (4.8), holding for an arbitrary wave speed $c(u)$, was constructed in Section 3.5.3, and exhibited in Figure 3.4.

We now classify nonlocal symmetries of the nonlinear wave equation $\mathbf{U}\{x, t\,; u\}$ (4.8) arising as point symmetries of any of the seven singlet potential systems **UA**$\{x,t\,;u,\alpha\}$, **UB**$\{x,t\,;u,\beta\}$, **UV**$\{x,t\,;u,v\}$, **UW**$\{x,t\,;u,w\}$, **TP**$\{u,v\,;t,p\}$, **TQ**$\{u,v\,;t,q\}$ and **TR**$\{u,v\,;t,r\}$ given by (3.81)–(3.84) and (3.92)–(3.94), respectively, or as point symmetries of the two nonlocally related subsystems **X**$\{u,v\,;x\}$ (3.86) and **T**$\{u,v\,;t\}$ (3.87) [Bluman & Cheviakov (2007); references therein].

In Tables 4.5a,b, for each of the nine above-mentioned nonlocally related systems, the situations are summarized where nonlocal symmetries arise for the nonlinear wave equation U$\{x,t\,;u\}$ (4.8) from point symmetries of any of these nine systems. The results are given modulo the equivalence transformations (4.9).

Table 4.5 (a) Cases for which nonlocal symmetries of the nonlinear wave equation U$\{x,t\,;u\}$ (4.8) arise

System	Potential(s)	Condition on $c(u)$	Symmetries; Remarks
UA (3.83)	α	No special cases	Nonlocal symmetries do not arise.
UB (3.84)	β	$c(u)=u^{-2/3}$	Linearizable by a point transformation.
		$\frac{F''(u)}{(F'(u))^2}=\frac{4F(u)+C_1}{(F(u)+C_2)^2+C_3}$, $(F(u)=\int c^2(u)du,\quad C_1,C_2,C_3=\text{const})$	One nonlocal symmetry.
UV (3.81)	v	Arbitrary	Infinite number of nonlocal symmetries; there exists an invertible mapping to linear system **XT** (3.85) (hodograph transformation).
		$\left[\frac{c'(u)}{c^3(u)}\left(\frac{c(u)}{c'(u)}\right)''\right]'=0$	One or two additional nonlocal symmetries.
UW (3.82)	w	$c(u)=(u+B)^{-2}$	Linearizable by a point transformation.
		$\frac{c'(u)}{c(u)}=-\frac{2u+C_1}{u^2+C_2}$ $(C_1,C_2=\text{const})$	One nonlocal symmetry.

The nonlocal symmetries for the cases listed in Tables 4.5a,b arise as follows.

(1) The potential system **UB**$\{x,t\,;u,\beta\}$
The potential system **UB**$\{x,t\,;u,\beta\}$ (3.84), i.e.,

Table 4.5 (b) Cases for which nonlocal symmetries of the nonlinear wave equation $\mathbf{U}\{x,t\,;u\}$ (4.8) arise

System	Poten-tial(s)	Condition on $c(u)$	Symmetries; Remarks
TP (3.92)	v, p	$\frac{-(2uc^2+u^2cc')c'''+2u^2c(c'')^2}{c^3(uc'+2c)^2}$ $+\frac{-(4c^2+u^2(c')^2-8ucc')c''+6(c')^2(c-uc')}{c^3(uc'+2c)^2}$ $=\lambda^2,\quad \lambda=\text{const}$	One or two nonlocal symmetries.
		$c(u)=u^{-2}$	Infinite number of nonlocal symmetries; there exists a point mapping to a linear system with constant coefficients.
TQ (3.93)	v, q	$c(u)=u^{-2/3};\quad c(u)=u^{-2}$	Two nonlocal symmetries.
TR (3.94)	v, r	$\frac{ucc''+c'(c-uc')}{(uc'+2c)^2}=\gamma^2=\text{const}$	Two nonlocal symmetries.
X (3.86)	v	$\frac{(-2cc''+5(c')^2)c^2c''''+3c^3(c''')^2+16c^2(c'')^3}{c^3(2cc''-5(c')^2)^2}$ $+\frac{-24c^2c'''c''c'+12c(c'c'')^2-10(c')^4c''}{c^3(2cc''-5(c')^2)^2}$ $=\sigma^2,\quad \sigma=\text{const}$	One or two nonlocal symmetries.
T (3.87)	v	$(\alpha'+H\alpha)'=\sigma^2\alpha c^2(u),\quad \sigma=\text{const.}$ $(H=c'(u)/c(u),\quad \alpha^2=(H^2-2H')^{-1})$	One or two nonlocal symmetries.
		$c(u)=u^{-2}$	Infinite number of nonlocal symmetries; there exists an invertible mapping to a linear system with constant coefficients.

$$\begin{aligned}\beta_x &= xu_t,\\ \beta_t &= xc^2(u)u_x - \textstyle\int c^2(u)du,\end{aligned}$$

has the group of equivalence transformations

$$\begin{aligned}&\widetilde{x}=a_1x,\quad \widetilde{t}=a_2t+a_4,\quad \widetilde{u}=a_3u+a_5,\\ &\widetilde{b}=a_1^2a_2^{-1}a_3b-a_2a_7t+a_6,\quad \widetilde{F}(\widetilde{u})=a_1^2a_2^{-2}a_3F(u)+a_7,\end{aligned}\tag{4.10}$$

where $F(u)=\int c^2(u)du$; $a_1,\ldots,a_7$ are arbitrary constants with $a_1a_2a_3\neq 0$.

For an arbitrary wave speed $c(u)$, the system $\mathbf{UB}\{x,t\,;u,\beta\}$ has three point symmetries given by

$$Y_1 = \frac{\partial}{\partial t}, \quad Y_2 = \frac{\partial}{\partial \beta}, \quad Y_3 = x\frac{\partial}{\partial x} + t\frac{\partial}{\partial t} + \beta\frac{\partial}{\partial \beta}.$$

These point symmetries project onto point symmetries of the nonlinear wave equation $\mathbf{U}\{x, t\,; u\}$ (4.8).

If the wave speed $c(u)$ satisfies the ODE

$$\frac{F''(u)}{(F'(u))^2} = \frac{4F(u) + C_1}{(F(u) + C_2)^2 + C_3}, \tag{4.11}$$

where $F(u) = \int c^2(u)du$ and C_1, C_2, C_3 are arbitrary constants, then the system $\mathbf{UB}\{x, t\,; u, \beta\}$ has an additional point symmetry

$$\begin{aligned} Y_4 = &\left(F(u) + \tfrac{1}{2}C_1\right)x\frac{\partial}{\partial x} + \beta\frac{\partial}{\partial t} + \frac{(F(u) + C_2)^2 + C_3}{F'(u)}\frac{\partial}{\partial u} \\ &+ \left(2C_2\beta - (C_2^2 + C_3)t\right)\frac{\partial}{\partial \beta}, \end{aligned}$$

which is a nonlocal symmetry of the nonlinear wave equation $\mathbf{U}\{x, t\,; u\}$ (4.8).

For $c(u) = u^{-2/3}$, the potential system $\mathbf{UB}\{x, t\,; u, \beta\}$ has an infinite number of point symmetries that lead to the linearization of the potential system $\mathbf{UB}\{x, t\,; u, \beta\}$ by a point transformation, and thus a linearization of the nonlinear wave equation $\mathbf{U}\{x, t\,; u\}$ (4.8) by a nonlocal transformation [Exercise 4.2.3].

(2) The potential system $\mathbf{UV}\{x, t\,; u, v\}$
The potential system $\mathbf{UV}\{x, t\,; u, v\}$ (3.81), i.e.,

$$\begin{aligned} v_x &= u_t, \\ v_t &= c^2(u)\, u_x, \end{aligned}$$

has the group of equivalence transformations

$$\begin{aligned} \widetilde{x} &= a_1 x + a_4 v + a_5, \quad \widetilde{t} = a_2 t + a_1^{-1} a_2 a_4 u + a_6, \\ \widetilde{u} &= a_3 u + a_7 t + a_8, \quad \widetilde{v} = a_1 a_2^{-1} a_3 v + a_1 a_2^{-1} a_7 x + a_9, \\ \widetilde{c}(\widetilde{u}) &= a_1 a_2^{-1} c(u), \end{aligned} \tag{4.12}$$

where $a_1, \ldots, a_9$ are arbitrary constants with $a_1 a_2 a_3 \neq 0$.

The nonlinear PDE system $\mathbf{UV}\{x, t\,; u, v\}$ is locally related to the linear PDE system $\mathbf{XT}\{u, v\,; x, t\}$ (3.85) through an interchange of dependent and independent variables in terms of the hodograph transformation $x = x(u, v)$, $t = t(u, v)$. Hence these two systems have the same point symmetries. In particular, the infinite number of point symmetries of the PDE system $\mathbf{XT}\{u, v\,; x, t\}$, due to its linearity, yields an *infinite number of nonlocal symmetries* of the nonlinear wave equation $\mathbf{U}\{x, t\,; u\}$.

The point symmetries of the PDE system $\mathbf{UV}\{x,t\,;u,v\}$ are summarized in Table 4.6.

For an arbitrary wave speed $c(u)$, in addition to the infinite number of point symmetries arising from the linearity of $\mathbf{XT}\{u,v\,;x,t\}$, the system $\mathbf{UV}\{x,t\,;u,v\}$ has four additional point symmetries that project onto the three point symmetries of the nonlinear wave equation $\mathbf{U}\{x,t\,;u\}$ (4.8) [Table 4.4]. Further point symmetries arise when $c(u)$ satisfies the ODE

$$\frac{c'(u)}{c^3(u)}\left(\frac{c(u)}{c'(u)}\right)'' = \lambda^2 = \text{const.} \tag{4.13}$$

For several classes of wave speeds $c(u)$ satisfying (4.13), these point symmetries yield nonlocal symmetries of $\mathbf{U}\{x,t\,;u\}$.

Table 4.6 Point symmetries of the potential system $\mathbf{UV}\{x,t\,;u,v\}$ of the nonlinear wave equation $\mathbf{U}\{x,t\,;u\}$ (4.8)

$c(u)$	#	Point Symmetries
Arbitrary	∞	Infinite number of point symmetries following from the linearity of the invertibly related system $\mathbf{XT}\{u,v\,;x,t\}$.
Arbitrary	4	$\mathrm{W}_1 = \frac{\partial}{\partial t}$, $\mathrm{W}_2 = \frac{\partial}{\partial x}$, $\mathrm{W}_3 = \frac{\partial}{\partial v}$, $\mathrm{W}_4 = t\frac{\partial}{\partial t} + x\frac{\partial}{\partial x}$.
u^ν $(\nu \neq 0, -1)$	6	W_1, W_2, W_3, W_4, $\mathrm{W}_5 = \nu t\frac{\partial}{\partial t} - u\frac{\partial}{\partial u} - (1+\nu)v\frac{\partial}{\partial v}$, $\mathrm{W}_6 = -((2\nu+1)tv + xu)\frac{\partial}{\partial t} - (tu^{1+2\nu} + xv)\frac{\partial}{\partial x}$ $+2uv\frac{\partial}{\partial u} + \left[(1+\nu)v^2 + \frac{u^{2+2\nu}}{1+\nu}\right]\frac{\partial}{\partial v}$.
e^u	6	W_1, W_2, W_3, W_4, $\mathrm{W}_7 = x\frac{\partial}{\partial x} + \frac{\partial}{\partial u} + v\frac{\partial}{\partial v}$, $\mathrm{W}_8 = -(2vt + x)\frac{\partial}{\partial t} - 2e^u t\frac{\partial}{\partial x} + 4v\frac{\partial}{\partial u}$ $+(4e^u + v^2)\frac{\partial}{\partial v}$.
u^{-1}	6	W_1, W_2, W_3, W_4, W_5 $(\nu = -1)$, $\mathrm{W}_9 = (tv - xu)\frac{\partial}{\partial t} - (tu^{-1} + xv)\frac{\partial}{\partial x} + 2uv\frac{\partial}{\partial u}$ $+2\log u\frac{\partial}{\partial v}$.
$c(u)$ satisfies (a), (b) or (c): (a) $c' = c^2\nu^{-1}\sinh(\nu\log c)$ (b) $c' = c^2\nu^{-1}\sin(\nu\log c)$ (c) $c' = c^2\nu^{-1}\cosh(\nu\log c)$	6	W_1, W_2, W_3, W_4, $\mathrm{W}_{10,11} = e^{\pm v}\Big\{((2+\Gamma')t \pm \Gamma x)\frac{\partial}{\partial t}$ $+(\Gamma' x \pm c^2\Gamma t)\frac{\partial}{\partial x} - 2\Gamma\frac{\partial}{\partial u} \mp 2(\Gamma'+1)\frac{\partial}{\partial v}\Big\}$, where $\Gamma = c/c'$.

The point symmetries $W_6, W_8, W_9, W_{10}, W_{11}$ of the potential system $\mathbf{UV}\{x,t\,;u,v\}$ correspond to nonlocal symmetries of the nonlinear wave equation $\mathbf{U}\{x,t\,;u\}$ (4.8).

(3) The potential system $\mathbf{UW}\{x,t\,;u,w\}$
The potential system $\mathbf{UW}\{x,t\,;u,w\}$ (3.82), i.e.,

$$\begin{aligned} w_x &= tu_t - u, \\ w_t &= tc^2(u)u_x, \end{aligned}$$

has the group of equivalence transformations that includes the transformations

$$\begin{aligned} &\widetilde{x} = a_1x + a_4, \quad \widetilde{t} = a_2t, \quad \widetilde{u} = a_3u + a_6t + a_7, \\ &\widetilde{w} = a_1a_3w - a_1a_7x + a_5, \quad \widetilde{c}(\widetilde{u}) = a_1a_2^{-1}c(u), \end{aligned} \tag{4.14}$$

where $a_1, \dots, a_7$ are arbitrary constants with $a_1a_2a_3 \neq 0$, and the projective transformation

$$\begin{aligned} &\widetilde{x} = x - bw, \quad \widetilde{t} = \frac{t}{1+bu}, \quad \widetilde{u} = \frac{u}{1+bu}, \quad \widetilde{w} = w, \\ &\widetilde{c}(\widetilde{u}) = (1+b\widetilde{u})^{-2}\, c\Big(\frac{\widetilde{u}}{1+b\widetilde{u}}\Big). \end{aligned} \tag{4.15}$$

For an arbitrary $c(u)$, the potential system $\mathbf{UW}\{x,t\,;u,w\}$ has the point symmetries

$$Z_1 = \frac{\partial}{\partial x}, \quad Z_2 = \frac{\partial}{\partial w}, \quad Z_3 = x\frac{\partial}{\partial x} + t\frac{\partial}{\partial t} + w\frac{\partial}{\partial w}.$$

These point symmetries project onto point symmetries of $\mathbf{U}\{x,t\,;u\}$.

If the wave speed $c(u)$ satisfies the ODE

$$\frac{c'(u)}{c(u)} = -\frac{2u + C_1}{u^2 + C_2}, \tag{4.16}$$

where C_1, C_2 are arbitrary constants, then the potential system $\mathbf{UW}\{x,t;\,u,w\}$ has an additional point symmetry

$$Z_4 = w\frac{\partial}{\partial x} + (u + C_1)t\frac{\partial}{\partial t} + (u^2 + C_2)\frac{\partial}{\partial u} - C_2x\frac{\partial}{\partial w},$$

which is obviously a nonlocal symmetry of $\mathbf{U}\{x,t\,;u\}$.

The general solution of (4.16) is found to be as follows:

$$
\begin{aligned}
C_2 = \omega^2 > 0: \quad & c(u) = \frac{c_0}{u^2+\omega^2}\exp\left\{-\frac{C_1}{\omega}\tan^{-1}\frac{u}{\omega}\right\};\\
C_2 = -\omega^2 < 0: \; & c(u) = \frac{c_0}{u^2-\omega^2}\left|\frac{u+\omega}{u-\omega}\right|^{C_1/2\omega};\\
C_2 = 0: \quad & c(u) = \frac{c_0}{u^2}e^{C_1/u}.
\end{aligned}
\tag{4.17}
$$

In (4.17), c_0 is an arbitrary constant of integration.

For $c(u) = (u+B)^{-2}$, where B is an arbitrary constant, the system $\mathbf{UW}\{x,t\,;u,w\}$ has an infinite number of point symmetries. One can show that here $\mathbf{UW}\{x,t\,;u,w\}$ is linearizable by a point transformation, and thus the nonlinear wave equation $\mathbf{U}\{x,t\,;u\}$ is linearizable by a nonlocal transformation [Exercise 4.2.2].

(4) The potential system $\mathbf{TP}\{u,v\,;t,p\}$
The potential system $\mathbf{TP}\{u,v\,;t,p\}$ (3.92), i.e.,

$$
\begin{aligned}
p_v &= ut_u - t,\\
p_u &= uc^2(u)t_v,
\end{aligned}
$$

has the group of equivalence transformations that includes the transformations

$$
\begin{aligned}
&\widetilde{u} = a_1 u, \quad \widetilde{v} = a_2 v + a_4, \quad \widetilde{t} = a_2^{-1}a_3 t + a_6 + a_7 u,\\
&\widetilde{p} = a_3 p + a_5 - a_2 a_6 v, \quad \widetilde{c}(\widetilde{u}) = a_1^{-1}a_2 c(u),
\end{aligned}
\tag{4.18}
$$

where $a_1, \ldots, a_7$ are arbitrary constants with $a_1a_2a_3 \neq 0$, and the projective transformation

$$
\begin{aligned}
&\widetilde{u} = \frac{u}{1+bu}, \quad \widetilde{v} = v, \quad \widetilde{t} = \frac{t}{1+bu}, \quad \widetilde{p} = p,\\
&\widetilde{c}(\widetilde{u}) = (1+b\widetilde{u})^{-2}\, c\Big(\frac{\widetilde{u}}{1+b\widetilde{u}}\Big)
\end{aligned}
\tag{4.19}
$$

similar to (4.15).

The point symmetry classification of the linear PDE system $\mathbf{TP}\{u,v\,;t,p\}$ (modulo its obvious infinite number of point symmetries due to its linearity) is as follows.

Case 1. For an arbitrary wave speed $c(u)$, the system $\mathbf{TP}\{u,v\,;t,p\}$ has the three point symmetries

$$
\mathrm{L}_1 = \frac{\partial}{\partial v}, \quad \mathrm{L}_2 = t\frac{\partial}{\partial t} + p\frac{\partial}{\partial p}, \quad \mathrm{L}_3 = \frac{\partial}{\partial p},
$$

that project onto point symmetries of the nonlinear wave equation $\mathbf{U}\{x,t\,;u\}$ (4.8).

Case 2. For $c(u) = u^{-2}$, the system $\mathbf{TP}\{u, v\,; t, p\}$ has an infinite number of point symmetries that are related to point symmetries of the system $\mathbf{UV}\{x, t\,; u, v\}$ with $c(u) = \text{const}$, since here the system $\mathbf{TP}\{u, v\,; t, p\}$ is mapped by the point transformation $y = -1/u, \gamma = t/u$ into the system with constant coefficients given by

$$\begin{aligned} p_v - \gamma_y &= 0, \\ p_y - \gamma_v &= 0. \end{aligned}$$

Remark 4.2.1. Note that the PDE system $\mathbf{TP}\{u, v\,; t, p\}$ is obviously not invariant under the translations $u \to u + B$, and thus it *does not* have an infinite number of point symmetries when $c(u) = (u+B)^{-2}$. However, by taking a linear combination of potential systems $\mathbf{TP}\{u, v\,; t, p\}$ and $\mathbf{XT}\{u, v\,; x, t\}$ (3.85) with weights 1 and B, and denoting a "combination" potential variable by $z = p + Bx$, one obtains a potential system $\mathbf{TZ}\{u, v\,; t, z\}$ which *does* have an infinite number of point symmetries when $c(u) = (u + B)^{-2}$

Case 3. For $c(u) \neq u^{-2}$, and $c(u)$ satisfying the ODE

$$\begin{aligned} &\frac{-(2uc^2 + u^2cc')c''' + 2u^2c(c'')^2 - (4c^2 + u^2(c')^2 - 8ucc')c''}{c^3(uc' + 2c)^2} \\ &\quad + \frac{6(c')^2(c - uc')}{c^3(uc' + 2c)^2} = \lambda^2, \end{aligned} \tag{4.20}$$

with λ a real or imaginary constant, the system $\mathbf{TP}\{u, v\,; t, p\}$ has additional point symmetries as follows.

Case 3a. When $\lambda \neq 0$ in (4.20), two additional point symmetries are given by

$$\begin{aligned} \mathrm{L}_{4,5} = e^{\pm\lambda v} \Bigg\{ &\Bigg[\pm \frac{\lambda^2 u^2 c}{2(2c + uc')} t \\ &- \left(\lambda \frac{c + uc'}{2c + uc'} - \frac{u^2(cc'' - 3(c')^2) - 4ucc' - 2c^2}{(2c + uc')^2} \right) p \Bigg] \frac{\partial}{\partial p} \\ &\pm \left[\frac{\lambda^2 c}{2(2c + uc')} p + \left(\frac{u^2(cc'' - 3(c')^2) - 4ucc' - 2c^2}{2(2c + uc')^2} \right) t \right] \frac{\partial}{\partial t} \\ &- \frac{\lambda uc}{2c + uc'} \frac{\partial}{\partial u} \pm \left[\frac{u^2(cc'' - 2(c')^2) - 2ucc' - 2c^2}{(2c + uc')^2} \right] \frac{\partial}{\partial v} \Bigg\} \end{aligned}$$

which yield *nonlocal symmetries* of the nonlinear wave equation $\mathbf{U}\{x, t\,; u\}$.

Case 3b. When $\lambda = 0$ in (4.20), the general solution of ODE (4.20) involves three distinguished classes given by

$$c(u) = Au^{\nu}(u+B)^{-2-\nu}; \tag{4.21}$$
$$c(u) = Au^{\nu}; \tag{4.22}$$
$$c(u) = Au^{-2}e^{B/u}; \tag{4.23}$$

A, B, ν are nonzero constants with $\nu \neq -2$.

From the equivalence transformations (4.18), it follows that a system $\mathbf{TP}\{u, v\,; t, p\}$ with wave speed (4.21) is invertibly equivalent to a system $\mathbf{TP}\{u, v\,; t, p\}$ with wave speed (4.22). Hence one only considers the non-equivalent cases (4.22), (4.23) (modulo the equivalence transformations (4.18)).

Case 3b(1). For wave speeds $c(u) = u^{\nu}$ with $\nu \neq -1$, the system $\mathbf{TP}\{u, v\,; t, p\}$ has two additional point symmetries given by

$$\begin{aligned}
\mathrm{L}_6 &= \nu t\frac{\partial}{\partial t} - u\frac{\partial}{\partial u} - (1+\nu)v\frac{\partial}{\partial v} - p\frac{\partial}{\partial p},\\
\mathrm{L}_7 &= -((2\nu+1)tv + p)\frac{\partial}{\partial t} + 2uv\frac{\partial}{\partial u}\\
&\quad + \left[(1+\nu)v^2 + \frac{u^{2+2\nu}}{1+\nu}\right]\frac{\partial}{\partial v} + (tu^{2+2\nu} - vp)\frac{\partial}{\partial p}.
\end{aligned}$$

Note that the symmetry L_7 is nonlocal for $\mathbf{U}\{x, t\,; u\}$ but local for the system $\mathbf{UV}\{x, t\,; u, v\}$; the symmetries L_6 and L_7 correspond to the symmetries W_5 and W_6, respectively, in Table 4.6.

Case 3b(2). For $c(u) = u^{-1}$, the system $\mathbf{TP}\{u, v\,; t, p\}$ again has two additional point symmetries given by

$$\mathrm{L}_6 \quad (\nu = -1), \quad \mathrm{L}_8 = (tv - p)\frac{\partial}{\partial t} + 2uv\frac{\partial}{\partial u} + 2\log u\frac{\partial}{\partial v} - (t - pv)\frac{\partial}{\partial p}.$$

The point symmetry L_8 is nonlocal for $\mathbf{U}\{x, t\,; u\}$ but local for the system $\mathbf{UV}\{x, t\,; u, v\}$. These symmetries correspond to W_5 $(\nu = -1)$ and W_7, respectively, in Table 4.6.

Case 3b(3). For $c(u) = u^{-2}e^{1/u}$, the system $\mathbf{TP}\{u, v\,; t, p\}$ has two additional point symmetries given by

$$\begin{aligned}
\mathrm{L}_9 &= (pu - 2tv(u+1))\frac{\partial}{\partial t} - 2u^2v\frac{\partial}{\partial u} + (u^2 + e^{2/u})\frac{\partial}{\partial v} + t\frac{e^{2/u}}{u}\frac{\partial}{\partial p},\\
\mathrm{L}_{10} &= t(u+1)\frac{\partial}{\partial t} + u^2\frac{\partial}{\partial u} - v\frac{\partial}{\partial v}.
\end{aligned}$$

The symmetries L_9 and L_{10} are nonlocal for both $\mathbf{U}\{x, t\,; u\}$ *and* the system $\mathbf{UV}\{x, t\,; u, v\}$.

(5) The potential system $\mathbf{TQ}\{u, v\,; t, q\}$
The potential system $\mathbf{TQ}\{u, v\,; t, q\}$ (3.93), i.e.,

$$\begin{aligned} q_v &= v t_u, \\ q_u &= c^2(u)(v t_v - t), \end{aligned}$$

has the group of equivalence transformations

$$\begin{aligned} &\widetilde{u} = a_1 u + a_4, \quad \widetilde{v} = a_2 v, \quad \widetilde{q} = a_3 q + a_5 + \tfrac{1}{3} a_2^2 a_7 v^3, \\ &\widetilde{t} = a_1 a_2^{-2} a_3 t + a_2 a_6 v + a_1 a_7 u v, \quad \widetilde{c}(\widetilde{u}) = a_1^{-1} a_2 c(u), \end{aligned} \tag{4.24}$$

where $a_1, \ldots, a_8$ are arbitrary constants with $a_1 a_2 a_3 \neq 0$.

The point symmetry classification of the linear potential system $\mathbf{TQ}\{u, v;$ $t, q\}$ is given in Table 4.7.

Table 4.7 Point symmetries of the potential system $\mathbf{TQ}\{u, v\,; t, q\}$ (3.93)

$c(u)$	#	Point Symmetries
Arbitrary	∞	Infinite number of point symmetries following from the linearity.
Arbitrary	2	$\mathrm{M}_1 = \frac{\partial}{\partial q}$, $\mathrm{M}_2 = t\frac{\partial}{\partial t} + q\frac{\partial}{\partial q}$.
u^ν	3	M_1, M_2, $\mathrm{M}_3 = (2\nu+1)t\frac{\partial}{\partial t} - u\frac{\partial}{\partial u} - (\nu+1)v\frac{\partial}{\partial v}$.
e^u	3	M_1, M_2, $\mathrm{M}_4 = 2t\frac{\partial}{\partial t} - \frac{\partial}{\partial u} - v\frac{\partial}{\partial v}$.
u^{-2}	5	M_1, M_2, M_3, $\mathrm{M}_5 = \frac{u^2}{u^2v^2-1}\left[t\frac{\partial}{\partial t} + u\frac{\partial}{\partial u} + v\frac{\partial}{\partial v}\right]$, $\mathrm{M}_6 = \frac{1}{u^2}\left[(4u^3q - 5tv^2u^2 - 3t)\frac{\partial}{\partial t} - (3u^2v^2+1)u\frac{\partial}{\partial u} + (u^2v^2+3)v\frac{\partial}{\partial v} + \frac{2}{u}(2tv^2 + (u^2v^2+1)uq)\frac{\partial}{\partial q}\right]$.
$u^{-2/3}$	5	M_1, M_2, M_3, M_7, M_8 [Exercise 4.2.4]

For $c(u) = u^{-2}$ or $c(u) = u^{-2/3}$, the system $\mathbf{TQ}\{u, v\,; t, q\}$ has five point symmetries; the symmetries $(\mathrm{M}_5, \mathrm{M}_6)$ and $(\mathrm{M}_7, \mathrm{M}_8)$, respectively, yield *nonlocal symmetries* of the nonlinear wave equation $\mathbf{U}\{x, t\,; u\}$ (4.8).

(6) The potential system $\mathbf{TR}\{u, v\,; t, r\}$
The potential system $\mathbf{TR}\{u, v\,; t, r\}$ (3.94), i.e.,

$$\begin{aligned} r_v &= v(u t_u - t), \\ r_u &= u c^2(u)(v t_v - t), \end{aligned}$$

has the group of equivalence transformations that includes the transformations

$$\begin{aligned}&\widetilde{u} = a_1 u, \quad \widetilde{v} = a_2 v, \quad \widetilde{r} = a_3 r + a_4 - \tfrac{1}{3} a_2^2 a_6 v^3, \\ &\widetilde{t} = a_2^{-2} a_3 t + a_5 uv + a_6 v^*, \quad \widetilde{c}(\widetilde{u}) = a_1^{-1} a_2 c(u),\end{aligned} \tag{4.25}$$

where $a_1, \ldots, a_6$ are arbitrary constants with $a_1 a_2 a_3 \neq 0$, and the projective transformation

$$\begin{aligned}&\widetilde{u} = \frac{u}{1+bu}, \quad \widetilde{v} = v, \quad \widetilde{t} = \frac{t}{1+bu}, \quad \widetilde{r} = r, \\ &\widetilde{c}(\widetilde{u}) = (1+b\widetilde{u})^{-2}\, c\Big(\frac{\widetilde{u}}{1+b\widetilde{u}}\Big).\end{aligned} \tag{4.26}$$

It follows that for the system $\mathbf{TR}\{u, v\,; t, r\}$, the wave speeds

$$c(u) = u^{\nu} \ \text{ and } \ c(u) = u^{\nu}(u+B)^{-2-\nu} \quad (\nu \neq -2), \tag{4.27}$$

are equivalent. In particular, the following wave speeds are equivalent.

1. $c(u) = u^{-1}$ and $c(u) = u^{-1}(Au+B)^{-1}$.
2. $c(u) = u^{-4/3}$ and $c(u) = u^{-4/3}(Au+B)^{-2/3}$.
3. $c(u) = 1$ and $c(u) = (Au+B)^{-2}$.

A, B are nonzero constants. The wave speed $c(u) = u^{-2}$ yields invariance under the equivalence transformation (4.27).

The point symmetry classification of the linear potential system $\mathbf{TR}\{u, v;$ $t, r\}$ (modulo its equivalence transformations (4.25), (4.26)) is given in Table 4.8.

Note that the system $\mathbf{TR}\{u, v\,; t, r\}$ has an additional point symmetry when $c(u)$ satisfies the ODE

$$\frac{ucc'' + c'(c - uc')}{(uc' + 2c)^2} = \gamma^2 = \text{const.} \tag{4.28}$$

The general solution of the ODE (4.28) for $\gamma \neq 0$ (modulo the equivalence transformations (4.25), (4.26)) consists of two families of solutions: (a) $c(u) = u^{\nu}$ ($\nu = \text{const}$) and (b) $c(u) = u^{-2} e^{1/u}$. For $c(u)$ satisfying the ODE (4.28) with $\gamma = 0$, i.e., $c(u) = u^{-4/3}$ (modulo the equivalence transformations (4.25), (4.26)), the system $\mathbf{TR}\{u, v\,; t, r\}$ has two additional point symmetries.

Comparing Tables 4.4 and 4.8, one observes that the point symmetries $\mathrm{N}_4, \ldots, \mathrm{N}_{10}$ of the potential system $\mathbf{TR}\{u, v\,; t, r\}$ yield nonlocal symmetries of the nonlinear wave equation $\mathbf{U}\{x, t\,; u\}$ (4.8). Of course, when $c(u) = 1$, $\mathbf{U}\{x, t\,; u\}$ is linear and $\mathbf{TR}\{u, v\,; t, r\}$ is a nonlinear system.

Note that at a first glance the symmetries N_4 and N_7 of the potential system $\mathbf{TR}\{u, v\,; t, r\}$ seem to project onto point symmetries of the nonlinear wave equation $\mathbf{U}\{x, t\,; u\}$. But since x is a nonlocal variable for the potential system $\mathbf{TR}\{u, v\,; t, r\}$, and the symmetry generators N_4 and N_7 do not con-

Table 4.8 Point symmetries of the potential system $\mathbf{TR}\{u, v\,; t, r\}$ (3.94)

$c(u)$	#	Point Symmetries
Arbitrary	∞	Infinite number of point symmetries following from the linearity.
Arbitrary	2	$\mathrm{N}_1 = \frac{\partial}{\partial r},\ \mathrm{N}_2 = t\frac{\partial}{\partial t} + r\frac{\partial}{\partial r}.$
$u^{\nu},\ \ \nu \neq -2$	3	$\mathrm{N}_1,\ \mathrm{N}_2,\ \mathrm{N}_3 = 2(\nu+1)t\frac{\partial}{\partial t} - u\frac{\partial}{\partial u} - (\nu+1)v\frac{\partial}{\partial v}.$
$u^{-2}e^{1/u}$	3	$\mathrm{N}_1,\ \mathrm{N}_2,\ \mathrm{N}_4 = (u+1)t\frac{\partial}{\partial t} + u^2\frac{\partial}{\partial u} - v\frac{\partial}{\partial v} - r\frac{\partial}{\partial r}.$
$u^{-4/3}$	4	$\mathrm{N}_1,\ \mathrm{N}_2,\ \mathrm{N}_5,\ \mathrm{N}_6$ [Exercise 4.2.4.]
u^{-2}	5	$\mathrm{N}_1,\ \mathrm{N}_2,\ \mathrm{N}_3\ (\nu=-2),\ \mathrm{N}_7 = tu\frac{\partial}{\partial t} + u^2\frac{\partial}{\partial u},$ $\mathrm{N}_8 = \frac{1}{u}\left[(tu^2v^2 + 2t - u^2 r)\frac{\partial}{\partial t} + 2v\frac{\partial}{\partial v} + (tv^2 + r)\frac{\partial}{\partial r}\right]$ $-(1+u^2v^2)\frac{\partial}{\partial u}.$
1	5	$\mathrm{N}_1,\ \mathrm{N}_2,\ \mathrm{N}_3\ (\nu=0),\ \mathrm{N}_9 = \frac{1}{u^2-v^2}\left(u\frac{\partial}{\partial u} - v\frac{\partial}{\partial v}\right),$ $\mathrm{N}_{10} = 2[t(u^2+v^2) + 2r]\frac{\partial}{\partial t} - u(u^2+3v^2)\frac{\partial}{\partial u} - v(3u^2+v^2)\frac{\partial}{\partial v}$ $+2[2tu^2v^2 - r(u^2+v^2)]\frac{\partial}{\partial r}.$

tain an explicit x-component, it turns out that the actual transformation of x is nonlocal under the actions of both N_4 and N_7.

(7) The nonlocally related subsystem $\mathbf{X}\{u, v\,; x\}$
The linear wave equation $\mathbf{X}\{u, v\,; x\}$ (3.86), i.e.,

$$x_{vv} = (c^{-2}(u)x_u)_u,$$

has the group of equivalence transformations

$$\begin{aligned} &\widetilde{u} = a_1 u + a_4, \quad \widetilde{v} = a_2 v + a_5, \quad \widetilde{x} = a_3 x + a_6 v + a_7, \\ &\widetilde{c}(\widetilde{u}) = a_1^{-1} a_2 c(u), \end{aligned} \tag{4.29}$$

where $a_1, \ldots, a_7$ are arbitrary constants with $a_1 a_2 a_3 \neq 0$.

For the wave speed $c(u) = u^{-2/3}$, the PDE $\mathbf{X}\{u, v\,; x\}$ has an infinite number of point symmetries [Exercise 4.2.5] (in addition to those due to its linearity), which suggests that it can be mapped by a point transformation into a constant coefficient linear PDE.

The PDE $\mathbf{X}\{u, v\,; x\}$ has two additional point symmetries when $c(u)$ satisfies the ODE

$$\begin{aligned} &\frac{(-2cc'' + 5(c')^2)c^2c'''' + 3c^3(c''')^2 + 16c^2(c'')^3 - 24c^2c'''c''c'}{c^3(2cc'' - 5(c')^2)^2} \\ &+ \frac{12c(c'c'')^2 - 10(c')^4c''}{c^3(2cc'' - 5(c')^2)^2} = \sigma^2 = \text{const.} \end{aligned} \tag{4.30}$$

The point symmetries of the PDE $\mathbf{X}\{u, v\,; x\}$ are summarized in Table 4.9.

Table 4.9 Point symmetries of the PDE $\mathbf{X}\{u, v\,; x\}$ (3.86)

$c(u)$	#	Point Symmetries
Arbitrary	∞	Infinite number of point symmetries following from the linearity.
Arbitrary	3	$J_1 = x\frac{\partial}{\partial x},\ J_2 = \frac{\partial}{\partial v},\ J_3 = \frac{\partial}{\partial x}$.
$u^{-2/3}$	∞	Exercise 4.2.5
(4.30) $(\sigma \neq 0)$	5	$J_1,\ J_2,\ J_3,\ J_{4,\,5} = e^{\pm\sigma v}\left\{\frac{1}{2}xFH\frac{\partial}{\partial x} + F(v)\frac{\partial}{\partial u} \pm\sigma^{-1}[F' + FH]\frac{\partial}{\partial v}\right\}$.
(4.30) $(\sigma = 0)$	5	$J_1,\ J_2,\ J_3,\ J_6 = v\left\{\frac{1}{2}xFH\frac{\partial}{\partial x} + F\frac{\partial}{\partial u}\right\} + \left\{K\frac{v^2}{2} + \int c^2 F du\right\}\frac{\partial}{\partial v}$, $J_7 = \frac{1}{2}xFH\frac{\partial}{\partial x} + F\frac{\partial}{\partial u} + Kv\frac{\partial}{\partial v}$.
Particular case (a) for $\sigma = 0$: $c(u) = u^{\nu}$ $(\nu = \text{const})$	5	$J_6^{(a)} = \nu(\nu+1)xv\frac{\partial}{\partial x} + 2(\nu+1)uv\frac{\partial}{\partial u} + [u^{2\nu+2} + v^2(\nu+1)^2]\frac{\partial}{\partial v}$, $J_7^{(a)} = u\frac{\partial}{\partial u} + (\nu+1)v\frac{\partial}{\partial v}$.
Particular case (b) for $\sigma = 0$: $c(u) = e^{u}$	5	$J_6^{(b)} = xv\frac{\partial}{\partial x} + 2v\frac{\partial}{\partial u} + [e^{2u} + v^2]\frac{\partial}{\partial v}$, $J_7^{(b)} = \frac{\partial}{\partial u} + v\frac{\partial}{\partial v}$.

In Table 4.9, $F(u) = (3H^2(u) - 2H'(u))^{-1/2}$, $H(u) = c'(u)/c(u)$. From the symmetry commutator relations, one can show that

$$(F' + HF)^2 - (\sigma c(u)F)^2 = K^2 = \text{const},$$

and hence for $\sigma = 0$, $F' + HF = K = \text{const}$ [Bluman & Kumei (1987)].

Comparing Tables 4.4 and 4.9, one observes that the symmetries J_4, J_5 and J_6 yield *nonlocal symmetries* of $\mathbf{U}\{x, t\,; u\}$ (4.8); the symmetry J_7 yields a nonlocal symmetry of $\mathbf{U}\{x, t\,; u\}$ except for the two listed particular cases. For the case $c(u) = u^{-2/3}$, the PDE $\mathbf{X}\{u, v\,; x\}$ has an infinite number of point symmetries that are nonlocal symmetries of the nonlinear wave equation $\mathbf{U}\{x, t\,; u\}$.

(8) The nonlocally related subsystem $\mathbf{T}\{u, v\,; t\}$
The linear wave equation $\mathbf{T}\{u, v\,; t\}$ (3.87), i.e.,

$$t_{uu} = c^2(u)t_{vv},$$

has the group of equivalence transformations that includes the transformations

$$\begin{aligned}&\widetilde{u} = a_1 u + a_4, \quad \widetilde{v} = a_2 v + a_5,\\ &\widetilde{t} = a_3 t + a_6 + a_7 u + a_8 v + a_9 uv, \quad \widetilde{c}(\widetilde{u}) = a_1^{-1} a_2 c(u),\end{aligned} \tag{4.31}$$

and the projective transformation

$$\widetilde{u} = \frac{u}{1+bu}, \quad \widetilde{v} = v, \quad \widetilde{t} = \frac{t}{1+bu}, \quad \widetilde{c}(\widetilde{u}) = (1+b\widetilde{u})^{-2}\, c\Big(\frac{\widetilde{u}}{1+b\widetilde{u}}\Big), \tag{4.32}$$

where $a_1, \ldots, a_9$ and b are arbitrary constants with $a_1 a_2 a_3 \neq 0$.

The point symmetry classification of the PDE $\mathbf{T}\{u, v\,; t\}$, modulo the equivalence transformations (4.31), (4.32), is given in Table 4.10.

Comparing with the point symmetry classification of the nonlinear wave equation $\mathbf{U}\{x, t\,; u\}$ [Table 4.4], one observes that the symmetries K_5, K_7, K_8, K_9, K_{10}, K_{11} and K_{12} yield *nonlocal symmetries* of the nonlinear wave equation $\mathbf{U}\{x, t\,; u\}$ (4.8).

Table 4.10 Point symmetries of the PDE $\mathbf{T}\{u,v\,;t\}$ (3.87) nonlocally related to the nonlinear wave equation $\mathbf{U}\{x,t\,;u\}$ (3.76)

$c(u)$	#	Symmetries
Arbitrary	∞	Infinite number of point symmetries following from the linearity.
Arbitrary	3	$\mathrm{K}_1 = t\frac{\partial}{\partial t}$, $\mathrm{K}_2 = \frac{\partial}{\partial v}$, $\mathrm{K}_3 = \frac{\partial}{\partial t}$.
u^ν, $\nu \neq 0,-1,-2$	5	K_1, K_2, K_3, $\mathrm{K}_4 = u\frac{\partial}{\partial u} + (1+C)v\frac{\partial}{\partial v}$, $\mathrm{K}_5 = -\frac{1}{2}Ctv\frac{\partial}{\partial t} + uv\frac{\partial}{\partial u} + \left[\frac{u^{2+2C}}{1+C} + \frac{1}{2}(1+C)v^2\right]\frac{\partial}{\partial v}$.
e^u	5	K_1, K_2, K_3, $\mathrm{K}_6 = \frac{\partial}{\partial u} + v\frac{\partial}{\partial v}$, $\mathrm{K}_7 = -\frac{1}{2}tv\frac{\partial}{\partial t} + v\frac{\partial}{\partial u} + \frac{1}{2}\left[e^{2u} + v^2\right]\frac{\partial}{\partial v}$.
u^{-1}	5	K_1, K_2, K_3, K_4 $(C=-1)$, $\mathrm{K}_8 = \frac{1}{2}tv\frac{\partial}{\partial t} + uv\frac{\partial}{\partial u} + (\log u)\frac{\partial}{\partial v}$.
u^{-2}	∞	Infinite number of nonlocal symmetries; there exists a point transformation into a linear PDE with constant coefficients [Exercise 4.2.6].
$\left[(Bu^2+C) \times \exp\{A\int (Bu^2+C)^{-1}du\}\right]^{-1}$ $(A,B,C = \text{const})$	5	K_1, K_2, K_3, $\mathrm{K}_9 = \frac{1}{2}t(A+2Bu)\frac{\partial}{\partial t} + (Bu^2+C)\frac{\partial}{\partial u} - Av\frac{\partial}{\partial v}$, $\mathrm{K}_{10} = \frac{1}{2}t(A+2Bu)v\frac{\partial}{\partial t} + (Bu^2+C)v\frac{\partial}{\partial u} + \left[-\frac{1}{2}Av^2 + \int c^2(u)(Bu^2+C)du\right]\frac{\partial}{\partial v}$.
$c(u)$ satisfies $(\alpha' + H\alpha)' = \sigma^2 c^2(u)\alpha$, where $\sigma = \text{const} \neq 0$, $H(u) = c'(u)/c(u)$, $\alpha^2(u) = (H^2(u) - 2H'(u))^{-1}$	5	K_1, K_2, K_3, $\mathrm{K}_{11,12} = e^{\pm\sigma v}\left[-\frac{1}{2}t\alpha H\frac{\partial}{\partial t} + \alpha\frac{\partial}{\partial u} \pm\sigma^{-1}(\alpha' + H\alpha)\frac{\partial}{\partial v}\right]$.

4.2.3 Classification of nonlocal symmetries of nonlinear telegraph equations arising from point symmetries of potential systems

Consider the nonlinear telegraph (NLT) equation $\mathbf{U}\{x,t\,;u\}$ given by

$$u_{tt} - (F(u)u_x)_x - (G(u))_x = 0. \tag{4.33}$$

The complete point symmetry classification of the PDE (3.60) with respect to the constitutive functions $F(u)$ and $G(u)$ [Kingston & Sophocleous (2001)], modulo the equivalence transformations (3.61), is presented in Table 4.11.

Table 4.11 Point symmetries of the nonlinear telegraph equation $\mathbf{U}\{x,t\,;u\}$ (3.60)

$F(u)$	$G(u)$	#	Point Symmetries
Arbitrary	Arbitrary	2	$X_1 = \frac{\partial}{\partial t}$, $X_2 = \frac{\partial}{\partial x}$.
$e^{(\alpha+1)u}$	e^u	3	$X_1, X_2, X_3 = (\alpha-1)t\frac{\partial}{\partial t} + 2\alpha x\frac{\partial}{\partial x} + 2u\frac{\partial}{\partial u}$.
u^α	$u^{\alpha+\beta+1}$	3	$X_1, X_2, X_4 = (\alpha+2\beta)t\frac{\partial}{\partial t} + 2\beta x\frac{\partial}{\partial x} - 2u\frac{\partial}{\partial u}$.
u^{-2}	u^{-1}	4	$X_1, X_2, X_4, X_5 = e^x\frac{\partial}{\partial x} - ue^x\frac{\partial}{\partial u}$.
u^α	$\ln u$	3	$X_1, X_2, X_6 = (\alpha+2)t\frac{\partial}{\partial t} + 2(\alpha+1)x\frac{\partial}{\partial x} + 2u\frac{\partial}{\partial u}$.
$e^{\alpha u}$	u	3	$X_1, X_2, X_7 = \alpha t\frac{\partial}{\partial t} + 2\alpha x\frac{\partial}{\partial x} + 2\frac{\partial}{\partial u}$.
u^{-4}	u^{-3}	4	$X_1, X_2, X_4, X_8 = t^2\frac{\partial}{\partial t} + ut\frac{\partial}{\partial u}$.

The complete point symmetry classification of the nonlocally related potential system $\mathbf{UV^1}\{x,t\,;u,v^1\}$ (3.62), i.e.,

$$\begin{aligned} v^1_x &= u_t, \\ v^1_t &= F(u)u_x + G(u), \end{aligned}$$

yielding nonlocal symmetries of the NLT equation (4.33), is presented in Table 4.12 for $G'(u) \neq 0$ [Bluman, Temuerchaolu & Sahadevan (2005)]. Part of this classification appears in Reid (1991b).

Observe that the point symmetries of the potential system $\mathbf{UV^1}\{x,t\,;u,v^1\}$ yield one nonlocal symmetry of the NLT equation $\mathbf{U}\{x,t\,;u\}$ for eight classes of constitutive functions. In the cases $F(u) = u^{-2}, G(u) = u^{-1}$, and $F(u)$ arbitrary, $G(u) = \text{const}$, the potential system $\mathbf{UV^1}\{x,t\,;u,v^1\}$ (3.62) is linearizable by a point transformation, and thus the corresponding NLT equation $\mathbf{U}\{x,t\,;u\}$ (4.33) is linearizable by a nonlocal transformation.

4.2.4 Nonlocal symmetries of nonlinear telegraph equations with power law nonlinearities

In this section, local conservation laws of the nonlinear telegraph equation $\mathbf{U}\{x,t\,;u\}$ (4.33) [Section 3.4.3] are used to construct extended trees of nonlocally related PDE systems for the three cases that arise. For the special situation of power law nonlinearities, $F(u) = u^\alpha$, $G(u) = u^\beta$, nonlocal sym-

Table 4.12 Point symmetries of the potential system $\mathbf{UV^1}\{x,t\,;u,v^1\}$ (3.62) that yield nonlocal symmetries of the NLT equation $\mathbf{U}\{x,t\,;u\}$ (4.33)

$F(u)$	$G(u)$	#	Point Symmetries Yielding Nonlocal Symmetries
Arbitrary	const	∞	Infinite number of symmetries; there exists a point mapping of the potential system $\mathbf{UV^1}\{x,t\,;u,v^1\}$ (3.62) to a linear system [Exercise 3.4.6].
u^{-2}	u^{-1}		
$\pm\frac{4u^{2\alpha+\beta-1}}{(u^{2\alpha}\pm 1)^2}$	$\frac{(u^{2\alpha}\mp 1)}{(u^{2\alpha}\pm 1)}$	1	$\mathrm{Y}_1=[(\beta+1)t+2\alpha v]\frac{\partial}{\partial t}+2[\beta x+\alpha\int F(u)du]\frac{\partial}{\partial x}$ $+2u\frac{\partial}{\partial u}+[2\alpha t+(\beta+1)v]\frac{\partial}{\partial v}$.
$u^{\beta-1}\sec^2(\alpha\ln u)$	$\tan(\alpha\ln u)$	1	$\mathrm{Y}_2=[(\beta+1)t-2\alpha v]\frac{\partial}{\partial t}+2[\beta x-\alpha\int F(u)du]\frac{\partial}{\partial x}$ $+2u\frac{\partial}{\partial u}+[2\alpha t+(\beta+1)v]\frac{\partial}{\partial v}$.
$-u^{\beta-1}(\ln u)^{-2}$	$(\ln u)^{-1}$	1	$\mathrm{Y}_3=[(\beta+1)t+2v]\frac{\partial}{\partial t}+2[\beta x+\int F(u)du]\frac{\partial}{\partial x}$ $+2u\frac{\partial}{\partial u}+(\beta+1)v\frac{\partial}{\partial v}$.
$e^{2\beta u}\sec^2 u$	$\tan u$	1	$\mathrm{Y}_4=(\beta t-v)\frac{\partial}{\partial t}+2[\beta x-\int F(u)du]\frac{\partial}{\partial x}+\frac{\partial}{\partial u}$ $+(t+\beta v)\frac{\partial}{\partial v}$.
$e^{2\beta u}\mathrm{sech}^2 u$	$\tanh u$	1	$\mathrm{Y}_5=(\beta t+v)\frac{\partial}{\partial t}+2[\beta x+\int F(u)du]\frac{\partial}{\partial x}+\frac{\partial}{\partial u}$ $+(t+\beta v)\frac{\partial}{\partial v}$.
$-e^{2\beta u}\mathrm{csch}^2 u$	$\coth u$		
$-u^{-2}e^{2\beta u}$	u^{-1}	1	$\mathrm{Y}_6=(\beta t+v)\frac{\partial}{\partial t}+2[\beta x+\int F(u)du]\frac{\partial}{\partial x}$ $+\frac{\partial}{\partial u}+\beta v\frac{\partial}{\partial v}$.

metries are classified that arise as point symmetries of nonlocally related PDE systems within these extended trees.

(1) Trees of nonlocally related systems for the NLT equation
The extended tree construction procedure [Section 3.5] is applied to the NLT equation $\mathbf{U}\{x,t\,;u\}$, through use of the local conservation laws obtained in Section 3.4.3. Note that the exclusion of dependent variables leads only to locally related subsystems. [Here there is no consideration of nonlocally related subsystems arising from interchanges of independent and dependent variables.] Three cases arise.

Case (a): Arbitrary $F(u),G(u)$. Here, the NLT equation (4.33) has two local conservation laws. The corresponding extended tree $\mathcal{T}_a$ consists of $2^2=4$ PDE systems.

- The NLT equation $\mathbf{U}\{x,t\,;u\}$ (4.33).
- Two singlet potential systems $\mathbf{UV^1}\{x,t\,;u,v^1\}$ (3.62) and $\mathbf{UV^2}\{x,t\,;u,v^2\}$ (3.63).
- One couplet $\mathbf{UV^1V^2}\{x,t\,;u,v^1,v^2\}$ [(3.62), (3.63)].

Case (b): $G'(u) = F(u)$, $F(u)$ arbitrary. Here, the NLT equation (4.33) has four local conservation laws. The corresponding extended tree $\mathcal{T}_b$ consists of 16 PDE systems.

- The NLT equation $\mathbf{U}\{x,t\,;u\}$ (4.33).
- Four singlet potential systems $\mathbf{UV^1}\{x,t\,;u,v^1\}$ (3.62), $\mathbf{UV^2}\{x,t\,;u,v^2\}$ (3.63), $\mathbf{UB^3}\{x,t\,;u,b^3\}$ (3.64) and $\mathbf{UB^4}\{x,t\,;u,b^4\}$ (3.65).
- Six couplets $\mathbf{UV^1V^2}\{x,t\,;u,v^1,v^2\}$ [(3.62), (3.63)], $\mathbf{UV^1B^3}\{x,t\,;u,v^1,b^3\}$ [(3.62), (3.64)], $\mathbf{UV^1B^4}\{x,t;u,v^1,b^4\}$ [(3.62), (3.65)], $\mathbf{UV^2B^3}\{x,t;u,v^2,b^3\}$ [(3.63), (3.64)], $\mathbf{UV^2B^4}\{x,t\,;u,v^2,b^4\}$ [(3.63), (3.65)] and $\mathbf{UB^3B^4}\{x,t\,;u,b^3,b^4\}$ [(3.64), (3.65)].
- Four triplets $\mathbf{UV^1V^2B^3}\{x,t\,;u,v^1,v^2,b^3\}$, $\mathbf{UV^1V^2B^4}\{x,t\,;u,v^1,v^2,b^4\}$, $\mathbf{UV^1B^3B^4}\{x,t\,;u,v^1,b^3,b^4\}$ and $\mathbf{UV^2B^3B^4}\{x,t\,;u,v^2,b^3,b^4\}$, given by the unions (3.62)–(3.64), [(3.62), (3.63), (3.65)], [(3.62), (3.64), (3.65)] and (3.63)–(3.65), respectively.
- One quadruplet $\mathbf{UV^1V^2B^3B^4}\{x,t\,;u,v^1,v^2,b^3,b^4\}$ (3.62)–(3.65), involving all four potentials.

Case (c): $G(u) = u$, $F(u)$ arbitrary. Here the NLT equation (4.33) again has four local conservation laws. The corresponding extended tree $\mathcal{T}_c$ of nonlocally related PDE systems consists of 16 PDE systems.

- The NLT equation $\mathbf{U}\{x,t\,;u\}$ (4.33).
- Four singlet potential systems $\mathbf{UV^1}\{x,t\,;u,v^1\}$ (3.62), $\mathbf{UV^2}\{x,t\,;u,v^2\}$ (3.63), $\mathbf{UC^3}\{x,t\,;u,c^3\}$ (3.66) and $\mathbf{UC^4}\{x,t\,;u,c^4\}$ (3.67).
- Six couplets $\mathbf{UV^1V^2}\{x,t\,;u,v^1,v^2\}$ [(3.62), (3.63)], $\mathbf{UV^1C^3}\{x,t\,;u,v^1,c^3\}$ [(3.62), (3.66)], $\mathbf{UV^1C^4}\{x,t\,;u,v^1,c^4\}$ [(3.62), (3.67)], $\mathbf{UV^2C^3}\{x,t;u,v^2,c^3\}$[(3.63), (3.66)], $\mathbf{UV^2C^4}\{x,t\,;u,v^2,c^4\}$ [(3.63), (3.67)] and $\mathbf{UC^3C^4}\{x,t;u,c^3,c^4\}$ [(3.66), (3.67)].
- Four triplets $\mathbf{UV^1V^2C^3}\{x,t\,;u,v^1,v^2,c^3\}$, $\mathbf{UV^1V^2C^4}\{x,t\,;u,v^1,v^2,c^4\}$, $\mathbf{UV^1C^3C^4}\{x,t\,;u,v^1,c^3,c^4\}$, and $\mathbf{UV^2C^3C^4}\{x,t\,;u,v^2,c^3,c^4\}$, given by the unions [(3.62), (3.63), (3.66)], [(3.62), (3.63), (3.67)], [(3.62), (3.66), (3.67)] and [(3.63), (3.66), (3.67)], respectively.
- One quadruplet $\mathbf{UV^1V^2C^3C^4}\{x,t\,;u,v^1,v^2,c^3,c^4\}$ [(3.62), (3.63), (3.66), (3.67)], involving all four potential variables.

(2) Symmetries of the NLT equation and nonlocally related systems for power law nonlinearities

Case (a): $F(u) = u^\alpha, G(u) = u^\beta$; $\alpha,\beta \neq 0$. The classification of the point symmetries of the four PDE systems within the tree $\mathcal{T}_a$ is presented in Table 4.13.

From the form of the point symmetries listed in Table 4.13, it follows that no nonlocal symmetries are obtained for the systems $\mathbf{U}\{x,t\,;u\}$ (4.33) and

Table 4.13 Point symmetries of the NLT equation (4.33) and nonlocally related systems in the general power law nonlinearity case (a): $F(u) = u^\alpha$, $G(u) = u^\beta$ ($\alpha, \beta \neq 0$)

System	#	Point Symmetries
$\mathbf{UV^1V^2}$ $\mathbf{UV^1}, \mathbf{UV^2}$, $\mathbf{U}$	5	$X_1 = (\alpha - \beta + 1)x\frac{\partial}{\partial x} + (\frac{\alpha}{2} - \beta + 1)t\frac{\partial}{\partial t} + u\frac{\partial}{\partial u}$ $+\frac{\alpha+2}{2}v^1\frac{\partial}{\partial v^1} + (\alpha - \beta + 2)v^2\frac{\partial}{\partial v^2}$, $X_2 = \frac{\partial}{\partial x}$, $X_3 = \frac{\partial}{\partial t} + v^1\frac{\partial}{\partial v^2}$, $X_4 = \frac{\partial}{\partial v^1}$, $X_5 = \frac{\partial}{\partial v^2}$.

$\mathbf{UV^1}\{x, t\,; u, v^1\}$ (3.62). The infinitesimal generator X_3 yields a nonlocal symmetry of the system $\mathbf{UV^2}\{x, t\,; u, v^2\}$ (3.63) (i.e., the system $\mathbf{UV^2}\{x, t\,; u, v^2\}$ is not invariant under translations in t) and a point symmetry of the other systems. All other infinitesimal generators define point symmetries of all systems in Table 4.13.

Case (b): $G'(u) = F(u)$, i.e., $F(u) = (\alpha+1)u^\alpha$, $G(u) = u^{\alpha+1}$, $\alpha \neq 0, -1, -2$. From the equivalence relation (3.61), this case is equivalent to the situation when $F(u) = u^\alpha$, $G(u) = u^{\alpha+1}$. The point symmetry classifications of the 16 PDE systems within the tree $\mathcal{T}_b$ are presented in Table 4.14.

Table 4.14 Point symmetries of the potential NLT systems for case (b): $F(u) = (\alpha + 1)u^\alpha$, $G(u) = u^{\alpha+1}$ ($\alpha \neq 0, -1, -2$)

System	$F(u)$	$G(u)$	#	Point Symmetries
$\mathbf{UV^1V^2B^3B^4}$, $\mathbf{UV^1V^2B^3}$, $\mathbf{UV^1V^2B^4}$, $\mathbf{UV^1B^3B^4}$, $\mathbf{UV^2B^3B^4}$,	$(\alpha+1)u^\alpha$	$u^{\alpha+1}$	7	$Y_1 = -\frac{\alpha}{2}t\frac{\partial}{\partial t} + u\frac{\partial}{\partial u} + v^2\frac{\partial}{\partial v^2} + \frac{\alpha+2}{2}v^1\frac{\partial}{\partial v^1}$ $+\frac{\alpha+2}{2}b^3\frac{\partial}{\partial b^3} + b^4\frac{\partial}{\partial b^4}$, $Y_2 = \frac{\partial}{\partial x} + b^3\frac{\partial}{\partial b^3} + b^4\frac{\partial}{\partial b^4}$, $Y_3 = \frac{\partial}{\partial t} + b^3\frac{\partial}{\partial b^4} + v^1\frac{\partial}{\partial v^2}$, $Y_4 = \frac{\partial}{\partial v^1}$, $Y_5 = \frac{\partial}{\partial v^2}$, $Y_6 = \frac{\partial}{\partial b^3}$, $Y_7 = \frac{\partial}{\partial b^4}$.
$\mathbf{UV^1V^2}, \mathbf{UV^1B^3}$, $\mathbf{UV^1B^4}, \mathbf{UV^2B^3}$, $\mathbf{UV^2B^4}, \mathbf{UB^3B^4}$, $\mathbf{UV^1}, \mathbf{UV^2}$, $\mathbf{UB^3}, \mathbf{UB^4}$, $\mathbf{U}$	$-3u^{-4}$	u^{-3}	8	Y_1, Y_2, Y_3, Y_4, Y_5, Y_6, Y_7, $Y_8 = t^2\frac{\partial}{\partial t} + tu\frac{\partial}{\partial u} - v^2\frac{\partial}{\partial v^1} - b^4\frac{\partial}{\partial b^3}$.
$\mathbf{UV^1V^2}$	$3u^2$	u^3	8	Y_1, Y_2, Y_3, Y_4, Y_5, Y_6, Y_7, $Y_9 = 3v^1\frac{\partial}{\partial x} + (tv^1 - v^2 + 3u)\frac{\partial}{\partial t} - uv^1\frac{\partial}{\partial u}$ $-(v^1)^2\frac{\partial}{\partial v^1} - v^1v^2\frac{\partial}{\partial v^2}$.

The case $\alpha = -2$ is not included in Table 4.14 since here the system $\mathbf{UV^1}\{x, t\,; u, v^1\}$ is linearizable by a point transformation [Bluman & Kumei (1989)] [Section 4.2.3].

From Table 4.14, it follows that for the case when $F(u) = 3u^2, G(u) = u^3$, the potential system $\mathbf{UV^1V^2}\{x, t\,; u, v^1, v^2\}$ [(3.62), (3.63)] has the point symmetry Y_9 which yields a nonlocal symmetry of the NLT equation $\mathbf{U}\{x, t;$ $u\}$ (4.33). Moreover, this is the only case yielding a nonlocal symmetry of the NLT equation $\mathbf{U}\{x, t\,; u\}$.

Note that the infinitesimal generator Y_3 yields a nonlocal symmetry of each of the systems $\mathbf{UV^1V^2B^4}\{x, t\,; u, v^1, v^2, b^4\}$ [(3.62), (3.63), (3.65)], $\mathbf{UV^2B^3B^4}\{x, t\,; u, v^2, b^3, b^4\}$ (3.63)–(3.65), $\mathbf{UV^1B^4}\{x, t\,; u, v^1, b^4\}$ [(3.62), (3.65)], $\mathbf{UV^2B^3}\{x, t\,; u, v^2, b^3\}$ [(3.63), (3.64)], $\mathbf{UV^2B^4}\{x, t\,; u, v^2, b^4\}$ [(3.63), (3.65)], $\mathbf{UV^2}\{x, t\,; u, v^2\}$ (3.63) and $\mathbf{UB^4}\{x, t\,; u, b^4\}$ (3.65), and a point symmetry of the other nine systems; the infinitesimal generator Y_8 yields a nonlocal symmetry of the systems $\mathbf{UV^1V^2B^3}\{x, t\,; u, v^1, v^2, b^3\}$ (3.62) – (3.64), $\mathbf{UV^1B^3B^4}\{x, t\,; u, v^1, b^3, b^4\}$ [(3.62), (3.64), (3.65)], $\mathbf{UV^1B^3}\{x, t\,; u, v^1, b^3\}$ [(3.62), (3.64)], $\mathbf{UV^1B^4}\{x, t\,; u, v^1, b^4\}$ [(3.62), (3.65)], $\mathbf{UV^2B^3}\{x, t\,; u, v^2, b^3\}$ [(3.63), (3.64)], $\mathbf{UV^1}\{x, t\,; u, v^1\}$ (3.62) and $\mathbf{UB^3}\{x, t\,; u, b^3\}$ (3.64), and a point symmetry of the other nine systems; the infinitesimal generator Y_9 yields a point symmetry of the system $\mathbf{UV^1V^2}\{x, t\,; u, v^1, v^2\}$ [(3.62), (3.63)] and a nonlocal symmetry of the other 15 listed nonlocally related systems.

Case (c): $F(u) = u^\alpha$, $G(u) = u$ $(\alpha \neq 0)$. The corresponding classification of the point symmetries is found in Table 4.15. The linear case $\alpha = 0$ is not considered. The entries in Table 4.15 for the triplets $\mathbf{UV^1C^3C^4}\{x, t\,; u, v^1, c^3, c^4\}$ [(3.62), (3.66), (3.67)], $\mathbf{UV^2C^3C^4}\{x, t\,; u, v^2, c^3, c^4\}$ [(3.63), (3.66), (3.67)], and the couplets $\mathbf{V^1C^4}\{x, t\,; u, v^1, c^4\}$ [(3.62), (3.67)], $\mathbf{UC^3C^4}\{x, t\,; u, c^3, c^4\}$ [(3.66), (3.67)] are missing since they are not known.

From the form of the known point symmetries listed in Table 4.15, it follows that no nonlocal symmetries arise for the systems $\mathbf{U}$ (4.33) and $\mathbf{UV^1}$ (3.62); the infinitesimal generator Z_2 yields a nonlocal symmetry of the systems $\mathbf{UV^2C^3}$ [(3.63), (3.66)], $\mathbf{UC^3}$ (3.66) and $\mathbf{UC^4}$ (3.67), and a point symmetry of the other listed systems; the infinitesimal generator Z_3 yields a nonlocal symmetry of the systems $\mathbf{UV^1V^2C^4}$ [(3.62), (3.63), (3.67)], $\mathbf{UV^1C^3}$ [(3.62), (3.66)], $\mathbf{UV^2C^3}$ [(3.63), (3.66)], $\mathbf{UV^2C^4}$ [(3.63), (3.67)], $\mathbf{UV^2}$ (3.63), $\mathbf{UC^3}$ (3.66) and $\mathbf{UC^4}$ (3.67), and a point symmetry of the other listed systems. All other infinitesimal generators yield point symmetries of each of the systems listed in Table 4.15.

Table 4.15 Point symmetries of the potential NLT systems for case (c): $F(u) = u^{\alpha}$, $G(u) = u$ $(\alpha \neq 0)$

System	Case	#	Point Symmetries
$\mathbf{UV^1V^2C^3C^4}$ $\mathbf{UV^1V^2C^3}$ $\mathbf{UV^1V^2C^4}$ $\mathbf{UV^1V^2}$, $\mathbf{UV^1C^3}$, $\mathbf{UV^2C^3}$, $\mathbf{UV^2C^4}$, $\mathbf{UV^1}, \mathbf{UV^2}$, $\mathbf{UC^3}, \mathbf{UC^4}$, $\mathbf{U}$	$\alpha \neq -1$	7	$\mathrm{Z}_1 = \frac{\alpha}{2} t \frac{\partial}{\partial t} + \alpha x \frac{\partial}{\partial x} + u \frac{\partial}{\partial u} + \frac{\alpha+2}{2} v_1 \frac{\partial}{\partial v_1}$ $+ v_2(a+1) \frac{\partial}{\partial v_2} + \frac{3\alpha+2}{2} c^3 \frac{\partial}{\partial c^3} + (2\alpha+1) c^4 \frac{\partial}{\partial c^4}$, $\mathrm{Z}_2 = \frac{\partial}{\partial x} + v_1 \frac{\partial}{\partial c^3} + v_2 \frac{\partial}{\partial c^4}$, $\mathrm{Z}_3 = \frac{\partial}{\partial t} + v_1 \frac{\partial}{\partial v_2} - v_2 \frac{\partial}{\partial c^3} + c^3 \frac{\partial}{\partial c^4}$, $\mathrm{Z}_4 = \frac{\partial}{\partial v_1}$, $\mathrm{Z}_5 = \frac{\partial}{\partial v_2}$, $\mathrm{Z}_6 = \frac{\partial}{\partial c^3}$, $\mathrm{Z}_7 = \frac{\partial}{\partial c^4}$.
	$\alpha = -1$	8	Z_2, Z_3, Z_4, Z_5, Z_6, Z_7, $\mathrm{Z}_8 = -\frac{1}{2} t \frac{\partial}{\partial t} - x \frac{\partial}{\partial x} + u \frac{\partial}{\partial u} + \frac{1}{2} v_1 \frac{\partial}{\partial v_1}$ $- \left(t + \frac{1}{2} c^3\right) \frac{\partial}{\partial c^3} - \left(\frac{t^2}{2} + c^4\right) \frac{\partial}{\partial c^4}$.
$\mathbf{UV^1C^3C^4}$, $\mathbf{UV^2C^3C^4}$ $\mathbf{UV^1C^4}, \mathbf{UC^3C^4}$			?

4.2.5 Nonlocal symmetries of the planar gas dynamics equations

In Section 3.5.4, an extended tree $\mathcal{T}_b$ of nonlocally related PDE systems was constructed for the planar gas dynamics equations. One should do a point symmetry classification for each PDE system in the tree $\mathcal{T}_b$ with respect to the constitutive function $B(p, q)$. In this section, it is shown that in many cases a point symmetry of one system in the tree yields a nonlocal symmetry of one or more other systems.

(1) A comparison of point symmetries of three nonlocally related PGD systems

In Table 4.16, for several representative classes of constitutive functions $B(p, q)$, there is a comparison of the point symmetries of three nonlocally related PGD systems: the Euler system $\mathbf{E}\{x, t\,; v, p, \rho\}$ (3.39), the Lagrange system $\mathbf{L}\{y, s\,; v, p, q\}$ (3.42), and the potential system $\mathbf{EA^1}\{x, t\,; v, p, \rho, \alpha^1\}$ (3.40) of the Euler system. [For a full classification, see Akhatov, Gazizov & Ibragimov (1991).]

Observe that the symmetry X_7 is local for the systems $\mathbf{E}\{x, t\,; v, p, \rho\}$ and $\mathbf{EA^1}\{x, t\,; v, p, \rho, \alpha^1\}$ but yields a nonlocal symmetry of the system $\mathbf{L}\{y, s\,; v, p, q\}$; the symmetries $\mathrm{Z}_7, \mathrm{Z}_8$ and Z_{Θ} are local for $\mathbf{L}\{y, s\,; v, p, q\}$ but yield nonlocal symmetries of the systems $\mathbf{E}\{x, t\,; v, p, \rho\}$ and $\mathbf{EA^1}\{x, t; v, p, \rho, \alpha^1\}$; the symmetries $\mathrm{Y}_8, \mathrm{Y}_{10}, \mathrm{Y}_{11}$ and Y_{12} are local for the systems $\mathbf{EA^1}\{x, t; v, p, \rho, \alpha^1\}$ and $\mathbf{L}\{y, s\,; v, p, q\}$ but yield nonlocal symmetries of the sys-

Table 4.16 A comparison of point symmetries of the PGD systems $\mathbf{E}\{x,t\,;v,p,\rho\}$, $\mathbf{EA}^1\{x,t\,;v,p,\rho,\alpha^1\}$ and $\mathbf{L}\{y,s\,;v,p,q\}$

$B(p,q)$	Point Symmetries		
	$\mathbf{E}\{x,t\,;v,p,\rho\}$	$\mathbf{EA}^1\{x,t\,;v,p,\rho,\alpha^1\}$	$\mathbf{L}\{y,s\,;v,p,q\}$
Arbitrary	$\mathrm{X}_1=\frac{\partial}{\partial x}$, $\mathrm{X}_2=\frac{\partial}{\partial t}$, $\mathrm{X}_3=t\frac{\partial}{\partial t}+x\frac{\partial}{\partial x}$, $\mathrm{X}_4=t\frac{\partial}{\partial x}+\frac{\partial}{\partial v}$.	$\mathrm{Y}_1=\frac{\partial}{\partial x}$, $\mathrm{Y}_2=\frac{\partial}{\partial t}$, $\mathrm{Y}_3=\mathrm{X}_3+\alpha^1\frac{\partial}{\partial \alpha^1}$, $\mathrm{Y}_4=\mathrm{X}_4$, $\mathrm{Y}_5=\frac{\partial}{\partial \alpha^1}$.	$\mathrm{Z}_1=\frac{\partial}{\partial s}$, $\mathrm{Z}_2=s\frac{\partial}{\partial s}+y\frac{\partial}{\partial y}$, $\mathrm{Z}_3=\frac{\partial}{\partial v}$, $\mathrm{Z}_4=\frac{\partial}{\partial y}$.
$3p/q$	$\mathrm{X}_1,\mathrm{X}_2,\mathrm{X}_3,\mathrm{X}_4$, $\mathrm{X}_5=x\frac{\partial}{\partial x}+v\frac{\partial}{\partial v}$ $-2\rho\frac{\partial}{\partial \rho}$, $\mathrm{X}_6=p\frac{\partial}{\partial p}+\rho\frac{\partial}{\partial \rho}$, $\mathrm{X}_7=t^2\frac{\partial}{\partial t}+tx\frac{\partial}{\partial x}$ $+(x-tv)\frac{\partial}{\partial v}$ $-3tp\frac{\partial}{\partial p}-t\rho\frac{\partial}{\partial \rho}$.	$\mathrm{Y}_1,\mathrm{Y}_2,\mathrm{Y}_3,\mathrm{Y}_4,\mathrm{Y}_5$, $\mathrm{Y}_6=\mathrm{X}_5-\alpha^1\frac{\partial}{\partial \alpha^1}$, $\mathrm{Y}_7=\mathrm{X}_6+\alpha^1\frac{\partial}{\partial \alpha^1}$, $\mathrm{Y}_8=\mathrm{X}_7$.	$\mathrm{Z}_1,\mathrm{Z}_2,\mathrm{Z}_3,\mathrm{Z}_4$, $\mathrm{Z}_5=-y\frac{\partial}{\partial y}+v\frac{\partial}{\partial v}$ $+2q\frac{\partial}{\partial q}$, $\mathrm{Z}_6=y\frac{\partial}{\partial y}+p\frac{\partial}{\partial p}$ $-q\frac{\partial}{\partial q}$. Nonlocal
$-p/q$	$\mathrm{X}_1,\mathrm{X}_2,\mathrm{X}_3,\mathrm{X}_4$, $\mathrm{X}_5,\mathrm{X}_6$. Nonlocal Nonlocal	$\mathrm{Y}_1,\mathrm{Y}_2,\mathrm{Y}_3,\mathrm{Y}_4$, $\mathrm{Y}_5,\mathrm{Y}_6,\mathrm{Y}_7$, Nonlocal Nonlocal	$\mathrm{Z}_1,\mathrm{Z}_2,\mathrm{Z}_3,\mathrm{Z}_4$, $\mathrm{Z}_5,\mathrm{Z}_6$, $\mathrm{Z}_7=\frac{\partial}{\partial p}+\frac{q}{p}\frac{\partial}{\partial q}$, $\mathrm{Z}_8=s\frac{\partial}{\partial v}-y\frac{\partial}{\partial p}$ $-\frac{yq}{p}\frac{\partial}{\partial q}$.
$pF(pe^q)$	$\mathrm{X}_1,\mathrm{X}_2,\mathrm{X}_3,\mathrm{X}_4$, Nonlocal	$\mathrm{Y}_1,\mathrm{Y}_2,\mathrm{Y}_3,\mathrm{Y}_4,\mathrm{Y}_5$, $\mathrm{Y}_8=t\frac{\partial}{\partial t}+2\alpha^1\frac{\partial}{\partial x}-v\frac{\partial}{\partial v}$ $-2p\frac{\partial}{\partial p}+2\rho^2\frac{\partial}{\partial \rho}$.	$\mathrm{Z}_1,\mathrm{Z}_2,\mathrm{Z}_3,\mathrm{Z}_4$, $\mathrm{Z}_9=s\frac{\partial}{\partial s}-v\frac{\partial}{\partial v}$ $-2p\frac{\partial}{\partial p}+2\frac{\partial}{\partial q}$.
$F(q)$	$\mathrm{X}_1,\mathrm{X}_2,\mathrm{X}_3,\mathrm{X}_4$, $\mathrm{X}_8=\frac{\partial}{\partial p}$. Nonlocal	$\mathrm{Y}_1,\mathrm{Y}_2,\mathrm{Y}_3,\mathrm{Y}_4,\mathrm{Y}_5$, $\mathrm{Y}_9=\frac{\partial}{\partial p}$, $\mathrm{Y}_{10}=\frac{t^2}{2}\frac{\partial}{\partial x}+t\frac{\partial}{\partial v}-\alpha^1\frac{\partial}{\partial p}$.	$\mathrm{Z}_1,\mathrm{Z}_2,\mathrm{Z}_3,\mathrm{Z}_4$, $\mathrm{Z}_{10}=\frac{\partial}{\partial p}$, $\mathrm{Z}_{11}=s\frac{\partial}{\partial v}-y\frac{\partial}{\partial p}$.
$F(p+nq)$ $n\neq 0$	$\mathrm{X}_1,\mathrm{X}_2,\mathrm{X}_3,\mathrm{X}_4$. Nonlocal Nonlocal	$\mathrm{Y}_1,\mathrm{Y}_2,\mathrm{Y}_3,\mathrm{Y}_4,\mathrm{Y}_5$, $\mathrm{Y}_{11}=n\alpha^1\frac{\partial}{\partial x}-\frac{\partial}{\partial p}-\rho^2\frac{\partial}{\partial \rho}$, $\mathrm{Y}_{12}=\frac{nt^2+(\alpha^1)^2}{2}\frac{\partial}{\partial x}+nt\frac{\partial}{\partial v}$ $-n\alpha^1\frac{\partial}{\partial p}-\rho^2\alpha^1\frac{\partial}{\partial \rho}$.	$\mathrm{Z}_1,\mathrm{Z}_2,\mathrm{Z}_3,\mathrm{Z}_4$, $\mathrm{Z}_{12}=\frac{\partial}{\partial q}-n\frac{\partial}{\partial p}$, $\mathrm{Z}_{13}=ns\frac{\partial}{\partial v}$ $-ny\frac{\partial}{\partial p}+y\frac{\partial}{\partial q}$.
$F(p)$	$\mathrm{X}_1,\mathrm{X}_2,\mathrm{X}_3,\mathrm{X}_4$. Nonlocal Nonlocal	$\mathrm{Y}_1,\mathrm{Y}_2,\mathrm{Y}_3,\mathrm{Y}_4,\mathrm{Y}_5$, $\mathrm{Y}_\Psi=\Psi(\alpha^1)\frac{\partial}{\partial x}-\rho^2\Psi'(\alpha^1)\frac{\partial}{\partial \rho}$. Nonlocal $\Psi(\alpha^1)$ arbitrary.	$\mathrm{Z}_1,\mathrm{Z}_2,\mathrm{Z}_3,\mathrm{Z}_4$, Nonlocal $\mathrm{Z}_\Theta=\Theta\Big(y,q$ $+\int\frac{dp}{F(p)}\Big)\frac{\partial}{\partial q}$, $\Theta(y,z)$ arbitrary.

tem $\mathbf{E}\{x,t\,;v,p,\rho\}$; the infinite number of symmetries Y_{Ψ} are local for the system $\mathbf{EA}^1\{x,t;v,p,\rho,\ \alpha^1\}$ but yield nonlocal symmetries of the systems $\mathbf{E}\{x,t\,;v,p,\rho\}$ and $\mathbf{L}\{y,s\,;v,p,q\}$.

(2) Nonlocal symmetries of polytropic PGD equations
Now consider symmetries of the nonlocally related PDE systems of planar gas dynamics equations in the tree $\mathcal{T}_b$ for the polytropic case $B(p,q)=\gamma p/q$, $\gamma\neq 0$. Comparisons are made for the complete point symmetry classifications of several such PDE systems: systems $\mathbf{E}\{x,t\,;v,p,\rho\}$ (3.39), $\mathbf{L}\{y,s\,;v,p,q\}$ (3.42) and $\underline{\mathbf{L}}\{y,s\,;p,q\}$ (3.46) [Table 4.17], as well as for the potential systems $\mathbf{LW}^1\{y,s\,;v,p,q,w^1\}$ (3.97), $\mathbf{LW}^4\{y,s\,;v,p,q,w^4\}$ (3.120) and $\underline{\mathbf{LW}}^4\{y,s\,;p,q,w^4\}$ (3.123) [Table 4.18].

Table 4.17 Point symmetries of the PGD systems $\mathbf{E}\{x,t\,;v,p,\rho\}$, $\mathbf{L}\{y,s\,;v,p,q\}$ and $\underline{\mathbf{L}}\{y,s\,;p,q\}$ in the polytropic case

γ	Point Symmetries		
	$\mathbf{E}\{x,t\,;v,p,\rho\}$	$\mathbf{L}\{y,s\,;v,p,q\}$	$\underline{\mathbf{L}}\{y,s\,;p,q\}$
Arbitrary	$\mathrm{X}_1=\frac{\partial}{\partial x}$, $\mathrm{X}_2=\frac{\partial}{\partial t}$, $\mathrm{X}_3=t\frac{\partial}{\partial t}+x\frac{\partial}{\partial x}$, $\mathrm{X}_4=t\frac{\partial}{\partial x}+\frac{\partial}{\partial v}$, $\mathrm{X}_5=x\frac{\partial}{\partial x}+v\frac{\partial}{\partial v}+p\frac{\partial}{\partial p}-\rho\frac{\partial}{\partial \rho}$, $\mathrm{X}_6=p\frac{\partial}{\partial p}+\rho\frac{\partial}{\partial \rho}$.	$\mathrm{Z}_1=\frac{\partial}{\partial s}$, $\mathrm{Z}_2=s\frac{\partial}{\partial s}+y\frac{\partial}{\partial y}$, $\mathrm{Z}_3=\frac{\partial}{\partial v}$, $\mathrm{Z}_4=v\frac{\partial}{\partial v}+p\frac{\partial}{\partial p}+q\frac{\partial}{\partial q}$, $\mathrm{Z}_5=y\frac{\partial}{\partial y}+p\frac{\partial}{\partial p}-q\frac{\partial}{\partial q}$, $\mathrm{Z}_6=\frac{\partial}{\partial y}$.	$\hat{\mathrm{Z}}_1=\mathrm{Z}_1$, $\hat{\mathrm{Z}}_2=\mathrm{Z}_2$, $\hat{\mathrm{Z}}_3=p\frac{\partial}{\partial p}+q\frac{\partial}{\partial q}$, $\hat{\mathrm{Z}}_4=\mathrm{Z}_5$, $\hat{\mathrm{Z}}_5=\mathrm{Z}_6$, $\hat{\mathrm{Z}}_6=y^2\frac{\partial}{\partial y}+yp\frac{\partial}{\partial p}-3yq\frac{\partial}{\partial q}$.
3	$\mathrm{X}_1,\mathrm{X}_2,\mathrm{X}_3,\mathrm{X}_4,\mathrm{X}_5,\mathrm{X}_6$, $\mathrm{X}_7=xt\frac{\partial}{\partial x}+t^2\frac{\partial}{\partial t}+(x-vt)\frac{\partial}{\partial v}-3tp\frac{\partial}{\partial p}-t\rho\frac{\partial}{\partial \rho}$.	$\mathrm{Z}_1,\mathrm{Z}_2,\mathrm{Z}_3,\mathrm{Z}_4,\mathrm{Z}_5,\mathrm{Z}_6$.	$\hat{\mathrm{Z}}_1,\hat{\mathrm{Z}}_2,\hat{\mathrm{Z}}_3,\hat{\mathrm{Z}}_4,\hat{\mathrm{Z}}_5,\hat{\mathrm{Z}}_6$, $\hat{\mathrm{Z}}_7=s^2\frac{\partial}{\partial s}-3sp\frac{\partial}{\partial p}+sq\frac{\partial}{\partial q}$.
−1	$\mathrm{X}_1,\mathrm{X}_2,\mathrm{X}_3,\mathrm{X}_4,\mathrm{X}_5,\mathrm{X}_6$.	$\mathrm{Z}_1,\mathrm{Z}_2,\mathrm{Z}_3,\mathrm{Z}_4,\mathrm{Z}_5,\mathrm{Z}_6$, $\mathrm{Z}_7=\frac{\partial}{\partial p}+\frac{q}{p}\frac{\partial}{\partial q}$, $\mathrm{Z}_8=-s\frac{\partial}{\partial v}+y\frac{\partial}{\partial p}+\frac{yq}{p}\frac{\partial}{\partial q}$.	$\hat{\mathrm{Z}}_1,\hat{\mathrm{Z}}_2,\hat{\mathrm{Z}}_3,\hat{\mathrm{Z}}_4,\hat{\mathrm{Z}}_5,\hat{\mathrm{Z}}_6$, $\hat{\mathrm{Z}}_8=\mathrm{Z}_7$, $\hat{\mathrm{Z}}_9=y\frac{\partial}{\partial p}+\frac{yq}{p}\frac{\partial}{\partial q}$, $\hat{\mathrm{Z}}_{10}=s\frac{\partial}{\partial p}+\frac{sq}{p}\frac{\partial}{\partial q}$, $\hat{\mathrm{Z}}_{11}=sy\frac{\partial}{\partial p}+\frac{syq}{p}\frac{\partial}{\partial q}$.

Observe that the symmetry $\widehat{\mathrm{Z}}_7$ yields nonlocal symmetries of each of the systems $\mathbf{L}\{y,s\,;v,p,q\}$ and $\mathbf{LW}^4\{y,s\,;v,p,q,w^4\}$ but yields local sym-

Table 4.18 Point symmetries of the PGD systems $\underline{\mathbf{LW}}^4\{y,s\,;p,q,w^4\}$, $\mathbf{LW}^4\{y,s;v,p,q,w^4\}$ and $\mathbf{LW}^1\{y,s\,;v,p,q,w^1\}$ in the polytropic case

γ	Point Symmetries		
	$\underline{\mathbf{LW}}^4\{y,s\,;p,q,w^4\}$	$\mathbf{LW}^4\{y,s\,;v,p,q,w^4\}$	$\mathbf{LW}^1\{y,s\,;v,p,q,w^1\}$
Arbitrary	$\hat{\mathrm{J}}_1=\frac{\partial}{\partial w^4}$, $\hat{\mathrm{J}}_2=\frac{\partial}{\partial s}$, $\hat{\mathrm{J}}_3=y\frac{\partial}{\partial y}+s\frac{\partial}{\partial s}+w^4\frac{\partial}{\partial w^4}$, $\hat{\mathrm{J}}_4=p\frac{\partial}{\partial p}+q\frac{\partial}{\partial q}+(\gamma+1)w^4\frac{\partial}{\partial w^4}$ $\hat{\mathrm{J}}_5=y\frac{\partial}{\partial y}+p\frac{\partial}{\partial p}-q\frac{\partial}{\partial q}+(2-\gamma)w^4\frac{\partial}{\partial w^4}$, $\hat{\mathrm{J}}_6=\frac{\partial}{\partial y}$.	$\mathrm{J}_1=\hat{\mathrm{J}}_1$, $\mathrm{J}_2=\hat{\mathrm{J}}_2$, $\mathrm{J}_3=\hat{\mathrm{J}}_3$, $\mathrm{J}_4=\frac{\partial}{\partial v}$, $\mathrm{J}_5=v\frac{\partial}{\partial v}+\hat{\mathrm{J}}_4$, $\mathrm{J}_6=\hat{\mathrm{J}}_5$, $\mathrm{J}_7=\hat{\mathrm{J}}_6$.	$\mathrm{Y}_1=\frac{\partial}{\partial w^1}$, $\mathrm{Y}_2=\hat{\mathrm{J}}_2$, $\mathrm{Y}_3=y\frac{\partial}{\partial y}+s\frac{\partial}{\partial s}+w^1\frac{\partial}{\partial w^1}$ $\mathrm{Y}_4=\frac{\partial}{\partial v}+s\frac{\partial}{\partial w^1}$, $\mathrm{Y}_5=v\frac{\partial}{\partial v}+p\frac{\partial}{\partial p}+q\frac{\partial}{\partial q}+w^1\frac{\partial}{\partial w^1}$, $\mathrm{Y}_6=y\frac{\partial}{\partial y}+p\frac{\partial}{\partial p}-q\frac{\partial}{\partial q}$, $\mathrm{Y}_7=\hat{\mathrm{J}}_6$.
3	$\hat{\mathrm{J}}_1,\hat{\mathrm{J}}_2,\hat{\mathrm{J}}_3,\hat{\mathrm{J}}_4,$ $\hat{\mathrm{J}}_5,\hat{\mathrm{J}}_6,$ $\hat{\mathrm{J}}_7=s^2\frac{\partial}{\partial s}-3sp\frac{\partial}{\partial p}+sq\frac{\partial}{\partial q}$.	$\mathrm{J}_1,\mathrm{J}_2,\mathrm{J}_3,\mathrm{J}_4,$ $\mathrm{J}_5,\mathrm{J}_6,\mathrm{J}_7$.	$\mathrm{Y}_1,\mathrm{Y}_2,\mathrm{Y}_3,\mathrm{Y}_4,$ $\mathrm{Y}_5,\mathrm{Y}_6,\mathrm{Y}_7,$ $\mathrm{Y}_8=\hat{\mathrm{J}}_7+(w^1-sv)\frac{\partial}{\partial v}+sw^1\frac{\partial}{\partial w^1}$.
-1	$\hat{\mathrm{J}}_1,\hat{\mathrm{J}}_2,\hat{\mathrm{J}}_3,\hat{\mathrm{J}}_4,$ $\hat{\mathrm{J}}_5,\hat{\mathrm{J}}_6,$ $\hat{\mathrm{J}}_7=\mathrm{Z}_7,$ $\hat{\mathrm{J}}_8=\mathrm{Z}_8,$ $\hat{\mathrm{J}}_9=\hat{\mathrm{Z}}_{10},$ $\hat{\mathrm{J}}_{10}=\hat{\mathrm{Z}}_{11}$.	$\mathrm{J}_1,\mathrm{J}_2,\mathrm{J}_3,\mathrm{J}_4,$ $\mathrm{J}_5,\mathrm{J}_6,\mathrm{J}_7,$ $\mathrm{J}_8=\mathrm{Z}_7,$ $\mathrm{J}_9=\mathrm{Z}_8$.	$\mathrm{Y}_1,\mathrm{Y}_2,\mathrm{Y}_3,\mathrm{Y}_4,$ $\mathrm{Y}_5,\mathrm{Y}_6,\mathrm{Y}_7$.
1	$\hat{\mathrm{J}}_1,\hat{\mathrm{J}}_2,\hat{\mathrm{J}}_3,\hat{\mathrm{J}}_4,$ $\hat{\mathrm{J}}_5,\hat{\mathrm{J}}_6,$ $\hat{\mathrm{J}}_{11}=\hat{\mathrm{Z}}_6$.	$\mathrm{J}_1,\mathrm{J}_2,\mathrm{J}_3,\mathrm{J}_4,$ $\mathrm{J}_5,\mathrm{J}_6,\mathrm{J}_7$.	$\mathrm{Y}_1,\mathrm{Y}_2,\mathrm{Y}_3,\mathrm{Y}_4,$ $\mathrm{Y}_5,\mathrm{Y}_6,\mathrm{Y}_7$.

metries of the other four considered systems $\mathbf{E}\{x,t\,;v,p,\rho\}$, $\underline{\mathbf{L}}\{y,s\,;p,q\}$, $\mathbf{LW}^1\{y,s\,;v,p,q,w^1\}$ and $\underline{\mathbf{LW}}^4\{y,s\,;p,q,w^4\}$; the symmetries Z_7 and Z_8 yield nonlocal symmetries of the systems $\mathbf{E}\{x,t\,;v,p,\rho\}$ and $\mathbf{LW}^1\{y,s;v,p,q,w^1\}$ but local symmetries of the other four considered systems $\mathbf{L}\{y,s;v,p,q\}$, $\underline{\mathbf{L}}\{y,s\,;p,q\}$, $\mathbf{LW}^4\{y,s\,;v,p,q,w^4\}$ and $\underline{\mathbf{LW}}^4\{y,s\,;p,q,w^4\}$; the symmetries $\hat{\mathrm{Z}}_{10}$ and $\hat{\mathrm{Z}}_{11}$ are local symmetries of the Lagrange subsystem $\underline{\mathbf{L}}\{y,s;p,q\}$ and the subsystem $\underline{\mathbf{LW}}^4\{y,s\,;p,q,w^4\}$ but yield nonlocal symmetries of the other four considered systems $\mathbf{E}\{x,t\,;v,p,\rho\}$, $\mathbf{L}\{y,s\,;v,p,q\}$, $\mathbf{LW}^1\{y,s;\ v,p,\ q,w^1\}$ and $\mathbf{LW}^4\{y,s\,;v,p,q,w^4\}$. Interestingly, the symmetry $\hat{\mathrm{Z}}_6$, a local symmetry of the Lagrange subsystem $\underline{\mathbf{L}}\{y,s\,;p,q\}$ for *any* value of the polytropic constant γ, yields a local symmetry of the subsystem $\underline{\mathbf{LW}}^4\{y,s\,;p,q,w^4\}$ *only in the case* $\gamma = 1$ (and yields a nonlocal symmetry otherwise), and is a nonlocal symmetry of the other four considered PGD systems $\mathbf{E}\{x,t\,;v,p,\rho\}$, $\mathbf{L}\{y,s\,;v,p,q\}$, $\mathbf{LW}^1\{y,s\,;v,p,q,w^1\}$ and $\mathbf{LW}^4\{y,s\,;v,p,q,w^4\}$ for all values of γ.

(3) Nonlocal symmetries of generalized polytropic PGD equations
As another example, consider a nonlocal symmetry classification problem for PGD equations with a generalized polytropic equation of state

$$B(p,q) = \frac{M(p)}{q}, \quad M''(p) \neq 0, \tag{4.34}$$

which excludes the polytropic case considered in the previous example.

For the sake of brevity, consideration is only given for the extended tree $\mathcal{T}_a$ (3.105) of PDE systems of planar gas dynamics equations. [These follow from local conservation laws of the Lagrange PGD system $\mathbf{L}\{y,s\,;v,p,q\}$ that arise from zeroth-order multipliers $\Lambda_i = \Lambda_i(y,s,v,p,q)$; see Section 3.5.4, Figure 3.5.]

The extended tree $\mathcal{T}_a'$ includes ten nonlocally related PDE systems.

- The Euler system $\mathbf{E}\{x,t\,;v,p,\rho\}$ (3.39).
- The Lagrange system $\mathbf{L}\{y,s\,;v,p,q\}$ (3.42).
- Three singlet potential systems $\mathbf{LW^1}\{y,s\,;v,p,q,w^1\}$ (3.97), $\mathbf{LW}^2\{y,s;v,p,q,w^2\}$(3.98), and $\mathbf{LW^3}\{y,s\,;v,p,q,w^3\}$ (3.99).
- Three couplets $\mathbf{LW^1W^2}\{y,s\,;v,p,q,w^1,w^2\}$ (3.100), $\mathbf{LW^1W}^3\{y,s;v,p,q,w^1,w^3\}$ (3.101), and $\mathbf{LW^2W^3}\{y,s\,;v,p,q,w^2,w^3\}$ (3.102).
- One triplet $\mathbf{LW^1W^2W^3}\{y,s\,;v,p,q,w^1,w^2,w^3\}$ (3.103).
- The nonlocally related subsystem $\underline{\mathbf{L}}\{y,s\,;p,q\}$ (3.46).

The point symmetry classification of each of the above seven potential systems (modulo the equivalence transformations (3.96)), i.e., the three singlets, three couplets and one triplet, yields Table 4.19 that lists point symmetries and nonlocal symmetries for the Lagrange PGD system $\mathbf{L}\{y,s\,;v,p,q\}$ with the equation of state (4.34).

In Table 4.19, the symmetries of each PDE system arise as projections of infinitesimal generators presented in the right-hand column on the space of variables of that system.

From Table 4.19, observe that the Euler system $\mathbf{E}\{x,t\,;v,p,\rho\}$ has the same symmetries for any $M(p)$. The infinitesimal generators $\mathrm{Z}_9,\ldots,\mathrm{Z}_{12}$ yield point symmetries of the systems $\underline{\mathbf{L}}\{y,s\,;p,q\}$, $\mathbf{L}\{y,s\,;v,p,q\}$ and $\mathbf{LW}^2\{y,s;v,p,q,w^2\}$, and nonlocal symmetries of all other systems; the infinitesimal generators $\mathrm{Z}_{13},\mathrm{Z}_{14}$ yield point symmetries of the systems $\underline{\mathbf{L}}\{y,s\,;p,q\}$ and

Table 4.19 Symmetries of the generalized polytropic planar gas dynamics equations

System	$M(p)$	Point Symmetries
$\mathbf{E}$	Arbitrary	$X_1 = \frac{\partial}{\partial t}$, $X_2 = \frac{\partial}{\partial x}$, $X_3 = t\frac{\partial}{\partial x} + \frac{\partial}{\partial v}$, $X_4 = x\frac{\partial}{\partial x} + v\frac{\partial}{\partial v} - 2\rho\frac{\partial}{\partial \rho}$, $X_5 = x\frac{\partial}{\partial x} + t\frac{\partial}{\partial t}$.
$\underline{\mathbf{L}}$, $\mathbf{L}$, $\mathbf{LW^1}$, $\mathbf{LW^2}$, $\mathbf{LW^3}$, $\mathbf{LW^1W^2}$, $\mathbf{LW^1W^3}$, $\mathbf{LW^2W^3}$, $\mathbf{LW^1W^2W^3}$	(i) Arbitrary	$Z_1 = \frac{\partial}{\partial s} + w^2\frac{\partial}{\partial w^3}$, $Z_2 = \frac{\partial}{\partial y} + w^1\frac{\partial}{\partial w^3}$, $Z_3 = \frac{\partial}{\partial v} + s\frac{\partial}{\partial w^1} + y\frac{\partial}{\partial w^2} + sy\frac{\partial}{\partial w^3}$, $Z_4 = -y\frac{\partial}{\partial y} + 2q\frac{\partial}{\partial q} + v\frac{\partial}{\partial v} + w^1\frac{\partial}{\partial w^1}$, $Z_5 = s\frac{\partial}{\partial s} + y\frac{\partial}{\partial y} + w^1\frac{\partial}{\partial w^1} + w^2\frac{\partial}{\partial w^2}$ $+2w^3\frac{\partial}{\partial w^3}$, $Z_6 = \frac{\partial}{\partial w^1}$, $Z_7 = \frac{\partial}{\partial w^2}$, $Z_8 = \frac{\partial}{\partial w^3}$.
$\underline{\mathbf{L}}$, $\mathbf{L}$, $\mathbf{LW^2}$	(ii) $-p\ln p$	$Z_9 = y\frac{\partial}{\partial y} + 2p\frac{\partial}{\partial p} + \frac{2q}{\ln p}\frac{\partial}{\partial q} + v\frac{\partial}{\partial v} + 2w^2\frac{\partial}{\partial w^2}$.
	(iii) $\gamma p + \delta p^{\frac{\gamma+1}{\gamma}}$ $\gamma \neq 0, -1$	$Z_{10} = \frac{(\gamma+1)y}{2\gamma}\frac{\partial}{\partial y} + p\frac{\partial}{\partial p} - \frac{q}{\delta p^{1/\gamma}+\gamma}\frac{\partial}{\partial q}$ $+\frac{(\gamma-1)v}{2\gamma}\frac{\partial}{\partial v} + w^2\frac{\partial}{\partial w^2}$.
	(iv) $1 + \alpha e^p$, $\alpha = \pm 1$	$Z_{11} = \frac{\partial}{\partial p} + \frac{\alpha e^p}{1+\alpha e^p} q\frac{\partial}{\partial q} - s\frac{\partial}{\partial w^2}$, $Z_{12} = y\frac{\partial}{\partial p} + \frac{\alpha e^p}{1+\alpha e^p} yq\frac{\partial}{\partial q} - s\frac{\partial}{\partial v} - sy\frac{\partial}{\partial w^2}$.
$\underline{\mathbf{L}}$, $\mathbf{LW^2}$	(ii) $-p\ln p$	$Z_{13} = y^2\frac{\partial}{\partial y} + yp\frac{\partial}{\partial p} - \left(3 - \frac{1}{\ln p}\right) yq\frac{\partial}{\partial q}$ $-(yv - w^2)\frac{\partial}{\partial v} + yw^2\frac{\partial}{\partial w^2}$.
	(iii) $\gamma p + \delta p^{\frac{\gamma+1}{\gamma}}$ $\gamma \neq 0, -1$	$Z_{14} = y^2\frac{\partial}{\partial y} + yp\frac{\partial}{\partial p} - \left(3 - \frac{\delta}{\gamma}\frac{p^{1/\gamma}}{\delta p^{1/\gamma}+\gamma}\right) yq\frac{\partial}{\partial q}$ $-(yv - w^2)\frac{\partial}{\partial v} + yw^2\frac{\partial}{\partial w^2}$.
$\underline{\mathbf{L}}$	(iii) with $\gamma = 3$: $3p + \delta p^{\frac{4}{3}}$	$\hat{Z}_{15} = \frac{1}{3}s^2\frac{\partial}{\partial s} - sp\frac{\partial}{\partial p} + \frac{1}{\delta p^{4/3}+3} spq\frac{\partial}{\partial q}$.

$\mathbf{LW^2}\{y, s\,; v, p, q, w^2\}$, and nonlocal symmetries of all other systems, including the Euler system $\mathbf{E}\{x, t\,; v, p, \rho\}$ and the Lagrange system $\mathbf{L}\{y, s\,; v, p, q\}$.

The point symmetries of the Lagrange subsystem $\underline{\mathbf{L}}\{y, s\,; p, q\}$ include all corresponding point symmetries of the system $\mathbf{LW^2}\{y, s\,; v, p, q, w^2\}$; for $M(p) = 3p + \delta p^{4/3}$, one additional symmetry $\hat{Z}_{15}$ is obtained that is a nonlocal symmetry of the Euler system $\mathbf{E}\{x, t\,; v, p, \rho\}$, the Lagrange system $\mathbf{L}\{y, s\,; v, p, q\}$ and all its seven potential systems considered in this example.

All other infinitesimal generators in Table 4.19 project onto point symmetries for both the Euler system $\mathbf{E}\{x, t\,; v, p, \rho\}$ and the Lagrange system $\mathbf{L}\{y, s\,; v, p, q\}$.

Exercises 4.2

4.2.1. Find equivalence transformations of the nonlocally related systems in the extended tree $\mathcal{T}_4$ for the nonlinear diffusion equation (3.18). Determine the point symmetry classifications of each of the nonlocally related subsystems $\mathbf{A}\{x, u\,; \alpha\}$, $\mathbf{V}\{u, t\,; v\}$ and $\mathbf{X}\{u, v\,; x\}$ within the extended tree $\mathcal{T}_4$ for the nonlinear diffusion equation (3.18).

4.2.2. Find the infinite set of point symmetries of the potential system $\mathbf{UW}\{x, t\,; u, w\}$ (3.82) of the nonlinear diffusion equation (3.18). Find a point transformation that maps $\mathbf{UW}\{x, t\,; u, w\}$ into a linear PDE system.

4.2.3. Show that the potential system $\mathbf{UB}\{x, t\,; u, \beta\}$ (3.84) of the nonlinear wave equation $u_{tt} = (c^2(u)u_x)_x$ (3.76), in the case $c(u) = u^{-2/3}$, has an infinite number of point symmetries. For this case, find an explicit form of the linearizing transformation. [Hint: In this case, instead of computing an infinite number of point symmetries and applying Theorem 2.4.2, one may start by introducing new independent variables $s = x^{-1}$, $\beta = x^3 u$. The resulting PDE system is linearizable by a hodograph transformation.]

4.2.4.

(a) Find the point symmetries M_7 and M_8 of the potential system $\mathbf{TQ}\{u, v;$ $t, q\}$ (3.93) of the wave equation $u_{tt} = (c^2(u)u_x)_x$ (3.76) [Table 4.7].
(b) Find the point symmetries N_5 and N_6 of the potential system $\mathbf{TR}\{u, v;$ $t, r\}$ (3.94) of the wave equation (3.76) [Table 4.8].

4.2.5. Find the point symmetries of the linear wave equation $\mathbf{X}\{u, v\,; x\}$ (3.86). Deduce whether this linear wave equation can be mapped by a point transformation into a constant coefficient linear PDE.

4.2.6. Calculate the components of the nontrivial infinite-parameter set of point symmetries of the linear wave equation

$$q_{tt} = x^2 q_{xx} \tag{4.35}$$

(equivalent to the equation (3.87) after a suitable renaming of the variables). Show that the scalar PDE (4.35) can be mapped into the constant coefficient linear wave equation $Q_{XT} = 0$ by the point transformation

$$X = 1/x + t, \quad T = 1/x - t, \quad Q = q/x + t$$

[Bluman (1983); Bluman & Kumei (1987)].

4.2.7. Show that the symmetry $\hat{\mathrm{Z}}_6$ [Table 4.17], which yields a nonlocal symmetry of both the polytropic Euler and Lagrange PGD systems and a local symmetry of the Lagrange subsystem $\underline{\mathbf{L}}\{y, s\,; p, q\}$, also yields a local symmetry of both the potential Lagrange system $\mathbf{LW^2}\{y, s\,; v, p, q, w^2\}$ (3.98) and the triplet potential Lagrange system $\mathbf{LW^1W^2W^3}\{y, s\,; v, p, q, w^1, w^2, w^3\}$ (3.103). Find the components of the infinitesimal symmetry generator corresponding to each of v, w^1, w^2, and w^3.

4.3 Construction of Non-invertible Mappings Relating PDEs

In this section, nonlocally related systems are used to extend the work presented in Sections 2.4–2.6 on the invertible mapping of a given PDE system to one of a simpler type that can draw on an arsenal of well-known solution techniques. In particular, it is shown how to find useful nonlocal mappings relating PDEs through the use of nonlocally related potential systems.

Firstly, the invertible mapping algorithm presented in Section 2.4 is extended to include nonlocal mappings of nonlinear PDEs to linear PDEs. Here, if a nonlocally related potential system has a point symmetry that satisfies the criteria of Theorems 2.4.1 and 2.4.2 and yields a nonlocal (potential) symmetry of a given PDE system, then one can construct an invertible mapping of the potential system to a linear system that in turn yields a nonlocal mapping of the given nonlinear PDE system to a linear PDE system. A similar extension occurs when such a nonlocal mapping exists of a nonlinear PDE system to a linear PDE system when the nonlocally related potential system of the nonlinear PDE system has a set of local conservation law multipliers that satisfies the criteria of Theorems 2.6.1 and 2.6.2.

Secondly, it is shown how to extend the invertible mapping algorithm presented in Section 2.4 to include nonlocal mappings of linear PDEs with variable coefficients to linear PDEs with constant coefficients. Here one starts from the observation that each solution set of the adjoint PDE system of a given linear PDE system is a set of conservation law multipliers of the given PDE system and correspondingly yields a nonlocally related linear potential system of the given PDE system. The aim is to find a particular solution set of the adjoint PDE system that yields an invertible mapping of its corresponding nonlocally related linear potential system to a constant coefficient linear system. In turn this yields a non-invertible (nonlocal) mapping of the given linear PDE with variable coefficients to a linear PDE with constant coefficients. As examples, we consider nonlocal transformations of Kolmogorov equations to the backward heat equation [Bluman & Shtelen (2004)]. There also exists related work on nonlocal transformations of Schrödinger equations to the free particle equation [Bluman & Shtelen (1996a)].

4.3.1 Non-invertible mappings of nonlinear PDE systems to linear PDE systems

Suppose a given nonlinear PDE system does not have local (point or contact) symmetries (or, equivalently, does not have local conservation law multipliers) that yield an invertible mapping to a linear PDE system. In particular, this means that its local symmetries do not satisfy the criteria of Theorems 2.4.1, 2.4.2 (or, equivalently, that its local conservation law multipliers do not satisfy the criteria of Theorems 2.6.1, 2.6.2) so that there does not exist an invertible mapping of the nonlinear PDE system to any linear PDE system. However, it could happen that a nonlocally related system has an infinite set of local symmetries (an infinite set of local conservation law multipliers) that yields an invertible mapping of the nonlocally related system to some linear PDE system. Consequently, through the invertible mapping of the nonlocally related system to a linear system, one obtains a nonlocal (non-invertible) mapping of the given nonlinear PDE system to a linear PDE system. Of course, the local symmetries (local conservation law multipliers) yielding such a linearization of the nonlocally related system, must have an essential dependence on nonlocal variables.

For illustration, the following examples are considered.

(1) Linearization of Burgers' equation
As a first example, consider Burgers' equation

$$u_t + uu_x - u_{xx} = 0. \tag{4.36}$$

One can show that equation (4.36) has at most a finite number of contact symmetries. Hence there exists no point or contact transformation that linearizes Burgers' equation. As written (for convenience, after multiplication by the factor 2), the PDE (4.36) can be expressed as the conservation law $\mathrm{D}_t(2u)+\mathrm{D}_x(u^2-2u_x)=0$. Correspondingly, one obtains the potential system

$$\begin{aligned} v_x &= 2u, \\ v_t &= 2u_x - u^2. \end{aligned} \tag{4.37}$$

The potential system (4.37) has an infinite number of point symmetries given by the infinitesimal generator

$$\mathrm{X} = e^{v/4}\left\{[2h(x,t) + g(x,t)u]\frac{\partial}{\partial u} + 4g(x,t)\frac{\partial}{\partial v}\right\}, \tag{4.38}$$

where $(g(x,t), h(x,t))$ is an arbitrary solution of the linear PDE system

$$\begin{aligned} h &= g_x, \\ h_x &= g_t \end{aligned} \tag{4.39}$$

[Vinogradov & Krasil'shchik (1984); Kersten (1987)]. Consequently, one can apply Theorems 2.4.1, 2.4.2 to obtain the well-known nonlocal Hopf–Cole transformation that linearizes Burgers' equation (4.36) [Exercise 2.4.4].

Note that from the form of the infinitesimal generator (4.38), one can immediately see that the locally related subsystem of (4.37), known as the integrated form of Burgers' equation, given by

$$v_t = v_{xx} - \tfrac{1}{4}(v_x)^2, \tag{4.40}$$

has the linearizing point symmetries

$$\mathrm{X} = e^{v/4} g(x,t) \frac{\partial}{\partial v},$$

where $g(x,t)$ is any solution of the linear heat equation

$$g_t - g_{xx} = 0.$$

(2) Linearization of a nonlinear heat conduction equation

The nonlinear heat conduction equation

$$u_t - (u^{-2} u_x)_x = 0, \tag{4.41}$$

which arises directly as a conservation law, does not have linearizing contact symmetries. However, one can show that the corresponding potential system given by

$$\begin{aligned} v_t &= u^{-2} u_x, \\ v_x &= u, \end{aligned} \tag{4.42}$$

has the infinite set of linearizing point symmetries

$$\mathrm{X} = g(t,v)\frac{\partial}{\partial x} - h(t,v) u^2 \frac{\partial}{\partial u}, \tag{4.43}$$

where $(g(t,v), h(t,v))$ is an arbitrary solution of the linear system

$$\begin{aligned} h &= g_v, \\ h_v &= g_t \end{aligned} \tag{4.44}$$

[Bluman, Kumei, & Reid (1988)]. See Exercise 2.4.3 for the corresponding transformation to a linear system.

Again, note that from the form of the infinitesimal generator (4.43), it follows that the locally related subsystem of (4.42), given by

$$v_t = (v_x)^{-2} v_{xx}, \tag{4.45}$$

has the infinite set of linearizing point symmetries

$$\mathrm{X} = g(t,v)\frac{\partial}{\partial x},$$

where $g(t,v)$ is an arbitrary solution of the linear heat equation

$$g_t = g_{vv}.$$

See Exercise 2.4.3 for the corresponding linearizing transformation.

(3) Linearization of the Thomas equations
As a third example, consider the nonlinear system of Thomas equations given by

$$\begin{aligned} v_t - u_x &= 0, \\ v_t - uv - u - v &= 0, \end{aligned} \tag{4.46}$$

that describes a fluid flow through a reacting medium [Thomas (1944); see also Whitham (1974)] and also can be related to the equations for two-wave interaction [Hasegawa (1974); Hashimoto (1974); Yoshikawa & Yamaguti (1974)]. Since the nonlinear PDE system (4.46) does not have an infinite number of point symmetries, it cannot be linearized by a point transformation. The first equation of (4.46) is written as a conservation law, which in turn leads directly to the corresponding potential system given by

$$\begin{aligned} w_x &= v, \\ w_t &= u, \\ v_t - uv - u - v &= 0. \end{aligned} \tag{4.47}$$

One can show [Bluman & Kumei (1990b)] that the potential system (4.47) has the infinite set of point symmetries

$$\begin{aligned} \mathrm{X} = e^{w}\Big\{&[F(x,t)u + H(x,t)]\frac{\partial}{\partial u} + [F(x,t)v + G(x,t)]\frac{\partial}{\partial v} \\ &+F(x,t)\frac{\partial}{\partial w}\Big\}, \end{aligned} \tag{4.48}$$

where $(F(x,t), G(x,t), H(x,t))$ is an arbitrary solution of the linear PDE system

$$\begin{aligned} F_x &= G, \\ F_t &= H, \\ G_t &= G + H. \end{aligned} \tag{4.49}$$

Applying Theorems 2.4.1 and 2.4.2, one obtains the point transformation

$$\begin{aligned} z^1 &= x, \\ z^2 &= t, \\ w^1 &= e^{-w}, \\ w^2 &= e^{-w}v, \\ w^3 &= e^{-v}u, \end{aligned} \tag{4.50}$$

that invertibly maps the nonlinear system (4.47) to the linear system given by

$$\begin{aligned} \frac{\partial w^1}{\partial z^1} &= w^2, \\ \frac{\partial w^1}{\partial z^2} &= w^3, \\ \frac{\partial w^2}{\partial z^2} &= w^2 + w^3. \end{aligned} \tag{4.51}$$

Consequently, any solution $(w^1(z^1,z^2), w^2(z^1,z^2), w^3(z^1,z^2))$ of the linear system (4.51) yields the solution

$$(u(x,t), v(x,t)) = -\left(\frac{w^3(x,t)}{w^1(x,t)}, \frac{w^2(x,t)}{w^1(x,t)}\right),$$

of the Thomas equations (4.47).

Note that from the form of the infinitesimal generator (4.48), it follows that the locally related subsystem of (4.47), given by

$$w_{xt} - w_t w_x - w_t - w_x = 0, \tag{4.52}$$

has the linearizing infinite set of point symmetries

$$\mathrm{X} = F(x,t)e^w \frac{\partial}{\partial w} \tag{4.53}$$

where $F(x,t)$ is any solution of the linear PDE

$$F_{xt} - F_t - F_x = 0.$$

In particular, one obtains the point transformation $W = e^{-w}$ that maps the nonlinear PDE (4.52) to the linear PDE $W_{xt} - W_t - W_x = 0$.

(4) Linearization of a nonlinear reaction-diffusion equation
Consider the nonlinear reaction-diffusion equation given by

$$u_t - u^2 u_{xx} - 2u^2 = 0. \tag{4.54}$$

One can show that the PDE (4.54) has no linearizing set of contact symmetries and hence cannot be linearized by an invertible transformation. Multiplying the PDE (4.54) by u^{-2} yields the conservation law

$$\mathrm{D}_t(u^{-1}) + \mathrm{D}_x(u_x + 2x) = 0,$$

and the corresponding potential system ($u \neq 0$)

$$\begin{aligned} v_x &= u^{-1}, \\ v_t &= -(u_x + 2x) = -(u + x^2)_x. \end{aligned} \tag{4.55}$$

The nonlinear PDE system (4.55) also has no linearizing set of point symmetries. However, since the second PDE in (4.55) is written as a conservation law, one can accordingly introduce a second potential variable w to obtain the nonlocally related potential system

$$\begin{aligned} v_x &= u^{-1}, \\ w_x &= v, \\ w_t &= -(u + x^2). \end{aligned} \tag{4.56}$$

On can show [Exercise 2.4.8] that the potential system (4.56) has an infinite number of linearizing point symmetries given by the infinitesimal generator

$$\begin{aligned} \mathrm{X} = e^{(w-xv)}\Big\{&(F(t,v) - xH(t,v))\frac{\partial}{\partial x} \\ &+(G(t,v) - 2xF(t,v) + (x^2 - u)H(t,v))\frac{\partial}{\partial u} \\ &+(vF(t,v) - (1 + xv)H(t,v))\frac{\partial}{\partial w}\Big\}, \end{aligned} \tag{4.57}$$

where $(F(t,v), G(t,v), H(t,v))$ is an arbitrary solution of the linear system

$$\frac{\partial H(t,v)}{\partial v} = F(t,v), \quad \frac{\partial H(t,v)}{\partial t} = G(t,v), \quad \frac{\partial F(t,v)}{\partial v} = G(t,v). \tag{4.58}$$

Consequently, one can show that the application of Theorems 2.4.1, 2.4.2 to the point symmetries (4.57) yields the point transformation

$$\begin{aligned} z^1 &= t, \\ z^2 &= v, \\ w^1 &= xe^{(xv-w)}, \\ w^2 &= (x^2 + u)e^{(xv-w)}, \\ w^3 &= e^{(xv-w)} - 1, \end{aligned}$$

that invertibly maps the nonlinear PDE system (4.56) to the linear system

$$\frac{\partial w^1}{\partial z^2} = w^2, \quad \frac{\partial w^3}{\partial z^2} = w^1, \quad \frac{\partial w^3}{\partial z^1} = w^2.$$

Correspondingly, one can show that any solution $(w^1, w^2, w^3) \neq (0, 0, -1)$ of this linear system yields the solution

$$u = \frac{w^2(w^3+1) - (w^1)^2}{(w^3+1)^2}$$

of the nonlinear reaction-diffusion equation (4.54).

(5) Linearization of a nonlinear telegraph equation
As a final example, consider the nonlinear telegraph equation [Varley & Seymour (1985)]

$$\phi_{tt} = (\phi_t)^2 \phi_{xx} + \phi_t(1 - \phi_t). \tag{4.59}$$

One can show that PDE (4.59) does not have contact symmetries yielding its linearization by an invertible point or contact transformation.

Let $u = \phi_t$, $v = \phi_x$. Then the corresponding PDE system

$$\begin{aligned} u &= \phi_t, \\ v &= \phi_x, \\ u_t &= u^2 v_x + u(1-u), \end{aligned} \tag{4.60}$$

is equivalent to and locally related to the scalar PDE (4.59), and hence (4.60) is also not linearizable by an invertible transformation.

Clearly, the nonlinear PDE system (4.60) has a nonlocally related subsystem given by

$$\begin{aligned} u_x &= v_t, \\ u_t &= u^2 v_x + u(1-u). \end{aligned} \tag{4.61}$$

As shown in Section 2.4.1, the nonlinear telegraph system (4.61) has an infinite set of point symmetries yielding its linearization by the point transformation (2.92) to the linear PDE system given by (2.93). In turn, this yields the linearization of the nonlinear telegraph equation (4.59) by a non-invertible (nonlocal) transformation.

Of course, one could consider the nonlinear PDE system (4.61) as the given PDE system with the nonlocally related potential system (4.60) arising from its first equation written as a conservation law. In turn, the scalar equation (4.59) is a locally related subsystem of the potential system (4.60).

4.3.2 Non-invertible mappings of linear PDEs with variable coefficients to linear PDEs with constant coefficients

In Section 2.5, there was consideration of the problem of determining whether a given linear PDE with variable coefficients can be mapped invertibly to a linear PDE with constant coefficients. The basis of the presented algorithm was the observation that a linear PDE with constant coefficients is completely characterized by its point symmetries connected with its linearity and invariance under the abelian group of translations of its independent variables. This led to a definitive answer to the posed problem and also to the construction of such an invertible mapping when one exists. Parabolic and hyperbolic equations were considered as specific examples.

Now suppose a given linear PDE with variable coefficients cannot be mapped invertibly to a linear PDE with constant coefficients. Using the linear parabolic PDE as a canonical example, it is shown how to construct non-invertible mappings to extend the class of linear PDEs with variable coefficients that can be mapped to linear PDEs with constant coefficients. This is accomplished through consideration of an appropriate potential system. In particular, for any given linear PDE, *any* solution of its adjoint equation is a multiplier for a conservation law that yields an equivalent nonlocally related potential system. The aim is to find such a multiplier so that the corresponding potential system can be mapped invertibly into a linear PDE system with constant coefficients. As a consequence, the given linear PDE could be mapped, non-invertibly, into an equivalent constant coefficient linear PDE. When the given PDE is a linear parabolic equation (without loss of generality, PDE (2.176)), then the constant coefficient PDE can be taken to be the backward heat equation.

The explicit relationship between the solutions of any given linear PDE system and its local conservation law multipliers (which satisfy the adjoint system of the given system) is exhibited by equations (2.219) and (2.220) in Section 2.6.

Now suppose the given PDE is the linear parabolic PDE in the standard form (see Section 2.5.1 and the discussion following equation (2.176)) given by

$$\mathrm{L}u = u_{xx} + u_y + V(x,y)u = 0. \tag{4.62}$$

The results presented in Section 2.5.1 can be summarized in terms of the following theorem [Bluman & Shtelen (2004)] which can be proven by direct calculation.

Theorem 4.3.1. *A linear parabolic PDE* (4.62) *can be mapped invertibly by a point transformation to the backward heat equation*

$$w_{z^1 z^1} + w_{z^2} = 0 \tag{4.63}$$

if and only if $V(x, y)$ *is of the form*

$$V(x, y) = a(y)x^2 + b(y)x + c(y) \tag{4.64}$$

for some functions $a(y), b(y), c(y)$. *The point transformation that yields the mapping is given by*

$$\begin{aligned} z^1 &= \sigma(y)x + \rho(y), \\ z^2 &= \int^y \sigma^2(\hat{y})d\hat{y}, \\ w &= u \exp \tfrac{1}{4}[\sigma^{-1}\sigma'(y)x^2 + 2\sigma^{-1}\rho'(y)x + \lambda(y)], \end{aligned} \tag{4.65}$$

where $(\sigma(y), \rho(y), \lambda(y))$ *is a solution of the nonlinear system of ODEs*

$$\begin{aligned} \sigma^{-2}(\sigma\sigma'' - 2\sigma'^2) &= 4a(y), \\ (\sigma\rho'' - 2\sigma'\rho') &= 2\sigma^2 b(y), \\ \lambda' &= \sigma^{-2}(\rho'^2 - 2\sigma\sigma') + c(y). \end{aligned} \tag{4.66}$$

The solution of ODE system (4.66) appears in Bluman & Shtelen (2004).

Now the result of Theorem 4.3.1 is extended to include nonlocal (non-invertible) transformations of linear parabolic equations of the form (4.62) to the backward heat equation (4.63), i.e., through nonlocal transformations arising from related potential systems, one can widen the class of functions $V(x, y)$ for which a linear PDE (4.62) can be mapped into the backward heat equation (4.63). The work presented here appears in Bluman & Shtelen (2004).

A multiplier $\phi(x, y)$ that yields a local conservation law of the linear parabolic PDE (4.62) is any solution $\phi(x, y)$ of its adjoint PDE

$$\mathrm{L}^*\phi = \phi_{xx} - \phi_y + V(x, y)\phi = 0. \tag{4.67}$$

In particular, for arbitrary functions $(U(x, y), \Phi(x, y))$, one has the relationship

$$\begin{aligned} &\Phi \mathrm{L}U - U\mathrm{L}^*\Phi \\ &\quad = \Phi[U_{xx} + U_y + V(x, y)U] - U[\Phi_{xx} - \Phi_y + V(x, y)\Phi] \\ &\quad = \mathrm{D}_x(\Phi U_x - \Phi_x U) + \mathrm{D}_y(\Phi U). \end{aligned} \tag{4.68}$$

Consequently, for any solution $\phi(x, y)$ of the adjoint equation (4.67), the given linear parabolic scalar PDE (4.62) is nonlocally equivalent to the corresponding linear potential system

$$\begin{aligned} v_x &= \phi u, \\ v_y &= \phi_x u - \phi u_x. \end{aligned} \tag{4.69}$$

By direct calculation, one can prove the following extended theorem.

Theorem 4.3.2. *Let $\psi(x,y)$ be any solution of the linear PDE*

$$\psi_{xx} + \psi_y + [a(y)x^2 + b(y)x + c(y)]\psi = 0, \tag{4.70}$$

for some specific coefficients $a(y), b(y), c(y)$. Let $\phi(x,y) = \psi^{-1}$. For the same coefficients $a(y), b(y), c(y)$, consider the linear parabolic PDE (4.62) with

$$V(x,y) = -2\frac{\partial^2}{\partial x^2}\log|\phi(x,y)| + a(y)x^2 + b(y)x + c(y). \tag{4.71}$$

The corresponding potential system (4.69) can be mapped invertibly by a point transformation to the backward heat potential system

$$\begin{aligned} \frac{\partial w^2}{\partial z^1} &= w^1, \\ \frac{\partial w^2}{\partial z^2} &= -\frac{\partial w^1}{\partial z_1}, \end{aligned} \tag{4.72}$$

for which each component satisfies the backward heat equation, i.e., $w^i_{z^1 z^1} + w^i_{z^2} = 0,\ i = 1,\ 2$. In particular, such a mapping is given by

$$\begin{aligned} z^1 &= \sigma(y)x + \rho(y), \\ z^2 &= \int^y \sigma^2(\hat{y})d\hat{y}, \\ w^1 &= \sigma^{-1}e^{g(x,y)}\left\{u + \left(\tfrac{1}{2}\sigma^{-1}(\sigma'(y)x + \rho'(y)) - \psi^{-1}\psi_x\right)\psi v\right\}, \\ w^2 &= e^{g(x,y)}\psi v, \end{aligned} \tag{4.73}$$

where $(\sigma(y), \rho(y), \lambda(y))$ is a solution of the corresponding nonlinear ODE system (4.66) and

$$g(x,y) = \tfrac{1}{4}[\sigma^{-1}\sigma'(y)x^2 + 2\sigma^{-1}\rho'(y)x + \lambda(y)].$$

The mapping (4.73) defines a point transformation acting on (x,t,u,v)-space that projects onto a nonlocal transformation acting on (x,t,u)-space if the coefficient of v is nonzero in the third equation of the mapping.

It is easy to see that the mapping (4.73) yields a nonlocal transformation of the linear PDE (4.62) to the backward heat equation if and only if $V(x,y)$ is of the form (4.71), $V(x,y)$ is not quadratic in x, and $\phi(x,y)$ satisfies the condition

$$\frac{\partial^5}{\partial x^5}\log|\phi(x,y)| \not\equiv 0.$$

Let $\hat{\psi}(z^1,z^2)$ be *any* solution of the backward heat equation $\hat{\psi}_{z^1z^1}+\hat{\psi}_{z^2}=0$. Then from the mapping equations (4.65) it follows that

$$\psi(x,y)=\hat{\psi}(z^1,z^2)\exp\{-\tfrac{1}{4}[\sigma^{-1}\sigma'(y)x^2+2\sigma^{-1}\rho'(y)x+\lambda(y)]\}$$

is a solution of the linear parabolic PDE (4.70), and accordingly, $V(x,y)$ given by the equation (4.71) becomes

$$V(x,y)=a(y)x^2+b(y)x+c(y)-2\sigma^2\left[\frac{\hat{\psi}_{z^2}}{\hat{\psi}}+\left(\frac{\hat{\psi}_{z^1}}{\hat{\psi}}\right)^2\right]-\frac{\sigma'(y)}{\sigma(y)}, \quad (4.74)$$

where $z^1=\sigma(y)x+\rho(y)$, $z^2=\int^y\sigma^2(\hat{y})d\hat{y}$, with $\sigma(y),\rho(y)$ related to $a(y),b(y)$ through the first two ODEs of the system (4.66). Hence *every* solution of the backward heat equation yields a coefficient $V(x,y)$ given by (4.74) for which the corresponding linear parabolic PDE (4.62) can be mapped to the backward heat equation. Moreover, one can prove the following theorem [Exercise 4.3.3].

Theorem 4.3.3. *Let $w=\hat{\psi}(z^1,z^2)$ be a solution of the backward heat equation $w_{z^1z^1}+w_{z^2}=0$. Such a solution yields a coefficient $V(x,y)$ given by* (4.74). *The corresponding linear parabolic PDE* (4.62) *can be mapped to the backward heat equation* only *through a nonlocal transformation if and only if $\hat{\psi}(z^1,z^2)$ is* not *one of the forms*

(I) $\hat{\psi}(z^1,z^2)=e^{(Pz^1-P^2z^2)}$,

(II) $\hat{\psi}(z^1,z^2)=\dfrac{1}{\sqrt{(z^2-\hat{z}^2)}}\exp\left\{\dfrac{(z^1-\hat{z}^1)^2}{4(z^2-\hat{z}^2)}\right\}$,

where $P,\hat{z}^1,\hat{z}^2$ are arbitrary constants.

In Bluman & Shtelen (2004), a recycling procedure [See also Bluman & Reid (1989).] is described that can further extend the class of linear parabolic equations that can be mapped into the heat equation by explicit nonlocal transformations. Interesting special cases include d-Bessel processes of the form

$$\frac{\partial u}{\partial t}+\frac{\partial^2 u}{\partial R^2}+\frac{(d-1)}{R}\frac{\partial u}{\partial R}=0, \quad d=2k+1, \quad k=1,2,\dots .$$

For related work on classes of Schrödinger equations that can be mapped into the free particle equation by nonlocal transformations, see Bluman & Shtelen (1996a).

Exercises 4.3

4.3.1. Consider the potential system $\mathbf{UW}\{x, t\,; u, w\}$ (3.82) of the nonlinear wave equation (4.8) in the case $c(u) = (u + B)^{-2}$.

(a) Find an infinite set of point symmetries of the potential system $\mathbf{UW}\{x, t\,; u, w\}$ (3.82).
(b) Find a point transformation that maps $\mathbf{UW}\{x, t\,; u, w\}$ into a linear PDE system.

4.3.2. Show that the potential system $\mathbf{UB}\{x, t\,; u, \beta\}$ (3.84) of the nonlinear wave equation $u_{tt} = (u^{-4/3}u_x)_x$ has an infinite number of point symmetries. Find the explicit form of a linearizing transformation.

4.3.3. Prove Theorem 4.3.3.

4.4 Discussion

Pucci & Saccomandi (1993) give some necessary conditions for the existence of potential symmetries that arise from the potential system for a given scalar PDE written as a conservation law.

For diffusion-convection equations of the form

$$u_t - [f(u)u_x + k(u)]_x = 0,$$

Sophocleous (1996) classifies all functions $f(u)$ and $k(u)$ for which there exist potential symmetries through analyzing the potential system that arises from the equation as written. He also finds the corresponding potential symmetries.

Chou & Qu (1999) consider the potential system and potential equation, respectively given by

$$\begin{aligned} v_x &= u, \\ v_t &= D(u)(u_x)^n + E(u) \end{aligned} \tag{4.75}$$

and

$$v_t = D(v_x)(v_{xx})^n + E(v_x) \tag{4.76}$$

for the class of diffusion-convection equations of the form

$$u_t - [D(u)(u_x)^n + E(u)]_x = 0. \tag{4.77}$$

They classify the cases when the potential system (4.75) yields a potential symmetry of (4.77) and classify the point symmetries of the potential equation (4.76). It is not noted in this paper that (1) each point symmetry of (4.76) yields a local symmetry of (4.75); and (2) each potential symmetry of

(4.77) that results from a point symmetry of (4.75), must yield a local symmetry (not necessarily a point symmetry) of (4.76). [It is easy to see that the potential system (4.75) and the potential equation (4.76) are locally related.]

Sophocleous (2005) finds potential symmetries of the class of nonlinear diffusion equations with variable coefficients of the form

$$u_t = [g(x)u^n u_x]_x \tag{4.78}$$

by considering the potential system that arises from the equation as written. In particular, he shows that such potential symmetries arise in two cases: (i) $n = -2$, $g(x) = x^2$; (ii) $n = -2$, $g(x) = x^{-2}$. For the first case, he obtains potential symmetries that yield the linearization of PDE (4.78) and also exhibits invariant solutions of (4.78), arising from potential symmetries.

Ivanova, Popovych & Sophocleous (2008a,b) classify potential systems and resulting nonlocal conservation laws and potential symmetries for variable coefficient diffusion-convection equations of the form

$$f(x)u_t - [g(x)f(u)u_x]_x - H(x)G(u)u_x = 0.$$

Ivanova & Sophocleous (2008) classify potential systems and find resulting potential symmetries of systems of diffusion equations of the form

$$\begin{aligned} u_t &= [f(u,v)u_x]_x, \\ v_t &= [g(u,v)u_x]_x. \end{aligned}$$

Senthilvelan & Torrisi (2000) find potential symmetries and resulting invariant solutions for a nonlinear PDE system representing a simplified model for reacting mixtures. Potential symmetries are exhibited that yield the linearization of the given PDE system by a nonlocal transformation.

Bluman, Cheviakov, & Ganghoffer (2008) consider the complete set of equations of nonlinear elasticity in a dynamical context. A tree of nonlocally related systems is constructed that includes both the Lagrange and Euler PDE systems. As a consequence, nonlocal symmetries are found for both systems. Invariant solutions are constructed from such a nonlocal symmetry of the Euler system.

Formally, nonlocal symmetries have been found for PDEs through infinitesimals depending on nonlocal variables that are integrals of the given dependent variables of a given PDE system [Konopelchenko & Mokhnachev [(1979), (1980)]; Kumei (1981); Kapcov (1982); Pukhnachev (1987)]. In these works nonlocal symmetries are not realized as local symmetries of potential systems.

A particular way of obtaining nonlocal symmetries of PDEs is to seek recursion operators, depending on inverse differentiation (integral) operators, that generate sequences of nonlocal symmetries from local symmetries. For further details, see Kapcov (1982), Bluman & Kumei (1989), and Guthrie (1994).

In Krasil'shchik & Vinogradov (1984) [see also Vinogradov & Krasil'shchik (1984); Kersten (1987); Vinogradov (1989); Krasil'shchik & Kersten (2000)], nonlocal symmetries are defined as local symmetries of an associated auxiliary PDE system whose integrability conditions yield the given PDE system. A rather general form is assumed for the auxiliary system which involves unspecified functions. In principle, these unspecified functions are determined by requiring that the integrability conditions of the auxiliary PDE system yield the given PDE system. In order to apply their method (related to an idea introduced by Wahlquist and Estabrook (1975)) it seems that one has to impose very strong assumptions on the form of the unspecified functions.

In the final chapter, the complexity in finding nonlocal symmetries and nonlocal conservation laws of a given PDE system in the case of three or more independent variables is considered. It is seen that in order that such nonlocal symmetries and/or nonlocal conservation laws can arise from local symmetries and/or local conservation laws, respectively, of a potential system, it is necessary to append the potential system with gauge constraints that relate the potential variables. On the other hand, it is shown that local symmetries of nonlocally related systems arising as subsystems of a given PDE system can yield nonlocal symmetries of the given PDE system as in the situation for two independent variables. Moreover, unlike potential systems arising from divergence-type conservation laws, potential systems arising from *lower-degree* (e.g., curl-type) conservation laws may require fewer or no gauge constraints in order to yield nonlocal symmetries and/or nonlocal conservation laws.

Chapter 5
Further Applications of Symmetry Methods: Miscellaneous Extensions

5.1 Introduction

In this chapter, we consider three further topics on symmetry methods for PDEs. In particular, it is shown how symmetry methods can be used and further adapted to *systematically* construct particular solutions of a PDE system, how to find nonlocally-related systems that could yield nonlocal symmetries and/or nonlocal conservation laws in the case of a PDE system with three or more independent variables, and how to find local symmetries and local conservation laws through symbolic manipulation software.

In Chapter 1, it was shown how to find local symmetries of a given PDE system. A local symmetry maps solutions of the PDE system into one-parameter families of solutions. However, there can exist solutions that map into themselves, i.e., are *invariant*, under the action of a local symmetry of the PDE system. Such solutions are called *invariant solutions* (*similarity solutions*) and include the well-known *self-similar solutions* (*automodel solutions*) that result from scaling symmetries. Invariant solutions of PDEs were first considered by Lie (1881) and then more extensively by Ovsiannikov [(1959), (1962), (1982)], Bluman (1967), Bluman & Cole [(1969), (1974)], Olver (1986), and Bluman & Kumei (1989). The method of finding invariant solutions is commonly referred to as the classical method [Bluman (1967); Bluman & Cole (1969)]. The construction of invariant solutions to solve boundary value problems (BVPs) posed for PDEs was presented in Bluman (1967). [See also Bluman & Cole [(1969), (1974)]; Bluman & Kumei (1989); Bluman & Anco (2002).]

Self-similar solutions (as well as traveling wave invariant solutions arising from translation symmetries in space and time) play an essential role in asymptotic analysis. Often, the asymptotic solution of a BVP for a nonlinear PDE is either a self-similar solution or traveling wave solution. Such a solution that also arises from reduction through a dimensional analysis argument is

G.W. Bluman et al., *Applications of Symmetry Methods to Partial Differential Equations*, Applied Mathematical Sciences 168, DOI 10.1007/978-0-387-68028-6_5,

called a *self-similar solution of the first kind* whereas one arising strictly as an invariant solution is called a *self-similar solution of the second kind* [Barenblatt & Zel'dovich (1972) and Barenblatt [(1979), (1987), (1996)]]. Comprehensive reviews of self-similar asymptotics appear in Newman (1984), Galaktionov et al. (1988), and Galaktionov & Svirshchevskii (2007). For applications of self-similar and traveling wave asymptotics to physical problems, see Barenblatt & Zel'dovich (1972) and Barenblatt [(1979), (1987), (1996)] and Goldenfeld (1992). Barenblatt & Zel'dovich (1972), and Barenblatt [(1979), (1987), (1996)] consider examples of "intermediate asymptotics" where, in an intermediate space-time domain, the solution of a BVP is approximated by a similarity solution that satisfies neither the given nor asymptotic boundary conditions. In such examples the similarity solution is not an equilibrium state. Kamin (1975) rigorously justified the evolution of the solution of a porous medium equation to a self-similar solution. There are many papers that rigorously justify self-similar asymptotics [e.g., Atkinson & Peletier (1974); Friedman & Kamin (1980); and Galaktionov & Samarskii (1984)]. A wealth of examples appears in the excellent book by Galaktionov & Svirshchevskii (2007).

An invariant solution of a given PDE system arises from invariants of a local symmetry of the PDE that satisfy an auxiliary PDE system (which plays the role of a constraint called the *invariant surface condition*). The resulting invariant solution satisfies both the invariant surface condition and the given PDE system. The classical method to find invariant solutions can be generalized to the nonclassical method. Here one considers an augmented system of PDEs consisting of the given PDE system and an unknown constraint system and seeks symmetries that leave invariant this augmented system such that the invariant surface condition is the unknown constraint system itself. In general, such "symmetries" do not map solutions of the given PDE system into one-parameter families of solutions but are useful to find further specific solutions beyond those obtained by the classical method. However, in the nonclassical method, the (over-determined) system of determining equations for symmetries is nonlinear unlike the situation in the calculations for local symmetries. By construction, solutions obtained by the nonclassical method include those obtained by the classical method. The nonclassical method was introduced in Bluman (1967) and Bluman & Cole (1969) with the restriction that the invariant surface condition is of the form arising for point symmetries. Further discussions of the nonclassical method appear in Levi & Winternitz (1989), Olver & Rosenau [(1986), (1987)], Nucci & Clarkson (1992) and Clarkson & Mansfield (1994a,b). In Fokas & Liu (1994), the nonclassical method is extended to include an invariant surface condition of the form that arises for local symmetries without restriction to the form arising for point symmetries.

It is shown that further solutions can arise for a given PDE system through consideration of invariant solutions of nonlocally related systems. In particular, an invariant solution of a nonlocally related system that arises from a nonlocal symmetry of the given PDE system could yield solutions of the given PDE system that arise neither as invariant solutions (classical method) nor from use of the nonclassical method. Moreover, a direct application of the nonclassical method to a nonlocally related system can yield still further solutions of a given PDE system [Bluman & Yan (2005)].

Other extensions that attempt to seek further solutions of a given PDE system include

- the use of the invariant form arising from a potential symmetry (without seeking specific solutions of the related potential system) to seek solutions of the given PDE system [Pucci & Saccomandi (1993)];
- the use of a one-to-one change of the variables in a potential system to the canonical coordinates arising from a potential symmetry of the potential system. Solutions of the given PDE system are then sought in which one of the dependent variables is allowed to have a dependence on the translated canonical coordinate (which is not an invariant of the potential symmetry). This modification has been shown to yield new solutions that are not obtainable by the above-listed procedures [Cheviakov (2008)]. A related method was suggested in Sjöberg & Mahomed (2004).

However, it often happens that a seemingly distinct solution arising from one of the above-mentioned methods is not distinct [Bluman & Yan (2005), Cheviakov (2008)].

A second important topic considered in this chapter is the study of nonlocally related systems in $n \geq 3$ dimensions (i.e., at least three independent variables). It is shown that the direct (and obvious) way of constructing potential systems in $n \geq 3$ dimensions, arising from local conservation laws, leads to PDE systems that are under-determined, i.e., subject to gauge freedom. It turns out that a useful potential system requires the appending of *gauge constraints* that relate the potential variables in such a way that any solution of the appended potential system yields a solution of the given PDE system and, vice versa, any solution of the given PDE system yields a solution of the appended potential system. In particular, it is impossible to obtain nonlocal symmetries or nonlocal conservation laws for such potential systems, without introducing gauge constraints [Anco & Bluman (1997b); Anco & The (2005)]. In general, for a given PDE system, the selection of appropriate gauge constraints remains an open problem.

In spite of being systematic, the computation of local symmetries and local conservation laws of PDE systems presents a significant computational challenge. For many real problems, the over-determined linear systems of determining equations for finding multipliers of conservation laws or infinitesimals of local symmetries often include hundreds (or thousands) of equations; those for the nonclassical method are even worse since here one has to deal with nonlinear systems of determining equations. Therefore various symbolic computation software packages have been developed.

A complete symbolic package for computations of symmetries and/or conservation laws naturally consists of two parts. The first part contains user routines that interpret equations, use a specified ansatz for symmetry components or conservation law multipliers, generate determining equations, and split them into an over-determined linear (or, in the nonclassical method, nonlinear) PDE system. The second part contains routines for effective symbolic reduction (and possibly the solution) of large over-determined PDE systems. Since PDE systems often contain arbitrary constitutive functions and/or parameters, it is highly beneficial when the reduction algorithm includes options for case splitting, i.e., the isolation of special forms of constitutive functions and/or parameters, for which additional symmetries or conservation laws arise.

Two popular approaches for the symbolic solution of large over-determined PDE systems are based on differential Gröbner bases and the characteristic set method, respectively. To name a few, the programs `DIFFGROB2` [Mansfield (1993)], `standard_form` [Reid (1991a)], `rif` [Reid, Wittkopf & Boulton (1996)], `CRACK` [Wolf & Brand (1992)] belong to the first class; a program developed for computer algebra system (CAS) `Mathematica` [Temuerchaolu (2003), see also Wu (1984)] and a package `diffalg` for `Maple` [Boulier et al. (1995)] belong to the second class. For a detailed review, see Hereman [(1996), (1997), (2005)].

To date, a variety of packages for the computation of symmetries and/or conservation laws has been developed for different computer algebra systems. For example, a set of programs `LiePDE`, `ApplySym` and `ConLaw` [Wolf (2002)] provides a user interface for local conservation law and symmetry computation in CAS `REDUCE`, subsequently using `CRACK` for the reduction and solution of linear over-determined systems. The package `GeM` [Cheviakov (2007)] for `Maple` offers a set of routines that generates the splitting of the over-determined systems of determining equations for the computations of local conservation laws, symmetries and approximate symmetries. It subsequently uses `rif` for solving over-determined systems, and contains another set of routines for the computations of fluxes of conservation laws and conservation law/symmetry output. In his programs [Temuerchaolu (2003)] for `Mathematica` for symmetry and conservation law computations, Temuerchaolu uses his own over-determined systems solver, which, unlike `rif` and `CRACK`, requires continuous user input.

5.2 Applications of Symmetry Methods to the Construction of Solutions of PDEs

Consider a PDE system $\mathbf{R}\{x\,;u\}$ of N PDEs of order k with n independent variables $x = (x^1, \ldots, x^n)$ and m dependent variables $u = (u^1, \ldots, u^m)$, given by

$$R^\sigma(x, u, \partial u, \ldots, \partial^k u) = 0, \quad \sigma = 1, \ldots, N, \tag{5.1}$$

that has the point symmetry with the infinitesimal generator

$$\mathrm{X} = \xi^i(x,u)\frac{\partial}{\partial x^i} + \eta^\mu(x,u)\frac{\partial}{\partial u^\mu}, \tag{5.2}$$

or, equivalently, in evolutionary form, the infinitesimal generator

$$\hat{\mathrm{X}} = (\eta^\mu(x,u) - \xi^i(x,u)u_i^\mu)\frac{\partial}{\partial u^\mu}. \tag{5.3}$$

Let $\xi(x,u) = (\xi^1(x,u), \ldots, \xi^n(x,u))$ and assume that $\xi(x,u) \not\equiv 0$.

Definition 5.2.1. $u = \Theta(x)$, with components $u^\nu = \Theta^\nu(x)$, $v = 1, \ldots, m$, is an *invariant solution* of the PDE system $\mathbf{R}\{x\,;u\}$ (5.1) resulting from the point symmetry (5.2) if and only if

(i) $u^\nu = \Theta^\nu(x)$ is an invariant surface of the point symmetry (5.2) for each $\nu = 1, \ldots, m$.

(ii) $u = \Theta(x)$ is a solution of $\mathbf{R}\{x\,;u\}$ (5.1).

It follows that $u = \Theta(x)$ is an invariant solution of the PDE system $\mathbf{R}\{x\,;u\}$ resulting from the point symmetry (5.2), if and only if $u = \Theta(x)$ satisfies

(i)

$$\mathrm{X}(u^\nu - \Theta^\nu(x)) = 0 \text{ when } u = \Theta(x), \quad \nu = 1, \ldots, m \tag{5.4a}$$

$$\leftrightarrow \mathrm{X}(u^\nu - \Theta^\nu(x))\big|_{u=\Theta(x)}, \quad \nu = 1, \ldots, m \tag{5.4b}$$

$$\leftrightarrow \eta^\nu(x, \Theta(x)) - \xi^i(x, \Theta(x))\frac{\partial \Theta^\nu(x)}{\partial x^i} = 0, \quad \nu = 1, \ldots, m \tag{5.4c}$$

$$\leftrightarrow \hat{\mathrm{X}}u^\nu\Big|_{u=\Theta(x)} = 0, \quad \nu = 1, \ldots, m; \tag{5.4d}$$

(ii)

$$R^\sigma(x,u,\partial u,\ldots,\partial^k u)=0 \text{ when } u=\Theta(x), \quad \sigma=1,\ldots,N \tag{5.5a}$$

$$\leftrightarrow R^\sigma(x,\Theta(x),\partial\Theta(x),\ldots,\partial^k\Theta(x))=0, \quad \sigma=1,\ldots,N \tag{5.5b}$$

$$\leftrightarrow R^\sigma(x,u,\partial u,\ldots,\partial^k u)\big|_{u=\Theta(x)}=0, \quad \sigma=1,\ldots,N. \tag{5.5c}$$

In (5.5b), $\partial^j\Theta(x)$ denotes the components $\partial^j\Theta^\mu(x)/(\partial x^{i_1}\ldots\partial x^{i_j})$, $\mu=1,\ldots,m$, for $i_j=1,\ldots,n$ with $j=1,\ldots,k$. The solutions of equations (5.4) are *invariant surfaces* of the point symmetry (5.2). Equations (5.4) and (5.5) define the *classical method* to obtain particular solutions of a PDE system $\mathbf{R}\{x\,;u\}$ (5.1).

5.2.1 The classical method

In summary, $u=\Theta(x)$ is a solution (invariant solution) of the PDE system $\mathbf{R}\{x\,;u\}$ (5.1) obtained through the *classical method* [Lie (1881)] if and only if there exists a Lie group of point transformations with infinitesimal generator X given by (5.2) [$\hat{\mathrm{X}}$ given by (5.3)], with its kth extension $\mathrm{X}^{(k)}$ given by (1.12), such that

(i)

$$\mathrm{X}^{(k)}\,R^\sigma(x,u,\partial u,\ldots,\partial^k u)\big|_{\{R^\lambda(x,u,\partial u,\ldots,\partial^k u)=0\}_{\lambda=1}^N}=0, \quad \sigma=1,\ldots,N; \tag{5.6}$$

(ii)

$$\hat{\mathrm{X}}u^\nu\Big|_{u=\Theta(x)}=0, \quad \nu=1,\ldots,m; \tag{5.7}$$

(iii)

$$R^\sigma(x,u,\partial u,\ldots,\partial^k u)\big|_{u=\Theta(x)}=0, \quad \sigma=1,\ldots,N. \tag{5.8}$$

Having found a point symmetry with infinitesimal generator X given by (5.2) through solving the linear system of determining equations (5.6), one can proceed in two ways to solve the systems of equations (5.7) and (5.8) to find an invariant solution $u=\Theta(x)$, as follows.

(1) Invariant form method
Here one first solves the invariant surface conditions (5.7) by explicitly solving the corresponding characteristic equations for $u=\Theta(x)$ given by

$$\frac{dx^1}{\xi^1(x,u)}=\cdots=\frac{dx^n}{\xi^n(x,u)}=\frac{du^1}{\eta^1(x,u)}=\cdots=\frac{du^m}{\eta^m(x,u)}. \tag{5.9}$$

If $z^1(x,u),\ldots,z^{n-1}(x,u),h^1(x,u),\ldots,h^m(x,u)$, are $n+m-1$ functionally independent constants of integration that arise from solving the characteristic system of ODEs (5.9) with the Jacobian $\partial(h^1,\ldots,h^m)/\partial(u^1,\ldots,u^m)\neq 0$, then the general solution $u=\Theta(x)$ of the invariant surface condition equations (5.7) is given implicitly by the invariant form

$$h^\nu(x,u)=H^\nu(z^1(x,u),\ldots,z^{n-1}(x,u)), \tag{5.10}$$

where H^ν is an arbitrary differentiable function of its arguments, $\nu=1,\ldots,m$. Note that $z^1(x,u),\ldots,z^{n-1}(x,u),h^1(x,u),\ldots,h^m(x,u)$, are $n+m-1$ functionally independent invariants of the one-parameter Lie group of point transformations with the infinitesimal generator X given by (5.2), and hence are $n+m-1$ canonical coordinates for the one-parameter Lie group of point transformations with the infinitesimal generator X given by (5.2). Let $z^n(x,u)$ be the $(n+m)$th canonical coordinate satisfying $\mathrm{X}z^n=1$. If the PDE system $\mathbf{R}\{x\,;u\}$ (5.1) is transformed by the corresponding invertible point transformation into a PDE system $\mathbf{S}\{z\,;h\}$ with independent variables $z=(z^1,\ldots,z^n)$ and dependent variables $h=(h^1,\ldots,h^m)$, then the transformed PDE system $\mathbf{S}\{z\,;h\}$ has the translation point symmetry given by

$$\begin{aligned}(z^*)^i &= z^i, \quad i=1,\ldots,n-1,\\ (z^*)^n &= z^n+\varepsilon,\\ (h^*)^\nu &= h^\nu, \quad \nu=1,\ldots,m.\end{aligned}$$

Thus the variable z^n does not appear explicitly in the transformed PDE system $\mathbf{S}\{z\,;h\}$, and hence the transformed PDE system has *particular* solutions of the form (5.10) that in turn define, implicitly, *specific* functions $u=\Theta(x)$ which are invariant solutions of the PDE system $\mathbf{R}\{x\,;u\}$ (5.1), i.e., the PDE system $\mathbf{R}\{x\,;u\}$ (5.1) has invariant solutions implicitly given by the invariant form (5.10). In particular, these invariant solutions are found by solving a reduced system of DEs with $n-1$ independent variables $z^1,\ldots,z^{n-1}$ and m dependent variables $h^1,\ldots,h^m$. The variables $z^1,\ldots,z^{n-1}$ are commonly called *similarity variables.* The reduced system of DEs is found by substituting the invariant form (5.10) into the given PDE system $\mathbf{R}\{x\,;u\}$ (5.1). It is assumed that this substitution does not lead to a DE system with a singular equation. Note that if $\partial\xi/\partial u\equiv 0$, as is commonly the case, then $z^i=z^i(x)$, $i=1,\ldots,n-1$. In the case when $\mathbf{R}\{x\,;u\}$ (5.1) has two independent variables, i.e., $n=2$, the reduced system of DEs is an ODE system with independent variable $z=z^1$.

(2) Direct substitution method

This procedure is essential if one is unable to solve explicitly the invariant surface condition equations (5.7), i.e., if one is unable to obtain the general solution of the characteristic ODE system (5.9). Without loss of generality, one can assume that $\xi^n(x,u) \neq 0$. Then the first-order PDE system (5.7) can be written as

$$\frac{\partial u^\nu}{\partial x^n} = \frac{\eta^\nu(x,u)}{\xi^n(x,u)} - \sum_{i=1}^{n-1} \frac{\xi^i(x,u)}{\xi^n(x,u)} \frac{\partial u^\nu}{\partial x^i}, \quad \nu = 1, \ldots, m. \tag{5.11}$$

From (5.11) and its differential consequences, it follows that any term involving derivatives of components of u with respect to the independent variable x^n can be expressed in terms of components of x and u as well as derivatives of components of u with respect to the independent variables $x^1, \ldots, x^{n-1}$. Hence, after directly substituting (5.11) and its differential consequences for any partial derivative with respect to x^n appearing in the given PDE system $\mathbf{R}\{x\,;u\}$ (5.1), one obtains a reduced DE system directly involving the m dependent variables $u^1, \ldots, u^m$, the $n-1$ independent variables $x^1, \ldots, x^{n-1}$, derivatives of $u^1, \ldots, u^m$ with respect to $x^1, \ldots, x^{n-1}$, and the *parameter* x^n. A solution $u = \Phi(x^1, \ldots, x^{n-1}; x^n)$ of this reduced DE system yields an invariant solution $u = \Theta(x)$ of the given PDE system $\mathbf{R}\{x\,;u\}$ (5.1) provided that the invariant surface condition equations (5.7) or, equivalently, the given PDE system $\mathbf{R}\{x\,;u\}$ (5.1) itself, are also satisfied. In the case when $\mathbf{R}\{x\,;u\}$ (5.1) has two independent variables, i.e., $n = 2$, the reduced system of DEs is an ODE system. Here the constants of integration that appear in the general solution of the reduced ODE system are arbitrary functions of the parameter x^n, and these arbitrary functions are then determined by substituting the general solution into either the invariant surface condition equations (5.7) or the given PDE system $\mathbf{R}\{x\,;u\}$ (5.1).

For examples of invariant solutions of PDEs, the reader is referred to the books of Ovsiannikov [(1962), (1982)], Bluman & Cole (1974), Olver (1986), Bluman & Kumei (1989), Stephani (1989), Hydon (2000), Bluman & Anco (2002) and Cantwell (2002).

Extension of the classical method to higher-order symmetries

The classical method to find invariant solutions can be easily extended to find invariant solutions arising from reductions due to higher-order symmetries of a given PDE system $\mathbf{R}\{x\,;u\}$ (5.1). Here it is assumed that $\mathbf{R}\{x\,;u\}$ is a scalar PDE of order k with two independent variables $(x,\, t)$ and dependent variable u, given by

$$R(x,t,u,\partial u, \ldots, \partial^k u) = 0, \tag{5.12}$$

that has the local symmetry, with the infinitesimal generator

$$\hat{X} = \eta(x, t, u, \partial u, \ldots, \partial^s u)\frac{\partial}{\partial u}. \tag{5.13}$$

Note that the authors are unaware of the existence of higher-order symmetries ($s \geq 2$) of scalar PDEs, or "non-degenerate" PDE systems, with more than two independent variables. It has been suggested by Vinogradov (1989) that higher-order symmetries cannot exist for PDEs with more than two independent variables.

Definition 5.2.2. $u = \Theta(x, t)$ is an *invariant solution* of the PDE $\mathbf{R}\{x\,;u\}$ (5.12) resulting from the local symmetry (5.13) if and only if

(i) $u = \Theta(x, t)$ is an invariant surface of the local symmetry (5.13).
(ii) $u = \Theta(x, t)$ is a solution of $\mathbf{R}\{x\,;u\}$ (5.12).

Hence, $u = \Theta(x, t)$ is a solution (invariant solution) of the PDE system $\mathbf{R}\{x\,;u\}$ (5.12) obtained through the extended classical method if and only if there exists a one-parameter group of local transformations with infinitesimal generator $\hat{X}$ given by (5.13) (with its extension $\hat{X}^\infty$ given by (1.38)), such that

(i)
$$\hat{X}^\infty\, R(x, t, u, \partial u, \ldots, \partial^k u)\big|_{R(x,t,u,\partial u,\ldots,\partial^k u)=0,} = 0; \tag{5.14}$$

(ii)
$$\hat{X}u\Big|_{u=\Theta(x,t)} = 0; \tag{5.15}$$

(iii)
$$R(x, t, u, \partial u, \ldots, \partial^k u)\big|_{u=\Theta(x,t)} = 0. \tag{5.16}$$

Note that equation (5.15) corresponds to the invariant surface condition equation

$$\eta(x, t, \Theta(x, t), \partial\Theta(x, t), \ldots, \partial^k\Theta(x, t)) = 0. \tag{5.17}$$

At first sight it would appear that if the local symmetry (5.13) is not a point symmetry, one is likely unable to solve (5.17) and hence the extended classical method would appear not to be useful to find invariant solutions resulting from local symmetries that are not point symmetries. However, as it is now seen, the situation is not so bleak in the case when the PDE (5.12) is an evolutionary PDE of the form

$$u_t = G(x, t, u, u^{(1)}, \ldots, u^{(k)}) \tag{5.18}$$

with $u^{(1)} = u_x$, $u^{(2)} = u_{xx}$, etc. Through the evolutionary PDE (5.18) and its differential consequences, it follows that the t-derivatives of u can be expressed as functions of t, x, u, and x-derivatives of u. Hence, without loss of generality, each local symmetry of the evolutionary PDE (5.18) can be represented by an infinitesimal generator of the form

$$\hat{\mathrm{X}} = \eta(x, t, u, u^{(1)}, \ldots, u^{(p)})\frac{\partial}{\partial u}, \tag{5.19}$$

for some integer p. Consequently, here the invariant surface condition equation (5.17) for an invariant solution $u = \Theta(x,t)$ becomes

$$\eta\left(x, t, u, \frac{\partial u}{\partial x}, \ldots, \frac{\partial^p u}{\partial x^p}\right) = 0. \tag{5.20}$$

Equation (5.20) is a pth-order ODE with respect to the dependent variable u and the independent variable x; t can be treated as a parameter in the ODE (5.20). The general solution of the ODE (5.20) is an invariant form

$$\Phi(x, t, u, c_1(t), \ldots, c_p(t)) = 0, \tag{5.21}$$

where Φ is a specific function of its arguments; the p integration constants $c_1(t), \ldots, c_p(t)$ are arbitrary functions of the parameter t. After solving (5.21) for u to obtain an expression

$$u = \phi(x, t, c_1(t), \ldots, c_p(t)), \tag{5.22}$$

one substitutes (5.22) for u in the evolution equation (5.18) to determine these p arbitrary functions: $c_1 = C_1(t), \ldots, c_p = C_p(t)$, to obtain the corresponding invariant solution $u = \Theta(x,t) = \phi(x, t, C_1(t), \ldots, C_p(t))$.

5.2.2 The nonclassical method

The nonclassical method, introduced in Bluman (1967) [cf. Bluman & Cole (1969)], generalizes and includes Lie's classical method *for obtaining solutions of PDEs.* Here one first seeks functions $\xi^i(x,u)$, $\eta^\mu(x,u)$, $i = 1, \ldots, n$, $\mu = 1, \ldots, m$, so that (5.2) is a "symmetry" ("*nonclassical symmetry*") of the *augmented PDE system* $\mathbf{A}\{x\,;u\}$ consisting of the given PDE system $\mathbf{R}\{x\,;u\}$ (5.1), the invariant surface condition equations

$$I^\nu(x, u, \partial u) = \eta^\nu(x,u) - \xi^i(x,u)\frac{\partial u^\nu}{\partial x^i} = 0, \quad \nu = 1, \ldots, m, \tag{5.23}$$

and the differential consequences of (5.23). Consequently, one obtains an overdetermined set of *nonlinear* determining equations for the unknown functions $\xi^i(x,u)$, $\eta^\mu(x,u)$, $i = 1, \ldots, n$, $\mu = 1, \ldots, m$. It is straightforward to show that, for any set of $\xi^i(x,u)$, $\eta^\mu(x,u)$, $i = 1, \ldots, n$, $\mu = 1, \ldots, m$, (5.2) is a symmetry of the invariant surface condition equations (5.23), and from this it follows that the nonclassical method includes Lie's classical method [Exercise 5.2-1]. The resulting set of determining equations is nonlinear due to the substitution of the equations (5.23) (each written in solved form with respect to some derivative term) and their differential consequences into the symmetry determining equations (5.6) that now hold only for solutions of the augmented PDE system. In the nonclassical method, the invariant surface condition equations (5.23) are essentially a set of constraint equations of a specific form. In particular, the nonclassical method is equivalent to seeking all solutions of the PDE system (5.1) of the form (5.23) for *any* possible set of $\xi^i(x,u)$, $\eta^\mu(x,u)$, $i=1,\ldots,n$, $\mu=1,\ldots,m$. The set of determining equations satisfied by $\xi^i(x,u)$, $\eta^\mu(x,u)$, $i=1,\ldots,n$, $\mu=1,\ldots,m$, are the compatibility conditions for the existence of solutions of the augmented PDE system $\mathbf{A}\{x\,;u\}$ that includes the PDE system $\mathbf{R}\{x\,;u\}$ and the constraint equations (5.23).

A "nonclassical symmetry" is not a symmetry of a given PDE system $\mathbf{R}\{x\,;u\}$ (5.1) unless the infinitesimals yielding an infinitesimal generator (5.2) yield a point symmetry of $\mathbf{R}\{x\,;u\}$. Otherwise, a mapping resulting from such an infinitesimal generator maps *no* solution of $\mathbf{R}\{x\,;u\}$ (5.1) into a different solution of $\mathbf{R}\{x\,;u\}$. It just maps the solution obtained by the nonclassical method into itself! This is why in the paper of Bluman & Cole (1969), the phrase "nonclassical symmetry" was never used in conjunction with the presentation of the nonclassical method. [Unfortunately, the phrase "nonclassical group" was used in the PhD thesis of Bluman (1967).] In other words, strictly speaking, the nonclassical method is not a "symmetry" method but an extension of Lie's symmetry method ("classical method") for the purpose of finding specific solutions of PDEs.

Definition 5.2.3. A solution of a given PDE system $\mathbf{R}\{x\,;u\}$ (5.12) is a *nonclassical solution* if it arises as an invariant solution of the augmented PDE system $\mathbf{A}\{x\,;u\}$ (consisting of the PDE system $\mathbf{R}\{x\,;u\}$ (5.12), the constraint equations (5.23) and their differential consequences), and does not arise as an invariant solution of the given PDE system $\mathbf{R}\{x\,;u\}$ (5.12) with respect to its local symmetries.

[Note that the existence of a "nonclassical symmetry" does not guarantee that the corresponding invariant solution is a nonclassical solution of the PDE system $\mathbf{R}\{x\,;u\}$ (5.12). The invariant solution might also arise as an invariant solution of $\mathbf{R}\{x\,;u\}$ (5.12) with respect to some point symmetry of $\mathbf{R}\{x\,;u\}$ (5.12).]

The situation for a scalar PDE with two independent variables

Now consider the situation of a scalar PDE (5.12) with two independent variables. Let $x^1=x$, $x^2=t$, $\xi^1=\xi(x,t,u)$, $\xi^2=\tau(x,t,u)$. Then the set of invariant surface condition equations (5.23) becomes the invariant surface condition equation

$$\xi(x,t,u)u_x + \tau(x,t,u)u_t = \eta(x,t,u). \tag{5.24}$$

For a specific set of $(\xi(x,t,u), \tau(x,t,u), \eta(x,t,u))$, the general solution of the invariant surface condition (5.24) can be represented in the form

$$z(x,t,u) = \text{const} = c_1, \tag{5.25a}$$

$$H(x,t,u) = \text{const} = c_2 = h(z), \tag{5.25b}$$

where $z(x,t,u)$ is a similarity variable. After solving equation (5.25b) for u, one obtains an *ansatz*

$$u = \phi(x,t,h(z(x,t,u))) \tag{5.26}$$

for solutions of the scalar PDE (5.12).

If a specific set of $(\xi(x,t,u), \tau(x,t,u), \eta(x,t,u))$ is a set of infinitesimals for a point symmetry of the PDE (5.12), then the dependence of ϕ on x, t, and $h(z)$ is explicit in the ansatz (5.26); $h(z)$ is an arbitrary function of the similarity variable z. Here, the substitution of the ansatz (5.26) into the scalar PDE (5.12) yields a reduced ODE of order at most k with independent variable z and dependent variable $h(z)$. Each solution of this ODE yields an invariant solution, obtainable by the classical method, of the PDE (5.12).

If $\xi_u = \tau_u \equiv 0$, then $z(x,t,u) \equiv z(x,t)$, and the ansatz (5.26) reduces to the form

$$u = \phi(x,t,h(z(x,t))). \tag{5.27}$$

If $\xi_u = \tau_u = \eta_{uu} \equiv 0$, the ansatz (5.26) further reduces to the form

$$u = A(x,t) + B(x,t)h(z(x,t)). \tag{5.28}$$

In the ansatz (5.28), the functions $A(x,t)$ and $B(x,t)$ are explicitly known for a specific set of functions $(\xi(x,t), \tau(x,t), \eta(x,t,u))$.

Now suppose that one is able to obtain the sets of all infinitesimals $(\xi(x,t,u), \tau(x,t,u), \eta(x,t,u))$ of the symmetries $\mathrm{X} = \xi(x,t,u)\,\partial/\partial x + \tau(x,t,u)\,\partial/\partial t + \eta(x,t,u)\,\partial/\partial u$ of the augmented system $\mathbf{A}\{x\,;u\}$ consisting of the PDE (5.12), the constraint invariant surface condition equation (5.24), and the differential consequences of (5.24). From the above discussion, it follows that the set of all solutions $u = \Phi(x,t)$ of the PDE (5.12), arising from the nonclassical method, includes the set of *all* solutions of PDE (5.12) that are of the form $u = \phi(x,t,h(z(x,t,u)))$ where $h(z)$ satisfies a reduced ODE.

Hence the solutions of the PDE (5.12), obtained by the nonclassical method, include all solutions of the PDE obtained by the *direct method* of Clarkson & Kruskal (1989) since the direct method aims to find all solutions of the PDE (5.12) that are of the ansatz (5.26) with the restriction that $z(x,t,u) \equiv z(x,t)$ [Nucci & Clarkson (1992)].

From the nature of the constraint invariant surface condition equation (5.24), without loss of generality, in using the nonclassical method, two simplifying cases need only be considered when solving the determining equations for $(\xi(x,t), \tau(x,t), \eta(x,t,u))$, namely, $\tau \equiv 1$; $\tau \equiv 0$, $\xi \equiv 1$. This follows from the observations that if $\tau \neq 0$, then the constraint invariant surface condition equation (5.24) can be divided through by τ, and hence, without loss of generality, one can set $\tau \equiv 1$, so that there are really only two independent infinitesimals; similarly if $\tau \equiv 0, \xi \neq 0$, then the constraint invariant surface condition equation (5.24) can be divided through by ξ, and hence, without loss of generality, one can set $\xi \equiv 1$, so that here there is really only one independent infinitesimal.

Note that for a given set of infinitesimals $(\xi(x,t), \tau(x,t), \eta(x,t,u))$ that satisfies the nonlinear determining equations, one can use either the invariant form or direct substitution method to find the resulting solutions of the scalar PDE (5.12).

Examples

Now consider examples of using the nonclassical method to obtain solutions of PDEs.

(1) Heat equation
The first PDE considered through the nonclassical method was the linear heat equation [Bluman (1967), Bluman & Cole (1969)]

$$u_t - u_{xx} = 0. \tag{5.29}$$

Case 1. The classical method
The classical method determining equation (5.6) that yields the point symmetries $\mathrm{X} = \xi(x,t,u)\,\partial/\partial x + \tau(x,t,u)\,\partial/\partial t + \eta(x,t,u)\,\partial/\partial u$ of the linear heat equation (5.29) is given by

$$\begin{aligned} &\tau_{uu} u_x^2 u_t + \xi_{uu} u_x^3 + 2\tau_u u_{xt} u_x + 2(\tau_{xu} + \xi_u) u_x u_t \\ &\quad + (2\xi_{xu} - \eta_{uu}) u_x^2 + 2\tau_x u_{xt} + (\tau_{xx} - \tau_t + 2\xi_x) u_t \\ &\quad + (\xi_{xx} - \xi_t - 2\eta_{xu}) u_x + (\eta_t - \eta_{xx}) = 0. \end{aligned} \tag{5.30}$$

In the determining equation (5.30), one treats $x, t, u, u_t, u_x, u_{xt}$ as independent variables. Since (5.30) has the form of a polynomial in u_t, u_x, u_{xt}, it consequently splits into nine equations that result from equating to zero the coefficients of like polynomial terms involving derivatives of u. This yields the well-known general solution of (5.30) given by

$$\begin{aligned} \xi(x,t,u) &\equiv \xi(x,t) = \alpha_1 + \alpha_2 x + \alpha_3 t + \alpha_4 xt, \\ \tau(x,t,u) &\equiv \tau(t) = 2\alpha_2 t + \alpha_4 t^2 + \alpha_5, \\ \eta(x,t,u) &= [-\tfrac{1}{2}\alpha_3 x - \alpha_4(\tfrac{1}{4}x^2 + \tfrac{1}{2}t) + \alpha_6]u + g(x,t), \end{aligned} \tag{5.31}$$

where $\alpha_1, \ldots, \alpha_6$ are arbitrary constants and, due to the linearity of PDE (5.29), $g(x,t)$ is an arbitrary solution of the heat equation, i.e., $g_t - g_{xx} = 0$. The resulting invariant solutions of the heat equation appeared in Bluman (1967) and Bluman & Cole (1969).

Case 2. The nonclassical method: $\tau \equiv 1$

If $u = \Phi(x,t)$ satisfies the augmented PDE system $\mathbf{A}\{x\,;u\}$ consisting of the linear heat equation (5.29), the corresponding constraint invariant surface condition

$$u_t = \eta(x,t,u) - \xi(x,t,u)u_x, \tag{5.32}$$

and the differential consequences of (5.32), it follows that all t-derivatives of u and all higher-order x-derivatives of u in the classical symmetry determining equation (5.30) can be expressed as polynomials in u_x, with coefficients that are functions of x, t, and u. In particular, after differentiating (5.32) with respect to x, and then replacing u_{xx} $(= u_t)$ by the right-hand side of (5.32), one obtains

$$u_{xt} = (\eta_x - \xi\eta) + (\eta_u - \xi_x + \xi^2)u_x - \xi_u u_x^2. \tag{5.33}$$

Consequently, after replacing u_t by the right-hand side of (5.32), and u_{xt} by the right-hand side of (5.33), the classical method determining equation (5.30) for the infinitesimals $(\xi(x,t,u), \tau(x,t,u), \eta(x,t,u))$ becomes the nonclassical method determining equation for the infinitesimals $(\xi(x,t,u), \eta(x,t,u))$ given by

$$\begin{aligned} &\xi_{uu}u_x^3 + (2\xi_{xu} - \eta_{uu} - 2\xi\xi_u)u_x^2 \\ &\quad +(\xi_{xx} - \xi_t - 2\eta_{xu} - 2\xi\xi_x + 2\eta\xi_u)u_x \\ &\quad +(\eta_t - \eta_{xx} + 2\eta\xi_x) = 0. \end{aligned} \tag{5.34}$$

Equation (5.34) is a polynomial equation in u_x, and hence splits into four equations whose solution is given by

$$\begin{aligned} \xi &= \xi(x,t), \\ \eta &= C(x,t)u + D(x,t), \end{aligned} \tag{5.35}$$

where $\{\xi(x,t),\ C(x,t),\ D(x,t)\}$ is any solution of the nonlinear system

$$\begin{aligned} \xi_t - \xi_{xx} + 2\xi\xi_x + 2C_x &= 0, \\ C_t - C_{xx} + 2\xi_x C &= 0, \\ D_t - D_{xx} + 2\xi_x D &= 0. \end{aligned} \tag{5.36}$$

This case was considered in more detail in Bluman & Cole (1969). Note that due to the form of (5.35), it follows that here all obtained solutions of the linear heat equation (5.29) are of the form (5.28).

Case 3. The nonclassical method: $\tau \equiv 0,\ \xi \equiv 1$
Here it is easy to show that after using the conditional invariant surface condition equation $u_x = \eta$ and its differential consequences, the classical method determining equation (5.30) for the infinitesimals $(\xi(x,t,u), \tau(x,t,u), \eta(x,t,u))$ becomes the nonclassical method determining equation for the infinitesimal $\eta(x,t,u)$ given by

$$\eta^2\eta_{uu} + 2\eta\eta_{xu} + \eta_{xx} - \eta_t = 0. \tag{5.37}$$

Note that

$$\eta = -\tfrac{1}{2}\sigma(x,t)u \tag{5.38}$$

solves the determining equation (5.37) if $v = \sigma(x,t)$ is any solution of the Burgers equation

$$v_t + vv_x - v_{xx} = 0. \tag{5.39}$$

Consequently, equation (5.38) together with the conditional invariant surface condition equation $u_x = \eta$, yields the Hopf–Cole transformation

$$v = -2\frac{u_x}{u}$$

that relates solutions of the Burgers equation (5.39) and the linear heat equation (5.29) through the nonclassical method. This case was first considered in Fushchich et al. (1992). [See also Appendix 7 in Fushchich, Shtelen & Serov (1993).]

Note that in this case, due to the form of an obtained infinitesimal that satisfies the determining equation (5.37), a resulting solution of the linear heat equation (5.29) is of the form

$$u = \phi(x,t,h(t)). \tag{5.40}$$

where $h(t)$ satisfies a reduced ODE.

(2) Boussinesq equation
The nonclassical method essentially lay dormant for two decades. A significant discussion of it appeared in the papers of Olver & Rosenau [(1986), (1987)] in the context of finding solutions of PDEs subject to differential constraints. A revived interest in the nonclassical method was ignited by the remarkable paper of Clarkson & Kruskal (1989), in which they exhibited solutions of the Boussinesq equation

$$u_{tt} + uu_{xx} + u_x^2 + u_{xxxx} = 0, \tag{5.41}$$

that are not obtainable by the classical method. In this paper, a direct method was introduced to find solutions of the Boussinesq equation (5.41) that are of the form (5.27). These were obtained by directly substituting the ansatz (5.27) into the Boussinesq equation (5.41) to find all cases leading to a reduced ODE for some $h(z(x,t))$. In a *tour de force*, Clarkson and Kruskal found all such solutions of (5.41). For example, their solutions of (5.41), given by

$$u(x,t) = t^2h(z) - t^{-2}(x+\lambda t^5)^2, \quad z = xt + \tfrac{1}{6}\lambda t^6, \quad \lambda = \text{const}, \tag{5.42}$$

with $h(z)$ satisfying the reduced ODE

$$(w''' + ww' + 5\lambda w - 50\lambda^2 z)' = 0, \tag{5.43}$$

are not obtainable by the classical method, i.e., as invariant solutions from the well-known point symmetries of the Boussinesq equation (5.41) [Nishitani & Tajiri (1982); Rosenau & Schwarzmeier (1986); Clarkson & Kruskal (1989)] with infinitesimals

$$\begin{aligned} \xi(x,t,u) &= \xi(x) = \alpha_1 x + \alpha_2, \\ \tau(x,t,u) &= \tau(t) = 2\alpha_1 t + \alpha_3, \\ \eta(x,t,u) &= \eta(u) = -2\alpha_1 u, \end{aligned} \tag{5.44}$$

where $\alpha_1, \alpha_2, \alpha_3$, are arbitrary constants.

As mentioned above, *all* solutions arising from the direct method must arise from the nonclassical method. In their seminal paper, Levi & Winternitz (1989) indeed showed how to use the nonclassical method to obtain all of the Clarkson and Kruskal solutions of the Boussinesq equation (5.41). In particular, they considered the nonclassical method for the case $\tau \equiv 1$, and showed that the resulting infinitesimals are given by

$$\begin{aligned} \xi(x,t) &= \alpha(t)x + \beta(t), \\ \eta(x,t,u) &= -2\alpha(t)u - [2\alpha(t)(\alpha'(t) + 2\alpha^2(t))x^2 + 2([\alpha(t)\beta(t)]' \\ &\quad +4\alpha^2(t)\beta(t))x + 2\beta(t)(\beta'(t) + 2\alpha(t)\beta(t))], \end{aligned} \tag{5.45}$$

where $\alpha(t)$ and $\beta(t)$ are solutions of the ODE system

$$\begin{aligned} \alpha'' + 2\alpha\alpha' - 4\alpha^3 &= 0, \\ \beta'' + 2\alpha\beta' - 4\alpha^2\beta &= 0. \end{aligned} \tag{5.46}$$

The general solution of the ODE system (5.46) is easily obtained, and for any solution of the ODE system (5.46), the general solution of the constraining invariant surface condition equation (5.24) is given by

$$u = K^2(t)h(z) - (\alpha(t)x + \beta(t))^2, \quad z = K(t)x - \int_0^t \beta(s)K(s)ds, \tag{5.47}$$

where

$$K(t) = \exp\left[-\int_0^t \alpha(s)ds \right].$$

The substitution of the form (5.47) into the Boussinesq equation (5.41) yields the reduced ODE

$$w^{(4)} + ww'' + w'^2 + (Az + B)w' + 2Aw = 2(Az + B)^2, \tag{5.48}$$

where

$$A = \frac{\alpha^2(t) - \alpha'(t)}{K^4(t)} = \text{const}, \quad B = \frac{\alpha(t)\beta(t) - \beta'(t)}{K^3(t)} + A\int_0^t \beta(s)K(s)ds = \text{const}.$$

Note that the solutions obtained by the classical method result from the two particular sets of solutions

$$\alpha(t) = \frac{1}{2t + C}, \quad \beta(t) = \frac{D}{2t + C}, \quad C = \text{const}, \quad D = \text{const},$$

and

$$\alpha(t) = 0, \quad \beta(t) = E = \text{const}$$

of the ODE system (5.46).

The reader is referred to Levi & Winternitz (1989) for further details.

Many other examples have illustrated the usefulness of the nonclassical method to obtain solutions of PDEs that are not obtainable by Lie's classical method. Nucci & Clarkson (1992) use the nonclassical method to obtain solutions of the Fitzhugh–Nagumo equation

$$u_t - u_{xx} + u(u-1)(u-a), \quad a = \text{const}, \tag{5.49}$$

that can be found by neither the classical nor direct methods. These solutions result from the ansatz (5.26) but not from the ansatz (5.27). Clarkson & Mansfield (1994a) apply the nonclassical method to the nonlinear heat equation

$$u_t - u_{xx} = f(u), \tag{5.50}$$

to find forms of the reaction term $f(u)$ that yield solutions of the PDE (5.50) not obtainable by Lie's classical method.

5.2.3 Invariant solutions arising from nonlocal symmetries that are local symmetries of nonlocally related systems

For a given PDE system $\mathbf{R}\{x\,;u\}$ (5.1), nonlocal symmetries arising as local symmetries (usually point symmetries) of nonlocally related systems can yield further solutions of the given PDE system that do not arise as invariant solutions from point symmetries of $\mathbf{R}\{x\,;u\}$ (5.1). To find further solutions arising from such nonlocal symmetries, one first constructs the invariant solutions resulting from Lie's classical method applied to the corresponding local symmetries of nonlocally related systems. From the relationship between $\mathbf{R}\{x\,;u\}$ (5.1) and the nonlocally related system, for a constructed invariant solution of the nonlocally related system, normally one can readily find the corresponding solution of $\mathbf{R}\{x\,;u\}$ (5.1). Such a solution of $\mathbf{R}\{x\,;u\}$ (5.1) is obtained directly through projection when the nonlocally related system results from a conservation law of $\mathbf{R}\{x\,;u\}$ (5.1). The situation is more complicated, and involves integration, when the nonlocally related system is a subsystem of $\mathbf{R}\{x\,;u\}$ (5.1).

In particular, let $\mathbf{R}\{x,t\,;u\}$ be a given PDE system, and let $\mathbf{S}\{y,z\,;w\}$ be a PDE system nonlocally related to $\mathbf{R}\{x,t\,;u\}$. Due to the relationship between their solution sets, to every solution $w = w(y,z)$ of the nonlocally related system $\mathbf{S}\{y,z\,;w\}$ there corresponds a solution $u = u(x,t)$ of $\mathbf{R}\{x,t\,;u\}$, and the converse is also true [Section 3.2]. Suppose Y is a point symmetry of $\mathbf{S}\{y,z\,;w\}$ that yields a nonlocal symmetry of $\mathbf{R}\{x,t\,;u\}$. If $w = \widetilde{w}(y,z)$ is an invariant solution of $\mathbf{S}\{y,z\,;w\}$ arising from its point symmetry Y, one can find a corresponding solution $u = \Theta(x,t)$ of the given PDE system $\mathbf{R}\{x,t\,;u\}$. In the case when $\mathbf{S}\{y,z\,;w\} = \mathbf{RV}\{x,t\,;u,v\}$ is a potential system of $\mathbf{R}\{x,t\,;u\}$ with potential variable v, the situation is straightforward since for any solution $(u,v) = (\Theta(x,t), \Xi(x,t))$ of $\mathbf{S}\{y,z\,;w\} = \mathbf{RV}\{x,t\,;u,v\}$, the corresponding solution $u = \Theta(x,t)$ of $\mathbf{R}\{x,t\,;u\}$ is found by projection.

For PDE systems that have nontrivial nonlocal symmetries, one can often construct exact solutions, with a transparent physical meaning, that do not arise as invariant solutions from local symmetries. From the point of view of applications, being able to find a previously unknown class of exact solutions for a nonlinear PDE system can be of great importance.

Mathematically, the following question is of significance. Suppose $u = \Theta(x,t)$ is a solution of $\mathbf{R}\{x,t\,;u\}$ arising from an invariant solution of a nonlocally related system with respect to a nonlocal symmetry Y. Is this solution directly obtainable as an invariant solution of $\mathbf{R}\{x,t\,;u\}$ from the point symmetries of $\mathbf{R}\{x,t\,;u\}$? [Note that two distinct point symmetries could yield the same invariant solution of a given PDE system.] A direct way to answer this question is to check if $u = \Theta(x,t)$ is invariant under some nontrivial linear combination of point symmetries $\{\mathrm{X}_1, \ldots, \mathrm{X}_q\}$ of $\mathbf{R}\{x, t\,; u\}$:

$$\left[\left(\sum_{i=1}^{q} \alpha_i \mathrm{X}_i\right)(u - \Theta(x,t))\right]\Big|_{u-\Theta(x,t)=0} = 0,$$

with at least one of the constants $\alpha_i \neq 0$. For an example, see Exercise 5.2.5.

Here, two examples are considered that illustrate the use of finding further solutions (beyond those obtainable as invariant solutions of local symmetries) of given PDE systems from nonlocal symmetries arising as point symmetries of nonlocally related PDE systems. In the first example, such solutions are found for the variable-coefficient linear wave equation $u_{tt} = c^2(x)u_{xx}$ for wave speeds $c(x)$ corresponding to two-layered media with smooth transitions from layer to layer [Bluman & Kumei (1988)]. Exact solutions are constructed for wave speeds $c(x)$ that have four free parameters to fit a given medium; in particular, solutions are obtained for initial value problems for data with compact support. In the second example, such solutions are obtained for the system of planar gas dynamics equations (3.42) in their Lagrange formulation, with a generalized polytropic equation of state [Bluman, Cheviakov & Ivanova (2006)]. In both examples, the obtained exact solutions do not arise as solutions invariant with respect to point symmetries of the corresponding given PDE systems.

(1) Linear wave equations for two-layered media with smooth transitions
Consider the linear wave equation $\mathbf{U}\{x, t\,; u\}$ given by

$$u_{tt} = c^2(x)u_{xx}, \quad -\infty < x < \infty. \tag{5.51}$$

The PDE (5.51) has the obvious conservation law $\mathrm{D}_t(c^{-2}(x)u_t) - \mathrm{D}_x(u_x) = 0$, which yields the potential system $\mathbf{UV}\{x, t\,; u, v\}$ given by

$$\begin{aligned} v_t &= u_x, \\ u_t &= c^2(x)v_x. \end{aligned} \tag{5.52}$$

[Note that through the substitution

$$x, t, u, v, c(x) \to u, v, t, x, c^{-1}(u),$$

equations (5.51) and (5.52), respectively, coincide with the PDE $\mathbf{T}\{u, v\,; t\}$ (3.87) and its potential system $\mathbf{XT}\{u, v\,; x, t\}$ (3.85) considered in Section 3.5.3.]

Consider the wave speeds $c(x)$ given by solutions of the first-order ODE

$$c'(x) = m \sin(\nu \log c(x)) \tag{5.53}$$

corresponding to Case (b) in Table 4.6. The corresponding wave equations (5.51) and (5.52) describe wave propagation in two-layered media with smooth transitions, with the properties

$$\lim_{x\to-\infty} c(x) = 1, \quad \lim_{x\to+\infty} c(x) = e^{\pi/\nu} = \gamma, \quad \gamma > 0, \tag{5.54}$$

$$\max_{x\in(-\infty,\infty)} c'(x) = m > 0, \tag{5.55}$$

where γ, m are independent parameters with γ representing the ratio of asymptotic wave speeds. [One can easily adapt the results presented here to the situation where

$$\lim_{x\to-\infty} c(x) = c_1 > 0, \quad \lim_{x\to+\infty} c(x) = e^{\pi/\nu} = c_2 \tag{5.56}$$

through appropriate scalings.] Typical profiles for $c(x)$ are shown in Figure 5.1. Without loss of generality $c'(0) = m$.

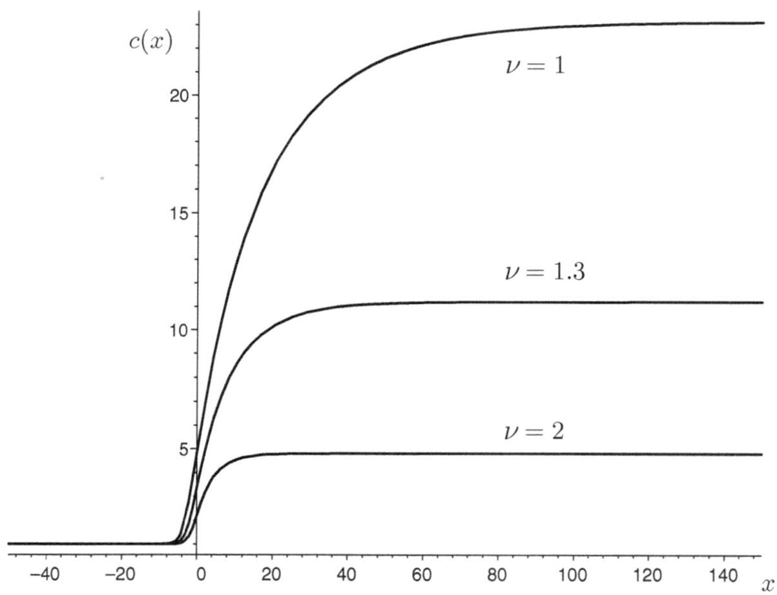

Fig. 5.1 Profiles of $c(x)$ for $m = 1/\nu$; $\nu = 1, 1.3, 2$ (top to bottom).

Let $\widetilde{\mathrm{W}}_{10,11}$ denote the infinitesimal symmetry generators $\mathrm{W}_{10,11}$ in Table 4.6 after the substitution $(u, v, t, x, c^{-1}(u)) \to (x, t, u, v, c(x))$. One can show that invariant solutions of the potential system $\mathbf{UV}\{x, t\,; u, v\}$ arising from its invariance under the point symmetry

$$\mathrm{W} = \widetilde{\mathrm{W}}_{10} + \widetilde{\mathrm{W}}_{11} + 4\nu ni\left[u\frac{\partial}{\partial u} + v\frac{\partial}{\partial v}\right]$$

are given by $(u,v) = (u_n(x,t), v_n(x,t))$ for $n = 0, 1, 2, \ldots$:

$$\begin{aligned}\begin{bmatrix} u_n(x,t) \\ v_n(x,t) \end{bmatrix} &= \sqrt{\sin y}\; e^{-2in \arctan \frac{\cot y}{\alpha(t)}} \begin{bmatrix} \sqrt{c(x)} & 0 \\ 0 & 1/\sqrt{c(x)} \end{bmatrix} \\ &\times \begin{bmatrix} \sqrt{\alpha(t)+\beta(t)\cos y} & \sqrt{\alpha(t)-\beta(t)\cos y} \\ \sqrt{\alpha(t)+\beta(t)\cos y} & -\sqrt{\alpha(t)-\beta(t)\cos y} \end{bmatrix} \begin{bmatrix} f_n(z) \\ g_n(z) \end{bmatrix},\end{aligned} \tag{5.57}$$

where

$$y = \nu \log c(x), \qquad \alpha(t) = \cosh m\nu t,$$

$$\beta(t) = \sinh m\nu t, \qquad z = \beta(t)\sin y,$$

$$\begin{bmatrix} f_n(z) \\ g_n(z) \end{bmatrix} = M_n(z) \begin{bmatrix} f_0(z) \\ g_0(z) \end{bmatrix},$$

with

$$\begin{bmatrix} f_0(z) \\ g_0(z) \end{bmatrix} = \frac{1}{\sqrt{z^2+1}} \begin{bmatrix} \cos\psi(z) & \sin\psi(z) \\ -\sin\psi(z) & \cos\psi(z) \end{bmatrix} \begin{bmatrix} P_n \\ Q_n \end{bmatrix},$$

$$\psi(z) = \frac{1}{2\nu}\log\left(z + \sqrt{z^2+1}\right),$$

$$M_n(z) = R_n(z) \times R_n(z) \times \cdots \times R_1(z) \times R_0(z),$$

$$R_0(z) = \begin{bmatrix} 1 & 0 \\ 0 & 1 \end{bmatrix},$$

and for $n \geq 1$,

$$R_n(z) = \begin{bmatrix} \left(n^2 - \frac{1}{4}\right)\left(\frac{z-i}{z+i}\right) - \frac{1}{4\nu^2} & \frac{i-2nz}{2\nu\sqrt{z^2+1}} \\ \frac{i+2nz}{2\nu\sqrt{z^2+1}} & \left(n^2-\frac{1}{4}\right)\left(\frac{z+i}{z-i}\right) - \frac{1}{4\nu^2} \end{bmatrix};$$

P_n and Q_n are arbitrary constants chosen separately for each invariant solution pair (u_n, v_n). For $n < 0$, it is convenient to define invariant solutions for the potential system (5.52) through

$$\begin{bmatrix} u_n(x,t) \\ v_n(x,t) \end{bmatrix} = \begin{bmatrix} \overline{u_{-n}}(x,t) \\ \overline{v_{-n}}(x,t) \end{bmatrix},$$

where a bar denotes the complex conjugate.

The solution of the potential system (5.52) with the initial conditions given by

$$u(x,0) = U(x), \quad v(x,0) = V(x) \tag{5.58}$$

can be represented formally in the form

$$\begin{bmatrix} u(x,t) \\ v(x,t) \end{bmatrix} = \sum_{n=-\infty}^{\infty} \begin{bmatrix} u_n(x,t) \\ v_n(x,t) \end{bmatrix} = \begin{bmatrix} u_0(x,t) \\ v_0(x,t) \end{bmatrix} + 2Re\left\{ \sum_{n=1}^{\infty} \begin{bmatrix} u_n(x,t) \\ v_n(x,t) \end{bmatrix} \right\}$$

with the constants P_n, Q_n determined from the initial conditions (5.58). In particular, since

$$u_n(x,0) = (-1)^n \sqrt{c(x)\sin y}\,(P_n + Q_n)e^{2iny},$$

$$v_n(x,0) = (-1)^n \sqrt{\frac{\sin y}{c(x)}}\,(P_n - Q_n)e^{2iny},$$

and $0 < y < \pi$, from the above Fourier series representation it follows that the Fourier coefficients are given by

$$P_n, Q_n = \frac{(-1)^n}{2\pi} \int_0^{\pi} \frac{e^{-2iny}}{\sqrt{\sin y}} \left[e^{-y/2\nu} U(x(y)) \pm e^{y/2\nu} V(x(y)) \right] dy.$$

For a given initial value problem, after determining the constants $\{P_n, Q_n\}$, one can directly compute the solution for any time t, $-\infty < t < \infty$. Note that no step-by-step marching in the time variable t is required as would be the case for numerical procedures based on the method of characteristics.

Full details of the derivation of these solutions and their properties are given in Bluman & Kumei (1988). It is easy to check that the corresponding projected solutions of the linear wave equation (5.51) do not arise as invariant solutions for any of its point symmetries.

(2) Generalized polytropic planar gas dynamics equations

Symmetries of nonlocally related systems of generalized polytropic PGD equations with the equation of state (4.34) were presented in Section 4.2.5. Here it is shown how to construct solutions of the Lagrange system arising from nonlocal symmetries that cannot arise as invariant solutions from point symmetries of the Lagrange system. In particular, the construction of such solutions is considered for the constitutive function $M(p) = -p \ln p$.

Among the potential systems of the Lagrange system $\mathbf{L}\{y, s\,; v, p, q\}$ (3.42) considered in Section 4.2.5, as seen in Table 4.19, the potential system $\mathbf{LW}^2\{y, s\,; v, p, q, w^2\}$ (3.98) has the largest number of point symmetries. [In particular, its point symmetries include all point symmetries of $\mathbf{L}\{y, s\,; v, p, q\}$.] Thus the potential system $\mathbf{LW}^2\{y, s\,; v, p, q, w^2\}$ (3.98) has the largest set of invariant solutions that can be constructed from the symmetries exhibited in Table 4.19. In particular, for $M(p) = -p\ln p$, the Lie algebra of symmetry generators for this constitutive function is spanned by the projections of the eight operators $Z_1, \ldots, Z_5, Z_7, Z_9, Z_{13}$ on the space of variables $\{y, s, v, p, q, w^2\}$ of $\mathbf{LW}^2\{y, s\,; v, p, q, w^2\}$. To find all solutions of $\mathbf{LW}^2\{y, s\,; v, p, q, w^2\}$ invariant with respect to these point symmetries, one proceeds as follows [Ovsiannikov (1982)].

1. Find optimal systems of one-dimensional invariant symmetry subalgebras, and construct invariant solutions with respect to each such subalgebra.
2. Use the transformation groups corresponding to the infinitesimal generators of the point symmetries to extend the set of solutions.

The optimal system of one-dimensional symmetry subalgebras for the point symmetries of the potential system $\mathbf{LW}^2\{y, s\,; v, p, q, w^2\}$ appears in Bluman, Cheviakov & Ivanova (2006).

Solutions of the Lagrange system $\mathbf{L}\{y, s\,; v, p, q\}$ are obtained from the solutions of the potential system $\mathbf{LW}^2\{y, s\,; v, p, q, w^2\}$ by simply excluding the potential variable w^2.

For the case $M(p) = -p\ln p$, a solution of the Lagrange system $\mathbf{L}\{y, s;$ $v, p, q\}$ (3.42) is constructed that arises from an invariant solution of the potential system $\mathbf{LW}^2\{y, s\,; v, p, q, w^2\}$ (3.98) with respect to the point symmetry

$$\mathrm{X} = Z_9 + Z_{13} \tag{5.59}$$

[Table 4.19]. The resulting solution does not arise as an invariant solution with respect to a local symmetry of $\mathbf{L}\{y, s\,; v, p, q\}$, since the symmetry (5.59) belongs to the invariant subalgebra

$$\mathcal{A}_8 = Z_{13} + \varepsilon_1 Z_1 + \varepsilon_2 Z_2 + \varepsilon_3 Z_7 + \alpha Z_9, \quad \varepsilon_i = 0, \pm 1, \ \alpha \in \mathbb{R},$$

which essentially involves the nonlocal symmetry Z_{13}.

In particular, the invariant solution has the form

$$\begin{aligned}
p(y,s) &= \frac{\beta^2\gamma}{\alpha^2}\frac{y^2}{y+\alpha}(1+\tan^2(\beta s)),\\
q(y,s) &= -\frac{\gamma}{(y+\alpha)^3}\ln\left[\frac{\gamma\beta^2}{\alpha^2}\frac{y^2}{y+\alpha}(1+\tan^2(\beta s))\right],\\
v(y,s) &= -\frac{\beta\gamma}{\alpha^2}\frac{y(y+2\alpha)}{(y+\alpha)^2}\tan(\beta s), \quad w(y,s) = -\frac{q_0\beta}{\alpha^2}\frac{y^2}{y+\alpha}\tan(\beta s),
\end{aligned} \tag{5.60}$$

where α, β, γ are arbitrary constants [Exercise 5.2.4]. One can show that the solution (5.60) does not arise as an invariant solution with respect to a point symmetry of the Lagrange system $\mathbf{L}\{y, s\,; v, p, q\}$ [Exercise 5.2.5].

Setting $\alpha = 1, \beta = i, \gamma = -2$ and applying the equivalence transformation (3.96) with $a_1 = a_2 = 1, a_3 = -1, a_4 = a_5 = a_6 = 0, a_7 = p_0, a_8 = q_0$, one obtains the solution

$$\begin{aligned} p'(y,s) &= p_0 - \frac{2y^2}{y+1}\frac{1}{\cosh^2 s}, \\ q'(y,s) &= q_0 - \frac{2}{(y+1)^3}\ln\left[\frac{2y^2}{y+1}\frac{1}{\cosh^2 s}\right], \\ v'(y,s) &= 2\frac{y(y+2)}{(y+1)^2}\tanh s \end{aligned} \tag{5.61}$$

of the Lagrange system $\mathbf{L}\{y, s\,; v, p, q\}$, for the constitutive function

$$B(p', q') = \frac{(p_0 - p')\ln(p_0 - p')}{q_0 - q'}.$$

The solution (5.61) is regular, bounded, and satisfies the physical conditions $p > 0, \rho > 0$ for all times $s \geq 0$ for the material space interval $0 \leq y \leq 5$. For full details of the derivation and properties of the presented solution, see Bluman, Cheviakov & Ivanova (2006).

5.2.4 Further extensions of symmetry methods for construction of solutions of PDEs connected with nonlocally related systems

In this section, other ansätze, based on symmetry extensions, are presented that could lead to further solutions of a given PDE system $\mathbf{R}\{x, t\,; u\}$. Each presented ansatz results from consideration of a nonlocally related system. The focus is on the situation when $\mathbf{R}\{x, t\,; u\}$ is a second-order scalar PDE. Extensions to higher-order PDE systems with two or more independent variables are straightforward.

Suppose $\mathbf{U}\{x, t\,; u\}$ is a given scalar PDE

$$R(x,t,u,\partial u,\partial^2 u) = \mathrm{D}_x f(x,t,u,u_x,u_t) - \mathrm{D}_t g(x,t,u,u_x,u_t) = 0 \tag{5.62}$$

that can be written as a conservation law. Correspondingly, the PDE (5.62) has the nonlocally related potential system $\mathbf{UV}\{x, t\,; u, v\}$ given by

$$\begin{aligned} v_t &= f(x,t,u,u_x,u_t), \\ v_x &= g(x,t,u,u_x,u_t). \end{aligned} \tag{5.63}$$

For many physical equations, one can eliminate u from the potential system $\mathbf{UV}\{x,t\,;u,v\}$ (5.63) to obtain a potential equation (which is a subsystem) $\mathbf{V}\{x,t\,;v\}$ given by

$$S(x,t,v,\partial v,\partial^2 v)=0. \tag{5.64}$$

By construction, the potential equation $\mathbf{V}\{x,t\,;v\}$ (5.64) is a nonlocally related system of $\mathbf{U}\{x,t\,;u\}$ (5.62) and may or may not be a nonlocally related subsystem of the potential system $\mathbf{UV}\{x,t\,;u,v\}$ (5.63). [One can show that the potential equation $\mathbf{V}\{x,t\,;v\}$ (5.64) is (1) locally related to the potential system $\mathbf{UV}\{x,t\,;u,v\}$ (5.63) if the given PDE $\mathbf{U}\{x,t\,;u\}$ (5.62) is parabolic; (2) nonlocally related to the potential system $\mathbf{UV}\{x,t\,;u,v\}$ (5.63) if the given PDE $\mathbf{U}\{x,t\,;u\}$ (5.62) is hyperbolic or elliptic [Exercise 5.2-7].]

For a point symmetry of the potential system $\mathbf{UV}\{x,t\,;u,v\}$ (5.63) with the infinitesimal generator

$$\mathrm{X}=\xi(x,t,u,v)\frac{\partial}{\partial x}+\tau(x,t,u,v)\frac{\partial}{\partial t}+\eta(x,t,u,v)\frac{\partial}{\partial u}+\zeta(x,t,u,v)\frac{\partial}{\partial v}, \tag{5.65}$$

the corresponding invariant surface condition (5.4) becomes

$$\begin{aligned}\xi(x,t,u,v)u_x+\tau(x,t,u,v)u_t&=\eta(x,t,u,v),\\ \xi(x,t,u,v)v_x+\tau(x,t,u,v)v_t&=\zeta(x,t,u,v).\end{aligned} \tag{5.66}$$

For a specific set of $\xi(x,t,u,v),\tau(x,t,u,v),\eta(x,t,u,v),\zeta(x,t,u,v)$, the general solution of the invariant surface condition (5.66) can be represented in the form

$$z(x,t,u,v)=\text{const}=c_1, \tag{5.67a}$$

$$\begin{aligned}H^1(x,t,u,v)&=\text{const}=c_2=h^1(z),\\ H^2(x,t,u,v)&=\text{const}=c_3=h^2(z),\end{aligned} \tag{5.67b}$$

where $z(x,t,u,v)$ is a similarity variable. After solving equations (5.67b) for u and v, one obtains the ansatz

$$u=\phi(x,t,h^1(z(x,t,u,v)),h^2(z(x,t,u,v))), \tag{5.68a}$$

$$v=\psi(x,t,h^1(z(x,t,u,v)),h^2(z(x,t,u,v))), \tag{5.68b}$$

for solutions of the potential system $\mathbf{UV}\{x,t\,;u,v\}$ (5.63).

If a specific set of $(\xi(x,t,u,v),\tau(x,t,u,v),\eta(x,t,u,v),\zeta(x,t,u,v))$ is a set of infinitesimals for a point symmetry of the potential system $\mathbf{UV}\{x,t\,;u,v\}$ (5.63), then the respective dependencies of ϕ and ψ on x, t, $h^1(z)$ and $h^2(z)$ are both explicit in the ansatz (5.68); $h^1(z)$ and $h^2(z)$ are both arbitrary functions of the similarity variable z. Here, the substitution of the ansatz (5.68) into the potential system $\mathbf{UV}\{x,t\,;u,v\}$ (5.63), yields a reduced ODE system of at most second order with independent variable z and dependent variables $h^1(z)$ and $h^2(z)$. Each solution of this ODE system yields an invariant solution,

$$u = \Phi(x, t, h^1(z(x, t, u, v)), h^2(z(x, t, u, v))), \tag{5.69a}$$

$$v = \Psi(x, t, h^1(z(x, t, u, v)), h^2(z(x, t, u, v))), \tag{5.69b}$$

obtainable by the classical method, of the potential system $\mathbf{UV}\{x, t\,; u, v\}$ (5.63). In turn, the invariant solution (5.69) of the nonlocally related PDE system (5.63), yields the solution

$$u = \Phi(x, t, h^1(z(x, t, u, v)), h^2(z(x, t, u, v))) \tag{5.70}$$

of the given scalar PDE $\mathbf{U}\{x, t\,; u\}$ (5.62). If the point symmetry (5.65) of the potential system $\mathbf{UV}\{x, t\,; u, v\}$ (5.63) is not a potential symmetry of the PDE (5.62), i.e., $\xi_v = \tau_v = \eta_v \equiv 0$, then the solution (5.70) is an invariant solution of the PDE (5.62) obtainable from reduction under the corresponding projected point symmetry of the PDE (5.62). [The converse does not necessarily hold, i.e., even if the point symmetry (5.65) is a potential symmetry of the given scalar PDE $\mathbf{U}\{x, t\,; u\}$ (5.62), it is still possible that the solution (5.70) can be obtained as an invariant solution resulting from a point symmetry of (5.62).]

In the following sections, other ansätze for solutions of a given scalar PDE $\mathbf{U}\{x, t\,; u\}$ (5.62) are presented that are based on further consideration of the nonlocally related system $\mathbf{UV}\{x, t\,; u, v\}$ (5.63) and the potential equation $\mathbf{V}\{x, t\,; v\}$ (5.64) (when it exists).

An ansatz based on a first refinement of the potential system ansatz (5.68)

The ansatz (5.68) is the invariant form for invariant solutions of the potential system $\mathbf{UV}\{x, t\,; u, v\}$ (5.63) obtained from its point symmetry (5.65). Instead of substituting this ansatz into $\mathbf{UV}\{x, t\,; u, v\}$ (5.63) to obtain an invariant solution (5.69) and hence the solution (5.70) of a given scalar PDE $\mathbf{U}\{x, t\,; u\}$ (5.62), more generally one could just directly substitute the ansatz (5.68) into $\mathbf{U}\{x, t\,; u\}$ (5.62). This method, introduced by Pucci & Saccomandi (1993), is a generalization of the classical method. Indeed, here both u and v are sought in the invariant form; u satisfies the given PDE $\mathbf{U}\{x, t\,; u\}$, but the pair (u, v) *may not* satisfy the potential system $\mathbf{UV}\{x, t\,; u, v\}$ (5.63).

As an example, Pucci & Saccomandi considered the Fokker–Planck equation

$$u_t - u_{xx} - (xu)_x = 0. \tag{5.71}$$

The scalar PDE (5.71) is directly written as a conservation law, hence one obtains the corresponding potential system

$$\begin{aligned} v_x &= u, \\ v_t &= u_x + xu. \end{aligned} \tag{5.72}$$

One can show that the potential system (5.72) has six nontrivial point symmetries including the point symmetry with the infinitesimal generator

$$\mathrm{X} = e^{2t}\left[x\frac{\partial}{\partial x} + \frac{\partial}{\partial t} - [(x^2+2)u + 2xv]\frac{\partial}{\partial u} - (x^2+1)v\frac{\partial}{\partial v}\right] \tag{5.73}$$

that yields a potential symmetry of (5.71). It is easy to show that the general solution of the corresponding invariant surface condition is given by

$$\begin{aligned} z(x,t) &= x^{-1}e^t = \text{const} = c_1, \\ H^1(x,v) &= e^{x^2/2}xv = \text{const} = c_2 = h^1(z), \\ H^2(x,u,v) &= e^{x^2/2}(x^2u + x^3v) = \text{const} = c_3 = h^2(z), \end{aligned} \tag{5.74}$$

which yields the ansatz

$$u = e^{-x^2/2}(x^{-2}h^2(z) - h^1(z)), \tag{5.75a}$$

$$v = e^{-x^2/2}x^{-1}h^1(z), \tag{5.75b}$$

for solutions of the potential system (5.72).

After substituting (5.75) into the potential system (5.72), one finds that $h^1(z)$, $h^2(z)$ satisfy the first order ODE system

$$\begin{aligned} h^2(z) + (zh^1(z))' &= 0, \\ z(h^2(z))' + 2h^2(z) &= 0, \end{aligned} \tag{5.76}$$

whose general solution yields the solution

$$u = e^{-x^2/2}[a_1(1-x^2)e^{-2t} + a_2xe^{-t}], \tag{5.77}$$

with arbitrary constants a_1, a_2 of the Fokker–Planck equation (5.71).

On the other hand, if one directly substitutes the ansatz (5.75a) into the Fokker–Planck equation (5.71), then one obtains the equation

$$\begin{aligned} &z^2[z^2(h^2(z))'' + 6z(h^2(z))' + 6h^2(z)] \\ &\quad -e^{2t}[z^2(h^1(z))'' + 2z(h^1(z))' - 2h^2(z)] = 0. \end{aligned} \tag{5.78}$$

Hence the ansatz (5.75a) yields a solution of the Fokker–Planck equation (5.71) if and only if $h^1(z)$, $h^2(z)$ satisfy the second-order ODE system

$$\begin{aligned} z^2(h^2(z))'' + 6z(h^2(z))' + 6h^2(z) = 0, \\ z^2(h^1(z))'' + 2z(h^1(z))' - 2h^2(z) = 0. \end{aligned} \tag{5.79}$$

From the previous remarks, it follows that each solution of the ODE system (5.76) must satisfy the ODE system (5.79). The general solution of the ODE system (5.79) yields the solution

$$u = e^{-x^2/2}[a_1(1-x^2)e^{-2t} + a_2 x e^{-t}] + e^{-x^2/2}[b_1(3x - x^3)e^{-3t} + b_2], \tag{5.80}$$

with arbitrary constants a_1, a_2, b_1, b_2 of the Fokker–Planck equation (5.71). The solution (5.77) corresponds to the situation where $b_1 = b_2 = 0$.

However, it is easy to check that the solution $u = e^{-x^2/2}$ can be obtained from the invariance of the PDE (5.71) under translations in t, and the solution $u = e^{-x^2/2}(3x - x^3)e^{-3t}$ can be obtained by applying the infinitesimal generator $e^{-t}\,\partial/\partial x$ of a point symmetry of the PDE (5.71) to the solution $u = e^{-x^2/2}(1-x^2)e^{-2t}$ in (5.77).

An ansatz based on a second refinement of the potential system ansatz (5.68)

Here the ansatz (5.68) is modified as follows [Cheviakov (2008)]. Let $z(x,t,u,v)$, $\hat{z}(x,t,u,v)$ be the canonical coordinates for the point symmetry X (5.65) of the potential system $\mathbf{UV}\{x,t\,;u,v\}$ (5.63), i.e, $\mathrm{X}z = 0$, $\mathrm{X}\hat{z} = 1$. Then the ansatz (5.68) is changed to allow one of $h^1(z)$, $h^2(z)$ to depend on the translated variable $\hat{z}$, i.e., say $h^2(z)$ is replaced by $h^2(z,\hat{z})$. Finally, one directly substitutes the corresponding ansatz into $\mathbf{U}\{x,t\,;u\}$ (5.62) to obtain a reduced PDE system satisfied by $h^1(z)$, $h^2(z,\hat{z})$. Any solution that can be found for this reduced PDE system yields a solution of the given scalar PDE $\mathbf{U}\{x,t\,;u\}$ (5.62). The second refinement ansatz directly generalizes both the classical ansatz [Section 5.2.1] and the first refinement of the potential system ansatz. In particular, in the second refinement of the potential system ansatz, u is a solution of PDE $\mathbf{U}\{x,t\,;u\}$ (5.62) that has an invariant form, whereas v *does not have an invariant form* (since it depends on $\hat{z}$), and the pair (u,v) is generally *not a solution* of the potential system $\mathbf{UV}\{x,t\,;u,v\}$ (5.63).

As an example, Cheviakov (2008) considered the Lagrange planar gas dynamics system $\mathbf{L}\{y,s\,;v,p,q\}$ (3.42) in the particular polytropic case $B(p,q) = 3p/q$. In this case, as seen from Table 4.18, this PDE system has the nonlocal symmetry (potential symmetry)

$$\mathrm{Y}_8 = s^2\frac{\partial}{\partial s} + (w^1 - sv)\frac{\partial}{\partial v} - 3sp\frac{\partial}{\partial p} + sq\frac{\partial}{\partial q} + sw^1\frac{\partial}{\partial w^1}, \tag{5.81}$$

which is a point symmetry of the potential system $\mathbf{LW}^1\{y, s\,; v, p, q, w^1\}$ (3.97). The invariants of the point symmetry (5.81) are given by

$$\begin{aligned} z = y &= \text{const} = c_1, \\ H^1(s,p) &= s^3 p = \text{const} = c_2 = h^1(z), \\ H^2(s,q) &= \frac{q}{s} = \text{const} = c_3 = h^2(z), \\ H^3(s,w^1) &= \frac{w^1}{s} = \text{const} = c_4 = h^3(z), \\ H^4(s,v,w^1) &= sv - w^1 = \text{const} = c_5 = h^4(z), \end{aligned} \tag{5.82}$$

and as the translated canonical coordinate, one can choose

$$\hat{z} = 1/s. \tag{5.83}$$

The corresponding invariant solutions of the potential system $\mathbf{LW}^1\{y, s\,;$ $v, p, q, w^1\}$(3.97) result from the ansatz

$$\begin{aligned} &p(y,s) = s^{-3}h^1(y), \quad q(y,s) = sh^2(y), \\ &v(y,s) = s^{-1}h^4(y) + h^3(y), \quad w^1(y,s) = sh^3(y). \end{aligned} \tag{5.84}$$

Substitution of (5.84) into $\mathbf{LW}^1\{y, s\,; v, p, q, w^1\}$ (3.97) yields the reduced ODE system

$$\begin{aligned} (h^1(y))' = 0, \quad (h^3(y))' = h^2(y), \quad h^4(y) = 0, \\ h^1(y)(h^4(y))' = 0, \quad h^1(y)h^2(y) = h^1(y)(h^3(y))' \end{aligned} \tag{5.85}$$

with its general solution ($h^1(y) \neq 0$) given by

$$h^1(y) = C, \quad h^2(y) = f'(y), \quad h^3(y) = f(y), \quad h^4(y) = 0, \tag{5.86}$$

where $f(y)$ is an arbitrary differentiable function and C is an arbitrary constant. The corresponding solutions of the Lagrange system $\mathbf{L}\{y, s\,; v, p, q\}$ (3.42) are given by

$$v(y,s) = f(y), \quad p(y,s) = Cs^{-3}, \quad q(y,s) = sf'(y). \tag{5.87}$$

On the other hand, using the second refinement of the potential system ansatz, if one directly substitutes [Without loss of generality one can replace $\hat{z} = 1/s$ by s.]) the ansatz

$$p(y,s) = s^{-3}h^1(y), \quad q(y,s) = sh^2(y), \quad v(y,s) = s^{-1}h^4(y) + h^3(y,s) \tag{5.88}$$

into the given Lagrange system $\mathbf{L}\{y, s\,; v, p, q\}$ (3.42), one obtains the reduced PDE system

$$s\frac{\partial h^3(y,s)}{\partial y} = sh^2(y) - (h^4(y))', \quad s^3\frac{\partial h^3(y,s)}{\partial s} = -sh^4(y) - (h^1(y))'. \quad (5.89)$$

The solution of the reduced PDE system (5.89) is given by

$$\begin{aligned} h^1(y) &= 2By + C, \quad h^2(y) = f'(y), \\ h^3(y,s) &= f(y) + Bs^{-2} + s^{-1}g(y), \quad h^4(y) = -g(y), \end{aligned} \quad (5.90)$$

where $f(y), g(y)$ are arbitrary differentiable functions and B, C are arbitrary constants. The corresponding solutions of the Lagrange system $\mathbf{L}\{y, s\,; v, p, q\}$ (3.42) are given by

$$v(y,s) = f(y) + Bs^{-2}, \quad p(y,s) = s^{-3}(2By + C), \quad q(y,s) = sf'(y), \quad (5.91)$$

generalizing the solutions (5.87).

Nonclassical potential solutions of PDEs

The nonclassical method is now applied to the potential system $\mathbf{UV}\{x, t\,; u, v\}$ (5.63) of a given scalar PDE $\mathbf{U}\{x, t\,; u\}$ (5.62) written as a conservation law. The situation is considered where a subsystem (potential equation) $\mathbf{V}\{x, t\,; v\}$ (5.64) exists.

Definition 5.2.4. A solution of a given scalar PDE $\mathbf{U}\{x, t\,; u\}$(5.62) obtained by applying the nonclassical method with $\tau \equiv 1$ to either a potential system $\mathbf{UV}\{x, t\,; u, v\}$ (5.63) or a potential equation $\mathbf{V}\{x, t\,; v\}$ (5.64) is called a *nonclassical potential solution* of (5.62) if it is neither a nonclassical solution obtained by directly applying the nonclassical method with $\tau \equiv 1$ to (5.62) nor a solution arising as an invariant solution from a point symmetry of $\mathbf{U}\{x, t\,; u\}$ (5.62), of the potential system $\mathbf{UV}\{x, t\,; u, v\}$ (5.63) or the potential equation $\mathbf{V}\{x, t\,; v\}$ (5.64).

The nonlinear heat conduction equation $\mathbf{U}\{x, t\,; u\}$ given by

$$u_t - (K(u)u_x)_x = 0, \quad K'(u) \not\equiv 0, \quad (5.92)$$

is considered as a prototypical example [Bluman & Yan (2005)]. The potential system $\mathbf{UV}\{x, t\,; u, v\}$ given by

$$\begin{aligned} v_x &= u, \\ v_t &= K(u)u_x, \end{aligned} \quad (5.93)$$

naturally arises from the conservation law (5.92). The dependent variable u can be eliminated from (5.93) to yield the subsystem (potential equation) $\mathbf{V}\{x,t\,;v\}$ given by

$$v_t - K(v_x)v_{xx} = 0, \tag{5.94}$$

which is nonlocally related to the given scalar PDE (5.92) but locally related to the potential system (5.93).

For arbitrary $K(u)$, the determining equations for the infinitesimals in the nonclassical method are set up for the given scalar PDE (5.92) in terms of the invariant surface condition (5.24), for the potential system (5.93) in terms of the invariant surface condition (5.66), and for the potential equation (5.94) in terms of the invariant surface condition

$$\xi(x,t,v)v_x + \tau(x,t,v)v_t = \zeta(x,t,v), \tag{5.95}$$

respectively, for $\tau \equiv 1$ and for $\tau \equiv 0$, $\xi \equiv 1$.

As a particular example,

$$K(u) = \frac{1}{u^2 + u} \tag{5.96}$$

is considered, following Bluman & Yan (2005).

Case 1. The nonclassical method applied to the given scalar PDE (5.92)

(1) $\tau \equiv 1$. Here the nonclassical method applied directly to the nonlinear heat conduction equation (5.92) yields the following system of four determining equations for the infinitesimals $\xi(x,t,u), \eta(x,t,u)$:

$$\begin{aligned}
&K'(u)\xi_u - K(u)\xi_{uu} = 0,\\
&\left[K(u)K''(u) - (K')^2(u)\right]\eta + K(u)K'(u)\eta_u\\
&\qquad +2K(u)\xi\xi_u + K^2(u)(\eta_{uu} - 2\xi_{xu}) = 0,\\
&K(u)(\xi_t - 2\xi_u\eta + 2\xi\xi_x) - K'(u)\xi\eta\\
&\qquad +K^2(u)(2\eta_{xu} - \xi_{xx}) + 2K(u)K'(u)\eta_x = 0,\\
&K(u)(2\xi_x\eta + \eta_t) - K'(u)\eta^2 - K^2(u)\eta_{xx} = 0.
\end{aligned} \tag{5.97}$$

One can show [Exercise 5.2.10] that for $K(u)$ given by (5.96), equations (5.97) yield one "nonclassical symmetry", which, however, only yields a classical solution of the PDE (5.92).

(2) $\tau \equiv 0$, $\xi \equiv 1$. Here the nonclassical method yields the following determining equation for the infinitesimal $\eta(x,t,u)$.

$$K(u)(\eta_{xx} + 2\eta\eta_{xu} + \eta^2\eta_{uu}) + K''(u)\eta^3 + K'(u)(3\eta\eta_x + 2\eta^2\eta_u) - \eta_t = 0. \tag{5.98}$$

In principle, any $K(u)$ yields solutions of the determining equation (5.98). In practice, for a given $K(u)$, one must use a specific ansatz for $\eta(x,t,u)$ to seek particular solutions of (5.98).

The systems of determining equations (5.97) and (5.98) appear in Bluman & Shtelen (1996b). To date it appears that the direct application of the nonclassical method to the PDE (5.97) has only yielded solutions that are obtainable by Lie's classical method.

Case 2. The nonclassical method applied to the potential system (5.93)

(1) $\tau \equiv 1$. Here the nonclassical method applied to the potential system (5.93) yields the following system of two determining equations for the infinitesimals $\xi(x,t,u,v), \eta(x,t,u,v), \zeta(x,t,u,v)$.

$$\begin{aligned}
&(\zeta - u\xi)(\zeta_u - u\xi_u) + [u(\zeta_v - \xi_x) - \xi_v u^2 - \eta + \zeta_x]K(u) = 0,\\
&(\zeta_u - u\xi_u)[\eta K(u) - \xi(\zeta - u\xi)]\\
&\quad +[(\zeta_v - u\xi_v)(\zeta - u\xi) - (\eta_v K(u) + \xi_t)u - \eta_x K(u)]K(u)\\
&\quad +(\zeta - u\xi)[(\xi_x + \xi_v u - \eta_u)K(u) - \eta K'(u)]\\
&\quad +\xi_u(\zeta_u - u\xi_u)^2 + \zeta_t K(u) = 0.
\end{aligned} \tag{5.99}$$

The set of determining equations (5.99) is clearly under-determined since it involves two equations for the three infinitesimals $\xi(x,t,u,v)$, $\eta(x,t,u,v)$, $\zeta(x,t,u,v)$. Consequently, in principle, any $K(u)$ yields an infinite number of solutions. So far no nonclassical potential solution has been found in this case.

(2) $\tau \equiv 0$, $\xi \equiv 1$. Here it is easy to show [Exercise 5.2.11] that the nonclassical method only yields solutions of the potential system (5.93) of the form $u = f(x)$, i.e., invariant solutions of the given scalar PDE (5.92) that are derivable from its invariance under translations in t.

Case 3. The nonclassical method applied to the potential equation (5.94)

(1) $\tau \equiv 1$. Here the nonclassical method applied to the potential equation (5.94) yields the following determining equation for the infinitesimals $\xi(x,t,v), \zeta(x,t,v)$.

$$\begin{aligned}
&[-\xi\xi_v v_x^3 + (\xi_v\zeta - \xi\xi_x + \xi\zeta_v)v_x^2 + (\xi_x\zeta - \zeta\zeta_v + \xi\zeta_x)v_x - \zeta\zeta_x]K'(v_x)\\
&\quad +[-2\xi\xi_v v_x^2 + (2\xi_v\zeta - 2\xi\xi_x - \xi_t)v_x + 2\zeta\xi_x + \zeta_t]K(v_x)\\
&\quad +[\xi_{vv}v_x^3 + (2\xi_{xv} - \zeta_{vv})v_x^2 + (\xi_{xx} - 2\zeta_{xv})v_x - \zeta_{xx}]K^2(v_x) \equiv 0.
\end{aligned} \tag{5.100}$$

Since the determining equation (5.100) must hold for all values of x, t, v, v_x, it follows that $K(u) = K(v_x)$ is restricted to satisfying a first-order Bernoulli equation (with variable coefficients) of the form

$$\begin{aligned}&[A_1v_x^3 + A_2v_x^2 + A_3v_x + A_4]K'(v_x) + [B_1v_x^2 + B_2v_x + B_3]K(v_x)\\&\quad+[C_1v_x^3 + C_2v_x^2 + C_3v_x + C_4]K^2(v_x) = 0,\end{aligned} \tag{5.101}$$

for some constants A_i, B_j, C_k. Consequently, $K(u)$ depends at most on 11 parameters. The solution of the determining equation (5.100) is further discussed in Bluman & Yan (2005). In particular, it can be shown [Exercise 5.2.12] that for $K(u)$ given by (5.96), equations (5.101) do yield nonclassical potential solutions of the given nonlinear heat conduction equation (5.92).

(2) $\tau \equiv 0,\ \xi \equiv 1$. Here the invariant surface condition (5.95) becomes $v_x = \zeta(x,t,v)$, and the nonclassical method applied to the potential equation (5.94) yields the following determining equation for the infinitesimal $\zeta(x,t,v)$.

$$[2\zeta\zeta_x\zeta_v + \zeta^2\zeta_v^2 + \zeta_x^2]K'(\zeta) + [\zeta^2\zeta_{vv} + 2\zeta\zeta_{xv} + \zeta_{xx}]K(\zeta) - \zeta_t = 0. \tag{5.102}$$

In principle, any $K(u) = K(\zeta)$ yields solutions of (5.102). In practice, one must use an ansatz for $\zeta(x,t,v)$ to seek particular solutions of (5.102).

Exercises 5.2

5.2.1. Show that any PDE solution obtained by Lie's classical method can be obtained by the nonclassical method.

5.2.2. Show that the solution of the infinitesimal equations (5.36) includes the classical case as a special case. In particular, find necessary and sufficient conditions so that $\xi(x,t), C(x,t), D(x,t)$ yield only solutions obtainable by Lie's classical method.

5.2.3. Consider the Euler PDE system $\mathbf{E}\{x,t\,;v,\sigma,\rho\}$ for the system of nonlinear one-dimensional elasticity equations in the Eulerian framework, given by

$$\begin{aligned}\rho_t + (\rho v)_x &= 0,\\ \sigma_x + \rho f(x) &= \rho(v_t + vv_x),\\ \sigma &= K(\rho).\end{aligned} \tag{5.103}$$

In (5.103), ρ is the density of a given material, σ is the Cauchy stress, v is the physical (material) velocity, $f(x)$ is the force per unit mass, and $K(\rho)$ is a constitutive function.

(a) Use the first equation of the PDE system (5.103) to introduce a nonlocal variable w. Show that the resulting nonlocally related PDE system $\mathbf{EW}\{x,t\,;v,\sigma,\rho,w\}$ is locally equivalent to the nonlinear elasticity system $\mathbf{L}\{y,t\,;v,\sigma,q,x\}$ in the Lagrangian framework, given by

$$\begin{aligned} q &= x_y, \\ v &= x_t, \\ v_t &= \sigma_y + f(x), \\ \sigma &= K(1/q), \end{aligned} \tag{5.104}$$

where $q = 1/\rho$, and $y = w$.

(b) Show that for $K(\rho) = (1/2)[\arctan(1/\rho) + \rho/(\rho^2 + 1)$, $f(x) = x$, the potential Euler (Lagrange) system $\mathbf{EW}\{x, t\,; v, \sigma, \rho, w\}$ (5.104) has the point symmetry

$$\mathrm{Y} = \frac{e^t}{\rho}\left[\frac{\partial}{\partial t} + (v + \rho w)\frac{\partial}{\partial x} + (x + \rho w)\frac{\partial}{\partial v} - \rho(\rho^2 + 1)\frac{\partial}{\partial \rho} - \rho(x - v)\frac{\partial}{\partial w}\right],$$

which yields a nonlocal symmetry of the Euler system $\mathbf{E}\{x, t\,; v, \sigma, \rho\}$ (5.103).

(c) Find invariant solutions of the Euler system $\mathbf{E}\{x, t\,; v, \sigma, \rho\}$ (5.103) arising from the nonlocal symmetry Y [Bluman, Cheviakov & Ganghoffer (2008)].

5.2.4. Derive the invariant solutions (5.60) and (5.61) of the Lagrange PGD system $\mathbf{L}\{y, s\,; v, p, q\}$ (3.42). In particular, show that the invariants of the symmetry (5.59) are given by

$$\begin{aligned} &I_1 = s, \quad I_2 \equiv P(s) = p(y, s)y^{-2}(y + \alpha), \\ &I_3 \equiv Q(s) = q(y, s)(y + \alpha)^3 \left(\ln \frac{P(s)y^2}{y + \alpha}\right)^{-1}, \\ &I_4 \equiv W(s) = w^2(y, s)y^{-2}(y + \alpha), \\ &I_5 \equiv V(s) = v(y, s)y^{-1}(y + \alpha)^2 - yW(s). \end{aligned} \tag{5.105}$$

After substitution of the dependent variables in terms of the invariants (5.105), show that the potential system $\mathbf{LW}^2\{y, s\,; v, p, q, w^2\}$ yields a system of ODEs

$$Q'(s) = 0, \quad W'(s) = -P(s), \quad P'(s) = 2\alpha^2 \frac{P(s)W(s)}{Q(s)}, \tag{5.106}$$

whose solution is given by (5.60).

5.2.5. Consider the invariant solution (5.60) of the potential system $\mathbf{LW}^2\{y, s;\ v, p, q, w^2\}$ (3.98) with $M(p) = -p \ln p$. Explicitly verify that the corresponding solution of the Lagrange system $\mathbf{L}\{y, s\,; v, p, q\}$ (3.42) does not arise as an invariant solution with respect to a point symmetry of $\mathbf{L}\{y, s\,; v, p, q\}$ [Bluman, Cheviakov & Ivanova (2006)].

5.2.6. Consider the symmetry Y_8 [Table 4.18], which yields a nonlocal symmetry of the polytropic Lagrange PGD system (3.42) ($\gamma = 3$), and a point symmetry of the potential system $\mathbf{LW^1}\{y, s\,; v, p, q, w^1\}$ (3.97).

(a) Find solutions of the potential system $\mathbf{LW^1}\{y, s\,; v, p, q, w^1\}$ invariant with respect to Y_8.
(b) Find the corresponding solution of the Lagrange system $\mathbf{L}\{y, s\,; v, p, q\}$.

5.2.7.

(a) Show that if a scalar PDE (5.62) is a linear parabolic PDE, then the potential equation (5.64) is locally related to the potential system (5.63).
(b) Show that if a scalar PDE (5.62) is either a linear elliptic or linear hyperbolic PDE, then the PDE (5.62), the potential equation (5.64), and the potential system (5.63) are three distinct nonlocally related systems.

5.2.8. Consider the wave equation in an inhomogeneous medium, given by

$$u_{tt} - xu_{xx} = 0. \tag{5.107}$$

(a) Show that x^{-1} is a multiplier for a conservation law of (5.107).
(b) Find the resulting potential system and find the potential symmetry of (5.107) that arises as a point symmetry of the potential system.
(c) Find the invariant solution of the potential system arising from the obtained potential symmetry and find the corresponding solution of the PDE (5.107).
(d) Use the first refinement of the potential system ansatz, in conjunction with the found potential symmetry, to find a wider class of solutions of the PDE (5.107).
(e) Show that all of the solutions of the PDE (5.107) obtained through (c) and (d) [Pucci & Saccomandi (1993)] arise from the invariant solutions of the PDE (5.107) resulting from its invariance under particular scalings and translations.

5.2.9. The Lagrange PDE system $\mathbf{L}\{y, s\,; v, p, q\}$ for the planar gas dynamics equations in the general polytropic case with the constitutive function $B(p, q) = \gamma p/q$ for some constant γ, is given by

$$\begin{aligned} q_s - v_y &= 0, \\ v_s + p_y &= 0, \\ p_s + \gamma \frac{p}{q} v_y &= 0. \end{aligned} \tag{5.108}$$

Consider the potential system $\mathbf{LW^2}\{y, s\,; v, p, q, w^2\}$ given by

$$\begin{aligned} q_s - v_y &= 0, \\ w^2_y &= v, \\ w^2_s &= -p, \\ p_s + \gamma \frac{p}{q} v_y &= 0. \end{aligned} \tag{5.109}$$

(a) Show that the potential system (5.109) has the point symmetry

$$\mathrm{Z} = y^2 \frac{\partial}{\partial y} + yp \frac{\partial}{\partial p} - 3yq \frac{\partial}{\partial q} + (w^2 - yv) \frac{\partial}{\partial v} + yw^2 \frac{\partial}{\partial w^2} \tag{5.110}$$

which yields a potential symmetry of the Lagrange PDE system (5.108).

(b) Show that invariants of the point symmetry (5.110) are given by

$$z = s, \quad H^1(y,p) = \frac{p}{y}, \quad H^2(y,q) = y^3 q,$$
$$H^3(y,w^2) = \frac{w^2}{y}, \quad H^4(y,v,w^2) = yv - w^2.$$

(c) Show that the invariant solutions of the potential system (5.109) corresponding to the point symmetry (5.110) arise from the ansatz

$$\begin{aligned} p(y,s) &= yh^1(s), \quad q(y,s) = \frac{h^2(s)}{y^3}, \\ v(y,s) &= \frac{h^4(s)}{y} + h^3(s), \quad w^2(y,s) = yh^3(s). \end{aligned} \tag{5.111}$$

(d) Find the solutions of the Lagrange PDE system (5.108) that arise from the ansatz (5.111).

(e) Use the second refinement of the potential system ansatz in which one replaces $h^3(s)$ by $h^3(y,s)$ in the ansatz (5.111). Find the corresponding families of solutions of the Lagrange PDE system (5.108) that arise from this refined ansatz, and do not arise as invariant solutions of the Lagrange PDE system (5.108) or the potential system (5.109) with respect to any of their point symmetries [Cheviakov (2008)].

5.2.10. Show that for $K(u)$ given by (5.96), the determining equations (5.97) yield only one "nonclassical symmetry" of the nonlinear heat conduction equation (5.92), given by

$$\mathrm{Y} = \frac{c_3 x + c_2}{2c_3 t + c_1} \frac{\partial}{\partial x} + \frac{\partial}{\partial t}.$$

Show that the corresponding invariant solution is a classical invariant solution of the PDE (5.92) [Bluman & Yan (2005)].

5.2.11. Consider the potential system (5.93) of the nonlinear heat conduction equation (5.92). In the case where $\tau \equiv 0$, $\xi \equiv 1$, show that the nonclassical method applied to the potential system (5.93) only yields solutions of (5.92) of the form $u = f(x)$.

5.2.12. Consider the potential equation (5.94) of the nonlinear heat conduction equation (5.92). For $K(u)$ given by (5.96), in the case where $\tau \equiv 1$, show that the determining equations (5.101) yield two "nonclassical symmetries"

$$\begin{aligned} \mathrm{Y}_1 &= \frac{\partial}{\partial t} + b\tanh\left[b\left(2\gamma - \tfrac{1}{2}\right)v + b\gamma x + c\right]\left(\frac{\partial}{\partial x} - 2\gamma\frac{\partial}{\partial v}\right), \\ \mathrm{Y}_2 &= \frac{\partial}{\partial t} + b\coth\left[b\left(2\gamma - \tfrac{1}{2}\right)v + b\gamma x + c\right]\left(\frac{\partial}{\partial x} - 2\gamma\frac{\partial}{\partial v}\right), \end{aligned} \tag{5.112}$$

where $b \neq 0, c$ and γ are arbitrary constants. Show also that invariant solutions arising from generators (5.112) include nonclassical potential solutions of the given nonlinear heat conduction equation (5.92). [See Bluman & Yan (2005). In this paper, an asymptotic analysis was used to show that the exhibited solutions were indeed nonclassical potential solutions.]

5.3 Nonlocally Related PDE Systems in Three or More Dimensions

For PDE systems with $n > 2$ independent variables, the situation for obtaining and using nonlocally related PDE systems is considerably more complex than in the two-dimensional case. In particular, every divergence-type conservation law gives rise to several potential variables, which are only defined to within arbitrary functions of the independent variables. The corresponding potential system is thus *under-determined*, and is said to have *gauge freedom*. Additional equations involving potential variables, called *gauge constraints*, are needed to make such potential systems determined [Section 5.3.1]. In contrast, nonlocally related subsystems do not require additional gauge constraints [Section 5.3.2].

In Section 5.3.3, trees of nonlocally related PDE systems are constructed for a given PDE system and several important theorems are presented. In particular, it is shown that *only determined* nonlocally related systems can yield nonlocal symmetries of a given PDE system.

An important difference between two-dimensional and higher-dimensional PDE systems is that in higher dimensions, there exist several different types of conservation laws. For example, in three dimensions, one may have a vanishing divergence or a vanishing curl; for $n > 3$, $n - 1$ types of conservation laws exist. The consideration of such conservation laws and their use for the construction of nonlocally related potential systems are both discussed in Section 5.3.4.

Finally, in Section 5.3.5, several instructive higher-dimensional examples are considered where the use of nonlocally related PDE systems leads to new results.

5.3.1 Divergence-type conservation laws and resulting potential systems

In the case of $n \geq 3$ independent variables, divergence-type conservation laws are given by expressions of the form

$$\operatorname{div} \Phi[u] \equiv \mathrm{D}_i \Phi^i(x, u, \partial u, \ldots, \partial^{k-1} u) = 0. \tag{5.113}$$

For a given PDE system, as shown in Chapter 1, such conservation laws can be constructed systematically, e.g. by finding multipliers through annihilations by Euler operators, as in the two-dimensional case.

A conservation law (5.113) directly yields potential equations. For example, consider a divergence-type conservation law in three-dimensional space

$$\operatorname{div} \Phi = \Phi^1_x + \Phi^2_y + \Phi^3_z = 0 \tag{5.114}$$

with a flux vector $\Phi = (\Phi^1(x,y,z), \Phi^2(x,y,z), \Phi^3(x,y,z))$ and independent variables (x,y,z). From (5.114) it immediately follows that $\Phi = \operatorname{curl} \Psi$, where $\Psi = (\Psi^1(x,y,z), \Psi^2(x,y,z), \Psi^3(x,y,z))$ is a vector potential, involving three scalar potential variables. Explicitly, the potential equations are given by

$$\begin{aligned} \Psi^3_y - \Psi^2_z &= \Phi^1, \\ \Psi^1_z - \Psi^3_x &= \Phi^2, \\ \Psi^2_x - \Psi^1_y &= \Phi^3. \end{aligned} \tag{5.115}$$

However, unlike in the two-dimensional situation, the system of potential equations (5.115) is *under-determined.* In particular, the system of potential equations (5.115) is invariant under the transformations

$$\Psi \to \Psi + \operatorname{grad} \phi(x,y,z), \tag{5.116}$$

where $\phi(x,y,z)$ is an arbitrary smooth function of its arguments. Thus the system of potential equations (5.115) has *gauge freedom.*

An additional equation involving the potential variables is required in order to complete the system of potential equations (5.115) to eliminate its gauge freedom. For example, one can have the gauges:

- divergence (Coulomb) gauge: $\operatorname{div} \Psi \equiv \Psi^1_x + \Psi^2_y + \Psi^3_z = 0$,
- spatial gauge: $\Psi^k = 0$, $k = 1$ or 2 or 3,
- Poincaré gauge: $x\Psi^1 + y\Psi^2 + z\Psi^3 = 0$,

or other gauges, provided that all solutions of (5.115) can be obtained from the solutions of the corresponding gauge-constrained (determined) potential system.

If one of the coordinates in a given PDE system is time t, special gauges are frequently used, such as

- Lorentz gauge (in (2+1) dimensions): $\Psi^1_t - \Psi^2_y - \Psi^3_z = 0$,
- Cronstrom gauge (in (2+1) dimensions): $t\Psi^1 - x\Psi^2 - y\Psi^3 = 0$.

For example, for the (2+1)-dimensional variable-coefficient wave equation

$$u_{tt} = (K(x,y)u_x)_x + (K(x,y)u_y)_y, \tag{5.117}$$

the determined potential system $\mathbf{S}\{x, y, t\,; u, \Psi^1, \Psi^1, \Psi^3\}$ with the Lorentz gauge is given by

$$\begin{aligned} \Psi^3_y - \Psi^2_t &= K(x,y)u_x, \\ \Psi^1_t - \Psi^3_x &= K(x,y)u_y, \\ \Psi^2_x - \Psi^1_y &= -u_t, \\ \Psi^1_t - \Psi^2_y - \Psi^3_z &= 0. \end{aligned} \tag{5.118}$$

Each of the above-listed gauge constraints eliminates gauge freedom, in the sense that the potential variables no longer depend on arbitrary functions of the independent variables. A choice of gauge constraint will depend on a particular application. In many cases the choice of gauge constraint is an open problem.

The same situation applies in the general case of $n \geq 3$ independent variables. Consider a PDE system $\mathbf{R}\{x\,; u\}$ of N PDEs of order k with n independent variables $x = (x^1, \ldots, x^n)$, $n \geq 3$, and m dependent variables $u = (u^1, \ldots, u^m)$, given by

$$R^\sigma(x, u, \partial u, \ldots, \partial^k u) = 0, \quad \sigma = 1, \ldots, N, \tag{5.119}$$

which has a divergence-type conservation law (5.113). From Poincaré's lemma it follows that there exist $\frac{1}{2}n(n-1)$ functions $v^{jk} = -v^{kj}(x)$ $(j, k = 1, \ldots, n)$, components of an $n \times n$ antisymmetric tensor, such that the system of n PDEs

$$\Phi^i[u] = \mathrm{D}_j v^{ij}, \quad i = 1, \ldots, n, \tag{5.120}$$

is equivalent to the divergence expression (5.113). The PDE system (5.120) generalizes the three-dimensional curl expression (5.115). Note that for $n > 3$, the number of potential variables is $\frac{1}{2}n(n-1) > n$. Hence here the PDE system (5.120) is even more under-determined than in the situation for the three-dimensional PDE system (5.115). In particular, here the gauge freedom is exhibited by invariance under the transformations

$$v^{ij} \to v^{ij} + \mathrm{D}_k w^{ijk}, \tag{5.121}$$

where w^{ijk} are $\frac{1}{6}n(n-1)(n-2)$ arbitrary functions that are components of a totally antisymmetric tensor. [In particular, for $n=3$, there is only one such free function, which corresponds to the gauge invariance condition (5.116) for curls.] In other words, the system of potential equations (5.120) has an infinite number of point symmetries (*gauge symmetries*)

$$\mathrm{X}_{\text{gauge}} = \mathrm{D}_k w^{ijk} \frac{\partial}{\partial v^{ij}}. \tag{5.122}$$

The corresponding *potential system* $\mathbf{S}\{x\,;u,v\}$ is given by

$$\begin{aligned} R^\sigma(x,u,\partial u,\ldots,\partial^k u) &= 0, \quad \sigma = 1,\ldots,N', \\ \Phi^i[u] &= \mathrm{D}_j v^{ij}, \quad i = 1,\ldots,n. \end{aligned} \tag{5.123}$$

[If all equations of $\mathbf{R}\{x\,;u\}$ (5.119) are in $\mathbf{S}\{x\,;u,v\}$ (5.123), one has $N=N'$. If one of the equations of $\mathbf{R}\{x\,;u\}$ (without loss of generality, $R^N[u]=0$) is a differential consequence of the system of potential equations (5.123) and thus not included in $\mathbf{S}\{x\,;u,v\}$ due to redundancy, then $N'=N-1$.]

As in the two-dimensional case, it follows that the solution sets of a given system $\mathbf{R}\{x\,;u\}$ and its potential system $\mathbf{S}\{x\,;u,v\}$ are equivalent. In general, the potential variables v are nonlocal variables relative to $\mathbf{R}\{x\,;u\}$, and the PDE system $\mathbf{S}\{x\,;u,v\}$ is nonlocally related to $\mathbf{R}\{x\,;u\}$.

As it stands, the potential system $\mathbf{S}\{x\,;u,v\}$ is *under-determined* due to its gauge freedom (5.121). A *determined potential system* is a union of a potential system $\mathbf{S}\{x\,;u,v\}$ and a set of one or more gauge constraints that eliminates the gauge freedom.

5.3.2 Nonlocally related subsystems

Another important way of finding nonlocally related PDE systems in higher dimensions is through subsystems that are obtained after elimination of dependent variables by differential operations. Similar to the situation in the two-dimensional case, in order to obtain a nonlocally related subsystem $\mathbf{U}\{x\,;u\}$ of a given PDE system $\mathbf{UV}\{x\,;u,v\}$, it is obviously necessary that a dependent variable v only occurs in $\mathbf{UV}\{x\,;u,v\}$ in terms of its derivatives.

It should be noted that the construction of nonlocally related subsystems requires no gauge constraints, which (as shown later) is of great value for nonlocal symmetry computations.

As a first example, consider the time-independent PDE system $\mathbf{VP}\{x, y, z; v^1, v^2, v^3, p\}$ of Euler equations of an inviscid, constant density fluid flow in three dimensions, which can be written as

$$\begin{aligned} \mathbf{v} \times (\operatorname{curl} \mathbf{v}) &= \operatorname{grad}\left(\frac{p}{\rho} + \tfrac{1}{2}|\mathbf{v}|^2\right), \\ \operatorname{div} \mathbf{v} &= 0. \end{aligned} \tag{5.124}$$

Here $\mathbf{v} = (v^1, v^2, v^3)$ is the fluid velocity vector and $\rho = \text{const}$ the fluid density. In the PDE system (5.124), one can exclude the pressure p by taking the curl of the vector equation. The resulting subsystem $\mathbf{V}\{x, y, z\,; v^1, v^2, v^3\}$ given by

$$\begin{aligned} \operatorname{curl}\Big[\mathbf{v} \times (\operatorname{curl} \mathbf{v})\Big] &= 0, \\ \operatorname{div} \mathbf{v} &= 0 \end{aligned} \tag{5.125}$$

is equivalent and nonlocally related to the Euler system $\mathbf{VP}\{x, y, z; v^1, v^2, v^3, p\}$.

As a second example, consider the PDE system $\mathbf{UV}\{x, y, t\,; u, v^1, v^2\}$ in one time and two space dimensions, given by

$$\begin{aligned} \mathbf{v}_t &= \operatorname{grad} u, \\ u_t &= K(|\mathbf{v}|) \operatorname{div} \mathbf{v}. \end{aligned} \tag{5.126}$$

In (5.126), $\mathbf{v} = (v^1, v^2)$ is a vector function, and $K(|\mathbf{v}|)$ is a constitutive function of the indicated scalar argument. The PDE system (5.126) has the nonlocally related subsystem $\mathbf{V}\{x, y, t\,; v^1, v^2\}$, given by

$$\mathbf{v}_{tt} = \operatorname{grad}\left[K(|\mathbf{v}|) \operatorname{div} \mathbf{v}\right]. \tag{5.127}$$

In Section 5.3.5, the subsystem (5.127) is used to obtain a nonlocal symmetry of the PDE system (5.126).

5.3.3 Tree construction, nonlocal conservation laws, and nonlocal symmetries

Similar to the situation for two independent variables, for a given PDE system with $n > 2$ independent variables, the construction of a tree of nonlocally related systems involves:

- Finding nonlocally related determined potential systems arising from conservation laws.
- Finding nonlocally related subsystems of the given PDE system and its determined potential systems by exclusion of dependent variables.

As in the case of two independent variables, a tree of nonlocally related systems can be further extended through:

- Obtaining subsystems following an interchange of dependent and independent variables (or a more general point transformation).
- Finding additional local or nonlocal conservation laws, arising as local conservation laws of determined potential systems, and introducing further potential variables as well as additional gauge constraints.

Two important theorems are now presented that are concerned with *nonlocal* conservation laws arising from potential systems in higher dimensions. The first theorem generalizes Theorems 3.4.1 and 3.5.1 to the higher-dimensional case [Cheviakov & Bluman (2009); see also Kunzinger & Popovych (2008) and Bluman, Cheviakov & Ivanova (2006)].

Theorem 5.3.1. *Suppose* $\mathbf{R}\{x\,;u\}$ (5.119) *is a PDE system for which* $K \geq 1$ *local conservation laws* (5.113) *are known. Let* $\mathbf{S}\{x\,;u,v\}$ *be the potential system given by the union of* $\mathbf{R}\{x\,;u\}$ *and the corresponding* K *systems of potential equations* (5.120). *Then each conservation law of the potential system* $\mathbf{S}\{x\,;u,v\}$*, arising from multipliers that do not depend on potential variables* v*, is linearly dependent on local conservation laws of* $\mathbf{R}\{x\,;u\}$.

Proof. For simplicity, we consider a singlet potential system $\mathbf{S}\{x\,;u,v\}$ (5.123) following from some divergence-type conservation law (5.113) of a $\mathbf{R}\{x\,;u\}$ (5.119). [The proof directly carries over to K-plet potential systems, $K > 1$.]

Consider a local conservation law

$$\mathrm{D}_k F^k[u,v] = 0 \tag{5.128}$$

of the potential system $\mathbf{S}\{x\,;u,v\}$, arising from multipliers independent of the potential variables v. A conservation law (5.128) corresponds to the divergence expression

$$\mathrm{D}_k F^k[U,V] = A_i[U](\mathrm{D}_j V^{ij} - \Phi^i[U]) + \sum_{\sigma=1}^{N'} \Lambda_\sigma[U] R^\sigma[U], \tag{5.129}$$

where $A_i[U]$ are multipliers of the potential equations, and $\Lambda_\sigma[U]$ are multipliers of the equations of $\mathbf{R}\{x\,;u\}$ (5.119) that are present in the potential system $\mathbf{S}\{x\,;u,v\}$. [Without loss of generality, summation in σ can be taken from 1 to N.]

Now apply Euler operators $\mathrm{E}_{V^{\alpha\beta}}$ (1.149) with respect to $V^{\alpha\beta}$ ($\alpha, \beta = 1, \ldots, n$, $\alpha \neq \beta$) to the equation (5.129). The divergence and terms not involving $V^{\alpha\beta}$ vanish identically. Using the antisymmetry of V^{ij}, one obtains

$$\frac{\partial A_\alpha[U]}{\partial x^\beta} - \frac{\partial A_\beta[U]}{\partial x^\alpha} \equiv 0.$$

By a basic lemma in variational calculus [Wald (1990)], it follows that

$$A_i[U] = \frac{\partial B[U]}{\partial x^i},$$

and the equation (5.129) can be rewritten as

$$\begin{aligned} &\mathrm{D}_i \left(F^i[U,V] - B[U](\mathrm{D}_j V^{ij} - \Phi^i[U])\right) \\ &\quad = B[U]\mathrm{D}_i \Phi^i[U] + \Lambda_\sigma[U] R^\sigma[U]. \end{aligned} \tag{5.130}$$

[Here the identity $\mathrm{D}_i \mathrm{D}_j V^{ij} \equiv 0$ has been used.]

Now consider the conservation law $\mathrm{D}_i \Phi^i[u] = 0$ (5.113). Assuming that $\mathbf{R}\{x\,; u\}$ (5.119) can be written in a solved form (1.152) with respect to some leading derivatives $\{u^{j_\sigma}_{i_{\sigma,1}\cdots i_{\sigma,s}}\}$, it follows that by adding trivial fluxes (of the first type), one can assume that each flux $\Phi^i[U]$ contains no leading derivatives nor their differential consequences. Thus the only leading derivatives (and their differential consequences) that can arise in the expression $\mathrm{D}_i \Phi^i[U]$ will come from subleading derivatives $\{U^{j_\sigma}_{i_{\sigma,1}\cdots i_{\sigma,s-1}}\}$ (and their differential consequences). Hence, the divergence expression $\mathrm{D}_i \Phi^i[U]$ must be a linear combination of $R^\sigma[U]$ and their differential consequences:

$$\mathrm{D}_i \Phi^i[U] = \Gamma_{(0)\,\sigma}[U] R^\sigma[U] + \Gamma^i_{(1)\,\sigma}[U] \mathrm{D}_i R^\sigma[U] + \cdots + \Gamma^{i_1 \ldots i_q}_{(q)\,\sigma}[U] \mathrm{D}_{i_1} \ldots \mathrm{D}_{i_q} R^\sigma[U].$$

for some coefficients $\Gamma_{(0)\,\sigma}[U], \ldots, \Gamma^{i_1 \ldots i_q}_{(q)\,\sigma}[U]$ which are non-singular functions of x, U and derivatives of U. Then one has

$$B[U]\mathrm{D}_i \Phi^i[U] = \Gamma_\sigma[U] R^\sigma[U] + \mathrm{D}_i Q^i[U],$$

where $\{Q^i[U]\}_{i=1}^n$ are linear combinations involving $R^\sigma[U]$ and its differential consequences. Moreover, each $Q^i[U]$ vanishes on all solutions $U(x) = u(x)$ of $\mathbf{R}\{x\,; u\}$. Therefore (5.130) becomes

$$\begin{aligned} &\mathrm{D}_i \left(F^i[U,V] - B[U](\mathrm{D}_j V^{ij} - \Phi^i[U]) - Q^i[U]\right) \\ &\quad = (\Lambda_\sigma[U] + \Gamma_\sigma[U]) R^\sigma[U]. \end{aligned} \tag{5.131}$$

Since expressions $B[U](\mathrm{D}_j V^{ij} - \Phi^i[U])$ and $Q^i[U]$ vanish on solutions $(U, V) = (u, v)$ of the potential system $\mathbf{S}\{x\,; u, v\}$ (5.123), the left-hand side of (5.131) is a divergence expression corresponding to a conservation law equivalent to the conservation law (5.128). The right-hand side of (5.131) is a linear combination of equations of the given system $\mathbf{R}\{x\,;u\}$ (5.119), with multipliers depending only on local variables of $\mathbf{R}\{x\,;u\}$. Therefore the conservation law (5.128) of the potential system $\mathbf{S}\{x\,;u,v\}$ (5.123) is equivalent to a local conservation law of the given system $\mathbf{R}\{x\,;u\}$. □

Note that Theorem 5.3.1 generally does not hold when a potential system includes a gauge constraint(s) [Exercise 5.3.1].

It is important to remark that nonlocal conservation laws can exist for both determined and under-determined potential systems. In particular, the following theorem holds [Anco & The (2005)].

Theorem 5.3.2. *Suppose a given PDE system* $\mathbf{R}\{x\,;u\}$ *has an under-determined potential system* $\mathbf{S}\{x\,;u,v\}$ *with gauge freedom given by the point symmetry* X_{gauge} (5.122). *Then all divergence-type conservation laws*

$$\operatorname{div}\Phi[u,v] = \mathrm{D}_i\Phi^i[u,v] = 0$$

of $\mathbf{S}\{x\,;u,v\}$ *are gauge-invariant under* (5.122). *In particular,* $\operatorname{div}(\mathrm{X}_{gauge}\Phi) \equiv 0$ *on solutions of* $\mathbf{S}\{x\,;u,v\}$.

Theorem 5.3.2 states the invariance of fluxes under gauge symmetries but does not rule out the possible explicit dependence of fluxes $\Phi^i[u,v]$ on potentials v. In particular, there exist examples of nonlocal conservation laws that arise from both determined and under-determined potential systems [Anco & The (2005)].

Unlike the situation for nonlocal conservation laws, *nonlocal symmetries can only arise from determined potential systems.* The following essential theorem holds [Anco & Bluman (1997b)].

Theorem 5.3.3. *Each local symmetry of an under-determined potential system* $\mathbf{S}\{x\,;u,v\}$ (5.123) *projects onto a local symmetry of the given system* $\mathbf{R}\{x\,;u\}$ (5.119).

Proof. Suppose the potential system $\mathbf{S}\{x\,;u,v\}$ (5.123) has a local symmetry that has a characteristic form

$$\hat{\mathrm{X}} = \hat{\eta}^\mu[u,v]\frac{\partial}{\partial u^\mu} + \hat{\zeta}^\nu[u,v]\frac{\partial}{\partial v^\nu}, \tag{5.132}$$

where $\hat{\eta}^\mu[u,v]$ and $\hat{\zeta}^\nu[u,v]$ are functions of x, u, v and derivatives of u, v. Since the potential system $\mathbf{S}\{x\,;u,v\}$ (5.123) is under-determined, it has an infinite number of point symmetries $\mathrm{X}_{\mathrm{gauge}}$ (5.122). The local symmetries form a Lie algebra, hence it follows that the commutator

$$[\mathrm{X}_{\mathrm{gauge}}, \hat{\mathrm{X}}] = \mathrm{X}^\infty_{\mathrm{gauge}}\hat{\mathrm{X}} - \hat{\mathrm{X}}\mathrm{X}_{\mathrm{gauge}} = \mathrm{X}^\infty_{\mathrm{gauge}}\hat{\mathrm{X}} \tag{5.133}$$

is also a local symmetry of $\mathbf{S}\{x\,;u,v\}$. [Here $\mathrm{X}^{\infty}_{\mathrm{gauge}}$ is a prolongation of $\mathrm{X}_{\mathrm{gauge}}$ given by (1.38).] The projection of (5.133), given by

$$\mathrm{Y}=\left((\mathrm{D}_k w^{ijk})\frac{\partial\hat{\eta}^{\mu}[u,v]}{\partial v^{ij}}+(\mathrm{D}_{k_1}\mathrm{D}_k w^{ijk})\frac{\partial\hat{\eta}^{\mu}[u,v]}{\partial v^{ij}_{k_1}}\right.$$
$$\left.+(\mathrm{D}_{k_1}\mathrm{D}_{k_2}\mathrm{D}_k w^{ijk})\frac{\partial\hat{\eta}^{\mu}[u,v]}{\partial v^{ij}_{k_1k_2}}+\cdots\right)\frac{\partial}{\partial u^{\mu}},$$

is a local symmetry of the PDE system $\mathbf{R}\{x\,;u\}$ (5.119). Since the PDE system $\mathbf{R}\{x\,;u\}$ (5.119) is determined, its symmetries do not include free functions of all variables. Moreover, derivatives of the functions w^{ijk} are linearly independent. Therefore one concludes that

$$\frac{\partial\hat{\eta}^{\mu}[u,v]}{\partial v^{ij}}=\frac{\partial\hat{\eta}^{\mu}[u,v]}{\partial v^{ij}_{k_1}}=\frac{\partial\hat{\eta}^{\mu}[u,v]}{\partial v^{ij}_{k_1k_2}}=\cdots=0,$$

which implies that the symmetry (5.132) is a local symmetry of the given PDE system $\mathbf{R}\{x\,;u\}$ (5.119). □

As mentioned above, in the case of $n\geq 3$ independent variables, potential systems arising from divergence-type conservation laws (5.113) are always under-determined so that gauge constraints are necessary in order to find nonlocal symmetries. In general, it remains an open question as to what type of gauge constraints one should choose in order to obtain nonlocal symmetries for a given PDE system. An example using the Lorentz gauge for Maxwell's equations is presented in Section 5.3.5.

In the following subsection, it is seen that potential systems arising from *lower-degree* (curl-type) conservation laws can require fewer or no gauge constraints to be determined. Examples of such potential systems are considered in Section 5.3.5.

Another way of finding nonlocal symmetries of a given PDE system is through the consideration of its nonlocally related subsystemsthat also require no gauge constraints. An example of a nonlocal symmetry arising from a nonlocally related subsystem is presented in Section 5.3.5.

5.3.4 Lower-degree conservation laws and related potential systems

In three or more dimensions, conservation laws are not limited to independent divergence expressions (5.113). For example, in three-dimensional space, a PDE system $\mathbf{R}\{x,y,z\,;u\}$ can have a conservation law

$$\operatorname{curl} \Psi[u] = 0, \tag{5.134}$$

where $\Psi = (\Psi^1, \Psi^2, \Psi^3)$ is some flux vector depending on independent variables (x, y, z) and dependent variables u. Such a curl-type conservation law is often referred to as a *lower-degree conservation law.*

Of course, a conservation law (5.134) can be viewed as three divergence-type conservation laws corresponding to the three projections of a curl. Accordingly, one can introduce a total of nine potential variables, with gauge constraints to be chosen. However, another (and in many ways more efficient) representation of a curl-free vector field is in terms of the gradient of a scalar function, i.e. the conservation law (5.134) is equivalent to the set of potential equations

$$\begin{aligned} \Psi^1[u] &= w_x, \\ \Psi^2[u] &= w_y, \\ \Psi^3[u] &= w_z. \end{aligned} \tag{5.135}$$

In (5.135), the nonlocal potential variable $w(x, y, z)$ is defined to within a constant. Therefore the corresponding potential system is determined and requires no gauge constraints.

For PDE systems with three independent variables, the only possible types of conservation laws are of divergence-type and curl-type. Divergence-type conservation laws occur more frequently, but curl-type conservation laws also arise in physical applications. Examples include PDE systems describing static electromagnetic fields, irrotational gas and fluid dynamics, and ideal plasma equilibria. An example of the application of a curl-type conservation law to generate a useful nonlocally related potential system is presented in Section 5.3.6.

In addition to divergence-type and curl-type conservation laws, PDE systems with $n > 3$ independent variables can have other types of *lower-degree conservation laws.* In particular, PDE systems with n independent variables can have $n - 1$ types of conservation laws. Similar to the conservation law (5.134), lower-degree conservation laws are expressed by several components, i.e., vanishing divergence expressions. It is important to note that lower-degree conservation laws can yield a smaller number of potential variables than divergence-type conservation laws, and thus require fewer gauge constraints. In particular, *conservation laws of degree one* (which generalize curl-type conservation laws in $n \geq 3$ dimensions) can be shown to always yield *determined potential equations*, requiring no gauge constraints. Several examples of potential systems following from lower-degree conservation laws that arise in applications are presented in Section 5.3.5.

For a detailed description of lower-degree conservation laws and arising potential systems, it is convenient to use differential form notation. For further details, see Anderson & Torre (1996).

5.3.5 Examples of applications of nonlocally related systems in higher dimensions

A nonlocal symmetry arising from a nonlocally related subsystem

The first example illustrates the use of nonlocally related subsystems to obtain nonlocal symmetries of PDE systems in higher dimensions.

Let the (2+1)-dimensional PDE system $\mathbf{UV}\{x, y, t\,; u, v^1, v^2\}$ (5.126) be a given PDE system with a scalar constitutive function $K(|\mathbf{v}|)$. The PDE system $\mathbf{UV}\{x, y, t\,; u, v^1, v^2\}$ (5.126) has the nonlocally related subsystem $\mathbf{V}\{x, y, t\,; v^1, v^2\}$ (5.127). Consider the one-parameter class of constitutive functions given by

$$K(|\mathbf{v}|) = |\mathbf{v}|^{2k} = \left((v^1)^2 + (v^2)^2\right)^k. \tag{5.136}$$

It is interesting to compare the symmetry classifications of the systems (5.126) and (5.127) with respect to the constitutive parameter $k \neq 0$.

For arbitrary k in (5.136), one can show that the given PDE system $\mathbf{UV}\{x, y, t\,; u, v^1, v^2\}$ (5.126) has seven point symmetries

$$\begin{aligned}
&\mathrm{X}_1 = \frac{\partial}{\partial t}, \quad \mathrm{X}_2 = \frac{\partial}{\partial x}, \quad \mathrm{X}_3 = \frac{\partial}{\partial y}, \quad \mathrm{X}_4 = \frac{\partial}{\partial u},\\
&\mathrm{X}_5 = t\frac{\partial}{\partial t} + x\frac{\partial}{\partial x} + y\frac{\partial}{\partial y},\\
&\mathrm{X}_6 = -y\frac{\partial}{\partial x} + x\frac{\partial}{\partial y} - v^2\frac{\partial}{\partial v^1} + v^1\frac{\partial}{\partial v^2},\\
&\mathrm{X}_7 = m\left(x\frac{\partial}{\partial x} + y\frac{\partial}{\partial y}\right) + (m+1)u\frac{\partial}{\partial u} + v^1\frac{\partial}{\partial v^1} + v^2\frac{\partial}{\partial v^2},
\end{aligned} \tag{5.137}$$

and the subsystem $\mathbf{V}\{x, y, t\,; v^1, v^2\}$ (5.127) has the corresponding six point symmetries

$$\begin{aligned}
&\mathrm{Y}_1 = \mathrm{X}_1, \quad \mathrm{Y}_2 = \mathrm{X}_2, \quad \mathrm{Y}_3 = \mathrm{X}_3, \quad \mathrm{Y}_4 = \mathrm{X}_5, \quad \mathrm{Y}_5 = \mathrm{X}_6,\\
&\mathrm{Y}_6 = m\left(x\frac{\partial}{\partial x} + y\frac{\partial}{\partial y}\right) + v^1\frac{\partial}{\partial v^1} + v^2\frac{\partial}{\partial v^2}.
\end{aligned} \tag{5.138}$$

In the case $k = -1$, one can show that both systems have an infinite number of point symmetries. Finally, in the case $k = -2$, one can show that the subsystem $\mathbf{V}\{x, y, t\,; v^1, v^2\}$ (5.127) has an additional point symmetry

$$\mathrm{Y}_7 = t^2\frac{\partial}{\partial t} + tv^1\frac{\partial}{\partial v^1} + tv^2\frac{\partial}{\partial v^2}, \tag{5.139}$$

whereas the given PDE system $\mathbf{UV}\{x, y, t\,; u, v^1, v^2\}$ (5.126) still has the same point symmetries (5.137). It follows that (5.139) yields a nonlocal symmetry of the PDE system $\mathbf{UV}\{x, y, t\,; u, v^1, v^2\}$ (5.126).

Nonlocal symmetries arising from a lower-degree conservation law

As a second example, consider the time-independent PDE system $\mathbf{H}\{x, y, z;\, h^1, h^2, h^3\}$ in three space dimensions, given by

$$\operatorname{curl}\Big(K(|\mathbf{h}|)(\operatorname{curl}\mathbf{h}) \times \mathbf{h}\Big) = 0, \quad \operatorname{div}\, \mathbf{h} = 0. \tag{5.140}$$

In (5.140), $\mathbf{h} = (h^1, h^2, h^3)$ is a vector of dependent variables. The first equation in PDE system (5.140) is a curl-type conservation law as it stands. The corresponding potential system $\mathbf{HW}\{x, y, z\,; h^1, h^2, h^3, w\}$ is given by

$$K(|\mathbf{h}|)(\operatorname{curl}\mathbf{h}) \times \mathbf{h} = \operatorname{grad} w, \quad \operatorname{div}\, \mathbf{h} = 0, \tag{5.141}$$

where $w(x, y, z)$ is a scalar potential variable. The potential system (5.141) is determined as written and hence needs no gauge constraints.

Now a comparison is made of the classifications of nonlocal symmetries of the PDE systems $\mathbf{H}\{x, y, z\,; h^1, h^2, h^3\}$ and $\mathbf{HW}\{x, y, z\,; h^1, h^2, h^3, w\}$ for the one-parameter family of constitutive functions $K(|\mathbf{h}|)$ given by

$$K(|\mathbf{h}|) = |\mathbf{h}|^{2k} \equiv \Big((h^1)^2 + (h^2)^2 + (h^3)^2\Big)^k, \tag{5.142}$$

where k is a parameter. For $k \neq -1$, the potential system $\mathbf{HW}\{x, y, z; h^1, h^2, h^3, w\}$ has nine point symmetries, corresponding to four translations (in x, y, z and w), three rotations, one scaling and one dilation [Exercise 5.3.2], spanned by $\mathrm{X}_1, \ldots, \mathrm{X}_9$. Correspondingly, the given PDE system $\mathbf{H}\{x, y, z\,; h^1, h^2, h^3\}$ has eight point symmetries (projections of $\mathrm{X}_1, \ldots, \mathrm{X}_9$). For $k = -1$, the point symmetries of $\mathbf{H}\{x, y, z\,; h^1, h^2, h^3\}$ remain the same, whereas the potential system $\mathbf{HW}\{x, y, z\,; h^1, h^2, h^3, w\}$ has an additional infinite number of point symmetries given by

$$\mathrm{X}_\infty = F(w)\Big(\frac{\partial}{\partial w} + h^1\frac{\partial}{\partial h^1} + h^2\frac{\partial}{\partial h^2} + h^3\frac{\partial}{\partial h^3}\Big) \tag{5.143}$$

depending on an arbitrary smooth function $F(w)$. The symmetries (5.143) are nonlocal symmetries of the given PDE system $\mathbf{H}\{x, y, z\,; h^1, h^2, h^3\}$ (5.140).

Note that the symmetries (5.143) cannot be used for the construction of invariant solutions since they do not involve spatial components. However, one can use the symmetries (5.143) to map any known solution of the PDE system (5.140) (with a corresponding potential variable w) to an infinite family of solutions of (5.140).

Nonlocal symmetries of the two-dimensional linear wave equation

Consider the linear wave equation $\mathbf{U}\{t, x, y\,; u\}$ given by

$$u_{tt} = u_{xx} + u_{yy}. \tag{5.144}$$

Equation (5.144) is a divergence-type conservation law as it stands. A vector potential $v = (v^0, v^1, v^2)$ is introduced and a Lorentz gauge is appended to the corresponding under-determined potential system. The resulting determined potential system $\mathbf{UV}\{t, x, y\,; u, v^0, v^1, v^2\}$ is given by

$$\begin{aligned} u_t &= v^2_x - v^1_y, \\ -u_x &= v^0_y - v^2_t, \\ -u_y &= v^1_t - v^0_x, \\ v^0_t - v^1_x - v^2_y &= 0. \end{aligned} \tag{5.145}$$

A comparison is now made of the point symmetries of the PDE systems $\mathbf{U}\{t, x, y\,; u\}$ (5.144) and $\mathbf{UV}\{t, x, y\,; u, v^0, v^1, v^2\}$ (5.145). Modulo the infinite number of point symmetries of any linear PDE system, the linear wave equation (5.144) has ten point symmetries:

- three translations $\mathrm{X}_1, \mathrm{X}_2, \mathrm{X}_3$ given by
$$\mathrm{X}_1 = \frac{\partial}{\partial t}, \quad \mathrm{X}_2 = \frac{\partial}{\partial x}, \quad \mathrm{X}_3 = \frac{\partial}{\partial y};$$
- one dilation given by
$$\mathrm{X}_4 = t\frac{\partial}{\partial t} + x\frac{\partial}{\partial x} + y\frac{\partial}{\partial y};$$
- one rotation and two space-time rotations (boosts) given by
$$\mathrm{X}_5 = x\frac{\partial}{\partial y} - y\frac{\partial}{\partial x}, \quad \mathrm{X}_6 = t\frac{\partial}{\partial x} + x\frac{\partial}{\partial t}, \quad \mathrm{X}_7 = t\frac{\partial}{\partial y} + y\frac{\partial}{\partial t};$$
- three additional conformal transformations given by

$$\begin{aligned}
\mathrm{X}_8 &= (t^2+x^2+y^2)\frac{\partial}{\partial t} + 2tx\frac{\partial}{\partial x} + 2ty\frac{\partial}{\partial y} - tu\frac{\partial}{\partial u},\\
\mathrm{X}_9 &= 2tx\frac{\partial}{\partial t} + (t^2+x^2-y^2)\frac{\partial}{\partial x} + 2xy\frac{\partial}{\partial y} - xu\frac{\partial}{\partial u},\\
\mathrm{X}_{10} &= 2ty\frac{\partial}{\partial t} + 2xy\frac{\partial}{\partial x} + (t^2-x^2+y^2)\frac{\partial}{\partial y} - yu\frac{\partial}{\partial u}.
\end{aligned}$$

The potential system $\mathbf{UV}\{t,x,y\,;u,v^0,v^1,v^2\}$ (5.145) has seven point symmetries $\mathrm{Y}_1,\ldots,\mathrm{Y}_7$ that project onto the point symmetries $\mathrm{X}_1,\ldots,\mathrm{X}_7$ of the wave equation (5.144). However the three additional conformal symmetries of the potential system (5.145) have the form

$$\begin{aligned}
\mathrm{Y}_8 = {} & \mathrm{X}_8 - (yv^1 - xv^2 + tu)\frac{\partial}{\partial u} - (2tv^0 + xv^1 + yv^2)\frac{\partial}{\partial v^0}\\
& -(xv^0 + 2tv^1 + yu)\frac{\partial}{\partial v^1} + (yv^0 + 2tv^2 - xu)\frac{\partial}{\partial v^2},\\
\mathrm{Y}_9 = {} & \mathrm{X}_9 - (-yv^0 - tv^2 + xu)\frac{\partial}{\partial u} - (2xv^0 + tv^1 + yu)\frac{\partial}{\partial v^0}\\
& -(xv^0 + 2tv^1 + yu)\frac{\partial}{\partial v^1} + (yv^0 + 2tv^2 - xu)\frac{\partial}{\partial v^2},\\
\mathrm{Y}_{10} = {} & \mathrm{X}_{10} - (xv^0 + tv^1 + yu)\frac{\partial}{\partial u} - (2yv^0 + tv^2 - xu)\frac{\partial}{\partial v^0}\\
& -(2yv^1 - xv^2 + tu)\frac{\partial}{\partial v^1} + (tv^0 + xv^1 + 2yv^2)\frac{\partial}{\partial v^2},
\end{aligned} \tag{5.146}$$

and, from their forms, clearly yield nonlocal symmetries of the wave equation $\mathbf{U}\{t,x,y\,;u\}$ (5.144). In addition, the potential system (5.145) has three point symmetries of duality-type given by

$$\begin{aligned}
\mathrm{Y}_{11} &= v^0\frac{\partial}{\partial u} - u\frac{\partial}{\partial v^0} + v^2\frac{\partial}{\partial v^1} - v^1\frac{\partial}{\partial v^2},\\
\mathrm{Y}_{12} &= v^1\frac{\partial}{\partial u} - v^2\frac{\partial}{\partial v^0} + u\frac{\partial}{\partial v^1} - v^0\frac{\partial}{\partial v^2},\\
\mathrm{Y}_{13} &= v^2\frac{\partial}{\partial u} + v^1u\frac{\partial}{\partial v^0} + v^0\frac{\partial}{\partial v^1} + u\frac{\partial}{\partial v^2},
\end{aligned} \tag{5.147}$$

that yield nonlocal symmetries of the wave equation $\mathbf{U}\{t,x,y\,;u\}$ (5.144). In summary, the potential system $\mathbf{UV}\{t,x,y\,;u,v^0,v^1,v^2\}$ (5.145) with the Lorentz gauge yields six nonlocal symmetries of the linear wave equation (5.144).

It turns out that no nonlocal symmetries of the wave equation arise from the potential system $\mathbf{UV}\{t,x,y\,;u,v^0,v^1,v^2\}$ if the Lorentz gauge is replaced by any one of the algebraic gauges $v^k=0$ for $k\in\{0,1,2\}$, the divergence gauge, the Poincaré gauge, or the Cronstrom gauge [Exercise 5.3.3].

Nonlocal symmetries and nonlocal conservation laws of Maxwell's equations in (2+1) dimensions

Maxwell's equations in a vacuum are given by

$$\begin{aligned} \operatorname{div}\mathbf{B} &= 0,\\ \operatorname{div}\mathbf{E} &= 0,\\ \mathbf{E}_t &= \operatorname{curl}\mathbf{B},\\ \mathbf{B}_t &= -\operatorname{curl}\mathbf{E}, \end{aligned} \tag{5.148}$$

where $\mathbf{B} = B^1\mathbf{e}_x + B^2\mathbf{e}_y + B^3\mathbf{e}_z$ is a magnetic field, $\mathbf{E} = E^1\mathbf{e}_x + E^2\mathbf{e}_y + E^3\mathbf{e}_z$ is an electric field, (x, y, z) are cartesian coordinates, and t is time.

In this example, nonlocal symmetries and nonlocal conservation laws of the PDE system (5.148) are sought in three-dimensional Minkowski space (x, y, t). It is assumed that $\mathbf{B} = B(x, y)\mathbf{e}_z$, $\mathbf{E} = E^1(x, y)\mathbf{e}_x + E^2(x, y)\mathbf{e}_y$. Then Maxwell's equations (5.148) can be written as the PDE system $\mathbf{M}\{t, x, y\,; B, E^1, E^2\}$ in terms of the four equations given by

$$E^1_x + E^2_y = 0, \tag{5.149a}$$
$$E^1_t = B_y, \tag{5.149b}$$
$$E^2_t = -B_x, \tag{5.149c}$$
$$B_t = -E^2_x + E^1_y. \tag{5.149d}$$

Note that each of the four equations (5.149) is a divergence expression as it stands. Hence for each equation in (5.149), one can introduce a three-component vector potential. This yields 12 potential variables. From Theorem 5.3.3, it follows that in order to obtain nonlocal symmetries of Maxwell's equations (5.148), gauge constraints are required. Since the form of gauge constraints that could yield nonlocal symmetries is not known a priori, a different approach is chosen. In particular, it is shown that the system of Maxwell's equations (5.149) is equivalent to the union of a divergence-type conservation law and a curl-type lower-degree conservation law, with the latter requiring no gauge constraints.

Let $\eta^{ij} = \eta_{ij} = \operatorname{diag}(-1, 1, 1)$ be the standard space-time metric tensor in three-dimensional Minkowski space. The electromagnetic field tensor F_{ij} and its raised version $F^{ij} = \eta^{ip}\eta^{jl}F_{pl}$ are, respectively, given by the antisymmetric matrices

$$F = (F_{ij}) = \begin{pmatrix} 0 & -E^1 & -E^2 \\ E^1 & 0 & B \\ E^2 & -B & 0 \end{pmatrix}, \quad (F^{ij}) = \begin{pmatrix} 0 & E^1 & E^2 \\ -E^1 & 0 & B \\ -E^2 & -B & 0 \end{pmatrix}. \tag{5.150}$$

In (5.150), the indices i, j take on the values $0, 1, 2$, with 0 corresponding to a t-component, and $1, 2$ to x- and y-components, respectively.

The dual of F_{ij} is the vector given by $*F_p = \frac{1}{2}\varepsilon_{ijp}F^{ij}$, where ε_{ijp} is the Levi–Civita symbol. In particular,

$$*F = (*F_k) = (B, -E^2, E^1).$$

As is well-known, using F and $*F$, one can express Maxwell's equations (5.149) in the elegant form

$$\partial_p F_{ij} + \partial_j F_{pi} + \partial_i F_{ip} = 0 \quad \Leftrightarrow \quad \eta^{ij}\partial_i {*F_j} = 0, \tag{5.151a}$$

$$\partial_i F^{ij} = 0 \quad \Leftrightarrow \quad \varepsilon^{pij}\partial_i {*F_j} = 0, \tag{5.151b}$$

where ∂_0, ∂_1 and ∂_2 denote partial derivatives with respect to t, x and y, respectively. Note that (5.151a) is equivalent to the scalar equation (5.149d), and (5.151b) is equivalent to the remaining three scalar equations (5.149a)–(5.149c). Consequently, equations (5.151) have the form of divergence and curl conservation laws, if one identifies $\varepsilon^{pij}\partial_i$ and $\eta^{ij}\partial_i$ as Minkowski space versions of curl and divergence operators that are taken with respect to the variables (t, x, y) [Exercise 5.3.4].

The conservation laws (5.151) are now used to construct potential systems for Maxwell's equations (5.149). The curl-type conservation law (5.151b) yields a scalar potential variable W satisfying the potential equations

$$\mathcal{P}^W: \quad \begin{cases} B = W_t, \\ -E^2 = W_x, \\ E^1 = W_y. \end{cases} \tag{5.152}$$

Consequently, one obtains the potential system

$$\mathbf{MW}\{t, x, y\,; B, E^1, E^2, W\} = \mathbf{M}\{t, x, y\,; B, E^1, E^2\} \cup \mathcal{P}^W, \tag{5.153}$$

which is determined and hence requires no gauge constraints.

Through the divergence conservation law (5.151a), one introduces a vector potential (A^0, A^1, A^2) satisfying the system of under-determined potential equations

$$\mathcal{P}^A: \quad \begin{cases} B = A^2_x - A^1_y, \\ E^2 = A^0_y - A^2_t, \\ -E^1 = A^1_t - A^0_x. \end{cases} \tag{5.154}$$

Using the Lorentz gauge, one obtains a determined potential system

$$\begin{aligned} &\mathbf{MA}\{t, x, y\,; B, E^1, E^2, A^0, A^1, A^2\} \\ &\quad = \mathbf{M}\{t, x, y\,; B, E^1, E^2\} \cup \mathcal{P}^A \cup \{A^0_t - A^1_x - A^2_y = 0\}. \end{aligned} \tag{5.155}$$

A couplet potential system $\mathbf{MAW}\{t,x,y\,;A^0,A^1,A^2,W\}$ is given by

$$\begin{aligned} W_t &= A^2_x - A^1_y, \\ -W_x &= A^0_y - A^2_t, \\ -W_y &= A^1_t - A^0_x, \\ A^0_t - A^1_x - A^2_y &= 0, \end{aligned} \tag{5.156}$$

where the components of the electric and magnetic fields are excluded through appropriate substitutions.

A nonlocally related subsystem $\mathbf{E}\{t,x,y\,;E^1,E^2\}$ is obtained directly from Maxwell's equations (5.149) by eliminating the magnetic field B:

$$\begin{aligned} E^1_x + E^2_y &= 0, \\ E^1_{tt} &= E^1_{xx} + E^1_{yy}, \\ E^2_{tt} &= E^2_{xx} + E^2_{yy}. \end{aligned} \tag{5.157}$$

In summary, a tree

$$\begin{aligned} \mathcal{T}_M = \{&\mathbf{M}\{t,x,y\,;B,E^1,E^2\}, \mathbf{MW}\{t,x,y\,;B,E^1,E^2,W\}, \\ &\mathbf{MA}\{t,x,y\,;B,E^1,E^2,A^0,A^1,A^2\}, \mathbf{MAW}\{t,x,y\,;A^0,A^1,A^2,W\}, \\ &\mathbf{E}\{t,x,y\,;E^1,E^2\}\} \end{aligned} \tag{5.158}$$

of nonlocally related PDE systems has been constructed for Maxwell's equations (5.149) in three-dimensional Minkowski space. This tree is presented in Figure 5.2.

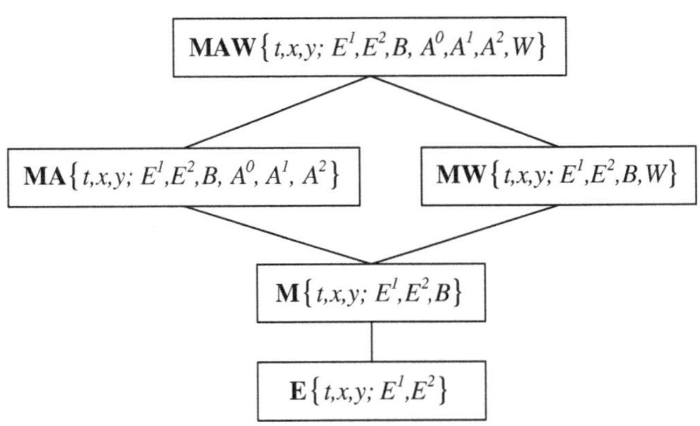

Fig. 5.2 A tree of nonlocally related systems for Maxwell's equations (5.149) in 3D Minkowski space.

Maxwell's equations (5.149) have eight point symmetries: three translations, one rotation, two space-time rotations (boosts), and two scalings [Exercise 5.3.5].

Using nonlocally related systems in the tree $\mathcal{T}_M$, one can find nonlocal symmetries of Maxwell's equations (5.149). In particular, the potential system $\mathbf{MW}\{t, x, y\,; B, E^1, E^2, W\}$ (5.153) has three additional conformal-type point symmetries given by

$$\begin{aligned}
\mathrm{Y}_1 &= (t^2+x^2+y^2)\frac{\partial}{\partial t} + 2tx\frac{\partial}{\partial x} + 2ty\frac{\partial}{\partial y} - (3tE^1+2yB)\frac{\partial}{\partial E^1} \\
&\quad -(3tE^2-2xB)\frac{\partial}{\partial E^2} - (2yE^1-2xE^2+3tB+W)\frac{\partial}{\partial B} - tW\frac{\partial}{\partial W}, \\
\mathrm{Y}_2 &= 2tx\frac{\partial}{\partial t} + (t^2+x^2-y^2)\frac{\partial}{\partial x} + 2xy\frac{\partial}{\partial y} - (3xE^1+2yE^2)\frac{\partial}{\partial E^1} \\
&\quad +(2yE^1-3xE^2+2tB+W)\frac{\partial}{\partial E^2} + (2tE^2-3xB)\frac{\partial}{\partial B} - xW\frac{\partial}{\partial W}, \\
\mathrm{Y}_3 &= 2ty\frac{\partial}{\partial t} + 2xy\frac{\partial}{\partial x} + (t^2-x^2+y^2)\frac{\partial}{\partial y} - (3yE^1-2xE^2+2tB+W)\frac{\partial}{\partial E^1} \\
&\quad -(2xE^1+3yE^2)\frac{\partial}{\partial E^2} - (2tE^1+3yB)\frac{\partial}{\partial B} - yW\frac{\partial}{\partial W},
\end{aligned}$$

that obviously yield nonlocal symmetries of Maxwell's equations (5.149).

Moreover, since the potential system $\mathbf{MAW}\{t, x, y\,; A^0, A^1, A^2, W\}$ (5.156) coincides with the potential system (5.145) for the wave equation (with $W = u, A^i = v^i$), it follows that it also has the duality-type symmetries (5.147). These symmetries yield three nonlocal symmetries of Maxwell's equations (5.149) [Anco & Bluman (1997b)].

A similar analysis can be done for Maxwell's equations (5.148) in four-dimensional Minkowski space. For this PDE system, local conservation laws and point symmetries of the determined PDE system consisting of the set of potential equations arising from a lower-degree conservation law and the four-dimensional Lorentz gauge, respectively yield nonlocal conservation laws and nonlocal symmetries of Maxwell's equations (5.148) [Anco & The (2005)].

5.3.6 Symmetries and exact solutions of the three-dimensional MHD equilibrium equations

Consider the PDE system of ideal magnetohydrodynamics (MHD) equilibrium equations in three space dimensions given by

$$\operatorname{div}(\rho\mathbf{V}) = 0, \quad \operatorname{div}\mathbf{B} = 0, \tag{5.159a}$$

$$\rho \mathbf{V} \times \operatorname{curl} \mathbf{V} - \mathbf{B} \times \operatorname{curl} \mathbf{B} - \operatorname{grad} P - \tfrac{1}{2}\rho \operatorname{grad} |\mathbf{V}|^2 = 0, \qquad (5.159b)$$

$$\operatorname{curl} \mathbf{V} \times \mathbf{B} = 0. \qquad (5.159c)$$

In (5.159), the dependent variables are the plasma density ρ, the plasma velocity $\mathbf{V} = (V^1, V^2, V^3)$, the pressure P and the magnetic field $\mathbf{B} = (B^1, B^2, B^3)$; the independent variables are the spatial coordinates (x, y, z). For closure, one must add an appropriate equation of state that relates pressure and density to the MHD equations (5.159).

It is now shown that an infinite number of nonlocal symmetries exist for the MHD equations (5.159) for two different equations of state. Moreover, the applications of these nonlocal symmetries are considered. For additional details and examples, see Bogoyavlenskij [(2001), (2002)] and Galas (1993).

(1) Incompressible MHD equilibria

As a first simplified example, consider the incompressible MHD equilibrium system $\mathbf{I}\{x, y, z\,; \mathbf{B}, \mathbf{V}, P\}$ with constant density (without loss of generality, $\rho = 1$), given by

$$\operatorname{div} \mathbf{V} = 0, \quad \operatorname{div} \mathbf{B} = 0, \qquad (5.160a)$$

$$\mathbf{V} \times \operatorname{curl} \mathbf{V} - \mathbf{B} \times \operatorname{curl} \mathbf{B} - \operatorname{grad} P - \tfrac{1}{2} \operatorname{grad} |\mathbf{V}|^2 = 0, \qquad (5.160b)$$

$$\operatorname{curl} \mathbf{V} \times \mathbf{B} = 0. \qquad (5.160c)$$

Using the lower-degree conservation law (5.160c), one introduces a potential variable Ψ:

$$\mathbf{V} \times \mathbf{B} = \operatorname{grad} \Psi. \qquad (5.161)$$

[Note that Ψ has the direct physical meaning of a function enumerating *magnetic surfaces*, i.e., two-dimensional surfaces to which streamlines and magnetic field lines are tangent. In general, every three-dimensional plasma domain is spanned by such surfaces.]

The resulting potential system $\mathbf{I\Psi}\{x, y, z\,; \mathbf{B}, \mathbf{V}, P, \Psi\}$ is determined (i.e., has no gauge freedom) and is given by

$$\operatorname{div} \mathbf{V} = 0, \quad \operatorname{div} \mathbf{B} = 0, \quad \mathbf{V} \times \mathbf{B} = \operatorname{grad} \Psi, \qquad (5.162a)$$

$$\rho \mathbf{V} \times \operatorname{curl} \mathbf{V} - \mathbf{B} \times \operatorname{curl} \mathbf{B} - \operatorname{grad} P - \tfrac{1}{2} \operatorname{grad} |\mathbf{V}|^2 = 0. \qquad (5.162b)$$

Now a comparison is made between the point symmetries of the PDE systems (5.160) and (5.162). The incompressible MHD equilibrium system (5.160) has 10 point symmetries: translations in pressure and two scalings, given, respectively, by

$$\mathrm{X_P} = \frac{\partial}{\partial P}, \quad \mathrm{X_D} = x\frac{\partial}{\partial x} + y\frac{\partial}{\partial y} + z\frac{\partial}{\partial z}, \quad \mathrm{X_S} = B^i \frac{\partial}{\partial B^i} + V^i \frac{\partial}{\partial V^i} + 2P\frac{\partial}{\partial P},$$

the interchange symmetry given by

$$\mathrm{X_I} = V^i \frac{\partial}{\partial B^i} + B^i \frac{\partial}{\partial V^i} - (\mathbf{B} \cdot \mathbf{V}) \frac{\partial}{\partial P},$$

and the Euclidean group (three space translations and three rotations) given by

$$\mathrm{X_E} = \boldsymbol{\zeta} \lrcorner \frac{\partial}{\partial \mathbf{x}} + (\mathbf{B} \cdot \mathrm{grad}) \boldsymbol{\zeta} \lrcorner \frac{\partial}{\partial \mathbf{B}},$$

where the hook symbol denotes summation over vector components, $\mathbf{x} = (x, y, z)$, $\boldsymbol{\zeta} = \mathbf{a} + \mathbf{b} \times \mathbf{x}$, and $\mathbf{a}$, $\mathbf{b}$ are arbitrary constant vectors in $\mathbb{R}^3$.

The first nine symmetries of the MHD system (5.160) directly yield point symmetries of the potential system (5.162). In addition, the potential system (5.162) has the obvious potential shift symmetry given by

$$\mathrm{X}_\Psi = \frac{\partial}{\partial \Psi},$$

as well as an infinite number of point symmetries given by

$$\mathrm{X}_\infty = M(\Psi) \left(V^i \frac{\partial}{\partial B^i} + B^i \frac{\partial}{\partial V^i} - (\mathbf{B} \cdot \mathbf{V}) \frac{\partial}{\partial P} \right), \tag{5.163}$$

where $M(\Psi)$ is an arbitrary smooth function of its argument. The point symmetries (5.163) yield nonlocal symmetries of the incompressible MHD equilibrium system (5.160). One can show that globally the symmetries (5.163) transform a given solution $(\mathbf{B}, \mathbf{V}, P)$ to a family of solutions $(\mathbf{B}', \mathbf{V}', P')$ given by

$$\begin{aligned} x' &= x, \quad y' = y, \quad z' = z, \\ \mathbf{B}' &= \mathbf{B} \cosh M(\Psi) + \mathbf{V} \sinh M(\Psi), \\ \mathbf{V}' &= \mathbf{V} \cosh M(\Psi) + \mathbf{B} \sinh M(\Psi), \\ P' &= P + \left(|\mathbf{B}|^2 - |\mathbf{B}'|^2 \right) /2 \end{aligned} \tag{5.164}$$

[Bogoyavlenskij (2001), Cheviakov (2004)].

Since the transformations (5.164) depend on an arbitrary function that is constant on magnetic surfaces, they can be used to obtain families of physically interesting solutions from a known MHD equilibrium solution. The transformations (5.164) preserve magnetic surfaces: $\mathbf{B}' \times \mathbf{V}'$ is parallel to $\mathbf{B} \times \mathbf{V}$.

As a simple example, consider the well-known simple “transverse flow” solution of the MHD equilibrium system (5.160) given by

$$\begin{aligned} \mathbf{B} = H(r)\mathbf{e}_z, \quad \mathbf{V} = \omega(r)(-y\mathbf{e}_x + x\mathbf{e}_y), \\ P(r) = F(r) - H^2(r)/2, \quad F(r) = \textstyle\int_0^r q\omega^2(q)dq \end{aligned} \tag{5.165}$$

depending on two arbitrary functions $H(r), \omega(r)$. This solution describes the differential rotation of a constant-density ideal gas plasma around the z-axis, for the vertical magnetic field; $r = \sqrt{x^2+y^2}$ is a cylindrical radius. The magnetic surfaces $\Psi = \text{const}$ are cylinders $r = \text{const}$ around the z-axis. In Figure 5.3(a), field lines of the solution (5.165) tangent to the cylinder $r = 1$ are shown for $H(r) = e^{-r}, \omega(r) = 2e^{-2r}$. Using the transformations (5.164) with an arbitrary function $M(\Psi) = f(r)$, one obtains an infinite family of solutions (5.165) for a noncollinear magnetic field and velocity

$$\begin{aligned} \mathbf{B} &= H(r)\cosh(f(r))\mathbf{e}_z + \mathbf{V}\sinh(f(r)), \\ \mathbf{V} &= \cosh(f(r))\mathbf{V} + H(r)\sinh(f(r))\mathbf{e}_z. \end{aligned} \tag{5.166}$$

Here the magnetic field lines and plasma streamlines are helices that are tangent to cylindrical magnetic surfaces $r = \text{const}$, with slopes depending on r. For $f(r) = e^{-r^2}$, original and transformed magnetic field lines and streamlines tangent to the cylinder $r = 1$ are shown in Figure 5.3.

One can show that for incompressible plasma equilibria with nonconstant plasma density, there exist infinite sets of transformations that generalize (5.164) [Bogoyavlenskij [(2001), (2002)]; see Exercise 5.3.7].

(2) Compressible adiabatic MHD equilibria

Now consider the system of compressible MHD equilibrium equations $\mathbf{C}\{x, y, z; \mathbf{B}, \mathbf{V}, P, \rho\}$ given by

$$\operatorname{div}(\rho\mathbf{V}) = 0, \quad \operatorname{div}\mathbf{B} = 0, \tag{5.167a}$$

$$\mathbf{V}\cdot \operatorname{grad} P + \gamma P \operatorname{div}\mathbf{V} = 0, \tag{5.167b}$$

$$\rho\mathbf{V}\times\operatorname{curl}\mathbf{V} - \mathbf{B}\times\operatorname{curl}\mathbf{B} - \operatorname{grad} P - \tfrac{1}{2}\rho\operatorname{grad}|\mathbf{V}|^2 = 0, \tag{5.167c}$$

$$\operatorname{curl}\ \mathbf{V}\times\mathbf{B} = 0. \tag{5.167d}$$

The PDE system (5.167) describes plasmas following the ideal gas equation of state and undergoing an adiabatic process. Here the entropy $S = P/\rho^\gamma$ is constant throughout the plasma domain.

A determined potential system $\mathbf{C\Psi}\{x, y, z; \mathbf{B}, \mathbf{V}, P, \rho, \Psi\}$ is obtained, as before, through replacing the conservation law (5.167d) by the potential equations (5.161).

The potential system $\mathbf{C\Psi}\{x, y, z; \mathbf{B}, \mathbf{V}, P, \rho, \Psi\}$ has an infinite number of point symmetries given by the infinitesimal generator

$$\mathrm{X}_C = N(\Psi)\left(V^i\frac{\partial}{\partial V^i} - 2\rho\frac{\partial}{\partial\rho}\right) + \left(\int N(\Psi)d\Psi\right)\frac{\partial}{\partial\Psi}, \tag{5.168}$$

where $N(\Psi)$ is an arbitrary smooth function [Galas (1993)]. The point symmetries (5.168) yield nonlocal symmetries of the compressible MHD equilibrium

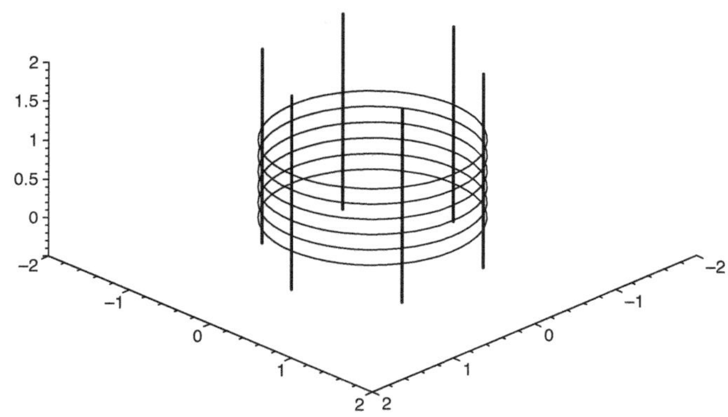

(a) original

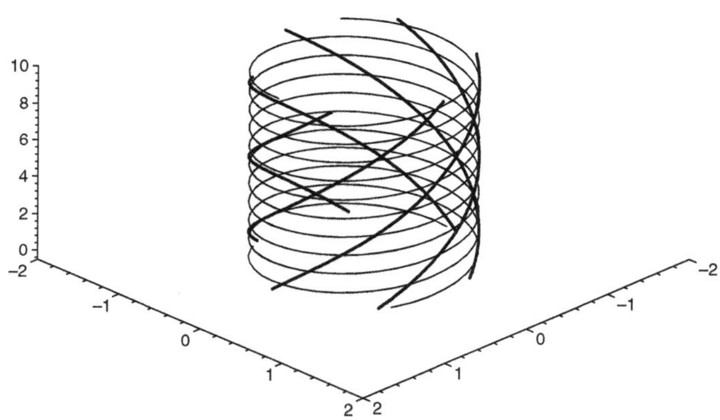

(b) transformed

Fig. 5.3 Magnetic field lines and streamlines of the "transverse flow" MHD equilibrium solution (5.165) (a) and its transformed version (5.166) (b). Magnetic field lines are shown with thick lines and plasma streamlines with thin lines.

system (5.167). The finite form of the transformations of physical variables is readily found to be given by

$$\begin{aligned} &x' = x, \quad y' = y, \quad z' = z, \quad \mathbf{B}' = \mathbf{B}, \quad P' = P, \\ &\mathbf{V}' = f(\Psi)\mathbf{V}, \quad \rho' = \rho/f^2(\Psi). \end{aligned} \tag{5.169}$$

Some generalizations of the symmetry transformations (5.169) are considered in Exercise 5.3.9.

Exercises 5.3

5.3.1. Show that, in general, Theorem 5.3.1 does not hold when a potential system is appended with a gauge constraint.

5.3.2. Classify the point symmetries of the PDE systems (5.140) and (5.141) with respect to the parameter k in the constitutive function (5.142).

5.3.3. Find the point symmetries of the potential system (5.145) of the (2+1) - dimensional wave equation (5.144), using the algebraic (spatial), divergence (Coulomb), Cronstrom, and Poincaré gauges, respectively.

5.3.4. Check that the PDE system (5.151) is equivalent to the PDE system

$$\operatorname{div}_{(t,x,y)}[B, E^2, -E^1] = 0, \tag{5.170a}$$

(5.170b)

$$\operatorname{curl}_{(t,x,y)}[B, -E^2, E^1] = 0, \tag{5.170c}$$

where both the curl and divergence are formally taken with respect to the variables (t, x, y). Show that the PDE system (5.170) in turn is equivalent to Maxwell's equations (5.149) in three-dimensional Minkowski space.

5.3.5.

(a) Find the point symmetries of Maxwell's equations (5.149) in three-dimensional Minkowski space.
(b) Find the point symmetries of the other PDE systems in the tree $\mathcal{T}_M$ (5.158). Isolate those symmetries that yield nonlocal symmetries of Maxwell's equations (5.149).

5.3.6. Show that other common choices of gauges such as spatial, divergence or Cronstrom gauges appended to the potential systems (5.155), (5.156) of Maxwell's equations in three-dimensional Minkowski space, yield no nonlocal symmetries of the Maxwell system (5.149).

5.3.7. The incompressible MHD equilibrium equations with non-constant density are given by (5.159) with $\operatorname{div}\mathbf{V} = 0$.

(a) Show that if the density ρ is constant on magnetic surfaces, i.e.

$$\operatorname{grad}\rho\cdot\mathbf{B}=\operatorname{grad}\rho\cdot\mathbf{V}=0,$$

then the infinite set of transformations

$$x'=x,\quad y'=y,\quad z'=z,$$

$$\mathbf{B}'=b(\Psi)\mathbf{B}+c\sqrt{\rho}\mathbf{V},\quad \mathbf{V}'=\frac{c(\Psi)}{a(\Psi)\sqrt{\rho}}\mathbf{B}+\frac{b(\Psi)}{a(\Psi)}\mathbf{V},\tag{5.171}$$

$$\rho'=a^2(\Psi)\rho, P'=CP+\left(C|\mathbf{B}|^2-|\mathbf{B}'|^2\right)/2,$$

maps a given solution $(\mathbf{B},\mathbf{V},P,\rho)$ of the PDE system (5.159) into a family of solutions $(\mathbf{B}',\mathbf{V}',P',\rho')$ with the same set of magnetic field lines. In (5.171), $a(\Psi)$, $b(\Psi)$ are arbitrary functions constant on magnetic surfaces $\Psi=\text{const}$, and $b^2(\Psi)-c^2(\Psi)=C=\text{const}$ [Bogoyavlenskij [(2001), (2002)]].

(b) Show that the function Ψ does not have to be defined as $\Psi=\mathbf{V}\times\mathbf{B}$, but may be more generally defined as a function constant on magnetic field lines and streamlines: $\operatorname{grad}\Psi\cdot\mathbf{V}=0$, $\operatorname{grad}\Psi\cdot\mathbf{B}=0$. The latter definition makes the transformations (5.171) usable for the field-aligned case $\mathbf{V}\parallel\mathbf{B}$.

5.3.8. Consider the incompressible MHD equilibrium equations with nonconstant density that are considered in Exercise 5.3.7. Show that solutions invariant with respect to the Bogoyavlenskij symmetries (5.171) are given by

$$\mathbf{B}=\pm\sqrt{\rho}\mathbf{V},\quad P+|\mathbf{B}|^2/2=\text{const}.\tag{5.172}$$

The solutions (5.172) are known as Chandrasekhar equipartition equilibria [Chandrasekhar (1956)].

5.3.9.

(a) Show that the symmetries (5.169) hold in the more general adiabatic case where the entropy $S(P,\rho)$ is constant on each plasma streamline. In particular, here the equations (5.167c) are replaced by the pair of equations

$$S=P/\rho^{\gamma},\quad \mathbf{V}\cdot\operatorname{grad}S=0.$$

(b) Show that the function Ψ does not have to be defined by $\Psi=\mathbf{V}\times\mathbf{B}$, but can be more generally defined by a function constant on magnetic field lines and streamlines, i.e., $\operatorname{grad}\Psi\cdot\mathbf{V}=0$, $\operatorname{grad}\Psi\cdot\mathbf{B}=0$. This definition takes care of the field-aligned case $\mathbf{V}||\mathbf{B}$ [Bogoyavlenskij [(2001), (2002)]].

5.4 Symbolic Software

In this section, several brief examples of the use of symbolic software for local symmetry and conservation law analysis and classification are given. The program sequence for local symmetry or conservation law analysis is similar for most symbolic packages and typically includes the following steps.

1. Declaration of variables and the given PDE system.
2. Construction of a set of symmetry or conservation law determining equations.
3. If necessary (e.g., for finding symmetries or solutions of the adjoint linearized system), in the set of determining equations, the dependent variables arising from the given PDE system are restricted to solutions of the given PDE system.
4. Simplification (e.g., elimination of redundancies, partial solution) of the over-determined set of determining equations.
5. Solution of the simplified set of determining equations. Output of the point symmetries or conservation law multipliers.
6. For conservation laws: generation of fluxes.

If the given PDE system contains constitutive function(s) and/or constant parameter(s), a classification and case splitting is performed at Step 4, and Steps 5 and 6 are performed separately for each case that arises.

As a sample package containing necessary routines for such analyses, the `GeM` (version 031) package for CAS `Maple` (version 12) is now described [Cheviakov (2008), (2009a)]. Note that similar to the situation for many other packages, the `GeM` package does not distinguish between PDEs and ODEs, and hence may be used for finding symmetries and integrating factors/first integrals of ODE systems.

In addition to local symmetry and conservation law analyses, some packages contain routines for further analysis, such as the computation of approximate and adjoint symmetries.

5.4.1 An example of symbolic computation of point symmetries

As a first example, consider the polytropic Euler PDE system $\mathbf{E}\{x,t\,;v,p,\rho\}$ of planar gas dynamics equations, given by

$$\begin{aligned} \rho_t + (\rho v)_x &= 0, \\ \rho(v_t + v v_x) + p_x &= 0, \\ p_t + v p_x + \gamma p v_x &= 0, \end{aligned} \tag{5.173}$$

for the particular case where the polytropic exponent $\gamma = 3$ [Section 4.2.5, Table 4.17]. First, the package is initialized using the command

```
with(GeM):
```

Second, one defines variables and differential equations, as follows.

```
gem_decl_vars(indeps=[x,t], deps=[V(x,t),P(x,t),R(x,t)]);
gem_decl_eqs([
  diff(R(x,t),t)+diff(R(x,t)*V(x,t),x),
  R(x,t)( diff(V(x,t),t)+V(x,t)*diff(V(x,t),x) )
  + diff(P(x,t),x)=0,
  diff(P(x,t),t)+V(x,t)*diff(P(x,t),x)
  + 3*P(x,t)*diff(V(x,t),x)=0
  ],
solve_for=[diff(R(x,t),t), diff(V(x,t),t), diff(P(x,t),t)]);
```

Note that it is necessary that a given PDE system can be written in a solved form with respect to a set of leading derivatives specified in the `solve_for` parameter. [The expressions for these leading derivatives will later be automatically substituted into the symmetry determining equations so that the determining equations are considered on the solutions of the given PDE system.] It is also important to note that if the differential orders of the PDEs in the given system differ, the software automatically computes differential consequences of the lower-order PDEs, up to the order equal to the maximal differential order of the PDEs in the given system. [For example, consider a PDE system $\mathbf{R}\{t,x\,;u,v\}$ in solved form, given by two equations $u_{tt} = xuv_x$, $v_t = v^2 + uu_x$. For this system, the maximal differential order is two. Hence for the second equation, differential consequences up to second order will be automatically computed: $v_{tt} = 2v(v^2 + uu_x) + u_tu_x + uu_{tx}$, $v_{tx} = 2vv_x + uu_{xx} + u_x^2$.]

Third, one generates the symmetry determining equations, using the command

```
det_eqs:=gem_symm_det_eqs([x,t, R(x,t),V(x,t),P(x,t)]);
```

The arguments of the command define the dependence of symmetry components. In the considered case of seeking point symmetries, symmetry components depend on x, t, ρ, v and p. The procedure `gem_symm_det_eqs` yields the over-determined system of symmetry determining equations obtained by setting up the determining equations (1.51), (1.52), restricting the dependent variables of the given PDE system (5.173) appearing in the determining equations to the solution set of the PDE system (5.173), and equating all coefficients of like partial derivative terms to zero (since symmetry components do not depend on derivatives). In the current example, the split over-determined system of symmetry determining equations stored in det_eqs contains 27 equations.

Next, the over-determined system is simplified, as follows.

```
sym_components:=gem_symm_components();
simplified_eqs:=DEtools[rifsimp](det_eqs, sym_components,
 mindim=1);
```

[In particular, the option mindim=1 forces the output of the number of linearly independent solutions of equations simplified_eqs, i.e., the number of point symmetries of the PDE system (5.173).] In this example, there are seven linearly independent solutions.

Finally, the determining equations are solved, using the command

```
symm_sol:=pdsolve(simplified_eqs[Solved]);
```

Here this yields symmetry components containing seven arbitrary constants. The command

```
gem_output_symm(symm_sol);
```

prints the seven point symmetries separately:

$$\mathrm{X}_1 = \frac{\partial}{\partial x}, \quad \mathrm{X}_2 = \frac{\partial}{\partial t}, \quad \mathrm{X}_3 = t\frac{\partial}{\partial t} + x\frac{\partial}{\partial x},$$

$$\mathrm{X}_4 = t\frac{\partial}{\partial x} + \frac{\partial}{\partial v}, \quad \mathrm{X}_5 = x\frac{\partial}{\partial x} + v\frac{\partial}{\partial v} + p\frac{\partial}{\partial p} - \rho\frac{\partial}{\partial \rho}, \quad \mathrm{X}_6 = p\frac{\partial}{\partial p} + \rho\frac{\partial}{\partial \rho},$$

$$\mathrm{X}_7 = xt\frac{\partial}{\partial x} + t^2\frac{\partial}{\partial t} + (x - vt)\frac{\partial}{\partial v}.$$

5.4.2 An example of point symmetry classification

As a second example, consider the classification of point symmetries of the nonlinear diffusion equation $\mathbf{U}\{x, t\,; u\}$ given by

$$u_t - (K(u)u_x)_x = 0 \tag{5.174}$$

with respect to the constitutive function $K(u)$, $K'(u) \neq 0$ [Section 4.2.1, Table 4.1]. One defines variables using the commands

```
with(GeM):
gem_decl_vars(indeps=[x,t], deps=[U(x,t)],
  freefunc=[K(U(x,t))]);
```

where the optional parameter freefunc=[...] contains arbitrary function(s). (Arbitrary constants are specified using another optional parameter freeconst=[...].)

The given equation (5.174) is defined using the command

```
gem_decl_eqs([diff(U(x,t),t)=diff(K(U(x,t))
  *diff(U(x,t),x),x)], solve_for=[diff(U(x,t),t)]);
```

The symmetry determining equations are generated using the command

```
det_eqs:=gem_symm_det_eqs([x,t, U(x,t)]);
```

which yields 10 determining equations.

Next, one performs the automatic reduction and case splitting of the overdetermined linear PDE system stored in det_eqs.

```
sym_components:=gem_symm_components();
split_eqs:=DEtools[rifsimp](det_eqs, sym_components,
    casesplit, mindim=1);
```

The Maple variable split_eqs now contains a table of different computed cases. The case tree can be plotted using the command

```
caseplot(split_eqs,pivots);
```

[Note that depending on the version of Maple that is used, case splitting can occur differently, and moreover, some cases can yield the same symmetries. However, the complete analysis of a tree always yields complete results.] For the PDE (5.174), the case tree is shown in Figure 5.4. The pivot expressions are given by

$$p_1 = K(u), \quad p_2 = K(u), \quad p_2 = K'(u), \quad p_3 = 4K(u)K''(u) - 7(K'(u))^2,$$
$$p_4 = K(u)K'(u)K'''(u) - 2K(u)(K''(u))^2 + (K'(u))^2K''(u).$$

In particular, at each pivot, the left branch of the case tree in Figure 5.4 corresponds to the case where the pivot expression vanishes, and the right branch to the case when the pivot expression is nonzero. Numbers below the branches denote case numbers.

For each case m in Figure 5.4 ($1 \leq m \leq 5$), the corresponding simplified set of determining equations is accessed by calling split_eqs[m][Solved], and the number of independent solutions is given by split_eqs[m][dimension].

Case 1. This is the most general case. Here one uses the commands

```
symm_sol:=pdsolve(split_eqs[1][Solved]);
gem_output_symm(symm_sol);
```

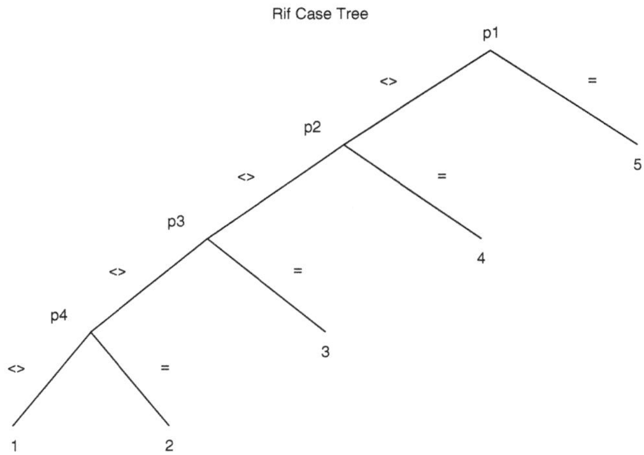

Fig. 5.4 The tree of cases in the classification of point symmetries of the nonlinear diffusion equation (5.174).

which yield the three symmetries

$$X_1 = \frac{\partial}{\partial x}, \quad X_2 = \frac{\partial}{\partial t}, \quad X_3 = x\frac{\partial}{\partial x} + 2t\frac{\partial}{\partial t}, \tag{5.175}$$

holding for an arbitrary constitutive function $K(u)$.

Case 2. In this case, the solution set has dimension four. This case is characterized by a restricted ODE satisfied by $K(u)$, contained in `split_eqs[2]` `[Solved]`:

$$K'''(u) = \frac{2K(u)(K''(u))^2 - (K'(u))^2K''(u)}{K(u)K'(u)}. \tag{5.176}$$

Modulo equivalence transformations (4.6), the equation (5.176) has two different solutions: $K(u) = u^\nu$ ($\nu = \text{const}$) and $K(u) = e^u$.

Case 2a. For $K(u) = u^\nu$, one obtains the corresponding point symmetries using commands

```
case2a_symm_sol:=pdsolve(subs(K(U)=U^nu,
  split_system[2][Solved]));
gem_output_symm(case2a_symm_sol);
```

This yields the three generic symmetries (5.175), and the additional symmetry

$$X_4 = x\frac{\partial}{\partial x} + \frac{2}{\nu}u\frac{\partial}{\partial u}. \tag{5.177}$$

Case 2b. For $K(u) = e^u$, one uses commands

```
case2b_symm_sol:=pdsolve(subs(K(U)=exp(U),
  split_system[2][Solved]));
gem_output_symm(case2b_symm_sol);
```

This yields the three generic symmetries (5.175), and the additional symmetry

$$X_5 = x\frac{\partial}{\partial x} + 2\frac{\partial}{\partial u}. \tag{5.178}$$

Case 3. In this case, the solution set has dimension five, and $K(u)$ is restricted to satisfying the ODE

$$K''(u) = \frac{7}{4}\frac{(K'(u))^2}{K(u)}. \tag{5.179}$$

Modulo equivalence transformations (4.6), the only solution of the equation (5.179) is $K(u) = u^{-4/3}$. The corresponding point symmetries are computed using commands

```
case3_symm_sol:=pdsolve(subs(K(U)=U^(-4/3),
  split_system[3][Solved]));
gem_output_symm(case3_symm_sol);
```

This yields the three symmetries (5.175), the symmetry (5.177) (with $\nu = -4/3$), and the additional symmetry

$$X_6 = x^2\frac{\partial}{\partial x} - 3xu\frac{\partial}{\partial u}. \tag{5.180}$$

Cases 4 and 5. These cases correspond to linear diffusion equations ($K(u) =$ const and $K(u) = 0$, respectively), and hence the PDE (5.174) has an infinite number of point symmetries in these cases. Indeed, `split_eqs[m][dimension]` $= \infty$ for $m = 4, 5$. This completes the classification of point symmetries of the nonlinear diffusion equation $\mathbf{U}\{x, t\,; u\}$ (5.174).

5.4.3 An example of symbolic computation of conservation laws

A symbolic implementation of the direct method is now used to compute several lower-order local conservation laws of the Korteweg–de Vries (KdV) equation

$$u_t + uu_x + u_{xxx} = 0 \tag{5.181}$$

[Section 1.3.5].

First, the local conservation law multipliers are computed. Let the multipliers have the dependence

$$\Lambda[U] = \Lambda(t, x, U, U_x, U_{xx}, U_{xxx}). \tag{5.182}$$

One defines variables and the PDE (5.181) using the commands

```
with(GeM):
gem_decl_vars(indeps=[x,t], deps=[U(x,t)]);
gem_decl_eqs([diff(U(x,t),t)=U(x,t)*diff(U(x,t),x)      (5.183)
  +diff(U(x,t),x,x,x)],
  solve_for=[diff(U(x,t),t)]);
```

[Note that in order to compute multipliers, the specification of the option `solve_for` in `gem_decl_eqs` is *not required*, since the conservation law multipliers are solutions of the conservation law determining equations (1.151) for arbitrary functions $U(x)$. However in (5.183), the option `solve_for` is specified. It is used later in the flux computation routine, which automatically verifies the correctness of flux computations by explicitly checking that $\mathrm{D}_i\Phi^i[U] = 0$ on solutions $U(x) = u(x)$ of the given PDE (5.181).]

The set of determining equations for the local conservation law multipliers is obtained and simplified using the routines

```
det_eqs:=gem_conslaw_det_eqs([x,t, U(x,t),
  diff(U(x,t),x), diff(U(x,t),x,x),
  diff(U(x,t),x,x,x)]):
CL_multipliers:=gem_conslaw_multipliers();              (5.184)
simplified_eqs:=DEtools[rifsimp](det_eqs,
  CL_multipliers, mindim=1);
```

The first command of (5.184) sets up the set of local conservation law multiplier determining equations (1.151), and splits them. [The splitting is done using the fact that the determining equations (1.151) are polynomial expressions in terms of derivatives of U of orders four and higher, and such derivatives are linearly independent.] This yields an over-determined system of 19 determining equations for the unknown local multiplier $\Lambda[U]$. After using the `rif` reduction algorithm, the system reduces to the nine equations

$$\frac{\partial^2\Lambda[U]}{\partial x^2} = \frac{\partial^2\Lambda[U]}{\partial x\partial U} = \frac{\partial^2\Lambda[U]}{\partial x\partial U_{xx}} = \frac{\partial^2\Lambda[U]}{\partial U\partial U_{xx}} = \frac{\partial^2\Lambda[U]}{\partial^2 U_{xx}} = \frac{\partial\Lambda[U]}{\partial U_x} = \frac{\partial\Lambda[U]}{\partial U_{xxx}} = 0,$$
$$\frac{\partial^2\Lambda[U]}{\partial U^2} = \frac{\partial\Lambda[U]}{\partial U_{xx}}, \quad \frac{\partial\Lambda[U]}{\partial t} = -U\frac{\partial\Lambda[U]}{\partial x}.$$

The four linearly independent solutions of these determining equations are obtained using the command

```
multipliers_sol:=pdsolve(simplified_eqs[Solved]);
```

and are given by

$$\Lambda_1[U] = 1, \quad \Lambda_2[U] = U, \quad \Lambda_3[U] = x - tU, \quad \Lambda_4[U] = \tfrac{1}{2}U^2 - U_{xx}.$$

The corresponding fluxes can be computed using one of the available methods. For example, using the homotopy method [Section 1.3.7] with the command

```
gem_get_CL_fluxes(multipliers_sol, method="Homotopy1");
```

one obtains fluxes corresponding to the conservation laws

$$\begin{aligned}
\mathrm{D}_t(u) + \mathrm{D}_x\left(\tfrac{1}{2}u^2 + u_{xx}\right) &= 0,\\
\mathrm{D}_t\left(\tfrac{1}{2}u^2\right) + \mathrm{D}_x\left(\tfrac{1}{3}u^3 + uu_{xx} - \tfrac{1}{2}u_x^2\right) &= 0,\\
\mathrm{D}_t\left(\tfrac{1}{2}tu^2 - xu\right) + \mathrm{D}_x\left(-\tfrac{1}{2}xu^2 + tuu_{xx} - \tfrac{1}{2}tu_x^2 - xu_{xx} + u_x\right) &= 0,\\
\mathrm{D}_t\left(\tfrac{1}{3}u^3 + uu_{xx}\right) + \mathrm{D}_x\left(\tfrac{1}{4}u^4u^2u_{xx} + u_{xx}^2 - uu_{xt} + u_xu_t\right) &= 0,
\end{aligned}$$

which are equivalent to the conservation laws given by (1.142), (1.144).

5.5 Discussion

Clarkson & Winternitz (1991) use the nonclassical method to obtain solutions of the Kadomtsev–Petviashvili equation

$$(u_t + uu_x + u_{xxx})_x \pm u_{yy} = 0$$

that do not result from reductions under Lie's classical method. Clarkson & Mansfield (1993) obtain nonclassical solutions for the nonlinear heat equation

$$u_t - u_{xx} = u^3$$

and the real Landau–Ginzburg PDE system

$$u_t + u_{xx} + 2u^2 v = \gamma u,$$
$$v_t + v_{xx} + 2v^2 u = \gamma v,$$

where γ is a real constant. Lou (1992) obtains nonclassical solutions for the system of dispersive wave equations in shallow water given by

$$v_t + (uv)_x = 0,$$
$$u_t + uu_x + v_x + \kappa v_{xxx} = 0, \quad \kappa = \text{ const.}$$

Clarkson & Mansfield (1994c) find nonclassical solutions for the generalized shallow water wave equation given by

$$u_{xxxt} + \alpha u_x u_{xt} + \beta u_t u_{xx} - u_{xt} - u_{xx} = 0, \quad \alpha, \beta = \text{ const.}$$

Arrigo & Hill (1995) obtain nonclassical solutions for nonlinear diffusion equations with source terms given by

$$u_t = [D(u)u_x]_x + Q(u)$$

for $D(u) = u^m,\ e^u$. Wiltshire & El-Kafri (2004) find nonclassical solutions for Richard's equation for water flow in an unsaturated uniform soil, i.e., for diffusion-convection equations of the form

$$u_t = [D(u)u_x + K(u)]_x.$$

Bradshaw-Hajek et al. (2007) use the nonclassical method to seek solutions, not obtainable by Lie's classical method, for reaction-diffusion equations with explicit spatial dependence of the form

$$u_t - u_{xx} = k(x)u^2(1-u). \tag{5.185}$$

In particular, they find that nonclassical solutions exist for (5.185) if and only if $k(x) = x^2$, $\tanh^2 x$, or $\tan^2 x$, modulo scalings given by $t \to \alpha^2 t$, $x \to \alpha x$ and translations in x.

The nonclassical method has been used to obtain solutions of the reaction-diffusion-convection equation [Cherniha & Serov (1998)]

$$u_t - u_{xx} = \lambda u u_x + u(\alpha - \beta u), \quad \lambda, \alpha, \beta = \text{const},$$

as well as [Cherniha (2007)] the generalized Fitzhugh–Nagumo equation

$$u_t - u_{xx} = \lambda u u_x + \alpha u(u-\delta)(1-u), \quad \lambda, \delta, \alpha = \text{ const with } 0 < \delta < 1,\ \alpha > 0,$$

and the generalized Kolmogorov–Petrovskii–Piskunov equation

$$u_t - u_{xx} = \lambda u u_x + \alpha u(1-u)^2, \quad \lambda, \alpha = \text{const with } \alpha > 0.$$

Clarkson & Priestley (1998) discuss the difficulties and extensions needed in using either the direct method or nonclassical method to obtain solutions for PDEs with nonlocal terms and, as examples, use various systems that represent the shallow water wave equation.

Burde (1996) uses the nonclassical method (and his own extension of it) to obtain further solutions of the boundary layer equations describing steady-state flow over a flat plate, given by

$$\begin{aligned} uu_x + vu_y - U(x)U'(x) &= \nu u_{yy}, \\ u_x + v_y &= 0. \end{aligned} \tag{5.186}$$

In particular, Burde obtains his solutions by applying the nonclassical method (and his generalization of it) to the subsystem (a scalar PDE) that one obtains after introducing the stream function (potential) $\psi(x,y)$ resulting from the conservation law expressed by the second equation of (5.186), namely the PDE

$$\psi_y \psi_{xy} - \psi_x \psi_{yy} - U(x)U'(x) + \nu \psi_{yyy} = 0. \tag{5.187}$$

Further nonclassical solutions of the PDE (5.187) are found by Saccomandi (2004). In particular, Saccomandi shows that the von Mises transformation that reduces the boundary layer equations (5.186) to a second-order evolution equation is a Bäcklund transformation related to the nonclassical method.

The nonclassical method extends to an ansatz related to contact symmetries (*conditional Lie–Bäcklund symmetries*; also called *generalized conditional symmetries*) [Fokas & Liu (1994); Liu & Fokas (1996)]. Such an ansatz is used by Qu, Ji & Wang (2007) to obtain solutions for quasi-linear diffusion equations with convection and source terms of the form

$$u_t = [u^m (u_x)^n]_x + P(u)u_x + Q(u),$$

and by Ji & Qu (2007) to obtain solutions for radially symmetric nonlinear diffusion equations with a spatially dependent source term of the form

$$u_t = [B(u)(u_r)^{m+1}]_r + \frac{(m+1)(n-1)}{r} B(u)(u_r)^{m+1} + Q(r,u).$$

A very good overview of several reduction methods, including the higher-order direct method (method of "nonlinear separation") due to Galaktionov (1990), to find special solutions of PDEs appears in Olver (1994). An ansatz technique (group foliation method) to solve a first-order PDE system whose independent variables and dependent variables are, respectively, the classical and differential invariants of a point symmetry of the PDE system appears in Ovsiannikov (1982), with interesting examples in the papers of Martina,

Sheftel & Winternitz (2001), Sheftel (2002), Nutku & Sheftel (2001), Golovin (2004) and Anco & Liu (2004).

Cheviakov & Bogoyavlenskij (2004) generalize the infinite set of transformations (5.171) to the PDE system of anisotropic (Chew–Goldberger–Low) plasma equilibrium equations in three dimensions, and use these transformations for the construction of families of exact solutions modelling anisotropic plasma equilibria. [See also Cheviakov (2005).]

King (1989) uses potential symmetries to construct solutions of boundary value problems for a class of nonlinear diffusion equations.

The Lorentz gauge is due to Ludvig Lorenz! It is commonly called the "Lorentz gauge" because of confusion with Hendrik Lorentz, after whom Lorentz invariance is named. We have used the common spelling in the text.

A computational way of checking for the possibility of the linearization of a given nonlinear PDE system [Sections 2.4 and 2.6], especially for classification problems, is to apply the work in Reid, Wittkopf & Boulton (1996) [See also Wittkopf (2004).] to find the size of the solution space of a given system of determining equations without actually obtaining any of its solutions. In particular, if the size of the solution space of the determining system either for local multipliers or for point (contact) symmetries is finite-dimensional, then no linearization by an invertible mapping is possible. But the converse (that the solution space is infinite-dimensional) is not sufficient for the existence of a linearization by a point or contact transformation. Specifically, the solution space must have a sufficiently large number of "parameters" such that the number of functions and independent variables, as well as the number of linear PDEs they satisfy, matches the corresponding number (i.e., the cardinality) in the given nonlinear PDE system. This counting is performed algorithmically [Reid, Wittkopf & Boulton (1996); Wittkopf (2004)] if a differential Gröbner basis is available.

References

Ablowitz, M.J., Kaup, D.J., Newell, A.C., and Segur, H. (1974). The inverse scattering transform-Fourier analysis for nonlinear problems. *Stud. Appl. Math.* **53**, 249–315.

Akhatov, I.S., Gazizov, R.K., and Ibragimov, N.H. (1987). Group classification of the equations of nonlinear filtration. *Sov. Math. Dokl.* **35**, 384–386.

Akhatov, I.S., Gazizov, R.K., and Ibragimov, N.H. (1988). Bäcklund transformations and nonlocal symmetries. *Soviet Math. Dokl.* **36**, 393–395. See also *CRC Handbook of Lie Group Analysis of Differential Equations,* Vol. I, 1994, Chap. 13, pp. 258–291 (N.H. Ibragimov, Ed.). CRC Press, Boca Raton, FL.

Akhatov, I.S., Gazizov, R.K., and Ibragimov, N.H. (1991). Nonlocal symmetries. Heuristic approach. *J. Sov. Math.* **55**, 1401–1450.

Ames, W.F., Lohner, R.J., and Adams, E. (1981). Group properties of $u_{tt} = [f(u)u_x]_x$. *Internat. J. Nonlinear Mech.* **16**, 439–447.

Anco, S.C. (2003). Conservation laws of scaling-invariant field equations. *J. Phys. A* **36**, 8623–8638.

Anco, S.C. (2009). Conservation laws of partial differential equations: theory, computation, and examples. *In preparation.*

Anco, S.C. and Bluman, G.W. (1996). Derivation of conservation laws from nonlocal symmetries of differential equations. *J. Math. Phys.* **37**, 2361–2375.

Anco, S.C. and Bluman, G.W. (1997a). Direct construction of conservation laws from field equations. *Phys. Rev. Lett.* **78**, 2869–2873.

Anco, S.C. and Bluman, G.W. (1997b). Nonlocal symmetries and nonlocal conservation laws of Maxwell's equations. *J. Math. Phys.* **38**, 3508–3532.

Anco, S.C. and Bluman, G.W. (1998). Integrating factors and first integrals for ordinary differential equations. *Eur. J. Appl. Math.* **9**, 245–259; Erratum (1999) *Eur. J. Appl. Math.* **10**, 223.

Anco, S.C. and Bluman, G.W. (2002a). Direct construction method for conservation laws of partial differential equations. Part I: Examples of conservation law classifications. *Eur. J. Appl. Math.* **13**, 545–566.

Anco, S.C. and Bluman, G.W. (2002b). Direct construction method for conservation laws of partial differential equations. Part II: General treatment. *Eur. J. Appl. Math.* **13**, 567–585.

Anco, S.C., Bluman, G.W., and Wolf, T. (2008). Invertible mappings of nonlinear PDEs to linear PDEs through admitted conservation laws. *Acta Appl. Math.* **101**, 21–38.

Anco, S.C. and Dar, A. (2009). Classification of conservation laws of compressible isentropic fluid flow in $n > 1$ spatial dimensions *Proc. Roy. Soc. A*, in press.

Anco, S.C. and Ivanova, N.M. (2007). Conservation laws and symmetries of semilinear wave equations. *J. Math. Anal. Appl.* **332**, 863–876.

Anco, S.C. and Liu, S. (2004). Exact solutions of semilinear radial wave equations in n dimensions. *J. Math. Anal. Appl.* **297**, 317–342.

Anco, S.C. and Pohjanpelto, J. (2001) Classification of local conservation laws of Maxwell's equations. *Acta Appl. Math.* **69**, 285–327.

Anco, S.C. and Pohjanpelto, J. (2003) Conserved currents of massless fields of spin $s \geq 1/2$. *Proc. R. Soc. London A* **459**, 1215–1239.

Anco, S.C. and Pohjanpelto, J. (2004) Symmetries and currents of massless neutrino fields, electromagnetic and gravitation fields. *CRM Proc. Lect. Notes* **34**, 1–12.

Anco, S.C. and Pohjanpelto, J. (2008) Generalized symmetries of massless free fields on Minkowski space. *SIGMA* **4**, 004.

Anco, S.C. and The, D. (2005). Symmetries, conservation laws, and cohomology of Maxwell's equations using potentials. *Acta Appl. Math.* **89**, 1–52.

Anderson, R.L., Kumei,S., and Wulfman, C.E. (1972) Generalization of the concept of invariance of differential equations. *Phys. Rev. Lett.* **28**, 988-991.

Anderson, I.M. and Torre, C.G. (1996). Asymptotic conservation laws in classical field theory. *Phys. Rev. Lett.* **77**, 4109–4113.

Anderson, R.L. and Ibragimov, N.H. (1979). *Lie–Bäcklund Transformations in Applications.* SIAM, Philadelphia.

Anderson, R.L., Kumei, S., and Wulfman, C.E. (1972). Generalization of the concept of invariance of differential equations. *Phys. Rev. Lett.* **28**, 988–991.

Arrigo, D.J. and Hill, J.M. (1995). Nonclassical symmetries for nonlinear diffusion and absorption. *Stud. Appl. Math.* **94**, 21–39.

Atkinson, F.V. and Peletier, L.A. (1974). Similarity solutions of the nonlinear diffusion equation. *Arch. Rat. Mech. Anal.* **54**, 373–392.

Bäcklund, A.V. (1876). Über Flächentransformationen. *Math. Ann.* **9**, 297–320.

Barenblatt, G.I. (1979). *Similarity, Self-Similarity, and Intermediate Asymptotics.* Consultants Bureau, New York.

Barenblatt, G.I. (1987). *Dimensional Analysis.* Gordon and Breach, New York.

Barenblatt, G.I. (1996). *Scaling, Self-Similarity, and Intermediate Asymptotics.* Cambridge University Press, Cambridge, UK.

Barenblatt, G.I. and Zel'dovich, Ya.B. (1972). Self-similar solutions as intermediate asymptotics. *Ann. Rev. Fluid Mech.* **4**, 285–312.

Benjamin, T.B., Bona, J.L., and Mahoney, J.J. (1972). Model equations for long waves in nonlinear dispersive media. *Phil. Trans. Roy. Soc. A* **272**, 47–48.

Bessel-Hagen, E. (1921). Über die Erhaltungssätze der Elektrodynamik. *Math. Ann.* **84**, 258–276.

Bluman, G.W. (1967). Construction of Solutions to Partial Differential Equations by the Use of Transformation Groups. Ph.D. Thesis, California Institute of Technology.

Bluman, G.W. (1974). Use of group methods for relating linear and nonlinear partial differential equations. *Proceedings of Symposium on Symmetry, Similarity and Group Theoretic Methods in Mechanics*, Calgary, AB. pp. 203–218.

Bluman, G.W. (1980). On the transformation of diffusion processes into the Wiener process. *SIAM J. Appl. Math.* **39**, 238–247.

Bluman, G.W. (1983). On mapping linear partial differential equations to constant coefficient equations. *SIAM J. Appl. Math.* **43**, 1259–1273.

Bluman, G.W. (1990). Simplifying the form of Lie groups admitted by a given differential equation. *J. Math. Anal. Appl.* **145**, 52–62.

Bluman, G.W. (1993). Use and construction of potential symmetries, *Math. Comput. Modelling* **8**, 1–14.

Bluman, G.W. (2005). Connections between symmetries and conservation laws, SIGMA **1**, Paper 011.

Bluman, G.W. and Anco, S.C. (2002). *Symmetry and Integration Methods for Differential Equations for Differential Equations.* Applied Mathematical Sciences, No. 154. Springer, New York.

Bluman, G.W. and Cheviakov, A.F. (2005). Framework for potential systems and nonlocal symmetries: Algorithmic approach. *J. Math. Phys.* **46**, 123506.

Bluman, G.W. and Cheviakov, A.F. (2007). Nonlocally related systems, linearizations and nonlocal symmetries for the nonlinear wave equation. *J. Math. Anal. Appl.* **333**, 93–111.

Bluman, G.W., Cheviakov, A.F., and Ganghoffer, J.-F. (2008). Nonlocally related PDE systems for one-dimensional nonlinear elastodynamics. *J. Eng. Math.* **62**, 203–221.

Bluman, G.W., Cheviakov, A.F., and Ivanova, N.M. (2006). Framework for nonlocally related partial differential equations systems and nonlocal symmetries: Extension, simplification, and examples. *J. Math. Phys.* **47**, 113505.

Bluman, G.W. and Cole, J.D. (1969). The general similarity solution of the heat equation. *J. Math. Mech.* **18**, 1025–1042.

Bluman, G.W. and Cole, J.D. (1974). *Similarity Methods for Differential Equations.* Applied Mathematical Sciences, No. 13. Springer-Verlag, New York.

Bluman, G.W. and Doran-Wu, P. (1995). The use of factors to discover potential systems or linearizations. *Acta Appl. Math.* **2**, 79–96.

Bluman, G.W. and Gregory, R.D. (1985). On transformations of the biharmonic equation. *Mathematika* **32**, 118–130.

Bluman, G.W. and Kumei, S. (1980). On the remarkable nonlinear diffusion equation $\frac{\partial}{\partial x}\left[a(u+b)^{-2}\frac{\partial u}{\partial x}\right] - \frac{\partial u}{\partial t} = 0$. *J. Math. Phys.* **21**, 1019–1023.

Bluman, G.W. and Kumei, S. (1987). On invariance properties of the wave equation. *J. Math. Phys.* **28**, 307–318.

Bluman, G.W. and Kumei, S. (1988). Exact solutions for wave equations of two-layered media with smooth transition. *J. Math. Phys.* **29**, 86–96.

Bluman, G.W. and Kumei, S. (1989). *Symmetries and Differential Equations.* Applied Mathematical Sciences, No. 81. Springer-Verlag, New York.

Bluman, G.W. and Kumei, S. (1990a). Symmetry-based algorithms to relate partial differential equations, I. Local symmetries. *Eur. J. Appl. Math.* **1**, 189–

216.

Bluman, G.W. and Kumei, S. (1990b). Symmetry-based algorithms to relate partial differential equations, II. Linearization by nonlocal symmetries. *Eur. J. Appl. Math.* **1**, 217–223.

Bluman, G.W., Kumei, S., and Reid, G.J. (1988). New classes of symmetries for partial differential equations. *J. Math. Phys.* **29**, 806–811; Erratum, *J. Math. Phys.* **29**, 2320.

Bluman, G.W. and Reid, G.J. (1989). Sequences of related linear PDE's, *J. Math. Anal. Appl.* **144**, 565–585.

Bluman, G.W. and Shtelen, V.M. (1996a). New classes of Schrodinger equations equivalent to the free particle equation through non-local transformations. *J. Phys. A* **29**, 4473–4480.

Bluman, G.W. and Shtelen, V.M. (1996b). Developments in similarity methods relating to pioneering work of Julian Cole, in *Mathematics is for Solving Problems* (edited by L.P. Cook, V. Roytbird and M. Tulin), SIAM, Philadelphia, pp. 105–117.

Bluman, G.W. and Shtelen, V.M. (2004). Nonlocal transformations of Kolmogorov equations into the backward heat equation. *J. Math. Anal. Appl.* **291**, 419–437.

Bluman, G.W. and Temuerchaolu (2005a). Conservation laws for nonlinear telegraph equations. *J. Math. Anal. Appl.* **310**, 459–476.

Bluman, G.W. and Temuerchaolu (2005b). Comparing symmetries and conservation laws of nonlinear telegraph equations. *J. Math. Phys.* **46**, 073513.

Bluman, G.W., Temuerchaolu, and Anco, S.C. (2006). New conservation laws obtained directly from symmetry action on a known conservation law. *J. Math. Anal. Appl.* **322**, 233–250.

Bluman, G.W., Temuerchaolu, and Sahadevan, R. (2005). Local and nonlocal symmetries for nonlinear telegraph equations. *J. Math. Phys.* **46**, 023505.

Bluman, G.W. and Yan, Z. (2005). Nonclassical potential solutions of partial differential equations. *Eur. J. Appl. Math.* **16**, 239–261.

Bogoyavlenskij, O.I. (2001). Infinite symmetries of the ideal MHD equilibrium equations. *Phys. Lett. A* **291**, 256–264.

Bogoyavlenskij, O.I. (2002). Symmetry transforms for ideal magnetohydrodynamics equilibria. *Phys. Rev. E* **66**, 056410.

Boulier, F., Lazard, D., Ollivier, F., and Petitot, M. (1995). Representation for the radical of a finitely generated differential ideal, In *ISSAC '95: Proceedings of the 1995 International Symposium on Symbolic and Algebraic Computation*, ACM, New York, pp. 158–166.

Boyer, T.H. (1967). Continuous symmetries and conserved currents. *Ann. Physics* **42**, 445–466.

Bradshaw-Hajek, B.H., Edwards, M.P., Broadbridge, P., and Williams, G.H. (2007). Nonclassical symmetry solutions for reaction-diffusion equations with explicit spatial dependence. *Nonlinear Anal.* **67**, 2541–2552.

Bragg, S.L. and Hawthorne, W.R. (1950). Some exact solutions of the flow through annular cascade actuator discs. *J. Aeronaut. Sci.* **17**, 243–249.

Broadbridge, P. and Tritscher, P. (1994). An integrable fourth-order nonlinear evolution equation applied to thermal grooving of metal surfaces. *IMA J. Appl.*

Math. **53**, 249–265.

Burde, G.I. (1996). New similarity reductions of the steady-state boundary layer equations. *J. Phys A* **29**, 1665–1683.

Cantwell, B.J. (2002). *Introduction to Symmetry Analysis.* Cambridge Texts in Applied Mathematics. Cambridge University Press, Cambridge, UK.

Carbonaro, P. (1991). Infinite Lie groups of point transformations leaving invariant the linear equation which describes in the hodograph plane the isentropic one-dimensional gas flow. *Internat. J. Non-linear Mech.* **26**, 531–536.

Chandrasekhar. S. (1956). The instability of a layer of fluid heated below and subject to the simultaneous action of a magnetic field and rotation. II. *Proc. Roy. Soc. Ser. A.* **237**, 476–484.

Cherniha, R. (2007). New Q-conditional symmetries and exact solutions of some reaction-diffusion-convection equations arising in mathematical biology. *J. Math. Anal. Appl.* **326**, 783–799.

Cherniha, R. and Serov, M.I. (1998). Symmetries, ansätze and exact solutions of nonlinear second-order evolution equations with convection term. *Eur. J. Appl. Math.* **9**, 527–542.

Cheviakov, A.F. (2004). Bogoyavlenskij symmetries of ideal MHD equilibria as Lie point transformations. *Phys. Lett. A* **321**, 34–49.

Cheviakov, A.F. (2005). Construction of exact plasma equilibrium solutions with different geometries *Phys. Rev. Lett.* **94**, 165001.

Cheviakov, A.F. (2007). GeM software package for computation of symmetries and conservation laws of differential equations. *Comput. Phys. Commun.* **176**, 48–61.

Cheviakov, A.F. (2008). An extended procedure for finding exact solutions of PDEs arising from potential symmetries. applications to gas dynamics. *J. Math. Phys.* **49**, 083502.

Cheviakov, A.F. (2009a). Symbolic package GeM (v.031) for automated symmetry and conservation law computation. *University of Saskatchewan Preprint.*

Cheviakov, A.F. (2009b). Computation of fluxes of conservation laws. To appear in *J. Eng. Math.*

Cheviakov, A.F. and Anco, S.C. (2008) Analytical properties and exact solutions of static plasma equilibrium systems in three dimensions. *Phys. Lett. A* **372**, 13631373.

Cheviakov, A.F. and Bluman, G.W. (2009). Potential systems and nonlocal symmetries in multi-dimensions. *University of Saskatchewan Preprint.*

Cheviakov, A.F. and Bogoyavlenskij, O.I. (2004). Exact anisotropic MHD equilibria. *J. Phys. A* **37**, 7593-7607.

Chou, K.S. and Qu, C. (1999). Symmetry groups and separation of variables of a class of nonlinear diffusion-convection equations. *J. Phys. A* **32**, 6271–6286.

Clarkson, P.A. and Kruskal, M.D. (1989). New similarity solutions of the Boussinesq equation. *J. Math. Phys.* **30**, 2201–2213.

Clarkson, P.A. and Mansfield, E.L. (1993). Nonclassical symmetry reductions and exact solutions of nonlinear reaction-diffusion equations, in P.A. Clarkson (Ed.), *Applications of Analytic and Geometric Methods to Nonlinear Differential Equations.* Kluwer, Dordrecht, pp. 375–389.

Clarkson, P.A. and Mansfield, E.L. (1994a). Symmetry reductions and exact solutions of a class of nonlinear equations. *Phys. D* **70**, 250–288.

Clarkson, P.A. and Mansfield, E.L. (1994b). Algorithms for the nonclassical method of symmetry reductions. *SIAM J. Appl. Math.* **54**, 1693–1719.

Clarkson, P.A. and Mansfield, E.L. (1994c). On a shallow water wave equation. *Nonlinearity* **7**, 7975–1000.

Clarkson, P.A. and Priestley, T.J. (1998). Shallow water wave systems. *Stud. Appl. Math.* **101**, 389–432.

Clarkson, P.A. and Winternitz, P. (1991). Nonclassical symmetry reductions for the Kadomtsev–Petviashvili equation. *Phys. D* **49**, 257–272.

Cole, J.D. (1951). On a quasilinear parabolic equation occurring in aerodynamics. *Quart. Appl. Math.* **9**, 225–236.

Fokas, A.S. and Liu, Q.M. (1994). Generalized conditional symmetries and exact solutions of nonintegrable equations. *Teoret. Mat. Fiz.* **99**, 263–277.

Friedman, A. and Kamin, S. (1980). The asymptotic behavior of gas in an n-dimensional porous medium. *Trans. Amer. Math. Soc.* **262**, 551–563.

Fushchich, W.I., Shtelen, V.M, and Serov, M.I. (1993). *Symmetry Analysis and Exact Solutions of Equations of Nonlinear Mathematical Physics.* Kluwer, Dordrecht.

Fushchich, W.I., Shtelen, V.M, Serov, M.I., and Popovych, R.O. (1992). Conditional symmetries of the linear heat equation. *Dokl. Ukr. Acad. Nauk A* **12**, 20–26.

Galaktionov, V.A. (1990). On new exact blow-up solutions for nonlinear heat conduction equations with source and applications. *Diff. Int. Eqs.* **3**, 863–874.

Galaktionov, V.A., Dorodnitsyn, V.A., Elenin, G.G., Kurdyumov, S.P., and Samarskii, A.A. (1988). A quasilinear heat equation with a source: Peaking, localization, symmetry, exact solutions, asymptotics, structures. *J. Soviet Math.* **41**, 1222–1292.

Galaktionov, V.A. and Samarskii, A.A. (1984). Methods of constructing approximate self-similar solutions of nonlinear heat equations, IV. *Math. USSR-Sb.* **49**, 125–150.

Galaktionov, V.A. and Svirshchevskii, S.R. (2007). *Exact Solutions and Invariant Subspaces of Nonlinear Partial Differential Equations in Mechanics and Physics.* Chapman & Hall, Boca Raton, FL.

Galas, F. (1993). Generalized symmetries for the ideal MHD equations. *Phys. D* **63**, 87–98.

Gardner, C.S., Greene, J.M., Kruskal, M.D., and Miura, R.M. (1967). Method for solving the Korteweg–de Vries equation. *Phys. Rev. Lett.* **19**, 1095–1097.

Goldenfeld, N. (1992). *Lectures on Phase Transitions and the Renormalization Group.* Addison-Wesley, Reading, MA.

Golovin, S.V. (2004). Applications of the differential invariants of infinite dimensional groups in hydrodynamics. Group analysis of nonlinear wave problems (Moscow, 2002). *Comm. Nonlin. Sci. Numer. Simul.* **9**, 35–51.

Gordon, T.J. (1986). On the symmetries and invariants of the harmonic oscillator. *J. Phys. A* **19**, 183–189.

Grad H. and Rubin H. (1958). Hydromagnetic equilibria and force-free fields. In *Proceedings of the Second United Nations International Conference on the Peaceful Uses of Atomic Energy,* pp. 190–204.

Guthrie, G.A. (1994). Recursion operators and non-local symmetries. *Proc. R. Soc. London A* **446**, 107–114.

Hasegawa, A. (1974). Propagation of wave intensity shocks in nonlinear interaction of waves and particles. *Phys. Lett. A* **47**, 165–166.

Hashimoto, H. (1974), Exact solutions of a certain semi-linear system of partial differential equations related to a migrating predation problem. *Proc. Japan Acad.* **50**, 623–627.

Hereman, W. (1996). *CRC Handbook of Lie Group Analysis of Differential Equations,* Vol. 3: *New Trends in Theoretical Developments and Computational Methods,* Chap. 13, pp. 367–413. (N.H. Ibragimov, Ed.). CRC Press, Boca Raton, FL.

Hereman, W. (1997). Review of symbolic software for Lie symmetry analysis. *Math. Comput. Modelling,* **25** (8-9), 115–132.

Hereman, W. (2005). Symbolic computation of conservation laws of nonlinear partial differential equations in multi-dimensions. *Int. J. Quant. Chem.* **106**, 278–299.

Hopf, E. (1950). The partial differential equation $u_t + uu_x = \mu u_{xx}$. *Comm. Pure Appl. Math.* **3**, 201–230.

Huang, D.J. and Ivanova, N.M. (2007). Group analysis and exact solutions of a class of variable coefficient nonlinear telegraph equations. *J. Math. Phys.* **48**, 073507.

Hydon, P.E. (2000). *Symmetry Methods for Differential Equations. A Beginner's Guide.* Cambridge University Press, Cambridge, UK.

Ibragimov, N.H. (1980). On the theory of Lie–Bäcklund transformation groups. *Math. USSR Sbornik* **37**, 205–226.

Ibragimov, N.H. (1985). *Transformation Groups Applied to Mathematical Physics.* Reidel, Boston.

Ibragimov, N.H. (1995). *CRC Handbook of Lie Group Analysis of Differential Equations,* Vol. 2, Chap. 11, pp. 269–293 (N.H. Ibragimov, Ed.). CRC Press, Boca Raton, FL.

Ibragimov, N.H., Torrisi, M., and Valenti, A. (1991). Preliminary group classification of equations $v_{tt} = f(x, v_x)v_{xx} + g(x, v_x)$. *J. Math. Phys.* **32**, 2988–2995.

Ivanova, N.M., Popovych, R.O., and Sophocleous, C. (2008a). Group analysis of variable coefficient diffusion-convection equations. III. Conservation laws. Preprint.

Ivanova, N.M., Popovych, R.O., and Sophocleous, C. (2008b). Group analysis of variable coefficient diffusion-convection equations. IV. Potential symmetries. Preprint.

Ivanova, N.M. and Sophocleous, C. (2008). Conservation laws and potential symmetries of systems of diffusion equations. Preprint.

Ji, L. and Qu, C. (2007). Conditional Lie–Bäcklund symmetries and solutions to $(n + 1)$-dimensional nonlinear diffusion equations. *J. Math. Phys.* **48**, 103509.

Kamin, S. (1975). Similarity solutions and the asymptotics of infiltration eqautions. *Arch. Rational Mech. Anal.* **60**, 171–183.

Kapcov, O.V. (1982). Extension of the symmetry of evolution equations. *Sov. Math. Dokl.* **25**, 173–176.

Kara, A.H. and Mahomed, F.M. (2002). A basis of conservation laws for partial differential equations. *J. Nonlin. Math. Phys.* **9**, 60–72.

Kersten, P.H.M. (1987). *Infinitesimal Symmetries: A Computational Approach.* CWI Tract No. 34. Centrum voor Wiskunde en Informatica, Amsterdam.

King, J.R. (1989). Exact solutions to some nonlinear diffusion equations. *Quart. J. Mech. Appl. Math.* **42**, 537–552.

Kingston, G. and Sophocleous, C. (1998). On form-preserving point transformations of partial differential equations. *J. Phys.* **A31**, 1597–1619.

Kingston, G. and Sophocleous, C. (2001). Symmetries and form-preserving transformations of one-dimensional wave equations with dissipation. *J. Non-Linear Mech.* **36**, 987–997.

Konopelchenko, B.G. (1987). *Nonlinear Integrable Equations.* Lecture Notes in Physics No. 270, Springer-Verlag, Berlin.

Konopelchenko, B.G. and Mokhnachev, V.G. (1979). On the group-theoretical analysis of differential equations. *Sov. J. Nucl. Phys.* **30**, 288–292.

Konopelchenko, B.G. and Mokhnachev, V.G. (1980). On the group theoretical analysis of differential equations. *J. Phys.* **A13**, 3113–3124.

Korobeinikov, V.P. (1982). Certain types of solutions of Korteweg–de Vries-Burgers' equations for plane, cylindrical, and spherical waves. In *Proceedings IUTAM Symposium on Nonlinear Wave Deformations*, Tallinn (in Russian).

Krasil'shchik, I.S. and Kersten, P.H.M. (2000). *Symmetries and Recursion Operators for Classical and Supersymmetric Differential Equations.* Kluwer, Dordrecht.

Krasil'shchik, I.S. and Vinogradov, A.M. (1984). Nonlocal symmetries and the theory of coverings: An addendum to A.M. Vinogradov's 'Local symmetries and conservation laws.' *Acta Appl. Math.* **2**, 79–96.

Krasil'shchik, I.S. and Vinogradov, A.M. (1989). Nonlocal trends in the geometry of differential equations: Symmetries, conservation laws, and Bäcklund transformations. *Acta Appl. Math.* **15**, 161–209.

Kumei, S. (1975). Invariance transformations, invariance group transformations, and invariance groups of the sine-Gordon equation. *J. Math. Phys.* **16**, 2461–2468.

Kumei, S. (1977). Group theoretical aspects of conservation laws on nonlinear dispersive waves: KdV-type equations and nonlinear Schrödinger equations. *J. Math. Phys.* **18**, 256–264.

Kumei, S. (1981). A Group Analysis of Nonlinear Differential Equations. Ph.D. Thesis, University of British Columbia, Vancouver, BC.

Kumei, S. and Bluman, G.W. (1982). When nonlinear differential equations are equivalent to linear differential equations. *SIAM J. Appl. Math.* **42**, 1157–1173.

Kunzinger, M. and Popovych, R.O. (2008). Potential conservation laws. Preprint.

Lax, P.D. (1968). Integrals of nonlinear equations of evolution and solitary waves. *Comm. Pure Appl. Math.* **21**, 467–490.

Levi, D. and Winternitz, P. (1989). Non-classical symmetry reduction: Example of the Boussinesq equation. *J. Phys.* **A22**, 2915–2924.

Lie, S. (1881). Über die Integration durch bestimmte Integrale von einer Klasse linearer partieller Differentialgleichungen. *Arch. Math.* **6**, 328–368; also *Gesammelte Abhandlungen,* Vol. III, pp. 492–523. B.G. Teubner, Leipzig, 1922.

Lie, S. (1890). *Theorie der Transformationsgruppen,* Vol. II. B.G. Teubner, Leipzig.

Lie, S. (1893). *Theorie der Transformationsgruppen,* Vol. III. B.G. Teubner, Leipzig.

Lisle, I. (1992). Equivalence Transformations for Classes of Differential Equations. Ph.D. Thesis, University of British Columbia, Vancouver, BC.

Liu, Q.M. and Fokas, A.S. (1996). Exact interaction of solitary waves for certain nonintegrable equations. *J. Math. Phys.* **37**, 324–345.

Lou, S-y. (1992). Nonclassical symmetry reductions for the dispersive wave equations in shallow water. *J. Math. Phys.* **33**, 4300–4305.

Ma, A. (1991). Extended Group Analysis of the Wave Equation. M.Sc. Thesis, University of British Columbia, Vancouver, BC.

Magri, F. (1978). A simple model of the integrable Hamiltonian equation. *J. Math. Phys.* **19**, 1156–1163.

Mansfield, E.L. (1996). The differential algebra package diffgrob2. *Maple Tech.* **3**, 33–37.

Mansfield, E.L. and Clarkson, P.A. (1997). Applications of the differential algebra package diffgrob2 to classical symmetries of differential equations. *J. Symb. Comput.* **23**, 517–533.

Martina, L., Sheftel, M.B., and Winternitz, P. (2001). Group foliation and non-invariant solutions of the heavenly equation. *J. Phys.* **A34**, 9243–9263.

Mayer, A. (1875). Directe Begründung der Theorie der Berührungstransformationen. *Math. Ann.* **8**, 304–312.

Meirmanov, A.M., Pukhnachev, V.V., and Shmarev, S.I. (1997). *Evolution Equations and Lagrangian Coordinates.* Walter de Gruyter, Berlin.

Mikhailov, A.V., Shabat, A.B., and Sokolov, V.V. (1991). The symmetry approach to classification of integrable equations. In *What Is Integrability?* (V.E. Zakharov, Ed.). Springer-Verlag, Berlin, pp. 115–184.

Mikhailov, A.V., Shabat, A.B., and Yamilov, R.I. (1987). The symmetry approach to the classification of non-linear equations. Complete lists of integrable systems. *Russ. Math. Surveys* **42**, 1–63.

Miller, W., Jr. (1968). *Symmetry and Separation of Variables.* Addison-Wesley, Reading, MA.

Miller, W., Jr. (1977). *Lie Theory and Special Functions.* Academic Press, New York.

Miura, R.M. (1968). Korteweg–de Vries equation and generalizations. I. A remarkable explicit nonlinear transformation. *J. Math. Phys.* **9**, 1202–1204

Miura, R. M. (1976). The Korteweg–de Vries equation: A survey of results. *SIAM Rev.* **18**, 412–459.

Miura, R.M., Gardner, C.S., and Kruskal, M.D. (1968). Korteweg–de Vries equation and generalizations. II. Existence of conservation laws and constants of motion. *J. Math. Phys.* **9**, 1204–1209.

Momoniat, E. (2007). A mapping of a generalization of a thin film equation on a moving substrate. *Nonlin. Anal.* **66**, 1980–1988.

Müller, E.A. and Matschat, K. (1962). Über das Auffinden von Ähnlichkeitslösungen partieller Differentialgleichungssysteme unter Benutzung von Transformationsgruppen, mit Anwendungen auf Probleme der Strömungsphysik. *Miszellaneen der Angewandten Mechanik,* Berlin, pp. 190–222.

Newman, W.I. (1984). A Lyapunov functional for the evolution of solutions to the porous medium equation to self-similarity, I. *J. Math. Phys.* **25**, 3120–3123.

Nishitani, T. and Tajiri, M. (1982). On similarity solutions of the Boussinesq equation. *Phys. Lett. A* **89** (8), 379–380.

Noether, E. (1918). Invariante Variationsprobleme. *Nachr. König. Gesell. Wissen. Göttingen, Math.-Phys. Kl.,* pp. 235–257.

Nucci, M.C. and Clarkson, P.A. (1992). The nonclassical method is more general than the direct method for symmetry reductions. An example of the Fitzhugh–Nagumo equation. *Phys. Lett. A* **164**, 49–56.

Nutku, Y. and Sheftel, M.B. (2001). Differential invariants and group foliation for the complex Monge-Ampère equation. *J. Phys. A* **34**, 137–156.

Oberlack, M., Wenzel, H., and Peters, N. (2001). On symmetries and averaging of the G-equation for premixed combustion. *Combust. Theory Modell.* **5**, 363–383.

Olver, P.J. (1977). Evolution equations possessing infinitely many symmetries. *J. Math. Phys.* **18**, 1212–1215.

Olver, P.J. (1983). Conservation laws and null divergences. *Math. Proc. Camb. Phil. Soc.* **94**, 529–540.

Olver, P.J. (1986). *Applications of Lie Groups to Differential Equations.* Graduate Texts in Mathematics, No. 107. Springer-Verlag, New York.

Olver, P.J. (1994). Direct reduction and differential constraints. *Proc. R. Soc. Lond. A* **444**, 509–523.

Olver, P.J. and Rosenau, P. (1986). The construction of special solutions to partial differential equations. *Phys. Lett. A* **114**, 107–112.

Olver, P.J. and Rosenau, P. (1987). Group-invariant solutions of differential equations. *SIAM J. Appl. Math.* **47**, 263–278.

Ovsiannikov, L.V. (1959). Group properties of the nonlinear heat conduction equation. *Dokl. Akad. Nauk USSR* **125**, 492–495 (in Russian).

Ovsiannikov, L.V. (1962). *Group Properties of Differential Equations.* Nauka, Novosibirsk (in Russian).

Ovsiannikov, L.V. (1982). *Group Analysis of Differential Equations.* Academic Press, New York.

Popovych, R.O. and Ivanova, N.M. (2004). New results on group classification of nonlinear diffusion-convection equations. *J. Phys. A* **37**, 7547–7565.

Popovych, R.O. and Ivanova, N.M. (2005a). Potential equivalence transformations for nonlinear diffusion-convection equations. *J. Phys. A* **38**, 3145–3155.

Popovych, R.O. and Ivanova, N.M. (2005b). Hierarchy of conservation laws of diffusion-convection equations. *J. Math. Phys.* **46**, 043502.

Popovych, R.O., Kunzinger. M., and Ivanova, N.M. (2008). Conservation laws and potential symmetries of linear parabolic equations. *Acta. Appl. Math.* **100**, 113–185.

Pucci, E. and Saccomandi, G. (1993). Potential symmetries and solutions by reduction of partial differential equations. *J. Phys. A* **26**, 681–690.

Pukhnachev, V.V. (1987). Equivalence transformations and hidden symmetry of evolution equations. *Sov. Math. Dokl.* **35**, 555–558.

Qu, C., Ji, L., and Wang, L. (2007). Conditional Lie–Bäcklund symmetries and sign-invariants to quasi-linear diffusion equations. *Stud. Appl. Math.* **119**, 355–391.

Reid, G.J. (1990). A triangularization algorithm which determines the Lie symmetry algebra of any system of PDEs. *J. Phys. A* **23**, L853–L859.

Reid, G.J. (1991a). Algorithms for reducing a system of PDEs to standard form, determining the dimension of its solution space and calculating its Taylor series solution. *Eur. J. Appl. Math.* **2**, 293–318.

Reid, G.J. (1991b). Finding abstract Lie symmetry algebra of differential equations without integrating determining equations *Eur. J. Appl. Math.* **2**, 319–340.

Reid, G.J., Wittkopf, A.D., and Boulton, A. (1996). Reduction of systems of nonlinear partial differential equations to simplified involutive forms. *Eur. J. Appl. Math.* **7**, 604–635.

Rosen, G. (1979). Nonlinear heat conduction in solid H_2. *Phys. Rev. B* **19**, 2398–2399.

Rosenau, P. and Schwarzmeier, J.L.. (1986). On similarity solutions of Boussinesq-type equations. *Phys. Lett. A* **115**, 75–77.

Saccomandi, G. (2004). A remarkable class of non-classical symmetries of the steady two-dimensional boundary-layer equations. *J. Phys. A* **37**, 7005–7017.

Sanders, J.A. and Wang, J.P. (1998). On the integrability of homogeneous scalar evolution equations. *J. Diff. Eqns.* **147**, 410–434.

Sanders, J.A. and Wang, J.P. (2000). On the integrability of non-polynomial scalar evolution equations. *J. Diff. Eqns.* **166**, 132–150.

Sarlet, W., Cantrijn, F., and Crampin, M. (1987). Pseudo-symmetries, Noether's theorem and the adjoint equations. *J. Phys. A* **20**, 1365–1376.

Sarlet, W., Prince, G.E., and Crampin, M. (1990). Adjoint symmetries for time-independent second-order equations. *J. Phys. A* **23**, 1335–1347.

Schiff, L.I. (1968). *Quantum Mechanics,* 3rd ed. McGraw-Hill, New York.

Senthilvelan, M. and Torrisi, M. (2000). Potential symmetries and new solutions of a simplified model for reacting mixtures. *J. Phys. A* **33**, 405–415.

Seyler, C.E. and Fenstermacher, D.L. (1984). A symmetric regularized long wave equation. *Phys. Fluids* **27**, 4–7.

Shafranov, V.D. (1958). Propagation of an electromagnetic field in a medium with spatial dispersion. *Soviet Phys. JETP* **34**, 1019–1029.

Sheftel, M.B. (2002). Method of group foliation and non-invariant solutions of partial differential equations. *Eur. Phys. J. B***29**, 203–206.

Sjöberg, A. and Mahomed, F.M. (2004). Non-local symmetries and conservation laws for one-dimensional gas dynamics equations. *Appl. Math. Comput.* **150**,

379–397.

Sophocleous, C. (1996). Potential symmetries of nonlinear diffusion-convection equations. *J. Phys. A* **29**, 6951–6959.

Sophocleous, C. (2005). Further transformation properties of generalized inhomogeneous nonlinear diffusion equations with variable coefficients. *Phys. A* **345**, 457–471.

Stephani, H. (1989). *Differential Equations: Their Solution Using Symmetries.* Cambridge University Press, Cambridge, UK.

Steudel, H. (1975a). Noether's theorem and higher conservation laws in ultrashort pulse propagation. *Ann. Physik* **32**, 205–216.

Steudel, H. (1975b). Noether's theorem and the conservation laws of the Korteweg–de Vries equation. *Ann. Physik* **32**, 445–455.

Storm, M.L. (1950). Heat conduction in simple metals. *J. Appl. Phys.* **22**, 940–951.

Sukharev, M.G. (1967). Invariant solutions of equations describing the motion of fluids and gases in long pipelines. *Dokl. Akad. Nauk USSR* **175**, 781-784 (in Russian).

Talman, J.D. (1968). *Special Functions; A Group Theoretic Approach.* W.A. Benjamin, New York.

Temuerchaolu. (2003). An algorithmic theory of reduction of differential polynomial systems. *Adv. Math.* **32**, 208–220 (in Chinese).

Thomas, H.C. (1944). Heterogeneous ion exchange in a flowing system. *J. Am. Chem. Soc.* **66**, 1664–1666.

Tsaousi, C. (2008). Differential Invariants of Hyperbolic Equations. Ph.D. Thesis, University of Cyprus, Lefkosia (Nicosia), Cyprus.

Tsaousi, C. and Sophocleous, C. (2008). On linearization of hyperbolic equations using differential invariants. *J. Math. Anal. Appl.* **338**, 762–773.

Vainberg, M.M. (1964). *Variational Methods for the Study of Nonlinear Operators.* Holden-Day, San Francisco.

Varley, E. and Seymour, B. (1985). Exact solutions for large amplitude waves in dispersive and dissipative systems. *Stud. Appl. Math.* **72**, 241–262.

Vilenkin, N.J. (1968). *Special Functions and the Theory of Group Representations.* American Mathematical Society, Providence, RI.

Vinogradov, A.M. (1989). Symmetries and conservation laws of partial differential equations: Basic notions and results. *Acta Appl. Math.* **15**, 3–21.

Vinogradov, A.M. and Krasil'shchik, I.S. (1984). On the theory of nonlocal symmetries of nonlinear partial differential equations. *Sov. Math. Dokl.* **29**, 337–341.

Volterra, V. (1913). *Leçons sur les Functions de Lignes.* Gauthier-Villars, Paris.

Wagener, A. (2004). Problem posted on The Math Form at Drexel.

Wahlquist, H.D. and Estabrook, F.B. (1975). Prolongation structures of nonlinear evolution equations. *J. Math. Phys.* **16**, 1–7.

Wald, R.M. (1990). On identically closed forms locally constructed from a field. *J. Math. Phys.* **31**, 2378–2384.

Wang, J.P. (2002). A list of 1+1 dimensional integrable equations and their properties. Recent advances in integrable systems (Kowloon 2000). *J. Nonlin.*

Math. Phys. **9**, 213–233.

Whitham, G.B. (1974). *Linear and Nonlinear Waves.* Wiley, New York, NY.

Wiltshire, R. and El-Kafri, M. (2004) Non-classical and potential symmetry analysis of Richard's equation for moisture flow in soil. *J. Phys. A* **37**, 823–839.

Winternitz, P., Smorodinsky, J.A., Uhlir, M., and Fris, I. (1967). Symmetry groups in classical and quantum mechanics. *Sov. J. Nucl. Phys.* **4**, 444–450.

Wittkopf, A.D. (2004). Algorithms and Implementations for Differential Elimination. Ph.D. Thesis. Simon Fraser University, Burnaby, BC. http://ir.lib.sfu.ca:8080/retrieve/205/etd0400.pdf

Wolf, T. (2002a). Investigating differential equations with CRACK, LiePDE, Applysymm and ConLaw. *Handbook of Computer Algebra, Foundations, Applications, Systems* (J. Grabmeier, E. Kaltofen, and V. Weispfenning, Eds.). Springer, New York, pp. 465–468.

Wolf, T. (2002b). A comparison of four approaches to the calculation of conservation laws. *Eur. J. Appl. Math.* **13**, 129–152.

Wolf, T. (2004). Applications of CRACK in the classification of integrable systems. In *CRM Proceedings and Lecture Notes.* Vol. 37, pp. 283–300. AMS, Providence, RI.

Wolf, T. and Brand, A. (1992). The computer algebra package CRACK for investigating PDEs. In *Proceedings of ERCIM Advanced Course on Partial Differential Equations and Group Theory*, Bonn.

Wu, W.T. (1984). Basic principles of mechanical theorem proving in elementary geometries. *J. Syst. Sci. Math. Sci.* **4**, 207–235.

Wybourne, B.G. (1974). *Classical Groups for Physicists.* Wiley, New York.

Yoshikawa, A. and Yamaguti, M. (1974). On further properties of solutions to a certain semi-linear system of partial differential equations. *Publ. Res. Inst. Math. Sci. Kyoto Univ.* **9**, 577–595.

Zakharov, V.E. and Shabat, A.B. (1972). Exact theory of two-dimensional self-focusing and one-dimensional self-modulation of waves in nonlinear media. *JETP* **34**, 62–69.

Zhdanov, R.Z. (1996). Conditional symmetries and spectrum of the one-dimensional Schrödinger equation. *J. Math. Phys.* **37**, 3198–3217.

Theorem, Corollary and Lemma Index

Chapter 1

Chapter 2

Chapter 3

Chapter 4

Chapter 5

Author Index

Subject Index

Boldface indicates a key reference. References to exercises are bracketed.

Kolibri 360 GmbH
Ferdinand-Jühlke-Straße 7
99095 Erfurt, Deutschland
produktsicherheit@kolibri360.de